8th China International Nanoscience and Technology Symposium 2009

(CINSTS 09)

Journal of Physics: Conference Series Volume 188

Xiangtan, China
23-27 October 2009

ISBN: 978-1-61738-288-8

Printed from e-media with permission by:

Curran Associates, Inc.
57 Morehouse Lane
Red Hook, NY 12571

proceedings.com

Some format issues inherent in the e-media version may also appear in this print version.

Printed by Curran Associates, Inc. (2010)

Institute of Physics
Dirac House, Temple Back
Bristol BS1 6BE UK

Phone: 44 1 17 929 7481
Fax: 44 1 17 920 0979

techtracking@iop.org

Additional copies of this publication are available from:

Curran Associates, Inc.
57 Morehouse Lane
Red Hook, NY 12571 USA
Phone: 845-758-0400
Fax: 845-758-2634
Email: curran@proceedings.com
Web: www.proceedings.com

8th China International Nanoscience and Technology Symposium 2009

(CINSTS 09)

Journal of Physics: Conference Series Volume 188

Xiangtan, China
23-27 October 2009

ISBN: 978-1-61738-288-8

IOP Institute of Physics

TABLE OF CONTENTS

The 8th China International NanoScience and Technology Symposium, Xiangtan (2009) – Nano-products Exposition, sponsored by Chinese Society of Miro-nanoTechnology and IEEE Nanotechnology Council, etc will be held on 23–27 October 2009 in Xiangtan, China. This symposium is held in order to promote the technology for the development of micro- and nano-scale, cross-scale integration, to share new micro/nano technologies, to exchange information and knowledge over all fields and promote the industrialization and development of nanotechnology. This is a leading professional and traditional conference with at least 400 participants every year. Famous experts, professors and government officials at home and abroad will give lectures during the symposium, which provides a good platform for delegates to discover the latest developments and dynamics of nanotechnology. Researchers, teachers and students in colleges, and technical personnel in the industrial community are welcome to contribute and actively participate in the symposium.

In our last symposium held in 2008, over 600 participants from all over the world attended, and we received over 570 abstract and paper submissions for the proceedings published in different languages in famous professional journals. And this year, we have already received over 400 submissions. After strict peer review, 60 of them are published in this volume of *Journal of Physics: Conference Series*. We are confident that the event will be even more successful this year. Consequently, the organizing committee and proceedings editorial committee would like to thank our colleagues at the IOP Publishing, the invited speakers, our sponsors and all the delegates for their great contributions in this conference.

Hailin Cong
Vice Chair of the proceedings editorial committee

All papers published in this volume of *Journal of Physics: Conference Series* have been peer reviewed through processes administered by the proceedings Editors. Reviews were conducted by expert referees to the professional and scientific standards expected of a proceedings journal published by IOP Publishing.

Preparation of Plate-shape Nano-Magnesium Hydroxide from Asbestos Tailings

DU Gao-xiang [1], ZHENG Shui-lin [2]

[1] School of Materials Science and Technology, China University of Geosciences (Beijing), Beijing, China

[2] Department of Chemical and Environment Engineering, China University of Mining and Technology (Beijing), Beijing, China

E-mail: dgx@cugb.edu.cn

Abstract. To prepare magnesium hydroxide is one of the effective methods to the comprehensive utilization of asbestos tailings. Nano-scale magnesium hydroxide was prepared and mechanisms of in-situ surface modification were characterized in the paper. Process conditions of preparation of magnesium hydroxide from purified hydrochloric acid leachate of asbestos tailings were optimized and in-situ surface modification of the product was carried out. Results showed that optimum process conditions for preparing nano-scale magnesium hydroxide were as follows: initial concentration of Mg^{2+} in the leachate was 22.75g/L, precipitant was NaOH solution (mass concentration 20%), reaction temperature was 50℃, and reaction time was 5min. The diameter and thickness of the plate nano-scale magnesium hydroxide powder prepared under optimal conditions were about 100 nm and10 nm, respectively. However, particle agglomeration was obvious, the particle size increased to micron-grade. Dispersity of the magnesium hydroxide powder could be elevated by in-situ modification by silane FR-693, titanate YB-502 and polyethylene glycol and optimum dosages were 1.5%, 1.5% and 0.75% of the mass of magnesium hydroxide, respectively. All of the modifiers adsorbed chemically on surfaces of magnesium hydroxide particles, among which Si-O-Mg bonds formed among silane FR-693 and the particle surfaces and Ti-O-Mg among titanate YB-502 and the surfaces.

Keywords. asbestos tailings; magnesium hydroxide; in-situ surface modification

Introduction

Asbestos tailings are hazardous solid wastes mainly consists of serpentine. Accumulation of asbestos tailings threatens healthiness of animals and human around mining area severely [1] because of large amount of serpentine fiber it contains. So, safe disposal of asbestos tailings is of great significance to improve the ecological environment of the mining area.

Presently researches on recycling asbestos tailings mainly focus on: (1) preparing building materials such as non-burnt brick, steam brick, ceramic and glass-ceramic [2,3,4]; (2) extracting chemicals of magnesium and silicon including magnesium hydroxide, light magnesium oxide, precipitated silica and sodium silicate [5]. The main chemical components of asbestos tailings are SiO_2 and MgO_2 comprising 80% of the total weight. Producing magnesium hydroxide is an effective way to recycle asbestos tailings as a kind of resource [6]. Extraction of chemicals of magnesium and silicon mainly focus on preparing magnesium oxide and metal magnesium. Preparation of magnesium hydroxide has not yet been reported.

Magnesium hydroxide is an important chemical, which has a wide application in areas such as flame retardant for polymers, electric and electronic products, lithium battery, neutralization of acidic wastewater and preparation of metal magnesium and magnesium oxide [7]. Preparation, dispersion and modification of nano-scale magnesium hydroxide have been popular in recent years because of its high flame retardance efficiency.

Taking purified hydrochloric acid leachate of asbestos tailings as raw material, by controlling reaction process conditions and in-situ surface modifications, flake nano-scale magnesium hydroxide was prepared in the paper.

1. Raw materials, reagents, instruments and methods

1.1. Experimental raw materials

The sketch map for preparing nano-magnesium hydroxide was as below:

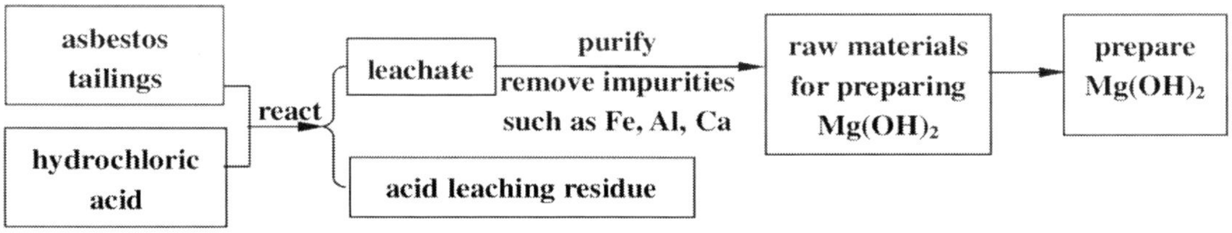

Concentrations of ions in purified acid leachate for preparing magnesium hydroxide were (g/L): Mg^{2+}, 45.50; Na^{+}, 20.54; Fe^{3+}, 0.06; Al^{3+}, 0.10; Ca^{2+}, 0.25.

1.2. Experimental reagents

NaOH (A.R.) from Beijing Chemical Plant; titanate coupling agent YB-502 produced by Changzhou Yabanyayu Promoter Co., Ltd; silane coupling agent FR-693 provided by America Compton Company; polyethylene glycol (A.R., n=8.2) from Beijing Chemical Plant; Di-n-butyl phthalate (A.R., $C_{16}H_{22}O_4$) from Beijing Yili Fine Chemicals Co., Ltd.

1.3. Experimental instruments

BT-1500 Particle Size Distribution Analysis produced by DanDong Better Instruments Co., Ltd.; ST-2000 BET Determinator manufactured by Beijing Beifen Instruments Company; D/MAX 2500-PC X-Ray Diffractometer from Japan Rigaku Company; WQF-410 Flourier Infrared Spectrometer produced by Beijing Second Optical Instrument Company; S-3500N Scanning Electron Microscope from Japan Hitachi Company.

1.4. Experimental methods

Purified acid leachate was taken to study the effect of process conditions on apparent particle size of nano-$Mg(OH)_2$ powder. Influences of dosages of different modifiers on apparent particle size were analyzed. The properties of nano-$Mg(OH)_2$ powder were characterized by SEM, XRD and other approaches.

2. Results and Discussions

2.1. Preparation and characterization of nano- $Mg(OH)_2$ powder

NaOH was considered as precipitant because particle size of prepared $Mg(OH)_2$ powder using NaOH as precipitant was smaller [8]. There was NaCl in the solution already, so NaOH is adopted as precipitator .

2.1.1. Influence of initial concentration of Mg^{2+} on apparent particle size

Experiment conditions: first measured 120mL Mg^{2+} solution, regulated the concentration of Mg^{2+} as 45.50g/L, 34.1 g/L, 27.30 g/L and 22.75 g/L, respectively, then put them into 500mL beakers, then titrated to pH=13 with 20% NaOH under 50℃, reacted for 8min and aged for 35min. Then filtrated, washed until no Cl^- was detected in the leachate. Then sampled from filter cakes and measured the particle sizes and distributions. Results were shown in figure.1.

As was shown in figure.1, the apparent particle size of $Mg(OH)_2$ powder decreased with decreasing concentration of Mg^{2+} solution. The $Mg(OH)_2$ powder prepared from lower initial concentration of Mg^{2+} had better dispersity. This was because much higher crystal nucleus formation rate than crystal nucleus growth rate was required in order to prepare superfine powder and the concentration of Mg^{2+} had a direct effect on supersaturation of $Mg(OH)_2$ powder during reaction. But the lower the initial concentration of Mg^{2+} was, the more amount of water was needed. So the optimum concentration of Mg^{2+} was 22.75g/L.

2.1.2. Influence of reaction temperature on apparent particle size

Test conditions: the reaction temperatures were 30℃, 50℃, 70℃ and 90℃, respectively, initial concentration of Mg^{2+} was 22.75g/L, other conditions were the same as that of section 2.1.1. Results were shown in figure.2.

As shown in figure.2, the prepared $Mg(OH)_2$ powder has minimum apparent particle size and distribution at temperature 50℃. So reaction temperature should be 50℃ in order to prepare superfine $Mg(OH)_2$ powder with good dispersity.

2.1.3. Influence of reaction time on apparent particle size

Test conditions: initial concentration of Mg^{2+} was 22.75g/L, reaction temperature was 50℃, and reaction times were 2min, 5min, 8min and 11min, respectively. Other conditions were the same as that of section 2.1.1. Results were shown in figure.3.

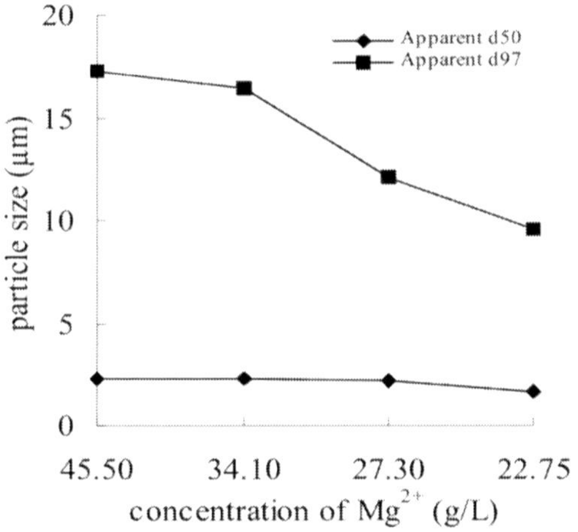

Figure.1 Effect of the concentration of Mg^{2+} on the properties of $Mg(OH)_2$ powder

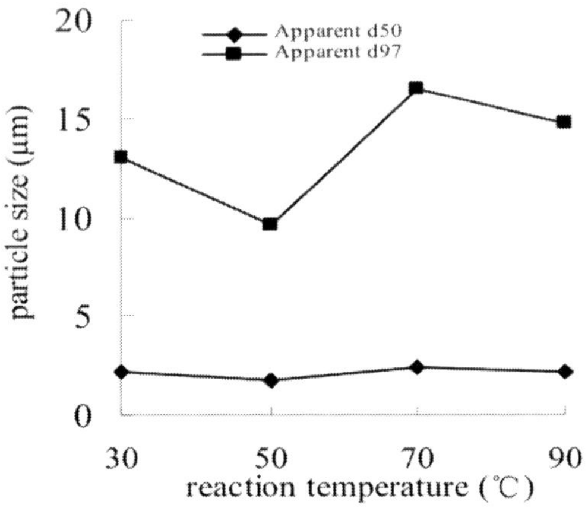

Figure.2 Effect of the reaction temperature on the apparent particle size of $Mg(OH)_2$ powder

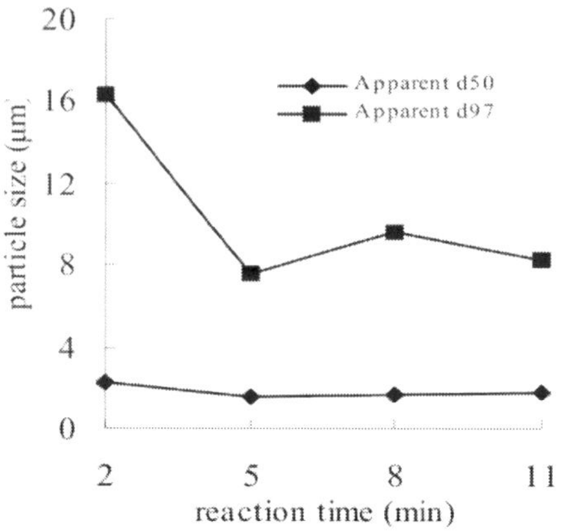

Figure.3 Effect of the reaction time on the properties of $Mg(OH)_2$ powder

Supersaturation of $Mg(OH)_2$ powder in the reaction system was directly affected by reaction time, as a result crystal nucleus formation and growth rate of $Mg(OH)_2$ powder were affected consequently and properties of $Mg(OH)_2$ were affected as well. As was shown in figure.3, minimum apparent particle size of $Mg(OH)_2$ powder is prepared when the reaction time is 5 min.

2.1.4. Characterization of $Mg(OH)_2$ powder

$Mg(OH)_2$ powder was prepared under the conditions of initial concentration of Mg^{2+} as 22.75g/L, reaction temperature as 50℃ and reaction time as 5min. The apparent particle size are: d_{50}=1.60μm, d_{97}=7.23μm. The properties were characterized by SEM and XRD approaches.

Results of chemical analysis of the prepared $Mg(OH)_2$ were (%): MgO, 66.59; SiO_2, 0.07; Fe_2O_3, 0.09; CaO, 0.18; Na_2O, 0.04; K_2O, 0.01; LOI, 31.62. It is calculated accordingly that $Mg(OH)_2$ has a relatively high purity of as high as 96.6%. The SEM photo was shown in figure.4 and the XRD pattern was shown in figure.5.

Figure.4 showed nano-scale flake particles with diameter of about 100nm and thickness of about 10nm and the phenomenon of particle agglomeration was obvious, the largest particle size was about 10μm. Figure.5 showed sharp narrow characteristic peaks of $Mg(OH)_2$ at d=4.795nm, d=2.369nm and d=1.799nm, indicating high crystallinity. The peak at d=2.728nm was characteristic peak of NaCl and d=3.042nm of $CaCO_3$, indicating trace NaCl and $CaCO_3$, which was consistent with chemical analysis.

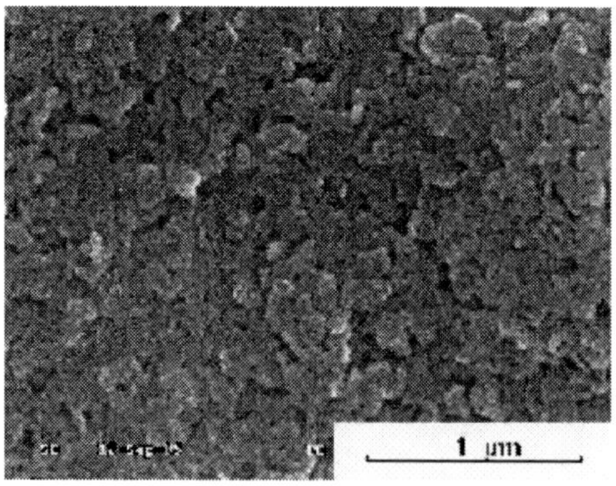

Figure.4 SEM photo of $Mg(OH)_2$ powder

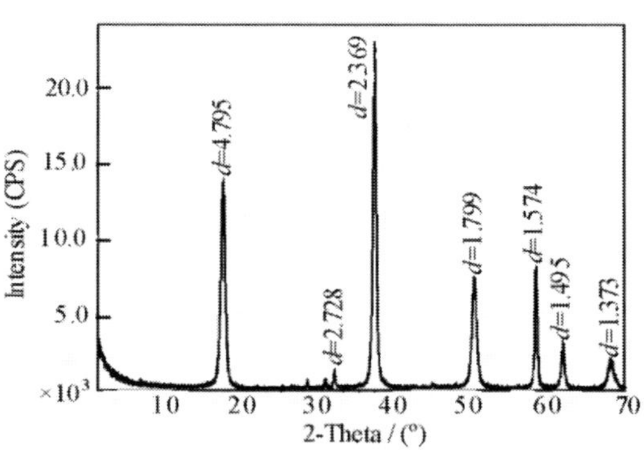

Figure.5 XRD pattern of $Mg(OH)_2$ powder

2.2. In-situ surface modification of $Mg(OH)_2$ powder

In-situ surface modification was carried out to improve dispersity of the $Mg(OH)_2$ powder. Research showed that dispersity of the powder in polymer matrix under dry condition could be improved by silane FR-693 and titanate TB-502 while dispersity in hydrophilic environment be improved by polyethylene glycol [9].

2.2.1. Modification by coupling agent

Conditions: first measured 120mL Mg^{2+} solution, regulated the concentration of Mg^{2+} as 22.75 g/L, then put it into 500mL beaker and added into certain amount of coupling agent (FR-693 or YB-502), then titrated to pH=13 with 20% NaOH under 50℃, reacted for 8min and aged for 35min. Then filtrated, washed until no Cl^- were detected in the leachate. Then sampled from filter cakes and measured the particle sizes and distributions. Results were shown in figure.6.

Figure.6 showed dispersity of the $Mg(OH)_2$ powder during the chemical preparation process could be improved to some extent by in-situ modification using silane FR-693 and the optimum dosage was 1.5% of the mass of $Mg(OH)_2$. YB-502 had similar function, the apparent particle size d_{50}=1.51μm and d_{97}=4.23μm when the dosage was 1.5% of the mass of $Mg(OH)_2$. But when the dosage was added up to 2.0% the apparent particle size of the powder increased slightly. So the optimum dosage of YB-502 was 1.5% of the mass of $Mg(OH)_2$.

2.2.2. Modification by polyethylene glycol

Experimental conditions were the same as that of section 2.2.1 with polyethylene glycol as modifier. Results were shown in figure.7.

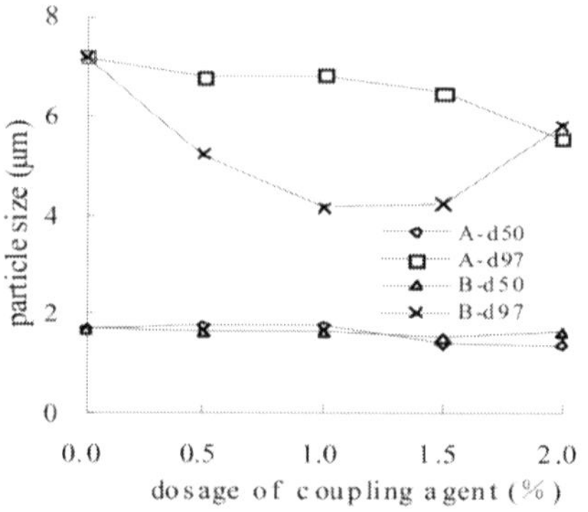

A-surface modification by FR-693,
B-surface modification by YB-502.

Figure.6 Effect of coupling agent treatment on the size distribution of $Mg(OH)_2$ powder

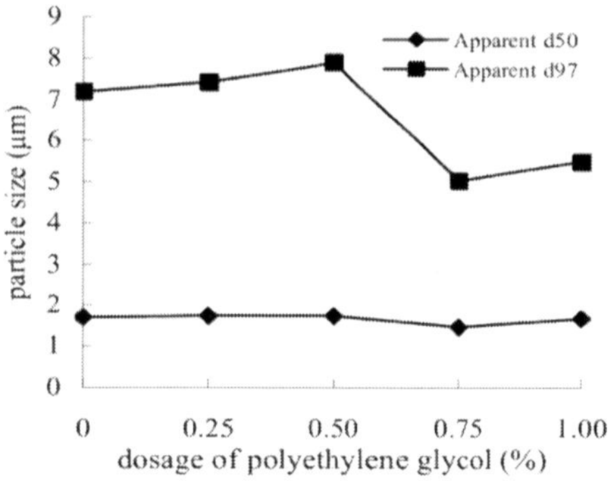

Figure.7 Effect of surface modification by polyglycol on particle size of $Mg(OH)_2$ powder

Figure.7 indicated that dispersity of the powder can be improved by in-situ modification using polyethylene glycol. When the dosage of modifier was 0.75% of the mass of $Mg(OH)_2$ powder the particle size d_{50}=1.46μm, d_{97}=5.02μm, but the apparent particle size of the powder increased when the dosage of modifier was added up to 1.0%. So optimum dosage of polyethylene glycol was 0.75% of the mass of $Mg(OH)_2$.

The SEM photos of the modified $Mg(OH)_2$ were shown in figure.8.

Comparison between figure.8 and figure.4 indicated that the dispersity of $Mg(OH)_2$ powder modified by silane was improved to some extent, and diameter-thickness ratio was smaller than that of non-modified powder, the modified powder had relatively good dispersive property. Dispersity of $Mg(OH)_2$ powder modified by titanate was improved significantly, flake micro-morphology of the particle was observed. Diameter-thickness ratio was a little bigger and the modified powder had relatively good dispersive property. Dispersity of the $Mg(OH)_2$ powder could be improved and agglomeration phenomenon was weakened by modification using polyglycol, the primary particle size of the $Mg(OH)_2$ powder was larger than non-modified powder and the dispersity was better. Of the three coupling agents dispersity of titanate and polyglycol modified powders were improved by a larger extent.

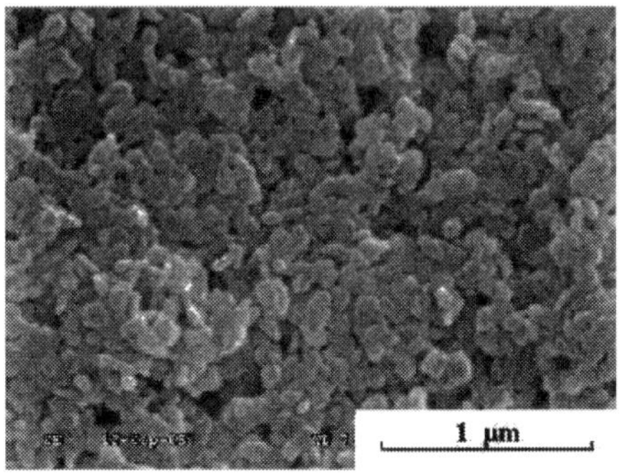

(a) surface modified by FR-693

(b)surface modified by YB-502

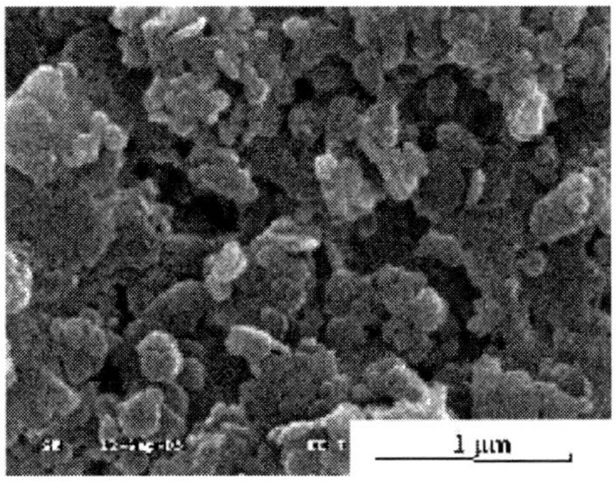

(c) surface modified by polyglycol

Figure.8 SEM photos of surface modified $Mg(OH)_2$ powder

2.2.4. Mechanism of in-situ modification of $Mg(OH)_2$

The FTIR pattern of non-modified $Mg(OH)_2$ powder was shown in figure.9. Differential patterns of modified and non-modified powders were shown in figure.10.

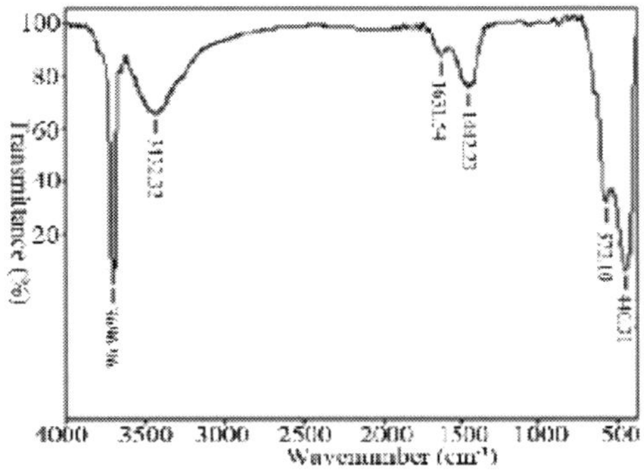

Figure.9 FTIR pattern of $Mg(OH)_2$ powder

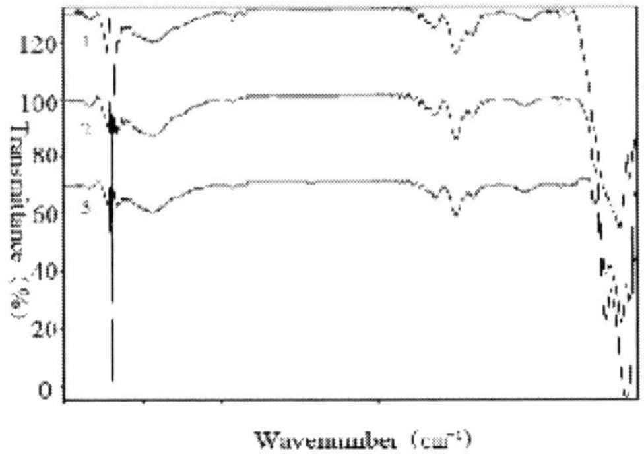

1—surface treated by FR-693;
2—surface treated by YB-502;
3—surface treated by polyglycol;

Figure.10 FTIR differential patterns of $Mg(OH)_2$ powders between non-modified and modified

As was seen in figure.9, wave numbers at 693.96cm^{-1}, 3432.32 cm^{-1} and 1442.23 cm^{-1} were absorption peaks of OH^-. Wave numbers at 440.31 cm^{-1} and 572.10 cm^{-1} were absorption peaks of Mg-O bond. Wave number at 1631 cm^{-1} was absorption peak of H_2O and 1442.23 cm^{-1} of CO_3^{2-}. Above peaks indicated that some of the prepared $Mg(OH)_2$ transformed into magnesium carbonate hydroxide or magnesium carbonate in air [10].

Differential peak at 3696 cm^{-1} in figure.10 indicated absorption peak of H_2O migration induced by adsorption of modifier on the powder surface. In curve 1 the vibration adsorption peak of –OH at 2921.67cm^{-1}, characteristic peak of $-CH_2$ at 2852.95 cm^{-1}, C-H bond in silane plane at 1398.03 cm^{-1} and C-H bond at 1488.34cm^{-1} indicated adsorption peak migration induced by chemical adsorption of silane on powder surface. It was seen that Si-O-Mg bond was formed between $Mg(OH)_2$ and silane on the powder surface. In curve 2 the vibration adsorption peak of –OH at 2921.10cm^{-1}, –C=O of titanate at 1743.36 cm^{-1}, C-N bond of amine at 1079.88cm^{-1} and characteristic peak of C-H bond at 1494.05 cm^{-1} indicated adsorption peak migration induced by titanate coating on surface of $Mg(OH)_2$ powder and adsorbed chemically. That was Ti-O-Mg bond was formed between $Mg(OH)_2$ and titanate on the powder surface. In curve 3 absorption peak migration induced by organic adsorption on powder surface at 3697.47 cm^{-1} and 3442.75 cm^{-1}, characteristic peak of C-O at 1743.13 cm^{-1} and characteristic peak of C-H bond at 1509.70 cm^{-1} indicated coating on $Mg(OH)_2$ surface by polyglycol and induced absorption peak migration.

3. Conclusions

(1) Optimum process conditions for preparing nano-scale magnesium hydroxide using Mg^{2+} solution from purified acid leachate of asbestos tailings were: initial concentration of Mg^{2+} solution 22.75g/L, precipitant 20% NaOH solution, reaction temperature and reaction time 50℃ and 5min respectively. It was observed from SEM photos that diameter of the prepared flake nano-scale $Mg(OH)_2$ powder was about 100nm and thickness was about 10nm. However, the phenomenon of agglomeration was obvious, the particle size increased to micron-grade.

(2) Dispersity of the $Mg(OH)_2$ powder could be improved by in-situ modification using silane FR-693, titanate YB-5002 and polyglycol as modifier. Optimum dosage of the three kinds of modifiers were 1.5%, 1.5% and 0.75% of the mass of $Mg(OH)_2$, respectively. All of the modifiers adsorbed chemically on $Mg(OH)_2$ particle surface, among which Si-O-Mg chemical bonds were formed between silane FR-693 and particle surface, Ti-O-Mg bonds were formed between titanate YB-502 and particle surface.

References

[1] Zhou Guihua and Wang Wenli 1999 Study on industrial status and prospect of asbestos in China *China Non-metallic Mining Industry Herald* **5** 23-25

[2] Lu Zhongyuan, Wang Haibin, Su Guanglan and Lv Shuzhen 1997 Study on hydrothermal synthesis of building materials by asbestos tailings *Environmental Sciences* **1** 68-70

[3] Lu Zhongyuan, Wang Pu and Li Heyu 1996 Study on preparation of autoclaved building materials by asbestos tailings *Non-metallic Mines* **4** 28-30

[4] Jiang Wenjiu 1997 Study on preparation of glass-ceramic decorated materials by asbestos tailings *Glass & Enamel* **26** 31-34

[5] Liao Qilong and Lu Zhongyuan 1997 Study on preparation of non-burnt brick by asbestos tailings *Non-metallic Mines* **6** 34-35

[6] DU Gaoxiang 2005 Preparation of superfine magnesium hydroxide and superfine amorphous silica from asbestos tailing *China University of Mining Science and Technology (Beijing)* **11** 75-79

[7] Du Gaoxiang, Zheng Shuilin and Li Yang 2005 Surface modification of ultra-fine brucite powder by silane coupling agent *Jouornal of the Chinese Ceramic Society* **33** 659-64

[8] Schofield, W.C.E., Hurst, S.J, Lees, G.C, Liauw, C.M and Rothon R.N 1998 Influence of surface modification of magnesium hydroxide on the processing and mechanical properties of composites of magnesium hydroxide and an ethylene vinyl acetate copolymer *Composite Interfaces* **5** 515-28

[9] Du Gaoxiang and Zheng Shuilin 2004 Surface modification of super-fine magnesium hydroxide powder and properties of PP with high contents of $Mg(OH)_2$ *China Plastics* **18** 75-79

[10] Qiu Deren 2003 *Industry Analysis Chemistry* (Shanghai: Fudan University Press) pp 154-64

Computer simulation of surfactant-mediated thin film growth *

X P Zheng[1], **P F Zhang**[2†], Z J Liu[1] and **E Y Wang**[1]
[1] Institute of Electronic Information Science and Technology, Lanzhou City University, Lanzhou 730070, PR China
[2] School of Mathematics and Statistics, Lanzhou University, Lanzhou 730000, PR China

E-mail: zxp@lztc.edu.cn and zpf519@sina.com

Abstract. A novel model consisting of basic micro-processes has been developed on the basis of the classic diffusion theory. It is first time that the concept of exchange rate has been introduced and the growing process of surfactant-mediated epitaxial thin-film growth has been simulated with Kinetic Monte Carlo (KMC) technique. The results of simulation found that the exchange reaction of RLA model is acombination of the basic micro-processes. The majorities of exchange are not complete position exchange and the exchange rate doesn't always equal one. Both surfactant atoms and adatoms would diffuse from one layer to another. The diffusion happens mostly between single atoms and the diffusing atoms increase with substrate temperature.

1. Introduction

During of thin film growing, it is found that the growth characteristics with surfactant-mediated would be completely opposite from the non-surfactant situation. The investigations of the abnormal effects become a focus of film growth field in recent years[1-3]. Up to date, the studies for the mechanism of thin film growth with surfactant-mediated are not perfect. Among the numerous models, Reaction Limited Aggregation (RLA) [4-6] model successfully explains abnormal effects found in experiments. However, the basic principle of RLA model is different from that of DLA (Diffusion Limited Aggregation)[7-9] model. In RLA model the nucleation is affected mainly by exchange reaction, but in DLA model the nucleation is affected by diffusion reaction.

The key of RLA model is the exchange reaction and there are three basic presuppositions in RLA model. First, the adatoms could become a stable nucleation centre after conquering a larger potential and exchanging position with the surfactant atoms. Secondly, to become a portion of the stable island, the adatoms that deposite subsequently conquer also a larger potential and exchange position with the surfactant atoms. The last assumption is that only exchanged atoms could constitiute form the island,

* Supported by No. 10574059 of the National Natural Science Foundation of China, No. 0710RJZA074 of the Natural Science Foundation of Gansu Province, No. 0711B-04 of the Second Scientific Research Project of Bureau of Gansu education, Key Project No. 209127 of the Science and technology Research of Ministry of Education andQing Lan' Talent Engineering Funds of Lanzhou Jiaotong University.
†Author/ Corresponding author (P. F. Zhang). E-mailzxp@lztc.edu.cn and zpf519@sina.com

namely the island locating in surfactant layer is stable. Basing on the first and second assumptions, exchange must be satisfied with the following qualifications:

(i) At the beginning, exchange must happen between nuclei (at least two atoms) and it is impossible that exchange happens between single atoms .

(ii) Exchange is the exchange of position, i.e. complete exchange.

(iii) The number of adatoms equals to one of surfactant atoms during exchange process.

The thin film growth is the result of atom diffusing and congregating at all, the micro-mechanism of exchange reaction in RLA model is not reported hitherto. In this paper, on the basis of diffusion theory we have simulated the mechanism of thin film growth with surfactant-mediated and investigated emphatically the essence of exchange reaction.

2. Description of simulator

It is known that the growth of thin film is due to atom diffusing and congregating. Basing upon diffusion theory and basic micro-processes, we advanced a novel three-dimensional model which consists of: (1) A atom deposits on the substrate. (2) A single atom diffuses on the surface of substrate. (3) The diffusing atom encounters another and then both of them nucleate. (4) The diffusing atom is captured by the existent island. (5) The atom locating at the edge of island detach island at a certain probability. (6) The atom locating at island edge retains bonding with island and removes along island edge. (7) After diffusing, the atom deposited directly on island falls on the substrate again. (8) The adatoms nucleate on a island. (9) The cluster which is composed of two or more atoms diffuses together, and so on.

The novel model does not comprise the exchange reactionof RLA model. By analyzing, we generalize the basic micro-processes of film growth into three classes:

(i) The incident atom deposits on the growing surface and isattached, called absorption event.

(ii) All kinds of atoms diffuse on the growing surface, called diffusion event.

(iii) The adatoms detach from the growing surface, called evaporation event. In fact, this process could be incorporated into a special kind of diffusion events.

After generalization, the novel model comprises not only all kinds of processes mentioned but also others, such as corner diffusion, layer-to-layer diffusion and so on.

The absorption rate of atoms is given by the number of incident atoms per unit time and the absorption event consists of:

(i) An incident atom deposits at a lattice site on the growing surface.

(ii) If the site has three atoms as nearest neighbors (stable site), the incident atom sticks and occupies this stable site, otherwise it will diffuse to one of the nearest or the next near vacant sites. If the site is not stable, the incident atom will continue to diffuse until it finds a stable site or evaporates from the growing surface. The whole process was regarded as one event.

The related effects caused by the atom diffusion were considered in the model. As illustrated in fig.1, since atom A has only two nearest neighbors, it will diffuse to a stable site. As a result, the site of atom B is no longer a stable site, it will diffuse to another stable site too, and so will atom C. Though different from the normal diffusion, it was regarded as one event rather than three events.

The diffusion rate of a single adatom was defined as the probability of a diffusion jump per unit time and is given by the Arrhenius-type expression

$$r = \nu_0 \exp\left(-\Delta E / k_B T\right), \qquad (1)$$

where ν_0 is the frequency of atom vibrations and it is assigned a value of $\nu_0 = (2k_B T)/h$, k_B the Boltzmann constant, T the substrate temperature and ΔE the activation energy.

The evaporation rate was calculated as the same as the diffusion rate, but the final target site is that without any neighbors.

We implemented the simulation onto face-centered cubic (100) surface, the substrate is 50×50, and we adopted the periodic border condition. The surface of substrate atom A is surfactant S and the incident atom is atom A.

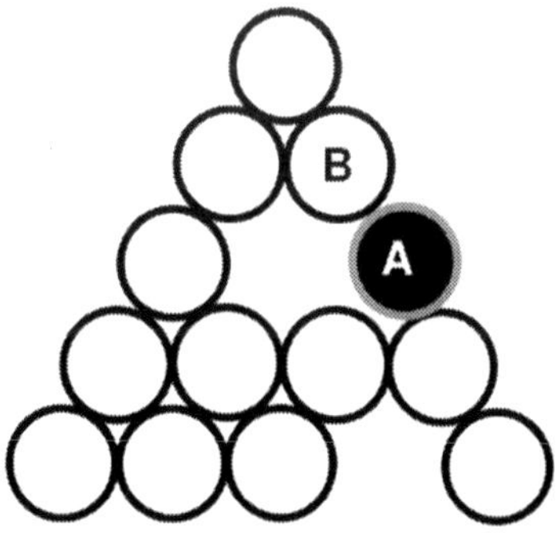

Figure 1. The sketch plot of related-effect

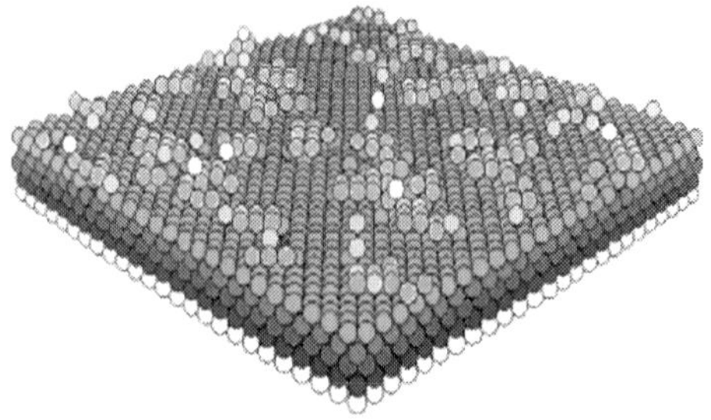

Figure 2. Film Morphology at the initial stages.

Morse potential was used to model the interaction between two atoms, where we considered not only the interaction between A and A, but also the interaction between A and S, S and S,

$$V(ij) = V_0\{\exp[-2\alpha(r_{ij}/r_0 - 1)] - 2\exp[-\alpha(r_{ij}/r_0 - 1)]\}, \quad (2)$$

where α is a constant depending on the interaction range of atoms and V_0 is the interactional potential between the nearest neighbour atoms. In the simulating, we referred to the interaction potentials reported[4] and took V_{AA}= 0.87ev, V_{AS}= 0.46ev and V_{SS}=0.13ev respectively, r_{ij} is the distance between atoms i and j, r_0 is the distance between the nearest neighbors, r_{ij}/r_0 is determined by the geometric relation of face-centered cubic structure. It has been demonstrated that Morse potential could reasonably represents the interaction between two atoms and easily calculates the potential change during the atom diffusion. The calculation of activation barrier was improved further on the basis of our previous methods[10-12]. In order to find the maximum activation barrier E_{max} between two atoms, we took 99 points P_i (i=1, 2, ···, 99) on the line connecting the initial lattice site to the next jump target. We took another two points P_0 and P_{100}, the distances from P_0 to the initial site and from P_{100} to the next jump target are 0.01a (a is the lattice constant). The respective potentials are E_i (i=0 to 100), E_{ini} (potential in the initial site) and E_{fin} (potential in the next jump site). The total potential for each point of interest can be obtained by summing up the contribution from all atoms within a sphere of radius of 1.5a. E_{max} is the maximum of E_i (i=0 to 99), and E_{fin}. The activation energy ΔE is determined by taking the difference between E_{max} and E_{ini}.

In this paper, we firstly introduced the concept of exchange rate in order to study the micro-mechanism of exchange reaction and defined it as

$$s = \mathbf{n}_2/\mathbf{n}_1, \quad (3)$$

where $\mathbf{n}_1$ is the numbers of surfactant atoms diffusing from surfactant layer to thin film surface, $\mathbf{n}_2$ is the numbers of incident atoms diffusing from the film surface to surfactant vacant sites. It is obvious that the exchange rate equals to one forever according to RLA model.

3. Results and discussion

Figure 2 shows a snapshot of film morphology at the initial stage with surfactant-mediated at temperature of 300K and deposition rate of 0.32 lay./s, where the layers from 1 to 3 are substrate atoms and the layers from 4 to 5 are surfactant atoms, white and grey spheres represent respectively the incident and surfactant atoms.

Seen from the figure 2, a part of surfactant atoms diffuse to the surface and simultaneously some adatoms diffuse to the surfactant during film growing, and the majority of diffusion is layer-to-layer diffusion of single atom. A trace file was created during the simulation, which could record the initial and the next jump positions of every jump. By analyzing the trace file, we found that both surfactant and adatoms diffuse in the way of layer-to-layer with the interactions of atom kinetic movement and interactions between atoms. On one hand some surfactant atoms diffuse up to the surface and leave vacant sites. Because of the existence of vacants, the surfactant atoms diffuse constantly in the surfactant layers so that the vacant sites move continually too. On the other hand adatoms diffuses on the film surface and might arrive at one of the nearest neighbours of vacant site (in the deposited surface). Because of lower energy of vacant site, the adatom diffuses more easily down and holds the vacant in the surfactant and then forms a filled site. At the same time, if the surfactant atom, which diffused into the surface, diffuses to other nearest neighbors of the filled site (in the surface, but not the position where the adatom located before filling), and the exchange is called quasi-exchange. If the vacant site just diffuses to the position where the adatom located before filling, and the exchange is called complete-exchange. It was found that the most of layer-to-layer diffusions happen between single atoms during simulation trace. Although there might exist a few quasi-exchange and few complete-exchange of position, these exchanges are the combination results of basic micro-processes. The possibility, in which the adatom nucleates firstly and then occurs the quasi-exchange or complete-exchange of nuclei, is comparatively small.

The dependence of the number of diffusing atoms on ubstrate temperature is presented in Fig. 3. In this figure, both the surfactant atoms which diffuse upwards to surface and the adatoms which diffuse downwards to surfactant layer increase sharply with temperature increasing. But the tendency becomes obviously slow down while the temperature is higher than the valve (approximately 320K), and the valve might be the maximum saddle energy of layer-to-layer diffusion. Because the atom can not be activated effectively at low temperature and the most of atoms possess larger barrier while diffusing, the atom can but diffuse in the same layer. However, the movement becomes more and more acute with temperature increasing so that many atoms can climb over the saddle and cross through layer when the temperature arrives at the valve.

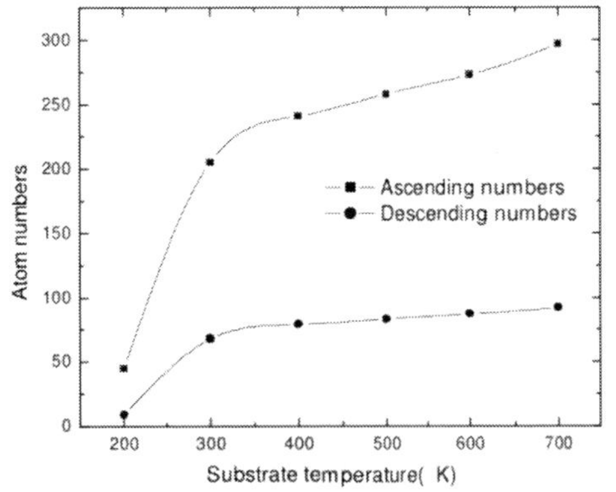

Figure 3. The dependence of atoms crossing layers on the substrate temperatures.

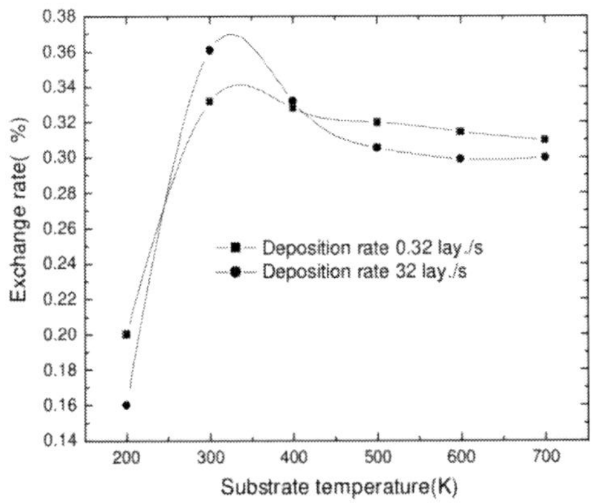

Figure 4. The dependence of exchange rate on the substrate temperature

4. Conclusion

A three-dimensional kinetic Monte Carlo (KCM) technique has been developed for simulating surfactant-mediated epitaxial growth of thin film Cu. The following results are drawn:

(1) The layer-to-layer diffusion is mostly the single atom diffusion during film growth. There might exist a few quasi-exchange and few complete-exchange of position, but these exchanges are combination results of basic micro-processes.

(2) Both surfactant atoms and adatoms diffuse from one layer to another. The majority of diffusion is layer-to-layer diffusion of single atom and the atoms of layer-to-layer diffusion increase with the temperature increasing.

(3) The layer-to-layer diffusion is not simple exchange of position, and not the exchange of 1: 1. The surfactant atoms which diffuse into the surface is greatly more than the adatoms which diffuse into the surfactant, namely the exchange rate is much less than one. The exchange rate increases rapidly with temperature increasing and reaches the maximum at the valve, and then begins to decrease and tend to be stable.

Reference

[1] Pomeroy J M, Jacobsen J, Hill C C, Cooper B H and Sethna J P, 2002*Phys. Rev. B* Kinetic Monte Carlo molecular dynamics investigations of hyper-thermal copper deposition on Cu(111)**66** 235412-1

[2] Combe N, Jensen P, and Pimpinelli 2000*Phys.Rew.Lett.* A Changing shapes in the Nanoworld**85(1)** 110

[3] Henkelman G and Jonsson H 2003 *Phys.Rew.Lett.* Time Scale Simulations of Metal Crystal Growth Reveal the Importance of Multiatom surface Processes**90(11)** 116101

[4] Wang E G 2003 *Physics Advance* Surface kinetic of film growth**23(1)** 1

[5] Zhu W G, Francesco Buatier de Mongeot, Valbusa U, Wang E G 2004*Phys.Rew.Lett.*Adatom Ascending at Stap Edges and Faceting on fcc Metal(110) Surfaces**92(10)** 106102

[6] Wu J, Wang E G, Kvarga, Liu B G, Pantelides S T, and Zhang Z Y 2002*Phys.Rew.Lett.* Island Shape Selection in Pt(111) submonolayer Homoepitaxy with or without CO as an Adsorbate**89(14)** 146103

[7] Zhang Z Y and Lagally M G 1997*Science* Atomistic Processes in the Early Stages of Thin-Film Growth**276** 377

[8] Wang L G and Clancy P2001 *Surf. Sci.* Kinetic Monte Carlo simulation of the growth of polycrystalline Cu film **473** 25

[9] Zhang P F, Zheng X P, and He D Y 2004 *Vacuum* Kinetic Monte carlo simulation of Cu film growth**72(4)** 405

[10] Zhang P F, Zheng X P, Wu S P, He D Y 2004 *Computational Materals Science* Application of computer simulation in thin-film epitaxy growth**30(3-4)** 331

[11] Zhang P F, Zheng X P, He D Y 2005.5 *J.Keroan Phys. Soc.* Monte carlo simulation of nucleation at the initial stage of thin film growth**46** 92

Journal of Physics: Conference Series **188** (2009) 012003 doi:10.1088/1742-6596/188/1/012003

Deflagration Products from Emulsion Explosive

Unconventional nano-powders from ion reaction

Xinghua Xie[1], Jing Zhu and Huisheng Zhou

Anhui University of Science and Technology, Huainan 232001, China

xxh1963@163.com

Abstract. The present situation of detonation synthesis and emulsion explosives is summarized. To solve the problems in research for lithium ion batteries, we suggested that zinc and lithium oxides should be used as cathode materials for lithium ion batteries. So, we design unconventional emulsion explosives, and synthesize zinc and lithium oxides by means of deflagration firstly. Also, microstructures and morphology of nanometer thin sheets in the deflagration soot of emulsion explosives are measured in our experiments; namely, we surveyed and analyzed products with the implement of XRD, TEM and FTIR. It is concluded that nanoparticles of lithium and zinc oxides can be synthesized through deflagration of the unconventional emulsion explosive. Main contents and research results are as follows： First, unconventional emulsion explosives for synthesis of zinc and lithium oxides are designed firstly, and we also discuss mechanics of deflagration. In the final analysis, we succeeded in synthesizing nano-sheets of zinc and lithium oxides by deflagrating at the fist time. Second, we tested microstructures of nanometer thin sheets of lithium and zinc composite oxides. Third, by contrast to industrial emulsion explosives, we analyzed the deflagration mechanics of the unconventional emulsion explosives. Last, zinc and lithium nanooxides are successfully collected and validated by XRD, TEM and FTIR exactly.

1. Introduction

A zinc and lithium composite oxide is the subject of growing interest for different applications in electronic components [1-4]. Especially, G.Ceder regarded that $LiZnO_2$ compound has a higher voltage than that of $LiCoO_2$ [5]. More recently, it was reported that these oxides nanomaterials have generated tremendous interests in both the scientific and engineering community, which has visibly led to rapid and intense growth in research focus [6-13].

[1] Xinghua Xie, Anhui University of Science and Technology, Huainan 232001, China.

In this paper, we report the synthesis of nanometer sheets of composite oxides, which we believe, has never been reported. We can grow these nanoscale structures in a cheap manner through the addition of nitrates using the emulsion explosive deflagration process. This novel method allows us to synthesize nanometer composite oxide sheets.Jianjun Liu et al [14] thought that the shock-induced chemical reactions leading to synthesis of compounds in powder mixtures occurred under conditions of the microsecond -scale duration of the high pressure, stress, strain- rate, and temperature states. Such high-rate chemical reactions can be advantageously utilized to synthesize materials with novel phases and unique microstructures, or to generate radically modified materials with physically interesting or technologically useful properties. I.J. Davidson [15] reported that using solid-state reactions method, usually, the powder preparation route is also quite complicated, for example, several times calcination and subsequent physical grindings. Moreover, its electrochemical properties are greatly dependent on its crystalline particle size [16].

Until now, there are a few reports concerning zinc and lithium composite oxide synthesized by deflagration of emulsion explosives, which is a kind of promising technique for synthesis of lithium transition metal oxide cathodes. Here, for the first time, zinc and lithium composite oxide was synthesized from $LiNO_3$ and $Zn(NO_3)_2 \cdot 6H_2O$ by using fast chemical reaction.

2. Experimental

The emulsions were prepared using a simple facility consisting of a thermostat and a container equipped with a stirrer. The solution of oxidizers was heated to 105℃ and then slowly added to the container, in which a preheated (90℃) mixture of fuels (oil and paraffine) with the emulsifier was agitated with the stirrer at a speed of 800 rpm. After adding the entire amount of the oxidizer, the agitation was continued for about three minutes to obtain fine particles of the emulsion matrices.

The explosive charges were put into a thin plastic bag with an approximate 1330 kg/m3 density and the mass was fixed at 0.800 kg of explosive matter. For each charge an electric No.8 initiator and a 0.038 kg plastic RDX booster were used. The deflagration experiments were performed in an explosive chamber (shown in Fig. 1). This method provides a very fast quenching space. The air surrounding the charge provides efficient cooling of deflagration products and thus reduces the reuniting of obtained nanoparticles. The deflagration experiments were done in a steel tank of 14.1 m3. The explosive charge was placed in a polyethylene bag, which was suspended at the tank center. The deflagration products contained some impurities such as fragments from the tank walls (Fe_2O_3, Al_2O_3), copper and steel from the detonator, and PE from the bag and the leg wires of the detonator. Large size impurities were eliminated by simple filtration of the suspension. The solid residue was washed thoroughly and dried. All the final products were analyzed by XRD [$\lambda CuK\alpha$=1.5406 A°]. So for all the experiments, the deflagration of emulsion explosives synthesized a black powder containing mainly ultradispersed composite oxides of zinc and lithium composite oxide.

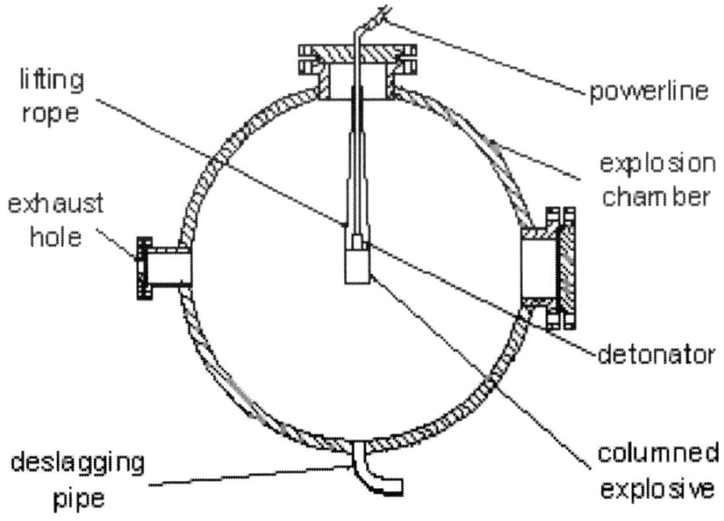

Figure 1 Schematic of explosive chamber

The deflagration soot was then studied by use of X-ray diffraction (XRD). Scanning electron microscopy (SEM) analysis was performed with JEDL JEM-1200EX (Japan) for deflagration synthesized Lithium zinc oxides. XRD analysis was carried out on an XRD-6000 Shimadzu (Japan) diffractometer using Cu Kα irradiation with input power of 50 kV and 150 mA. The divergence slit angle, scattering slit angle and receiving slit height were selected as 2, 2°, and 0.3 mm. The diffraction intensities were measured every 0.028 step for 1 s in the wide 2θ range from 10 to 100° at room temperature (293 K). After sintered at 600℃ for 1 hour, the shape and size of the as-obtained particles were observed by transmission electron microscope (TEM, Tecnai G^2 20 S-twin).

3. Results & Discussion

3.1. X-ray diffraction (XRD) analysis

The exact mechanism of the formation of such nano-grains in the emulsion explosive derived so far is poorly understood. Fig. 2 shows the XRD patterns of the dynamically synthesized zinc and lithium composite oxide. We obtained a nanosized diamond texture containing zinc and lithium composite oxide. It is obvious that the Bragg reflection peaks of the dynamically synthesized zinc and lithium composite oxide are broadened, which may result form small grain size and/or presence of microstrain. Here the mean grain size for deflagration synthesized zinc and lithium composite oxide refers to the mean size of crystallites of crystalline particles. XRD analyses were conducted at a fixed temperature in the present study; a precise determination of the structural parameters need more experiments including both high temperature and low temperature experiments. Explosive deflagration is nonequilibrium processes, generating a short duration of high pressure and high temperature.

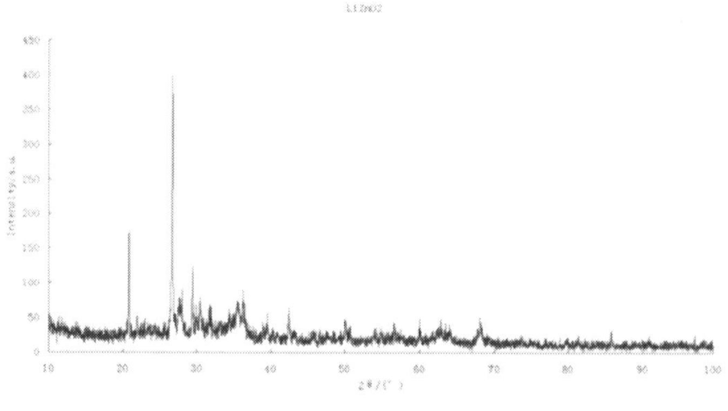

Figure 2 XRD pattern of lithium and zinc oxides.

The structures for the solid solutions were identified by means of powder X-ray diffraction (XRD) on an XRD-6000 diffractometer with Cu Kα radiation at room temperature. The average grain size (D) was measured from the XRD peak using the Scherrer formula [10]:

$$D = 0.9\frac{\lambda}{\beta\cos\theta}$$

(1)

Where λ=0.15406 nm, θ is the diffraction angle of the most intense peak, and β is defined as the experimental half-width.

The calculated mean grain sizes were 51.42 nm for deflagration synthesized zinc and lithium composite oxide. Here the mean grain size for deflagration synthesized zinc and lithium composite oxide refers to the mean size of crystallites of crystalline particles.

3.2. Analysis of TEM

A distinguishing feature of emulsion explosives is that in the aqueous solution, the lithium nitrate, zinc nitrate and ammonium nitrate are present in the form of tiny droplets covered with a very thin fuel layer. As a result, the interfacial surface of the emulsion is very large, the unconventional emulsion explosive in a higher density can be deflagrated completely when initiated.

Fig. 3 (a) and (b) are morphologies by TEM. Transmission electron microscopy was used to characterize the products. Zinc and lithium composite oxide with smaller primary sheets of widths from 50 to 100 nm, thickness 5 nm, and a variety of morphologies were found.

Although TEM picture shows smaller particles, which are aggregated further to form agglomeration, varied sizes, it is necessary to obtain the information regarding smallest primary particles. For this purpose, the TEM study and selected area electron diffraction (SAED) were shown in Fig 3. The SAED pattern of lithium and zinc oxide nanometer sheets shows a clear diffraction ring and varied points corresponding to the crystal planes of phase.

P.W. Chen et al [17] regarded that due to the active surface and strong adsorption activity, UFD particles were aggregated into clusters of dimensions from hundreds of nanometers to several micrometers. The clusters could exhibit round or polyhedral shapes. Highly dispersed and stabilized suspensions of UFD could be obtained through selecting the appropriate liquid media and dispersing agents. UFD surface was covered by a large amount of carbonyl, carboxyl, methyl and nitryl groups. UFD was characterized by having considerable microstress and larger crystal lattice parameters than bulk diamond. Such a high level of lattice deformation could be explained either by the inhomogeneous deformation of UFD particles during the strongly non-equilibrium deflagration process, or by the effect of other inserted atoms in the crystal lattice. The value of UFD microstress did not change noticeably when the UFD was heated up to 1000° C in an Ar atmosphere, which can be explained by the high potential barrier hindering the annealing of defects.

Troyanov et al [18] reported that for a mixture of 20% gibbsite and 80% hexogen the theoretical temperature of the explosion is about 2000°C. Hexogen C3H6N6O6 decomposes into CO_2, CO, H_2O, H and N under explosion. The positive oxygen balance governs the sufficient supply of oxygen in the final effluxes of the explosion.

3.3. Infrared Analysis

The Fourier-IR spectrum Fig. 4 shows that the sheet surface was covered by a large amount of carbonyl, carboxyl, methyl and nitryl groups. Due to this fact, it was possible to chemically modify the sheet surface. The synthesis and purification conditions could have affected the type and amount of groups on the surface of sheet. The presence of these functional groups could cause the deviation of the lattice parameter and the density of sheet from the corresponding values for normal oxide crystals. Due to the active surface and strong adsorption activity, lithium and zinc oxide particles were aggregated into clusters of dimensions from hundreds of nanometers to several micrometers. The clusters could exhibit sheet polyhedral shapes. Highly dispersed and stabilized suspensions of sheet could be obtained through selecting the appropriate liquid media and dispersing agents.

4. Conclusions

The zinc and lithium composite oxides with a thin sheet structure formed after deflagration wave treatment due to the high quenching rate are different from that of the normal combustion products. It might also provide a cheap large-scale synthesis method. Explosive deflagration is strongly nonequilibrium processes, generating a short duration of high pressure and high temperature. Free metal atoms are first released with the decomposition of explosives, and then theses metal and oxygen atoms are rearranged, coagulated and finally crystallized into zinc and lithium composite oxide during the expansion of deflagration process. For emulsion explosive deflagration, the pressure, the temperature and the adiabatic gamma were lower than those of deflagration process, but higher than those of combustion or sintering processes. The inherent short duration, high heating rate and high cooling rate prevent the zinc and lithium composite oxide crystallites from growing into larger sizes and induce considerable lattice distortion.

We can control the morphology and composition of zinc and lithium composite oxide by using deflagration of the unconventional emulsion explosives. The nanoagglomerations, and high surface area of these unique nano structures make them a potential candidate of cathode active materials for Lithium ion batteries.

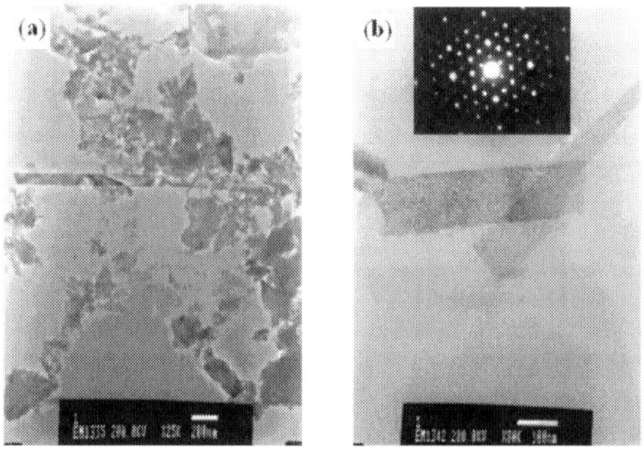

Figure 3 TEM image of deflagration synthesized lithium and zinc oxides.

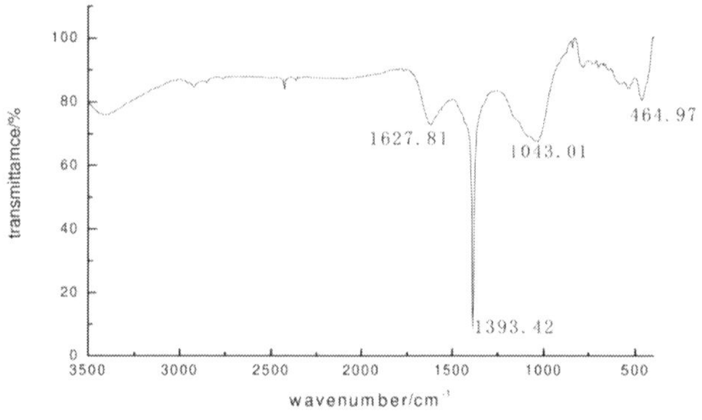

Figure 4 Infrared image of deflagration synthesized lithium and zinc oxides

Acknowledgements

This work was supported by the Anhui Provincial Program for the Tackling Key Problem (07010302189), the Program for the Provincial Innovation Research and Development Team of Anhui Provincial Universities (TD200705) and the 2007 Municipal Natural Science Foundation of Huainan and the Doctoral Fund of Anhui University of Science and Technology.

References

[1] N.V. Borissova, M. M. Pivovarov, V. M. Ushakov, M. M. Schultz, J. Chem. Thermodynamics, 28 (1996) 115-124.
[2] C. J. Rydh, B. A. Sandén, Energy Conversion and Management, 46 (2005) 1957-1979.
[3] T. R. Jensen, R. G. Hazell, A. N. Christensen, J. C. Hanson, Journal of Solid State Chemistry, 166 (2002) 341-351.
[4] A. Apelblat, E. Korin, J. Chem. Thermodynamics, 34 (2002) 1621-1637.
[5] G. Ceder, M.K. Aydinol, A.F. Kohan, Computational Materials Science, 8 (1997) 161-169.
[6] Y.J. He, Powder Technology, 147 (2004) 59-63.
[7] J. Wang, J.M. Cao, B.Q. Fang, P. Lu, S.G. Deng, H.Y. Wang, Materials Letters, 59 (2005) 1405-1408.
[8] Q.Y. Zhang, C.S. Xie, S.P. Zhang, A.H. Wang, B.L. Zhu, L. Wang, Z.B. Yang, Sensors and Actuators B, 110 (2005) 370-376.

[9] T. Fang, J. Duh, S. Sheen, Thin Solid Films, 469–470 (2004) 361-365.
[10] C.C. Hwang, T.Y. Wu, J. Wan, J.S. Tsai, Materials Science and Engineering B, 111 (2004) 49-56.
[11] J. Zhang, Y.D. Yang, B.L. Xu, F.H. Jiang, J.P. Li, Journal of Crystal Growth, 280 (2005) 509-515.
[12] H.H. Wang, C.S. Xie, D.W. Zeng, Journal of Crystal Growth, 277 (2005) 372-377.
[13] P.X. Gao, C.S. Lao, W.L. Hughes, Z. L. Wang, Chemical Physics Letters, 408 (2005) 174-178.
[14] J.J. Liu, H.L. He, X.G. Jin, Z.P. Hao, Z.Q. Hu, Materials Research Bulletin, 36 (2001) 2357-2363.
[15] I.J. Davidson, R.S. McMillan, H. Slegr, B. Luan, I. Kargina, J.J. Murray, I.P. Swainson, Journal of Power Sources, 81–82 (1999) 406-411.
[16] S. Komaba, S.T. Myung, N. Kumagai, T. Kanouchi, K. Oikawa, T. Kamiyama, Solid State Ionics, 152–153 (2002) 311-318.
[17] P.W. Chen, Y.S. Ding, Q. Chen, F.L. Huang, S.R. Yun, Diamond and Related Materials, 9 (9-10) (2000) 1722-1725.
[18] S. I. Troyanov, A. N. Tsvigunov, V. G. Khotin, and T. B. Puzyreva, Glass and Ceramics, 57(7-8) (2000) 241-242.

Easily preparation of high yield anatase TiO_2 crystals by adding a wonderful additive

Yibing Zhang [1,2†], **Lei Jiang** [2,3]
[1]Shangrao Normal University, Shangrao ,Jiangxi 334001 P.R.China.
[2]National Center for Nanoscience and Nanotechnology, Beijing 100080,P.R.China.
[3]Institute Chemistry Chinese Academy of Sciences, Beijing 100080,P.R.China.

E-mail: srxb@sina.com

Abstract. TiO_2 exists in three principal crystallographic forms (anatase, brookite and rutile) but only the anatase form is photocatalytically active and it is changes easily to a rutile-type one, so the technique to easily prepare high yield anatase TiO_2 is difficulty. In this present work, high yield anatase TiO_2 nanocrystals were easily prepared by adding a additive(the key technique) via hydrothermal treatment under optimum conditions (0.15M $Ti(SO_4)_2$ sulfate aqueous solution, the Ti^{4+}/additive molar ratio 1/1, the heat temperature 170℃ and reaction time 4h). The as-prepared TiO_2 was characterized by SEM, TEM and XRD.

TiO_2 exists in three principal crystallographic forms (anatase, brookite and rutile) but only the anatase form is photocatalytically active. Anatase, which has high oxidizing power, is widely used as a photocatalyst in extensive fields such as environmental purification, sterilization and deodorization[1]. TiO2 is also utilized in electrodes of electrochemical and dye-sensitized solar cells[2]. Thus, a technique that easily controls a polymorph of TiO2 in a synthetic process is industrially important. It was known that anatase-type structure changes easily to a rutile-type one, so the technique to easily prepare high yield anatase TiO2 is difficulty and to prepare pure anatase crystals is the hardest work. There are many methods of producing anatase, such as sol–gel method[3], hydrolysis[4], hydrothermal method[5], chemical coprecipitation peptization method[6], electron-beam evaporation[7], pulsed laser deposition (PLD) method[8]. In this present work, high yield anatase TiO_2 crystals were easily synthesized from titanium sulfate aqueous solution by adding additive via hydrothermal treatment, large scale anatase TiO2 nanocrystals self-assembled into microspheres.

The additive (cheaply, Ti^{4+}/additive molar ratio 1/1) was slowly added into the $Ti(SO_4)_2$ aqueous solution under stirring to form 0.15M $Ti(SO_4)_2$, then the solution was placed in Teflon-lined stainless steel autoclaves to heat 170℃ for 4h, the depositions on glass substrates were cooled to room temperature in stove after shut power then removed from the solution, rinsed with deionized water and absolute ethanol and dried in air. Figures 1a, b and c are typical field-emission scanning electron microscope (FE-SEM) top-images of the as prepared microspheres at low and high magnifications, respectively, showing a large scale self-assembly of anatase TiO_2 microspheres which were made of nanocrystals and densely packed. The anatase TiO_2 crystals are octahedron with well-developed faceted surfaces appeared flat triangle planes 100-200nm in side and 1μm approximately in length(Fig.2b), the diameter of microspheres is 2-3μm. From image (Fig.1c) we know the top surface of grains appears to be a gathering of octahedron crystals, while the side surface is relatively smooth. Fig. 1d shows the XRD patterns of depositions, all the diffraction peaks agree with those of TiO_2 in the anantase form and exist a remarkably enhanced (101) peak that is stronger and sharper than the other peaks, which indicates that TiO_2 microspheres have the anatase single crystallinity(supported by inset of Fig.2a).

The samples were then dispersed onto a transmission electron microscopy (TEM) grid for further detailed structure analysis. A careful look at the image in Figure 2 reveals that the morphology and sizes of anatase TiO_2 nanocrystals agree well with the FE-SEM examinations. The electron diffraction pattern (inset in Fig. 2a) indicates that the crystals were single crystals, the rings of the sample were indexed to diffraction from the (101), (004), (200), (105), (204) planes of anatase, which were consistent with the XRD patterns depicted above (Fig.1d).

Figure 1. (a, b and c) FE-SEM top-images of the as-prepared anatase TiO2 microshperes made of nanocrystals at low and high magnifications, respectively. (d) XRD pattern of the as-prepared microspheres.

Fig.2 **(a)**Transmission electron microscopy (TEM) image and selective area diffraction (SAD) pattern (inset) of the partly TiO2;(b) Cross-sectional view of the microshperes deposited for 4 h/170℃. (c)TEM image of the as prepared microspheres at high magnifications

In order to prepare large scale self-assembly of anatase TiO_2 microspheres, we study the optimal reaction conditions (including: concentration of $Ti(SO_4)_2$ aqueous solution, temperature, time and quantity of additive, see supporting information).

In conclusion, high yield anatase TiO_2 crystals were easily prepared by adding additive(key technique) via hydrothermal treatment. Future work involves the investigation of application of anatase in photo-catalysis under visible light.

Acknowledgment. The authors thank the State Key Project for Fundamental Research (G199064504) and the Special Research Foundation of the Natural Science Foundation of Chain(2992530,20/25/02) for continuing financial support.

References

[1] Mattsson A, Leideborg M, Larsson K, Westin G, Qsterlund L 2006 *J. Phys. Chem. B* **110** 1210
[2] Alarco´n H, Boschloo G, Mendoza P, Solis J L, Hagfekdt A J2005 *Phys. Chem. B* **109** 18483
[3] Li G S, Li L P, Goates J B, Woodfield B F 2005 *J. Am. Chem. Soc.* **127** 8659
[4] Hu Y, Yuan C 2005 *J. Cryst. Grow.* **274** 563
[5] Yang Y, Wu Q Y, Guo Y H 2005 *J Molecular Catal. A: Chem.* **225** 203
[6] Xie Y B, Yuan C W, Li X Z 2005 *Colloids and Surf. A: Physicochem. Eng. Aspects* **252** 87
[7] Yang M C, Yang T S, Wong M S 2004 *Thin Solid Films* **221** 469
[8] Suda Y, Kawasaki H, Ueda T 2004 *Thin Solid Films* **453** 162

Supporting Information

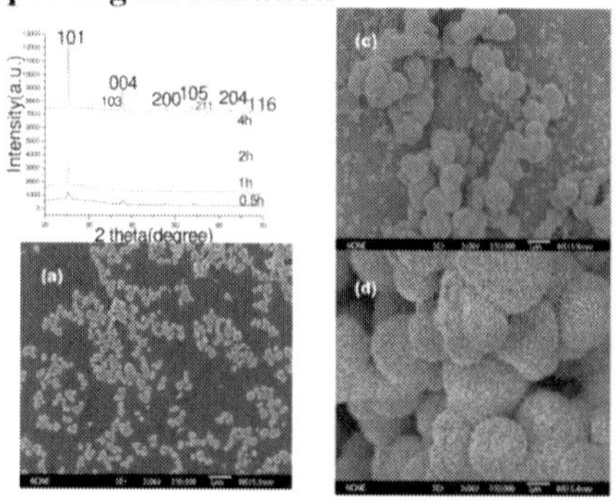

Fig. 1XRD pattern and SEM images of the as-prepared TiO2 microshperes for: (a)0.5h,(b)1h, (c)2h and (d)4h

Fig.3. XRD pattern and SEM images of the as-prepared TiO_2 microshperes in: (a)0.15M, (b)0.10M, (c)0.05M ,(d)0.01M

Fig.2. XRD pattern and SEM images of the as-prepared TiO_2 microshperes usingTi4+/additive: (a)1/1, (b)1/2, (c)2/1, and (d)1/4

Fig.4. XRD pattern and SEM images of the as-prepared TiO2 microshperes at: (a)90℃,(b)130℃ and (c)170℃

Preparation of $LaTiO_2N$ Photocatalyst with a Large Surface Area

Liu Shuiping[1,2], Tan Lianjiang[1,2], Zhang Yu[1,2], Chen Yanmo[1,2*]

Abstract

In this following communication, a new nanocrystalline photocatalyst $LaTiO_2N$ with a large surface area of 27.5 m^2/g was synthesized by nitridation of amorphous La_2O_3/TiO_2 composite powder at 900 °C for 8 h using NH_3 as the reactant gas, and the co-precipitate method was used to synthesize the high active amorphous La_2O_3/TiO_2 composite powder. The X-ray powder diffraction (XRD) and transmission electron microscopy (TEM) results revealed that the as-prepared $LaTiO_2N$ nanocrystals had a mean diameter of about 30 nm. With increasing the nitrided temperature from 700 °C to 900 °C, it can be seen that the color of the oxynitride vary from yellow to red, and the specific surface area of the sample became lower since the high temperature can lead to the coalescence and aggregation of the nanoparticles. Both a higher nitridation temperature and a longer nitridation time can lead to a higher nitrogen content of the oxynitride. The TEM of the oxynitride obtained at 700 °C, 800 °C revealed some voids are in the oxynitride nanoparticles, which may be attribute to the density difference between oxide and oxynitride. The band gap energy of $LaTiO_2N$ was estimated to be 2.1 eV for its absorption band was founded to extend to 600 nm, it indicated that the synthesized oxynitrides displayed good light absorption properties not only in the ultraviolet light but also in the visible light region. Its high specific surface area and narrow band gap energy may guarantee photocatalytic activity under visible light excitation.

Keywords: Red-shifted; Visible-Light; oxynitride; photocatalyst

1. Introduction

In recent years, there has been great concern over many serious environmental problems and a lack of natural energy resources we face on a global scale. The photocatalytic reaction has attracted much interest as a potential route to solve the energy and environmental issues [1-4], but most of photocatalysts as found only respond to UV light excitation. Of them, TiO_2 is known as one of the most effective photocatalysts for the degradation of organic pollutants, and its photocatalytic behavior has been studied extensively [5-7]. However titanium dioxide is effective only under UV excitation (λ < 387 nm) because of its wide band gap energy of 3.2 eV, which limits its further application in the visible-light region (λ > 400 nm). Therefore there is an urgent need to develop new types of photocatalysts responding to visible light irradiation. Recently, some (oxy)nitrides and oxysulfides such as Ta_3N_5 [8], TaON [9], $LaTiO_2N$ [10], $Sm_2Ti_2O_5S_2$ [11], $BaNbO_2N$ [12] and $(Ga_{1-x}Zn_x)(N_{1-x}O_x)$ [13] have been reported as potential visible-light-driven photocatalysts. These stable inorganic materials were usually synthesized by nitriding the corresponding metal oxides under NH_3 [10, 12], and they could absorb visible light up to wavelengths of 500-600 nm. However, the corresponding metal oxides were most prepared by high temperature solid-state reaction, which lead to a surface area lower than 5 m^2g^{-1} [9, 10, 12]. In this communication, wet chemical method was used to synthesize amorphous metal oxides as a highly reactive raw material for the nitridation reaction. Since its high reactive activity, the $LaTiO_2N$ with large surface area is expected to be synthesized by ammonolysis of a homogenous nano-sized La_2O_3/TiO_2 composite powder at relatively low temperature.

2. Experimental

Synthesis of La_2O_3/TiO_2 and $LaTiO_2N$

In a typical synthesis, 0.02 mol lanthanum nitrate and 0.02 mol tetrabutyl orthotitanate were separately dissolved in 50 ml absolute alcohol in each case. After being stirred for 20 min, two solutions were mixed under vigorous stirring, yielding a clear, yellow solution. Then the mixed solution was

*Correspondence to Prof. Chen Yanmo, State Key Laboratory for Modification of Chemical Fibers and Polymer Materials, Donghua University, Shanghai 201620, P. R. China. E-mail: ymc@dhu.edu.cn or liushuiping123@163.com.

slowly dropped into violently stirring 250 ml 4.0 M ammonia solution, which contained 0.5 g polyethylene glycol. The precipitate was collected by the filtration, and then washed with the distilled water and ethanol. After vacuum-oven drying at 60 °C for 24 h, the dried precipitate was calcined at 500 °C for 2 h. Finally amorphous La_2O_3/TiO_2 composite powder was obtained.

For the preparation of the $LaTiO_2N$, La_2O_3/TiO_2 composite powder was put into a tube furnace and subjected to the nitridation in the flow of ammonia gas. The flow volume of the NH_3 gas was 0.2 $Lmin^{-1}$, and the nitridation temperature ranged from 700 to 900 °C, while the nitridation time was fixed at 8 h. Samples were taken from the furnace after cooling to room temperature in the flow of NH_3 gas.

The powder phase composition was identified by X-ray diffraction (XRD) equipment (model D/max 2550pc, Rigaku Co., Tokyo, Japan), using Cu Kα (λ=1.5406 Å) radiation. Elemental analysis for carbon (C), hydrogen (H), and nitrogen (N) contents in the nitride powder was carried out on an Elmentar CHN analyzer (Vario EL III, Germany), The morphology and size of the resultant oxynitrides were observed using transmission electron microscopy (TEM; H-800, Hitachi, Japan) and the BET specific surface area measurement was performed on a nitrogen adsorption apparatus (model ASAP 2010, Micromeritics Instruments, Norcross, GA). UV–vis absorption spectra were recorded on a Perkin Elmer UV Lambda 900 PC instrument.

3. Results and Discussion

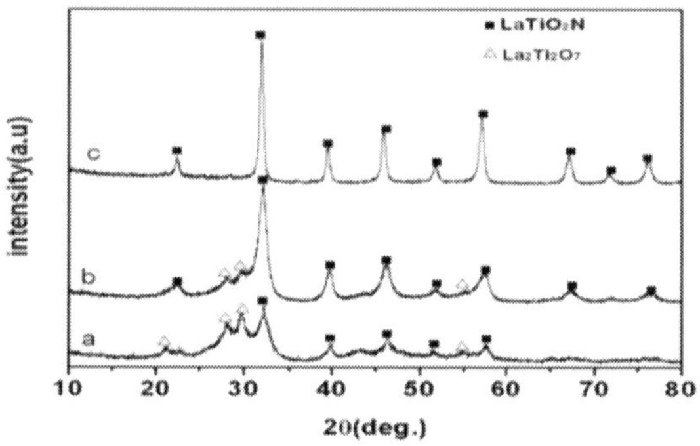

Fig. 1. XRD pattern of oxynitride powder obtained by nitridation of the amorphous oxide at (a) 700 °C, (b) 800 °C and (c) 900 °C.

Fig. 1 shows the XRD patterns of the nitrided powders derived from amorphous nanoparticles annealed at 500 °C. The whole nitridation reaction can be expressed as:

$$La_2Ti_2O_7 + 2NH_3 = 2LaTiO_2N + 3H_2O \quad (1)$$

For the oxide nitrided at 700 °C for 8 h, orthorhombic $LaTiO_2N$ (JCPDS 48-1230 card) crystalline phases can be detected, while a large number of the monoclinic $La_2Ti_2O_7$ (JCPDS 28-0571 card) crystalline phases still can be observed. When the amorphous oxide was nitrided at 800 °C for 8 h, only a little monoclinic $La_2Ti_2O_7$ crystalline phase can be detected. The complete nitridation of amorphous $La_2Ti_2O_7$ into $LaTiO_2N$ was achieved at 900 °C. All the diffraction peaks could be perfectly indexed to the orthorhombic $LaTiO_2N$ structure (JCPDS 48-1230 card). And the average size of the $LaTiO_2N$ particles could be estimated from the Scherrer' equation to be of about 25 nm.

Table 1 Surface area, crystalline phase, nitrogen content and color of the resulting powders

	Sample	T, °C	t, h	Sa, m^2g^{-1}	Phase	N, wt%	Color
Oxynitride	**A**	700	8	32.3	$LaTiO_2N$+$La_2Ti_2O_7$	2.36	Yellow
	B	800	8	30.4	$LaTiO_2N$+$La_2Ti_2O_7$	4.32	Orange
	C	900	8	27.5	$LaTiO_2N$	5.45	Red
Oxide	**D**	800	2	42.3	$La_2Ti_2O_7$	none	White
	E	500	2	93.8	Amorphous	none	White

Note: T = temperature, t = nitridation time, Sa = the specific surface area, and N = nitrogen content in weight.

As listed in Table 1, the oxynitride powder obtained after nitridation varied in color from yellow to red as increasing nitridation temperature. The nitrogen contents of oxynitride determined by combustion method on a CHN elemental analyzer are also listed in Table 1. The average nitrogen content of sample C is 5.45 wt %, which is very close to the theoretical value 5.98 wt % for $LaTiO_2N$. As listed in Table 1, both a higher temperature and a longer time can lead to a higher nitrogen content. The specific surface area of the sample was also listed in the Table 1. As the nitridation temperature varied from 700 °C to 900 °C, the specific surface area of the sample became lower since the high temperature can lead to the coalescence and aggregation of the nanoparticles.

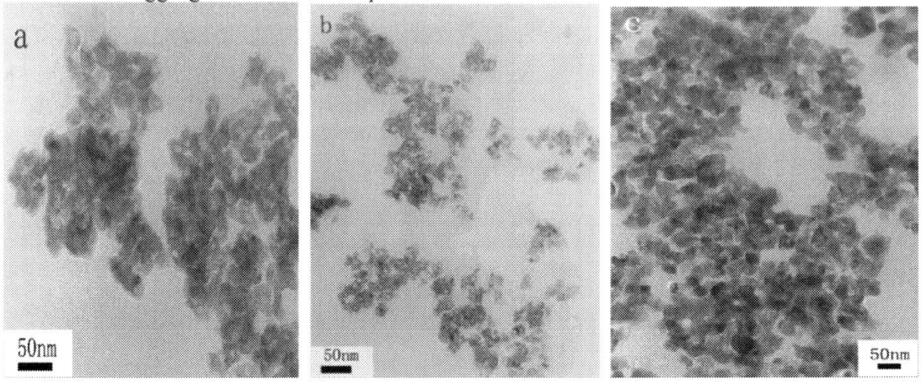

Fig. 2. TEM micrograph of the oxynitride powder obtained by nitridation of the amorphous oxide at (a) 700 °C, (b) 800 °C and (c) 900 °C.

TEM micrographs of the powder nitrided at different temperatures are given in Fig. 2. For the oxynitride powder by ammonolysis of the amorphous La_2O_3/TiO_2 at 900 °C for 8 h, it can be seen that the particle size of the oxynitride is about 30 nm, the shape is spherical, and little agglomerate can be observed. The size observed by TEM is consistent with the calculated crystallite size by Scherrer' equation. The TEM of the oxynitride obtained at 700 °C, 800 °C are also given in the Fig. 2(a) and 2(b) respectively, which reveal the nitridation process of the amorphous oxides. It can be seen that some voids are in the oxynitride nanoparticles, which may be attribute to the density difference between oxide and oxynitride. The $LaTiO_2N$ has higher density than the $La_2Ti_2O_7$. Herein, as La_2O_3/TiO_2 converted to one particle of $LaTiO_2N$, a volume shrinkage will occur during the nitridation. But for the higher-temperature nitridation at 900 °C, these voids are lost, which can be explained as the completely nitridation. This is consisted with the result that no other crystalline phases besides those of $LaTiO_2N$ were detected by the XRD (Fig. 1c).

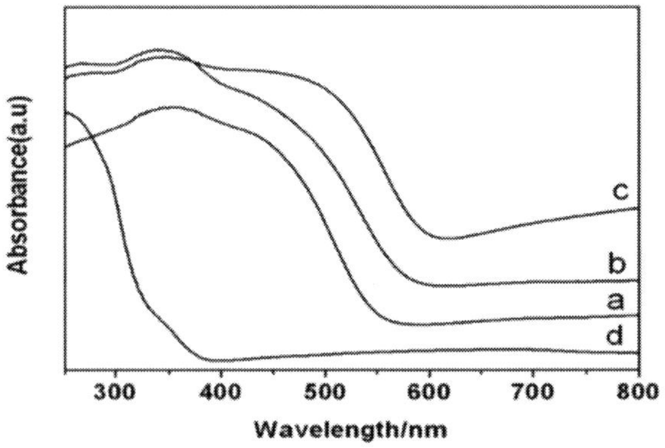

Fig. 3. UV-vis absorption spectra of the oxynitride powder obtained by nitridation of the amorphous oxide at (a) 700 °C, (b) 800 °C, (c)900 °C, curve (d) obtained by annealing of the amorphous oxide at 800 °C for 2 h.

The UV-vis absorption spectra of the oxynitride nanocrystals synthesized at different temperatures are shown in Fig. 3. It can be seen that the absorption edge of the oxynitride shifted towards the red region as increasing the nitridation temperature. The band gap energy of $LaTiO_2N$ (sample C) was estimated to be 2.1 eV[14], which is in a good agreement with the Domen's reports [10, 15]. This result can be implied from the analogous to the $La_2Ti_2O_7$ photocatalyst that the top of the valence band (i.e., highest occupied molecular orbital, HOMO) for the $La_2Ti_2O_7$ consists of O2p orbitals. When N atoms are partially substituted for O atoms, the HOMO of the $LaTiO_2N$ is expected to be shifted higher compared to the $La_2Ti_2O_7$ since it consists predominantly of N 2p orbitals with a small contribution by O 2p orbitals, without affecting the level of the bottom of the conduction band (i.e., lowest unoccupied molecular orbital, LUMO), which is made up entirely of empty Ti 3d orbitals.

4. Conclusion

In summary, the preparation of nanocrystalline $LaTiO_2N$ powder based on the nitridation of amorphous La_2O_3/TiO_2 composite oxide nanoparticles is presented. The quality of the resulting oxynitride powder was very sensitive to the preparative parameters, especially the nitriding temperature. When the amorphous La_2O_3/TiO_2 composite oxide was nitridated at 900 °C for 8 h, the perfect crystalline $LaTiO_2N$ powder can be obtained. The band gap energy of $LaTiO_2N$ was estimated to be 2.1 eV for its absorption band was founded to extend to 600 nm. The crystalline $LaTiO_2N$ powder was of the size of about 30 nm with a specific surface area up to 27.5 m^2g^{-1}. Its high specific surface area and narrow band gap energy may guarantee photocatalytic activity under visible light excitation.

References

[1] A. Fujishima, K. Honda, Nature 238 (1972) 37.
[2] R. Asahi, T. Morikawa, T. Ohwaki, K. Aoki, Science 293 (2001) 269.
[3] Z. G. Zou, J. H. Ye, K. Sayama, H. Arakawa, Nature 414 (2001) 625.
[4] M. Kitano, M. Matsuoka, M. Ueshima, M. Anpo, Appl. Catal. A: General. 325 (2007) 1.
[5] M. R. Hoffmann, S. T. Martin, W. Choi, D. W. Bahnemann, Chem. Rev. 95 (1995) 69.
[6] A. L. Linsebigler, G. Lu, J. T. Yates, Chem. Rev. 95 (1995) 735.
[7] S. Anandan, M. Yoon, J. Photochem. Photobiol. C: Photochem. Rev. 4 (2003) 5.
[8] Q. H. Zhang, L. Gao, Langmuir 20 (2004) 9821.
[9] G. Hitoki, A. Ishikawa, T. Takata, J. N. Kondo, M. Hara, K. Domen, Chem. Commun. 16 (2002) 1698.
[10] A. Kasahara, K. Nukumizu, G. Hitoki, T. Takata, J. N. Kondo, M. Hara, et al., J. Phys. Chem. B 107 (2003) 791.
[11] A. Ishikawa, T. Takata, J. N. Kondo, M. Hara, H. Kobayashi, K. Domen, J. Am. Chem. Soc.124 (2002) 13547.
[12] K. Maeda, K. Domen, J. Phys. Chem. C 111 (2007) 7851.
[13] K.Maeda, K.Teramura, D. Lu, T. Takata, N.Saito, Y. Inoue,et al., Nature 440 (2006) 295.
[14] A. Hagfeldt, M. Gratzel, Chem. Rev. 95 (1995) 49.
[15] A. Kasahara, K. Nukumizu, G. Hitoki, T. Takata, J. Kondo, M. Hara, et al., J. Phys. Chem. A 106 (2002) 6750.

Effect of Ti doping on microstructure of SiO_2-CTAB mesoporous films

YUE Chunxiao[1,2], YAO Lanfang[2], TIAN Linlin[2] and FANG Xueling[2],
1 Shanghai Jianqiao College, shanghai 201319,
2 University of Shanghai for Science and Technology, Shanghai 200093, China

E-mail:chunxiao.yue@gmail.com

Abstract. $n$$SiO_2$-$TiO_2$-CTAB ($n$ is the molar ratio of Si to Ti, n=2, 10, 20, 40, 60, 80) composite films were synthesized by sol-gel process of titanium isopropoxide (TPOT) and tetraethyl orthosilicate (TEOS) hydrolyzation using surfactant cetyltrimethyl ammonium bromide (CTAB) as the template. The microstructure of the films was investigated by Transmission electron microscopy (TEM), atomic force microscope (AFM), X-ray diffractometer (XRD) and UV- infrared spectrophotometer. The results show that Ti ions homogeneously disperse in the mesoporous SiO_2 matrix forming Si-O-Ti bonding when n is greater than 20, and Ti ions disrupt the uniformity of the pores in the composite mesoporous films and reduce the structural order when n is lesser than 10. SiO_2 and TiO_2 of SiO_2-TiO_2-CTAB mesoporous films is amorphous calcined at 450℃, 600℃, 700℃, 800℃.

1 Introduction

Mesoporous silica films have many unusual properties such as an adjustable refractive index, high porosity and low thermal conductivity and dielectric constants. Therefore, it has become increasingly attractive to use the films in optical and microelectronic applications, thermal insulation, catalyzed sorption and so on[1,2]. Currently significant progress has been made in the fabrication and characterization of ordered mesoporous SiO_2 films with rigidly uniform pores[3]. For pure SiO_2 material, its skeleton is absence of acid sites, and catalytic activity is low. One way to do this problem is dopant of activated element to prepare composite mesoporous material. TiO_2 is a wide bandgap (E_g≈3.2eV) material exhibiting excellent photoelectric efficiency and photocatalysis properties[4]. The addition of TiO_2 to SiO_2 induces higher acidic properties to a great extent, and enhances the photocatalytic performance and mechanical property as well[5]. Ti-doped SiO_2 mesoporous material can also be considered as attractive materials for high thermal stability, superior adsorption capability with active components and good redox properties[6~8]. In syntheses through conventional sol–gel processes, however, the preparation of mesoporous TiO_2–SiO_2 mixed oxides due to the differences in hydrolysis and condensation rates of Ti- and Si-alkoxides was identified as a major problem.

In the present paper, the fabrication of SiO_2-TiO_2-CTAB mesoporous films via sol-gel processing using CTAB as template and acetylacetone as hydrolysis inhibitor of TPOP was described, and the effect of Ti doping on microstructure of SiO_2-CTAB mesoporous films were investigated. Such composite mesoporous materials would not only take advantage of both TiO_2 and SiO_2, but would also extend their applications through the generation of new catalytic active sites due to the interaction of TiO_2 with SiO_2.

2 Experimental

2.1 Sol preparation

In the preparation of the sols for dip coating, TEOS and TPOT were used as silica and titanium sources, respectively. For the SiO_2 precursor solution, TEOS, ethanol, hydrochloric acid and deionized water were mixed at room temperature under magnetic stirring. Dilute hydrochloric acid was then added dropwise to the above sol to adjust the PH-value about 2. After aged at 50°C for several hours, the sol was added to CTAB(2.5 wt. %) and then stirred for 2h. For TiO_2 precursor solution, TPOT was mixed with acetylacetone (AcAc, as chelating agent of the titanium alkoxide) at a molar ratio of 1:1. Ethanol, deionized water and HAc were then added dropwise. According to the desired titanium contents in the final sol-gel films, $n$$SiO_2$-$TiO_2$-CTAB composite sols were formed. n is the molar ratios of Si to Ti, n=2, 10, 20, 40, 60 and 80. The final composite sols were aged for about 5~7 days before coating the films.

2.2 Samples preparation

Both silicon wafers and microscope slide glass were used as substrates. The films were performed by the dip-coating method in a clean room with a relative humidity <60% at 20℃~25℃. The dip-speed was optimized at 8inch/min. The samples were baked in a SiO_2 tube with an inner diameter of 10 mm.

2.3. Characterization

X-ray diffraction analysis was done on the films by using RigakuD/max 2550 Series XRD operated with CuKa radiation, at 40 KV, 100 mA. A FTIR (FTS-40 Bio-Rad) spectrometer was employed to distinguish properties of the films with different TiO_2 content. The resolution is 2 cm^{-1}, and an uncoated silicon wafer was used as a reference. The UV-visible absorbance spectra (measured by a JASCO V-570 UV/vis spectrophotometer) of the films deposited on microscope glass slides were examined in the wavelength range of 200~800 nm. TEM image was recorded on a JEM-2100F HR-TEM operated at 200kV. Surface image was observed by AFM (PSIA Corporation XE-100 advanced Scanning Probe Miscrope).

3 Results and discussion

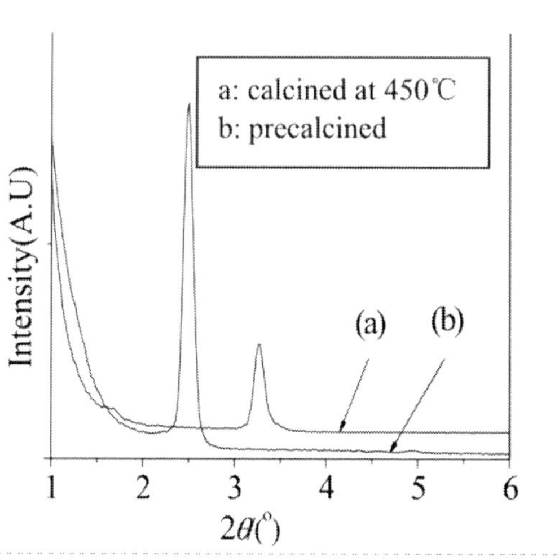

Figure 1. SXRD patterns of SiO_2-CTAB mesoporous films

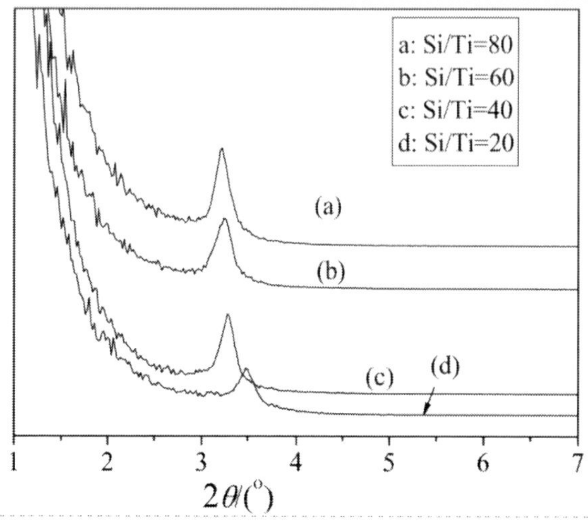

Figure 2. SXRD patterns of SiO_2-TiO_2-CTAB mesoporous films.

As seen in figure 1, a sharp diffraction peak indexed as (100) reflection of a hexagonal structure presents in small angle region, which illuminates that the structure of SiO_2-CTAB films is long-term ordered mesostructure because of the orientation of the surfactant template. The diffraction peak of SiO_2-CTAB film precalcined locates at about 2.5°, and interplanar crystal spacing $d_{(100)}$ is 3.54 nm. After calcined at 450℃, the diffraction peak shifts to 3.3° and $d_{(100)}$ is 2.68 nm because the framework of SiO_2-CTAB film shrinks. Full width at half maximum has few changes and ordered mesostructure not spoiled which hole spacing is about 3.12 nm as supported by TEM image.

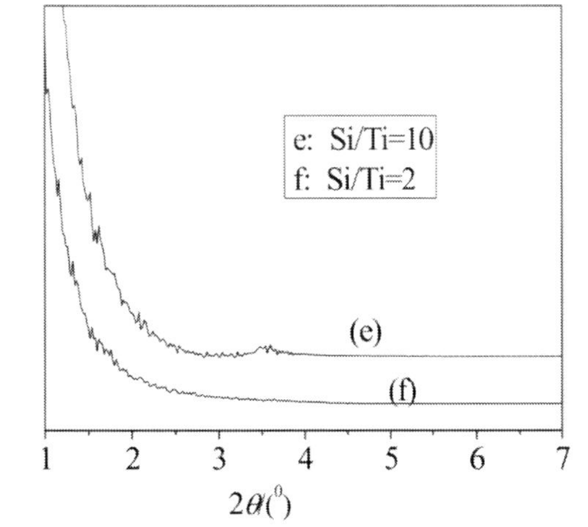

Figure 3. SXRD patterns of SiO_2-TiO_2-CTAB composite films.

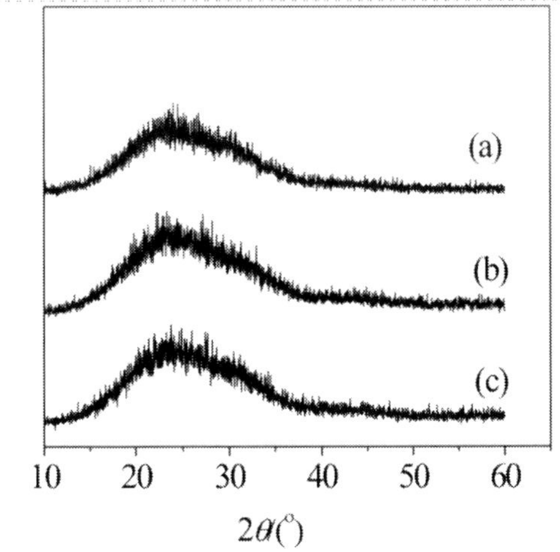

Figure 4. XRD patterns of SiO_2-TiO_2-CTAB mesoporous film calcined at 600℃, 700℃, 800℃.

Figure 2 and figure 3 display SXRD patterns of $n$$SiO_2$-$TiO_2$-CTAB ($n$=2, 10, 20, 40, 60 and 80) composite films calcined at 450℃. The diffraction peak (100) can be observed distinctly as shown in figure 2, which indicates that the films are long-term ordered mesostructure. Additionally, it can be noted that the diffraction peak (100) shifts slightly towards the higher value and $d_{(100)}$ decreases with the increase content of titanium. These results suggest that Ti ions incorporate into the silicon-oxide skeleton and compose the mesoporous structures with SiO_2 uniformly. With the increase of titanium content, the intensity of t the diffraction peak (100) decreases gradually. Especially, it becomes quite weak when $n \leq 10$ (shown in figure 3). It indicates that Ti ions disrupt the uniformity of the pores in the hybrid mesoporous materials and reduce the structural order. More Ti ions could change the silica charge density and interfere with the formation of the mesoporous structures.

As regards the microstructural evolution of the films, XRD patterns of all samples calcined at 450℃ show a diffraction peak pronounced at about 24° respectively, which is characteristic of no crystalline phase. XRD patterns of SiO_2-TiO_2-CTAB mesoporous film (Si/Ti=40) calcined at 600℃, 700℃, 800℃ are shown in figure 4. It indicates the channels' walls of the SiO_2-TiO_2-CTAB mesoporous films are essentially amorphous structures, and TiO_2 particles homogeneously disperse in the amorphous SiO_2 matrix.

As shown in Figure 5, the most intense absorption peak located at 1075cm^{-1} corresponds to a large amount of Si-O-Si stretching vibrational bonding, which constitutes the skeleton structure of films. The wide band at around 3430 cm^{-1} corresponds to Si-OH stretching. An shoulder peak at around 953 cm^{-1} is due to Si-O^- stretching vibration in Si-O-Ti^{+4} sequences involving tetrahedrally coordinated Ti^{+4} cations, together with Si-O vibration in Si-OH groups. And the intensity of the absorption band for Si-O-Ti bonding can be seen as an indication of degree of inclusion of titanium ions in the silica structure.

Journal of Physics: Conference Series **188** (2009) 012006 doi:10.1088/1742-6596/188/1/012006

Figure 6 shows the optical absorption spectra of SiO_2-TiO_2-CTAB mesoporous films coated on microscope slide glass after subtracted the effect of substrates. A relatively strong absorption at about 220 nm can be clearly seen and it becomes stronger with the increase of Ti content. It suggests that Si-O-Ti tetrahedral structure has formed via distribution of Ti ions in SiO_2 framework of mesoporous films. In addition, it must be noted that the adsorption edges of SiO_2-TiO_2-CTAB mesoporous films shift to red with the increase of Ti content. Owing to agglomeration, band-gap energy decreases.

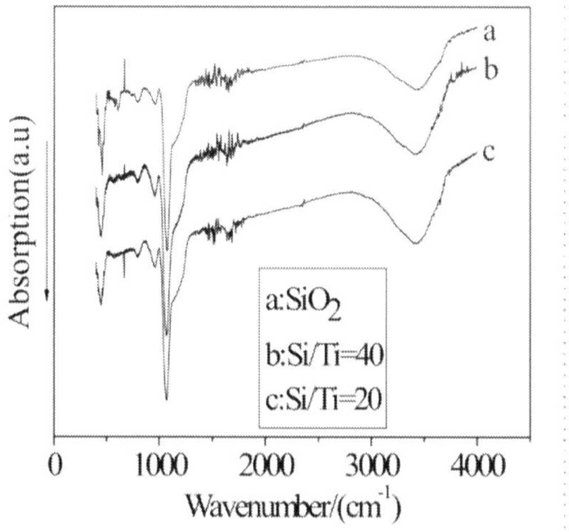

Figure 5. FTIR absorption spectrum of mesoporous films

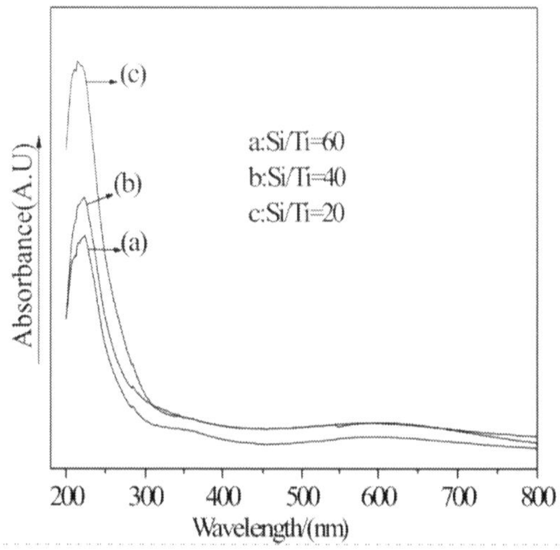

Figure 6. UV-Vis absorption spectrum of mesoporous films.

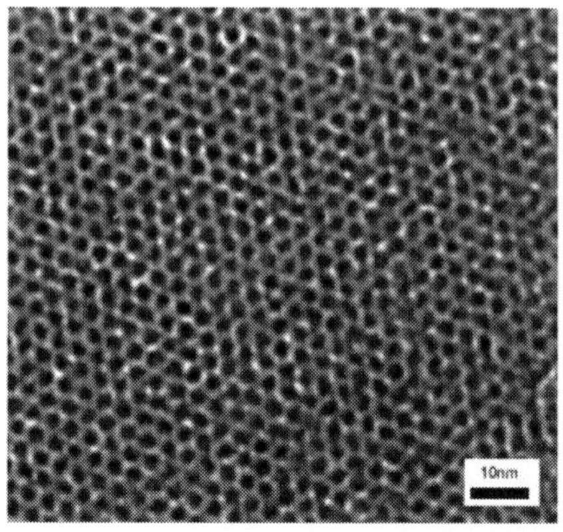

Figure 7. TEM image of SiO_2-CTAB mesoporous film.

Figure 8. AFM image of SiO_2-TiO_2-CTAB mesoporous film.

Figure 7 is TEM image of SiO_2-CTAB mesoporous film. It clearly shows the well-ordered periodic hexagonal mesostructure, and the hole spacing being about 3.2 nm. AFM image of SiO_2-TiO_2-CTAB mesoporous film (Si/Ti=80, calcined at 450℃) is shown in figure 8. its pore channels parallel to the substrate surface which is character of hexagonal mesostructure.

4 conclusion

SiO_2-TiO_2-CTAB mesoporous hybrid thin films were successfully synthesized by a modified sol-gel process. The results show that Ti ions homogeneously disperse in the mesoporous SiO_2 matrix forming Si-O-Ti bonding when the molar ratio of Si to Ti is greater than 20, and Ti ions disrupt the uniformity of the pores in the hybrid mesoporous materials and reduce the structural order when the molar ratio of to is lesser than 10. The adsorption edges of SiO_2-TiO_2-CTAB mesoporous films shift to red with the increase of Ti content .In addition, XRD analysis suggest that SiO_2 and TiO_2 of the framework is amorphous calcined at 450℃～800℃.

5 References

[1] Yamane T, Nagai N, Katayama S and Todoki M 2002 *J. Appl. Phys.* **91** 9772
[2] Mark E and Davis 2002 *J. Nature.* **417** 813
[3] Zhang J L, Li W, Meng X K, Wang L and Zhu L 2003 *J. Memb. Sci.* **222** 219
[4] Sabataitytė J, Oja I, Lenzmann F, Volobujeva O and Krunks M 2006 *J. C.R . Chimie.* **9** 708
[5] Kwon C H, Kim J H, Jung I S, Shin H and Yoon K H 2003 *J. Ceramics International.* **29** 851
[6] Belhekar A A, Awate S V and Anand R 2002 *J. Catalysis Communications.* **3** 453
[7] Li Z J, Hou B, Xu Y, Wu D, Sun Y H, Hu W and Deng F 2005 *J. Solid State Chemistry* **178** 1395
[8] Lafond V,Mutin and Vioux A 2004 *J.Chem. Mate.* **16** 5380

Journal of Physics: Conference Series **188** (2009) 012007 doi:10.1088/1742-6596/188/1/012007

Enhanced photocatalytic properties of quantum-sized ZnO induced by La^{3+}-Nd^{3+} co-doping

S X Lu*, T Zhu and W G Xu*
Institute of Chemical Physics, Beijing Institute of Technology, Beijing 100081, China

E-mail: S X Lu: shixianglu@bit.edu.cn; W G Xu: xuwg60@bit.edu.cn

Abstract. Nanoparticles of ZnO doped with La^{3+}, Nd^{3+}, co-doped with La^{3+} and Nd^{3+} were prepared using the sol-gel method. The samples were characterized by means of X-ray diffraction (XRD), UV-Vis spectroscopy (UV). The photocatalytic reactivities were evaluated by photodegradation of unsymmetrical dimethylhydrazine solution. Nanocrystalline ZnO co-doped with La^{3+} and Nd^{3+} at optimal concentration (2 at.% La^{3+} and 1.5 at.% Nd^{3+}) shows a better synergistic effect, which significantly increases the photodegradation activity of nano-ZnO. The factors influencing photoreactivity of co-doped nanocrystalline ZnO have been discussed.

1. Introduction

Nanoparticles zinc oxide and doped zinc oxide have been used as effective, inexpensive, nontoxic semiconductor for the degradation of a wide range of organic chemicals[1-5]. The present work focuses on the use of ZnO and and solar radiation for effective elimination of organic chemicals at ambient temperature[6-7]. However, ZnO has some drawbacks including the fast recombination rate of photogenerated electron-hole pair and a low quantum yield in the photocatalytic reactions in aqueous solutions, which obstruct commercialization of the photocatalytic degradation process [10]. Therefore, various methods have been developed to reduce the e^-/h^+ recombination rate of ZnO in the photocatalytic processes. The challenge of this work is how to obtain different types of oxygen defects in ZnO nanocrystals [8]. One of the interesting approaches is to dope the transition, alkaline and rare earth metal atoms in semiconductors [9]. It has been known that the advantage of the doping of the metal ions in semiconductor particles is the temporary trapping of the photogenerated charge carriers by the dopant and the inhibition of their recombination during migration from inside of the material to the surface or the enhanced association of the functionalized organic pollutants to the doping ion surface sites. The efficiency of photocatalytic reactions is limited by the high recombination rate of photoinduced electron-hole pairs formed in photocatalytic processes and by the absorption capability for visible light of photocatalysts [10, 11]. Ping Yang et al [10] found that TiO_2 co-doped with Fe^{3+} and Eu^{3+} ions could lead to superior photocatalysis. A pronounced synergistic effect in catalytic activity was noted. The facts indicate that introducing two or more proper metals into nanocrystalline semiconductor particles will improve the photocatalytic effect.

We believe that co-doping rare earth metal ions onto the nanocrystalline ZnO may achieve the same effect as co-deposited onto TiO_2. To the best of our knowledge, there are few studies carried out on ZnO semiconductor particles co-doped with two kinds of rare earth metal ions for photo-oxidative degradation of toxic compounds. Here, in order to research the photocatalytic property of nanocrystalline ZnO co-doped with La^{3+} and Nd^{3+} metal ions, the samples synthesized by sol-gel technique was prepared, its characterizations were described in detail, and its photocatalytic property was investigated with unsymmetrical dimethylhydrazine solution as the model.

2. Experimental

2.1 Photocatalyst Determination

The crystalline structure was investigated by XRD on x'Pert. Pro. MPD X-ray diffractmeter with Ni filtered Cu K_α radiation of 0.1542 nm at the scan rate of $0.05°s^{-1}$. The UV of ZnO were carried out on HITACHI U-3310 UV-Vis spectrophotometer. All the chemicals and solvent used in present study were of AR grade. Anhydrous ethanol were refluxed over magnesium strips and then distilled.

2.2 Sample preparation

The preparation procedure of ZnO nanoparticles was similar to a method described elsewhere [12], with some modifications. Lithium hydroxide monohydrate (0.29 g, 7 mmol) was dissolved in 50 ml of absolute ethanol at room temperature in an ultrasonic bath. Zinc acetate dihydrate (1.10 g, 5 mmol) was dissolved in 50 ml of refluxing ethanol. The stirred zinc acetate solutions were cooled to 50°C and held isothermal while the lithium hydroxide monohydrate solution was added. The resulting mixtures were stirred for 40 minutes as ZnO sol, and then 200 ml of cold hexane was added to the ZnO sol and kept them at the temperature 4°C for 24 h to prevent particles of ZnO growing rapidly. The white powder would be achieved by evaporated at room temperature. The powder was calcined for 2 h in muffle at 300°C as the best calcined temperature [12]. Doped and co-doped ZnO nanoparticles were synthesized using almost the same method. The appropriate amount of $LaCl_3$ and/or $NdCl_3$ was dissolved in 50 ml anhydrous ethanol which was added prior to the $Zn(CH_3COO)_2{\cdot}6H_2O$ The remaining procedures were the same as described above. For convenience, the samples were labeled as La(X) for La^{3+}doped ZnO, Nd(Y) for Nd^{3+}doped ZnO, and La(X)Nd(Y) for La^{3+} and Nd^{3+} co-doped ZnO, respectively, where X and Y refer to the nominal atomic concentration of La^{3+} and Nd^{3+}, respectively, which is based upon the assumption of quantitative incorporation of the dopants.

2.3 Photocatalytic activity

Unsymmetrical dimethylhydrazine solution was chosen as the reaction to quantify the photocatalytic reactivity of each sample. The experiments were carried out with 100 ml reactive unsymmetrical dimethylhydrazine solution prepared with distilled water in 200 ml beakers. The beakers were put under the ultraviolet light maintaining the distance between the light source and the surface of the solution about 10 cm in the appropriate stirring. The known amount of ZnO was added to the solution. After a known period of time, the solution was centrifuged at 2000 rpm for 10 min and then filtered to remove the catalyst particles completely. $Na_3[Fe(CN)_5NH_3{\cdot}3H_2O]$ (TPF) was obvious reagent of unsymmetrical dimethylhydrazine solution. The absorbency was measured with 722 UV-Vis spectrophotometer (λ_{max} = 500 nm), and the degradation rate (η) was calculated according to the equation below:

$$\eta = (A_0 - A)/A_0 \times 100\%$$

A_0 and A is the initial and absorbency of unsymmetrical dimethylhydrazine solution; A is the absorbency of unsymmetrical dimethylhydrazine solution after degradation.

3. Results and discussion

Figure 1 shows the XRD spectra. X-ray diffraction measurements show that all samples prepared have the wurtzite structure attached to hexagonal system according to the standardized JCPDS card. From the Figure 1 (b) we can see that the position of peak having a little remotion to the left. The particle characteristics and the average grain size calculated from the broadening of the XRD peak of wurtzite using Scherrer's equation were listed in table 1. From table 1 we can see that co-doping makes the XRD peak getting wider compared with pure ZnO, the grain size is in the range from 10.69

to 32.55 nm. Compared with pure ZnO unit cell dimensins, the unit cell dimensins of doped sample changed: a getting longer and b getting shorter. But unit cell volume has little changed.

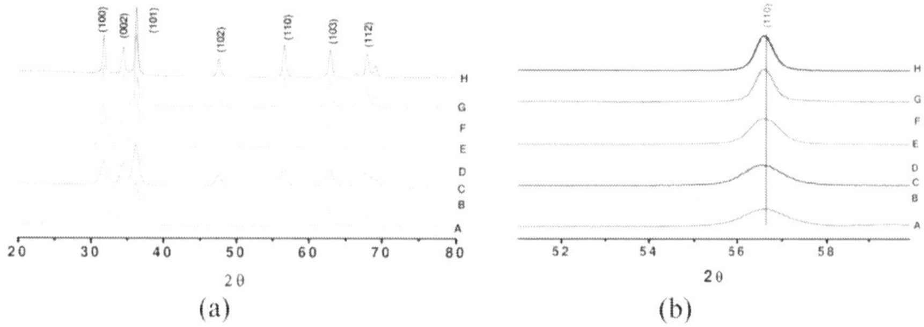

Figure 1. X-ray diffraction patterns of Samples calcined at 300 °C.A: ZnO; B: Nd(2.5); C: La(2); D: La(1)Nd(2.5); E: La(2)Nd(2.5); F: La(2)Nd(1.5); G: La(3)Nd(2.5); H: La(2)Nd(3.5)

Table 1. The calculated unit cell parameterof Samples by XRD. A, B, C, D, E, F, G and H is the same sample as that in figure 1, respectively.

sample	unit cell dimensin (nm) a	unit cell dimensin (nm) b	grain size (nm)	unit cell volume (nm^3)
A	**3.2491**	**5.2350**	**10.69**	**143.57**
B	**3.2545**	**5.2219**	**11.58**	**143.69**
C	**3.2552**	**5.2122**	**25.27**	**143.49**
D	**3.2497**	**5.2267**	**18.54**	**143.40**
E	**3.2559**	**5.1990**	**22.79**	**143.19**
F	**3.2594**	**5.2282**	**32.55**	**143.34**
G	**3.2518**	**5.2095**	**21.30**	**143.11**
H	3.2663	5.1990	21.39	144.10

Figure 2 shows the UV-Vis absorption spectra of colloidal solutions of pure, doped, and co-doped ZnO. The onset of the absorption spectrum of the ZnO colloid appears at about 360 nm. The absorption profile of doped ZnO appears a blue-shift about 10 nm. The absorption profile of co-doped ZnO shows the same blue-shift. The threshold value of semiconductor photoabsorption λ_g has a relation with the band gap E_g, the relation formula is E_g (eV) = 1240/λ_g (nm). From the Fig.2 we can see that a smaller λ_g lead to a wider band gap, and this makes photogenerated holes have a stronger reductive electric potential which may result in the increasing of photocatalysis of semiconductor.

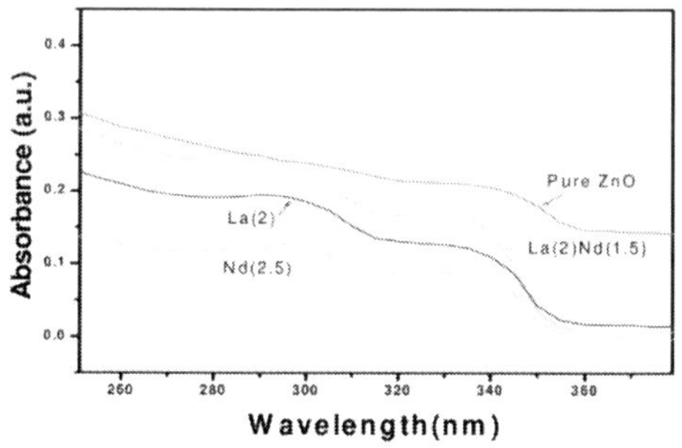

Figure 2. UV-Vis spectra of samples

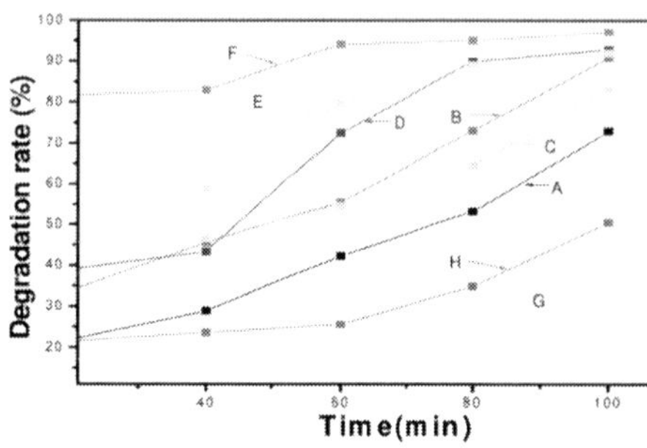

Figure 3. Degradation rate of UDMH using different La,Nd-doped ZnO samples. A, B, C, D, E, F, G and H is the same sample as that in figure 1, respectively.

The photoreactivities of pure ZnO, Nd(Y), La(X), and La(X)Nd(Y) samples are presented in Figure 3. Curve B (Nd(2.5)) and curve C (La(2)) are the best degradation effect for mono-doped ZnO. Curve F (sample La(2)Nd(1.5)) is the best degradation effect for co-doped ZnO. The degradation effect of samples La(1)Nd(2.5) (curve D) and La(2)Nd(2.5) (curve E) is better than mono-doped ZnO. The degradation effect of samples La(3)Nd(2.5) (curve G) and La(2)Nd(3.5) (curve H) are worse than mono-doped ZnO. So ZnO co-doped with La^{3+} and Nd^{3+} ions could lead to superior photocatalysis. Using the sample La(2), Nd(2.5), La(2)Nd(1.5), the degradation rate of unsymmetrical dimethylhydrazine could reach 83%, 91%, 97% respectively by catalyst of 0.1 g/L in 100 min at room temperature.

The effect may be explained by the cooperative operation of the two dopants in trapping charge carriers and mediating interfacial charge transfer process. Co-dopant of La^{3+} and Nd^{3+}could effectively inhibit the recombination of the photoinduced electrons and holes. We also notice that the proportion of dopants beyond the optimal proportion would lead to the decreasing of photodegradation efficiency which coincides with the situation of doping La^{3+} or Nd^{3+} alone such as sample La(2.5)Nd(2.5), its grain size is smaller than ZnO, but photodegradation rate do not increase than ZnO without dopant.

Nanocrystalline ZnO co-doped with La^{3+} and Nd^{3+} at optimal concentration (2 at.% La^{3+} and 3 at.% Nd^{3+}) shows a better synergistic effect, which significantly increases the photodegradation activity of pure ZnO. The separation of the charge carriers is attributed to such trappings. The co-doped ZnO photocatalysts with appropriate content of La^{3+} and Nd^{3+} possessed abundant electron traps so as to be favorable for the separation of the photoinduced electron-hole pairs, which could greatly enhance the activity of the photocatalysts. The efficiency of photocatalytic reactions is limited by the high recombination rate of photoinduced electron-hole pairs formed in photocatalytic processes and by the absorption capability for visible light of photocatalysts.

4. Conclusion

We prepared quantum-sized ZnO of doped with La^{3+} or $Nd^{3+}$and co-doped with La^{3+} and Nd^{3+} using sol-gel method. An investigation on the photodegradation of unsymmetrical dimethylhydrazine solution has been revealed that the photocatalytic reactivity of optimal co-doped ZnO was significantly increased as compared with ones over nanoparticles of ZnO no doped or monodoped with La^{3+} or Nd^{3+} alone. The photodegradation rate of unsymmetrical dimethylhydrazine solution reaches 97% by catalysis of 0.1g/L La^{3+}-Nd^{3+} co-doped ZnO at room temperature in 100 minutes. The synergistic effect of two dopants leads to significant enhancement of photodegradation. It can be expected that photocatalytic activity can be further improved by choosing proper dopants.

Reference

[1] Hoffmann M R, Martin S T, Choi W Y and Bahnemann D W 1995 *Chem. Rev.* **95** 69-96
[2] Kormann C, Bahnemann DW and Hoffmann M R 1988 *Envlron. Sci. Technol.* **22** 798-806
[3] Szacilowski K, Macyk W, Drzewiecka-Matuszek A, Brindell M and Stochel G 2005 *Chem. Rev.* **105** 2647-94
[4] Peng F , Wang H J, Yu H and Chen S H 2006 *Materials Research Bulletin* **41** 2123-29
[5] Chakrabarti S and Dutta B K 2004 *Journal of Hazardous Materials* **B112** 269-78
[6] Lam S W, Chiang K, Lim T M, Amal R and Low G K-C 2007 *Appl.Catal. B:Environ.* **72** 363-72
[7] Pardeshi S K and Patil A B 2009 *Journal of Hazardous Materials* **163** 403-09
[8] Zheng Y H ,Chen C Q, Zhan Y Y, Lin X Y, Zheng Q, Wei K M, Zhu J F and Zhu Y J 2007 *Inorg. Chem.* **46** 6675-82
[9] Anandan S, Vinu A, Mori T, Gokulakrishnan N, Srinivasu P, Murugesan V and Ariga K 2007 *Catalysis Communications* **8** 1377-82
[10] Yang P, Lu C, Hua N P and Du Y K 2002 *Materials Letters* **57** 794-801
[11] Qamar M, Saquib M and Muneer M 2005 *Dyes & Pigments* **65** 1-9
[12] Su S, Lu S X and Xu W G 2008 *Materials Research Bulletin* **43** 2172-78

Fabricating Arrays of Si(110)/Si(100) Microstructures by Atom Lithography using Organosilane Self-assembled Monolayers

J W Zhang

Department of Physics, University of Science and Technology of China, Hefei, Anhui 230026, P.R. China

E-mail: zjw@ustc.edu.cn

Abstract. We reported a monolayer-derived fabrication of Si(110)/Si(100) microstructures arrays over an area of about 2 cm^2 by atom lithography, in which self-assembled monolayer (SAM) of organosilane (octadecyltrichlorosilane, ODTS; dodecyltrichlorosilane, DDTS) and metastable helium beam (He-MAB) were used to pattern the surface of silicon substrates without coating intermediate layer. ODTS and DDTS SAMs were formed on Si(110)/Si(100) surfaces as resists for exposure to MAB. The hydrophobicity and the durability of the outermost surface of the SAM in the regions exposed to MAB were altered by varying MAB dose of exposure and the length of alkyl chains of SAM molecule. The negative/positive patterns of Si(110)/Si(100) surfaces with depths of 10~220nm and edge step resolutions of 40~500nm were obtained successfully. The effect of a wet etching process on the resolutions was discussed in brief. Our results suggested that the top-down atom lithography combined with bottom-up chemical assembly techniques could contribute significantly to the development of more monolayer-derived micro- and nanofabrication.

1. Introduction

The arrays of micro- and nanostructures of silicon show interesting characteristics in applications such as photonics, electronics, optoelectronics, and sensing, etc. [1, 2]. The array properties can be tuned further by varying the geometry, the doping, the periodicity, and the size of the micro- and nanostructures [2]. These controllable and elaborate arrays are commonly fabricated by X-ray, electron-beam or ion-beam lithography [3-5]. However, there are inherent limitations to pattern over large areas at nanoscale using these lithographical techniques due to the light diffraction, the long-range inter-particle interactions or the proximity effects. The conventional photolithographic techniques currently used in IC industry and micro-/nano- devices fabrication have also reached their ultimate spatial resolution because of both the light diffraction and the photoresist restrictions [5]. To overcome these limitations, alternative techniques with a shorter wavelength and a thinner resist for traditional optical lithography, including extreme ultraviolet lithography, atom lithography, microcontact printing and soft lithography, etc., have been proposed and developed to meet the requirements of a high spatial resolution of sub-100nm as well as the capability to fabricate an arbitrary structure and achieve mass production [5-9].

The atom lithography using metastable atoms beam (MAB) and self-assembled monolayer (SAM) has shown great potential to fabricate arrays of micro- and nanostructures, which is a major goal in nanoscience and technology [9-14]. This new fabrication technique, which bridges the gap between the bottom-up chemical self-assembly techniques and the modern top-down lithography, can overcome the intrinsic resolution limitations of traditional photolithographic techniques [9, 10]. In principle, the atom lithography based on neutral MAB and SAM ideally meets the required conditions for sub-100nm fabrication [9]. MAB with a de Broglie wavelength of less than 0.1nm eliminates diffractive resolution limitations, and an ultrathin organic SAM resist with a thickness of 1~3nm ensures the sharpness of the edge profile within the resist. Moreover, the neutral metastable atoms are insensitive to the electric and magnetic fields, and the long-range inter-particle interactions are weak. Therefore, this novel method permits the direct and large area manufacturing of micro- and nanostructures on a silicon wafer, avoiding some inherent complications of electron-beam, ion-beam or photolithography. With these unique advantages, the atom lithography with neutral MAB and high-resolution SAM resists makes it possible to achieve nanolithography and provides a potential way in manufacturing structures at a large-scale based on micro- and nanoscale features.

Previous reports have investigated the mechanism of the metastable atom lithography and the microfabrication with different resolutions on various substrates, such as gold, silicon, silicon oxide, and mica [9-14]. The typical micropatterning with sub-100 nm resolution on a silicon substrate covered with a SiO_2 or Au layer has been achieved successfully [9, 11, 14]. Most of these studies were conducted through multistep etching processes using the cover layers as intermediate masks. For example, in the He-MAB lithography with dodecanethiol-SAM [9, 11], a Au/Ti coating layer was used to create an intermediate mask on silicon substrate by the first etching process, and then, the intermediate mask pattern is transferred to silicon substrate by the second etching process. It is very interesting to directly transfer negative and positive patterns of SAM induced by metastable helium atoms beam (He-MAB) into a silicon substrate to fabricate arrays of micro- and nanostructures, which did not use metal/oxide coating layer. Recently, the He-MAB lithography of Si with SAM, which formed directly on silicon substrate by direct covalent linkage instead of metal/oxide coating layers, has been accomplished [10]. The latent image formed in SAM by He-MAB passing through a stencil was directly transferred into the underlaying Si substrate by KOH etching, which was a single step etching process. Undoubtedly, to realize and understand this direct transfer process will open a novel way in the practical application of the atom lithography in micro- and nanofabrication of silicon, especially in arrays of micro- and nanostructures of silicon.

In this paper, we reported a micropatterning of Si(110)/Si(100) wafer surfaces by atom lithography using He-MAB, SAMs of organosilanes (ODTS, DDTS) and a single step wet etching process, which indicated a potential application in fabricating arrays of silicon micro- and nanostructures. The internal energy (~20eV) of the metastable helium atom was used to selectively damage the SAM bonded directly on the Si(110)/Si(100) surfaces through a stencil masks to generate a latent pattern in SAM, and then a wet etching process was employed to transfer and develop the latent pattern into the underlying silicon substrate. Negative/positive patterns of Si(110)/Si(100) surfaces with a depth of 10~220nm and an edge resolution of 40~500nm were obtained, and arrays of silicon microstructures were fabricated successfully on Si(110) and Si(100) substrates. The mechanism of positive-tone and negative-tone patterning of silicon substrates were investigated and discussed briefly. Our results suggested that this versatile process, combining bottom-up chemical assembly techniques with top-down atom lithography, could contribute significantly to the development of more monolayer-derived micro- and nanofabrication.

2. Experimental procedure

2.1 *Preparation of Si(110)/Si(100) substrates with SAM*

The p-type Si(110) /Si(100) wafers with a native oxide layer (Φ=100mm, thickness=300μm, ρ=1~3 $\Omega\cdot$cm) were used and cut into pieces of 10mm×20mm as Si substrates. The Si substrates were first rinsed using toluene, ethanol, and deionized water sequentially, then immersed into a 2vol.% aqueous HF solution for 5min to remove the native oxide layer and hydrogenate the silicon substrates, and finally dried. Protected from air by a flowing argon gas, the Si substrates were immersed into a 10mM organosilanes (ODTS and DDTS) toluene solution at room temperature for 24 hours, and then rinsed in toluene and dried at 100°C for 30min. The mechanism of forming SAM on silicon substrate is illustrated in details in Scheme 1. Thus, through direct covalent linkage, i.e., silicon-carbon, fine organosilanes (ODTS and DDTS) SAMs were formed as a resist on silicon surface by introducing a hydrogenation process into the pretreatment of silicon wafer and a controlled argon atmosphere at elevated temperature.

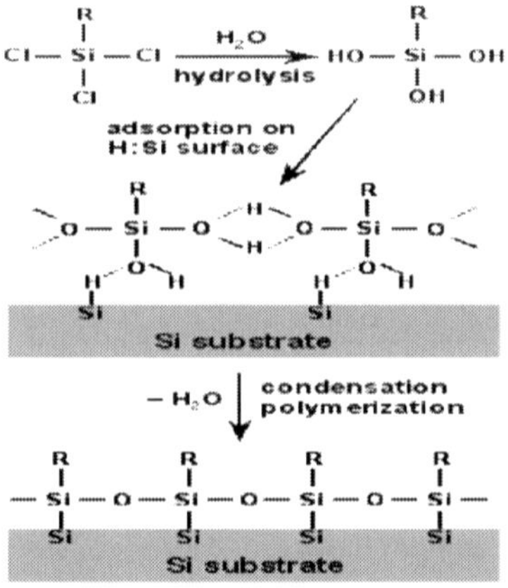

Scheme 1. The illustration of the stepwise process to form SAM.

2.2 *Exposure to MAB*

The experimental setup for the sample exposure to He-MAB was depicted in Scheme 2. The Si substrate with organosilane SAM was covered by a mask with TEM Cu grids. As shown in the inset of Scheme 2, six TEM Cu grids with a pitch size of 12.5μm (5.5μm for wires and 7.0μm for spaces) were well distributed on a nickel sheet (with a thickness of 0.1mm and a square area of 1cm×2cm) and fixed over the opening holes with a diameter of 2mm using Ag conductive paste. The assembly was immediately inserted into the sample chamber of the MAB source (as shown in Scheme 2) and exposed to He-MAB for 5~120 min, producing latent patterns in organosilanes SAM on Si substrates. The experiments were performed with the existing apparatus, as described in detail elsewhere [10, 13]. Briefly, the He* source was operated in a dc discharge mode, and a stable pressure (~5×10^{-7} Pa for the ultimate pressure and ~3×10^{-2} Pa for the working pressure) of chambers was retained by turbo-molecule pump (T.M.P) vacuum systems. A transverse electrical field with a deflection voltage of ~800 V was applied to an atom beam for removing ions or electrons that can cause damage to the SAM resist during the whole exposure. The helium beam through a skimmer opening with a diameter of 0.7 mm was directed toward a rotatable thin tantalum (Ta) sheet sample holder at a 26.5 cm distance. This Ta holder had a triangle arrangement along the rotation axis, on which two faces were used for installing samples and the rest for monitoring the intensity of the primary beam (shown in the inset of Scheme 2). The thickness of the sheet defines the gap between masks and sample surfaces (~100um), and avoids the mechanical contact that may cause possible artifacts. The metastable flux was estimated by measuring metastable-atominduced electron emission on both the sample and Ta plate. The typical values of the measured current on both the Ta plate and sample were 100 and 50 nA, which corresponded to an electron emission rate of ~1×10^{12} and ~2×10^{12} $s^{-1}\cdot cm^{-2}$, respectively.

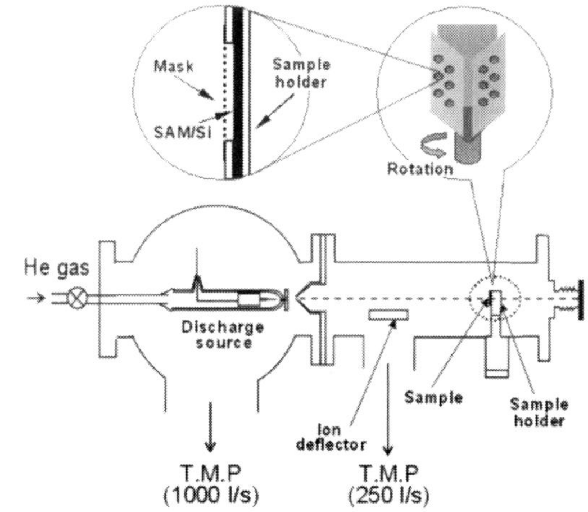

Scheme 2. Schematic setup of the sample exposure to MAB for atom lithography.

2.3 *Etching processing and pattern transferring*
After exposure, the samples with latent patterns in SAM were taken out from the vacuum chamber and rapidly dipped into an etching solution in which a magnetic bar was used for stirring, and were etched for 5~60min at room temperature in an aqueous solution of 0.1M KOH, rinsed by deionized water, and then dried in air.

2.4 *Characterization*
AFM images were taken with a multimode Nanoscope IV atomic force microscope (Veeco Metrology Group, Santa Barbara, CA), operated at ambient conditions. For all images we recorded the retrace direction of the tip using a scan angle of 0° or 90°. Substrates decorated with organosilanes SAMs were imaged in tapping mode using silicon cantilevers (NanoWorld, Neuchâtel, Switzerland) with a spring constant of 42 N m^{-1}. All images were recorded at a rate of 1.0 Hz, and with a pixel resolution of 512. SEM inspection of patterned Si substrates and the masks was carried out with a JSM-6500/SG scanning electron microscope operated at 30 keV.

3. Results and discussion
The process used to create the arrays of silicon micro- and nanostructures on a Si substrate was illustrated in Scheme 3. By introducing a hydrogenation process into the pretreatment of silicon substrates (as described in (I) and (II) in Scheme 3), we successfully formed fine organosilanes SAMs as a resist on silicon surface under a controlled argon gas atmosphere at elevated temperature. The chemical modification of non-oxidized silicon surfaces utilizing monolayers could be achieved by neutralizing the silicon surfaces with alkyl chains through direct covalent linkage, i.e., silicon-carbon [15]. This provided us an opportunity to directly use an ultrathin hydrophobic SAM with a thickness of 1~3nm as a super resist on silicon surfaces. The samples with the stencil masks were exposed to the irradiation of He-MAB at thermal energy, and the SAM was damaged through the interaction between the outermost surface of the SAM and the metastable atoms (as described in (III) in Scheme 3). When the metastable atoms hit the outermost surface of the SAM on Si substrate, they transferred their internal energy (~20eV) to the chemisorbed hydrocarbons of SAM molecules and caused chemical changes that might result in either the formation of a durable crossing-linking polymerization material or the loss of hydrophobicity of the SAM (as indicated in (IV) and (V) in Scheme 3). After exposure, the pattern was transferred into a latent pattern in SAM. The SAM exhibited the positive-tone and negative-tone sensitivity due to largely different dosage of MAB and the length of the alkyl chains of SAM molecule. The negative patterning indicated more He-MAB irradiation dose than the positive

patterning, corresponding to the polymerization induced by He-MAB. Meanwhile, the positive patterning corresponded to the hydrophilic region induced by He-MAB. Furthermore, the longer alkyl chains are easier to undergo cross-linking polymerization than the shorter alkyl chains under the MAB irradiation. Consequently, these could affect the etching process in the fabrication of the micro- or nanostructure (i.e., the hydrophilic exposure region should be easier to etch using KOH solution than the hydrophobic unexposed region, whereas the cross-linking polymerization region should be more durable to KOH etching than the hydrophobic unexposed region), resulting in the positive and negative pattern transfers (as described in (VI) and (VII) in Scheme 3), respectively. Our previous investigations of the pattern transfer process have confirmed this mechanism for the contrast inversion, i.e., the competition between the loss of hydrophobicity and the cross-linking polymerization of SAM molecules determined the polarity [10]. In our experiments, the lithographic patterns were obtained in samples with a relatively large area of about 2 cm^2. In these areas, the mask pattern was reproduced with high fidelity, and the repetition of lithographic patterns was consistent in different runs. To obtain better contrast patterns, the experimental parameters were carefully optimized. After etching, the resulting arrays fabricated on the silicon substrate were clearly observed under the optical microscope. Furthermore, the samples were investigated under the atomic force microscope (AFM) and the scanning electron microscope (SEM).

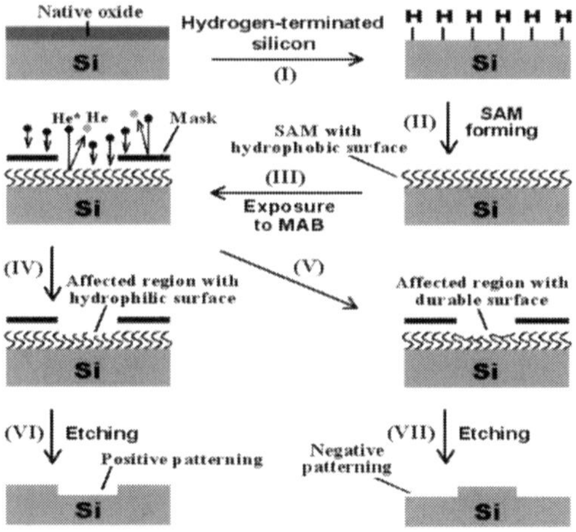

Scheme 3. Schematic illustration of the procedure used for the fabrication of periodic arrays of Si(110)/Si(100) by He-MAB lithography (excited helium atom: He*; ground-state helium atom: He).

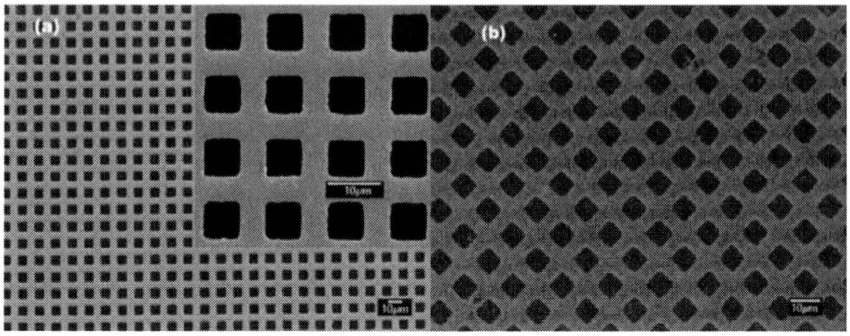

Figure 1. (a) SEM image of TEM Cu grid used as a mask (Note: the pitch size of patterns is 12.5 um, 5.5 um for wires, and 7.0 um for spaces) and (b) SEM image of negative patterning on Si(100) substrate with the ODTS SAM (exposure time: 60min; etching time: 5min).

Figure 1 showed the SEM images of the mask and one sample. The pattern from the stencil mask (as shown in Figure 1(a)) was transferred into the silicon substrate well, and the resulting micromesas array of silicon fabricated on the substrate were very intact and clear (as shown in Figure 1(b)). The preliminary SEM observations revealed that the direct fabrication of the array on the surfaces of silicon substrates by He-MAB and SAM was viable, and our results indicated that the patterning also appeared on either kind of surface of the Si(110) or Si(100) substrate with various organosilane SAM mentioned above.

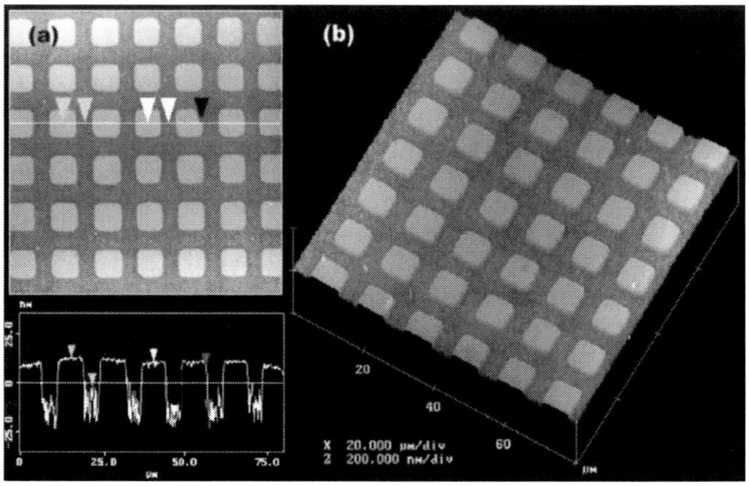

Figure 2. Tapping-mode AFM images of periodic arrays of square silicon micromesa fabricated on Si(110) substrate with ODTS SAM (exposure time: 20min, etching time: 5min): (a) AFM image with sectional analysis of the arrays of silicon micromesas with a height of about 26 nm ; (b) 3D views of the arrays of silicon micromesa over a 80 um×80 um area and with a periodicity of 12.5 um.

The AFM images of the square micromesa arrays formed on the Si(110) substrate with ODTS SAM was shown in Figure 2, which corresponded to negative patterning with an exposure time of 20 min and an etching time of 5 min. A square silicon micromesa arrays with a micromesa size of ~7×7 μm^2 (e.g., the size of one mesh of TEM stencil mask) was fabricated successfully on the substrate. The average height of the step was 27 nm, indicating an etching rate at 5 nm/min. The sharpest step edge had a width of 1 pixel (41 nm) and a height of 26 nm, indicating a steep sidewall with an aspect ratio of 1:1.7. The sectional analysis shown in Figure 2(a) indicates that the rms roughness of the top area of the micromesa and the pedestal base area were 2 nm and 15 nm, respectively. The flat top area of the micromesas could be observed clearly, suggesting the strong durability of the exposed SAM resist against chemical etching. The flaw pedestal base area of the micromesa was also very conspicuous, and this rough characteristic could be attributed to the imperfect SAM formation on the silicon substrate, which induced a local fluctuation in the etching rate of transferring the latent image in SAM resist onto the underlying silicon substrate and, consequently, produced a rough surface. As described in scheme 3, the negative-tone sensitivity of the SAM might be mainly due to the cross-linking polymerization of ODTS molecules induced by the irradiation of the metastable helium beam [10]. In addition, it was noteworthy that the etching time had a notable influence on the production of high-quality patterning. In our practical runs, the following etching times were chosen: 5, 10, and 20 min with identical exposure times. The sectional analysis of AFM images of these samples indicated that the rms roughness of top area of the micromesa in the case of overetching was bigger than that in the case of underetching.

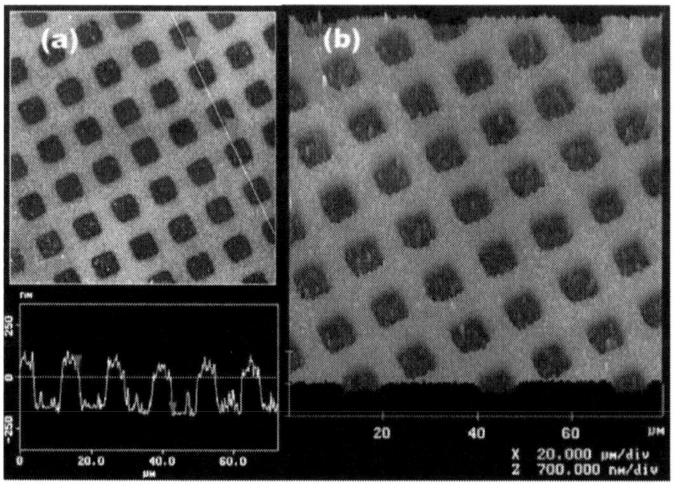

Figure 3. Tapping-mode AFM images with sectional analysis of periodic arrays of square silicon microwell fabricated on Si(100) substrate with DDTS SAM (exposure time: 20min, etching time: 10min): (a) AFM images with sectional analysis of arrays of square microwell with a depth of about 220 nm; (b) 3D views of the arrays of silicon microwell over a 80 um×80 um area and with a periodicity of 12.5 um.

Compared with the negative patterning discussed above, a positive patterning could be obtained by adjusting the dosage of He-MAB irradiation and the length of alkyl chains of SAM molecule. Figure 3 showed the AFM image of the square microwell arrays fabricated on the Si(100) substrate with DDTS SAM, corresponding to the positive patterning (exposure time: 20 min; etching time: 10 min). A square microwell arrays with a microwell size of ~7×7 μm^2 was fabricated on the Si(100) substrate. The depth of the microwells was about 220 nm. The sharpest wall edge had a width of 1 pixel (41 nm) and a depth of 100 nm, indicating a steep sidewall with an aspect ratio of 2.5:1. The sectional analysis shown in Figure 3(a) indicated that the rms roughness of the bottom area of the microwell and the top area of the wall were 31 nm and 22 nm, respectively. The rough characteristic could also be attributed to the imperfect SAM formation on the silicon substrate, as mentioned above. The positive-tone sensitivity of the SAM should be mainly due to the loss of hydrophobicity of DDTS SAM (with shorter alkyl chains) induced by the irradiation of the metastable helium atoms. The topography and line profile of the microwells array indicated a much rougher top surface of the sidewall with an rms roughness of 22 nm. This large rms value suggested that the hydrophobic area might be much weaker in resisting the KOH etchant, and the DDTS SAM would not be a good resist for atom lithography.

For the wet etching of Si in KOH solution, the anisotropy of etching rate of silicon have been studied in details in the case of the photolithography. The etching rates for the silicon wafer in the KOH solution are known to depend strongly on its orientation (etching ratio: {110}>{100}>>{111}) [16]. The anisotropy can induce obvious difference of the spatial resolution of patterning of Si along different crystal orientation. In our experiments, the anisotropic etching rate seemed to improve the edge resolution, and the anisotropic etching process of silicon strongly affected the patterning results including edge step resolution. It was easier to obtain a large aspect ratio and a better resolution for Si(110) rather than for Si(100). The sharpest step edge resolution obtained presently in Si(110) samples was only 41nm with 1 pixel and the step edge resolution of about 120 nm, indicating an aspect ratio of 3:1. In principle, the estimated value of geometrical blurring is ~240 nm with the geometry described in Scheme 2, and the spatial resolution could be improved by reducing the distance between the source and the sample, which was related to the divergence of the metastable helium beam. The achievable width in our experiments was much smaller than the calculated one because of the hardening effect in the process of chemical etching. If the experimental parameters in both the exposure and etching processes (e.g., the exposure time, etching time, etc.) were optimized further, higher resolution of the patterning onto the silicon substrates could be obtained.

The interaction between the outermost surface of the SAM and the irradiation of metstable atoms has been investigated for about ten years [17-20]. However, the detailed change in surface chemistry of the outermost of the SAM molecules under the irradiation of helium atom beam is still difficult to predict and remains to be explored. What we demonstrated in the present study was that both the positive-tone and the negative-tone sensitivity of the SAM were related to largely different dosage of MAB and the length of the alkyl chains of SAM molecule, as illustrated in Scheme 3. When we optimized the experimental parameters we accidentally obtained a novel patterning with both negative and positive pattern on one Si(100) substrate with DDTS SAM, which was very different from previous reports. In this case, a novel array of square micromesas with microholes in it was fabricated successfully. The mechanism to induce this dual patterning has remained to be explored at present. This dual array indicated some interesting flexibility of atom lithography in micro- and nanofabrication.

4. Conclusions

We reported a monolayer-derived fabrication of Si(110)/Si(100) microstructures arrays over a large area (greater than 1cm×2cm) by atom lithography, in which SAM and MAB were used to pattern the surface of silicon substrates without coating intermediate layer. ODTS and DDTS SAMs can not only be formed on Si(110)/Si(100) surfaces but also be used as resists for exposure to metastable-atom beams. Negative/positive patterns of Si(110)/Si(100) surfaces with a depth of 10~220nm and an edge step resolution of 40~500nm were obtained. We were able to alter the hydrophobicity and the durability of the outermost surface of the SAM in the regions exposed to MAB by varying MAB dose of exposure and the length of alkyl chains of SAM molecule, subsequently realize positive-tone or negative-tone patterning of silicon, and finally fabricate arrays of silicon microstructures. Our strategy, which was essentially a top-down integrating with bottom-up approach, involved the formation of SAM on the surface of silicon substrate and pattern transferring from physical mask to silicon substrate. The improved approach presented here could be commonly used and extended to the micropatterning of a variety of other metal oxides or materials, in which SAM could be formed directly onto the substrate surfaces without an oxide or metal intermediate layer. Our results suggested that this versatile process, combining chemical assembly schemes with top-down atom lithography, could contribute significantly to the development of more monolayer-derived micro- and nanofabrication.

Acknowledgments

The author gratefully acknowledges support from the National Natural Science Foundation of China (Grant No. 20777072), and would like to thank Prof. Y. Yamauchi, Dr. M. Kurahashi and Dr. X. Sun for their help with experiments and their helpful comments. The author also notes that part of the work was completed while he was at the NIMS.

References

[1] a) Dong B, Lu N, Zelsmann M, Kehagias N, Fuchs H, Torres C M S and Chi L 2006 *Adv. Funct. Mater.* **16** 1937. b) Jang J H, Ullal C K, Maldovan M, Gorishnyy T, Kooi S, Koh C Y and Thomas E L 2007 *Adv. Mater.* **17** 3027.

[2] a) Henzie J, Lee M H and Odom T W 2007 *Nature Nanotech.* **2** 549. b) Tao A, Sinsermsuksakul P and Yang P D 2006 *Nature Nanotech.* **2** 435. c) McAlpine M C, Ahmad H, Wang D W and Heath J R 2005 Nature **434** 285. d) Lei Y, Cai W P and Wilde G 2007 *Progress in Materials Science* **52** 465.

[3] a) Shin S W, Lee S G, Lee J, Whang C N, Lee J H, Choi I H, Kim T G and Song J H 2005 *Nanotechnol.* **16** 1392. b) Tseng A A 2005 *Small* **1** 924.
[4] Carr D W, Lercel M J, Whelan C S, Craighead H G, Seshadri K and Allara D L 1997 *J. Vac. Sci. Technol.* B **15** 1446.
[5] Xia Y, Rogers J A, Paul K E and Whitesides G M 1999 *Chem. Rev.* **99** 1823.
[6] a) Campbell C J, Smoukov S K, M Bishop K J, Baker E and Grzybowski B A 2006 *Adv. Mater.* **18** 2004. b) Ahn S J, Kaholek M, Lee W K, LaMattina B, LaBean T H and Zauscher S 2004 *Adv. Mater.* **16** 23. c) Wu N, Pease L F and Russel W B 2006 *Adv. Funct. Mater.* **16** 1992.
[7] a) Palmer A J, Baker M and Sang R T 2006 *Nanotechnol.* **17** 1166. b) Sawada S, Masuda Y, Zhu P and Koumoto K 2006 *Langmuir* **22** 332.
[8] a) Harnett C K, Satyalakshmi K M and Craighead H G 2000 *Appl. Phys. Lett.* **76** 2466. b) Kan J A V, A A Bettiol and F Watt 2006 *Nano Lett.* **6** 579. c) Saito N, Kadoya Y, Hayashi K, Sugimura H and Takai O 2003 *Jpn. J. Appl. Phys.* **42** 2534.
[9] Berggren K K, Bard A, Wilbur J L, Gillaspy J D, Helg A G, McClelland J J, Rolston S L, Phillips W D, Prentiss M and Whitesides G M 1995 *Science* **269** 1255.
[10] Zhang J W, Kurahashi M, Suzuki T, Sun X and Yamauchi Y 2006 *Jpn. J. Appl. Phys.* **45** 8020.
[11] a) Hill S B, Haich C A, Dunning F B, Walters G K, McClell J J, Celotta R J and Craighead H G 1999 *Appl. Phys. Lett.* **74** 2239. b) Younkin R, Berggren K K, Johnson K S, Prentiss M, Ralph D C and Whitesides G M 1997 *Appl. Phys. Lett.* **71** 1261.
[12] a) Lu W, Baldwin K G H, Hoogerland M D, Buckman S J, Senden T J, Sheridan T E and Boswell R W 1998 *J. Vac. Sci. Technol.* B **16** 3846. b) O'Dwyer C, Gay G, Viaris de Lesegno B, Weiner J, Camposeo A, Tantussi F, Fuso F, Allegrini M and Arimondo E 2005 *Nanotechnol.* **16** 1536.
[13] a) Ju X, Kurahashi M, Suzuki T and Yamauchi Y 2004 *Thin Solid Films* **464** 420. b) Ju X, Kurahashi M, Suzuki T and Yamauchi Y 2003 Jpn. J. Appl. Phys. **42** 4767. c) Ju X, Kurahashi M, Suzuki T and Yamauchi Y 2005 *Appl. Surf. Sci.* **241** 241.
[14] a) Brezger B, Schulze T, Drodofsky U, Stuhler J, Nowak S, Pfau T and Mlynek J 1997 *J. Vac. Sci. Technol.* B **15** 2905. b) Hill S B, Haich C A, Dunning F B, Walters G K, McClelland J J, Celotta R J, Craighead H G, Han J and Tanenbaum D M 1999 *J. Vac. Sci. Technol.* B **17** 1087.
[15] a) Kulkarni S A, Mirji SA, Mandale A B, Gupta R P and Vijayamohanan K P 2005 *Mater. Lett.* **59** 3890. b) Shirahata N, Masuda Y, Yonezawa T and Koumoto K 2004 Eur. Ceram. Soc. **24** 427. c) Heid S and Effenberger F 1996 *Langmuir* **12** 2118.
[16] Sato K, Shikida M, Matsushima Y, Yamashiro T, Asaumi K, Iriye Y and Yamamoto M 1998 *Sens. and Actuators* A **64** 87.
[17] a) Yamauchi Y, Noro T, Kurahashi M, Suzuki T and Ju X 2005 *Appl. Surf. Sci.* **241** 141. b) Setoyama H, Kera S, Okudaira K K, Hara M, Harada Y and Ueno N 2003, *Jpn. J. Appl. Phys.* **42** 597.
[18] a) Thywissen J H, Johnson K S, Dekker N H, Prentiss M, Wong S S, Weiss K and Grunze M 1998 *J. Vac. Sci. Technol.* B **16** 1155. b) Johnson K S, Thywissen J H, Dekker N H, Berggren K K, Chu A P, Younkin R and Prentiss M 1998 *Science* **280** 1583.
[19] a) Johnson K S, Berggren K K, Black A T, Chu A P, Dekker N H, Ralph D C, Thywissen J H, Younkin R, Thinkham M, Prentiss M and Whitesides G M 1996 *Appl. Phys. Lett.* **69** 2773. b) Rehse S J, Glueck A D, Lee S A, Goulakov A B, Menoni C S, Ralph D C, Johnson K S and Prentiss M 1997 *Appl. Phys. Lett.* **71** 1427.
[20] a) Close J D, Baldwin K G H, Hoffmann K and Quaas N 2000 *Appl. Phys.* B **70** 651. b) Nowak S, Pfau T and Mlynek J 1996 *Appl. Phys.* B **63** 203.

Effect of Mg-doping on pyrolysis evolution and phase constitutes of $Ba_{0.7}Sr_{0.3}TiO_3$ (BST) system derived from sol-gel

Hengcheng LIAO, Qichang WU, Liang XU

School of Materials Science and Engineering, Southeast University, Nanjing 211189, China; Jiangsu Key Laboratory for Advanced Metallic materials, Nanjing 211189, China

E-mail: hengchengliao@seu.edu.cn

Abstract. Gels and powders of dielectric $Ba_{0.7}Sr_{0.3}(Ti_{1-x}Mg_x)O_3$ (BST) were prepared by sol-gel technique. Effect of Mg-doping of on the pyrolysis evolution and phase constitutes of BSTs was investigated using DSC-TG and XRD. Results show that Mg-doping leads to a remarkable change of crystallizing process of BSTs from a double phase-transformation of non-doping sample into a single phase-transformation and quickens the crystallizing reaction of formation of BST perovskite. The maximum Mg-doping level should be less than x= 0.15 in $Ba_{0.7}Sr_{0.3}(Ti_{1-x}Mg_x)O_3$ system. The amount of $(Ba,Sr)CO_3$ phase in the BST powders increases considerably with the Mg-doping level, indicating Mg atoms should occupy the A positions of ABO_3 structure of BST, instead of the B position.

1. Introduction

In recent years, barium strontium titanate (BST) has received considerable attention. Due to their high dielectric nonlinearity, high dielectric constant, good pyroelectric properties and composition-dependant Curie temperature adjustability, BST ferroelectric thin films have many potential applications, such as in tunable microwave devices, dynamic random access memories and pyroelectric far-infrared detectors etc [1-4]. The Cure temperature and electric properties of BST could be controlled by adjusting the Ba/Sr ratio. It has been reported [5-6] that the Cure temperature of BST is near room temperature when Ba/Sr is 7/3 and the best electric pyroelectric properties of it could be obtained operating d.c. bias voltages.

While the relatively high dielectric loss of BST thin films compared to their bulk counterparts limits its applications. Physics processes controlling leakage current density, resistance degradation, and dielectric loss of BST still need to be further studied to provide better performance. Doping with small amounts of dopants such as Mg, Ni, Mn, La, and Nb was found to suppress leakage current [7-10]. The changed electrical properties are attributed to the charged defects, which are created to maintain charge neutrality.

It has been thought that Mg-doping, as an acceptor-dopant occupying the B sites of the ABO3 perovskite structure in BST system [9,11-14], on the one hand, could restrain coarsening of grains and condense the surface microstructure, and on the other hand, of importance, could enhance the potential barrier at grain boundary and suppress the leakage current, and therefore decreases the dielectric loss [11,15,16]. Sol-gel technique is one of the best preparing processes of doped BST system. Adopting this method, Joshi et al [8], Yoon et al [9], Zhang et al [11] and Cole et al [17] investigated Mg-doping effect on leakage current and its mechanisms of BSTs. Gel should be pyrolyzed and crystallized to produce BST ferroelectric thin films. It is well known that the pyrolyzing and crystallizing processes have an important influence on the microstructure and dielectric properties of BSTs [18-21]. However, the influence of dopants on pyrolysis evolution and phase constitutes of BST system, involving in Mg, Ni, Mn and B etc, has not been well established. In this paper, the aim is to have a highlight in the Mg-doping effect on pyrolysis evolution and phase constitutes of BST derived by Sol-gel technique.

2. Experimental Procedure

It has been thought Mg atoms into ABO_3 lattice of BST should occupy B sites (namely Ti site). But, substituting Ti by Mg during composition design of the precursor of BST did not be considered [16]. In this paper $Ba_{0.7}Sr_{0.3}(Ti_{1-x}Mg_x)O_3$ (x=0, 0.05, 0.10 and 0.15) precursor solutions were confected using barium acetate, strontium acetate and $Ti(OC_4H_9)_4$ to form BST, acetic acid and 2-methoxyethanol as solvents, and magnesium acetate as the dopant precursor. Figure1. shows the process of preparing $Ba_{0.7}Sr_{0.3}(Ti_{1-x}Mg_x)O_3$ sol-gels. The wet gels were analyzed by DSC-TG (SDTQ600) among a temperature range from 25°C to 900°C with a rate of 20°C /min and high purity of N_2 gas as protecting atmosphere. According the results of DSC-TG analysis, the gels were heated at 210°C for 0.5 hour and 340°C for 2 hours for pyrolyzing, and then crystallizing-annealed at 750°C for 0.5 hour at air atmosphere to produce BST powder. Then BSTs crystallizing-annealed were analyzed by XRD (XD-3AX) for determining phase constitutes, with CuKα radiation (λ=0.154nm) at 40 kV and 30mA. The scanning range was 2θ=20°-60° and rate was 2°/min.

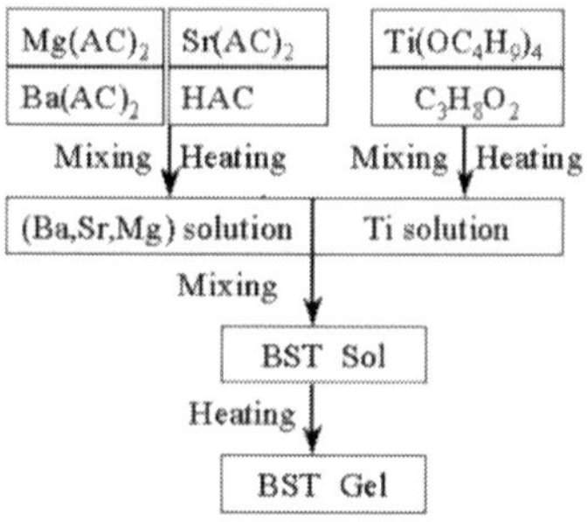

Figure 1. Preparation of $Ba_{0.7}Sr_{0.3}(Ti_{1-x}Mg_x)O_3$ sol-gels

3. Results and discussion

3.1. Thermal analysis (DSC-TG)

Figure 2 shows the pyrolysis evolution of $Ba_{0.7}Sr_{0.3}(Ti_{1-x}Mg_x)O_3$ (x=0, 0.05, 0.10 and 0.15) gels. It can be divided into three regions: evaporating (temperature about 25°C-250°C), pyrolyzing (temperature about 250°C-600°C) and crystallizing (temperature above 600°C). The first suggests evaporation of residual free organic solvents. In the pyrolyzing process, residual organic matters begin to burn. Perovskite BST begins to form in the last crystallizing process.

Pyrolyzing process of all gel samples is made up of exothermal peaks and endothermal peaks, as shown in Figure 2. Endothermal peaks indicate pyrolysis of organics, and exothermal peaks suggest the breaking-down of gels' framework. From Figure 2, it is seen that Mg-doping has an obvious influence on the pyrolyzing process of BST gels, even the details occurred are not understood clearly now.

Of more interest, Mg-doping has a considerable influence on the crystallization of BSTs. With increase in the doping level of Mg, temperature of the last exothermic peak (crystallizing peak) forming perovskite BST first rises a little and then drops remarkably, and the intensity of heat flow, representing the reaction rate, increases from 1.25 W g^{-1} of the sample with x=0 to larger than 2.0Wg^{-1}

of the sample with x=0.10, although when x =0.15 the intensity of heat flow decreases down to 1.08W g^{-1}. It indicates Mg-doping accelerates the crystallizing reaction of BST system. According to the regulation of International Union of Thermal Analysis and Calorimetry, the onset temperature of crystallizing reaction is the onset temperature of the last exothermic peak (crystallization peak). Therefore the onset temperatures of crystallizing reaction of $Ba_{0.7}Sr_{0.3}(Ti_{1-x}Mg_x)O_3$ (x=0, 0.05, 0.10 and 0.15) are 662.7°C, 659.3°C, 627.3°C and 609.1°C respectively. It suggests that Mg-doping makes the crystallizing reaction occurred easily.

Figure 2b-c show the crystallizing process of Mg-doped BST samples has only one exothermal peak, namely one reaction occurring, and it is worth to note the exothermal peak of un-doped BST sample seems to be a superposition of two exothermal peaks, indicating there are two crystallizing reactions occurred in the crystallizing process, the first one feeble begun at 623.2°C and the latter strong and dominant. So it is rational to think that Mg-doping might lead to a remarkable change of the crystallizing process of BST system from a double phase-transformation of non-doping sample into a single phase-transformation.

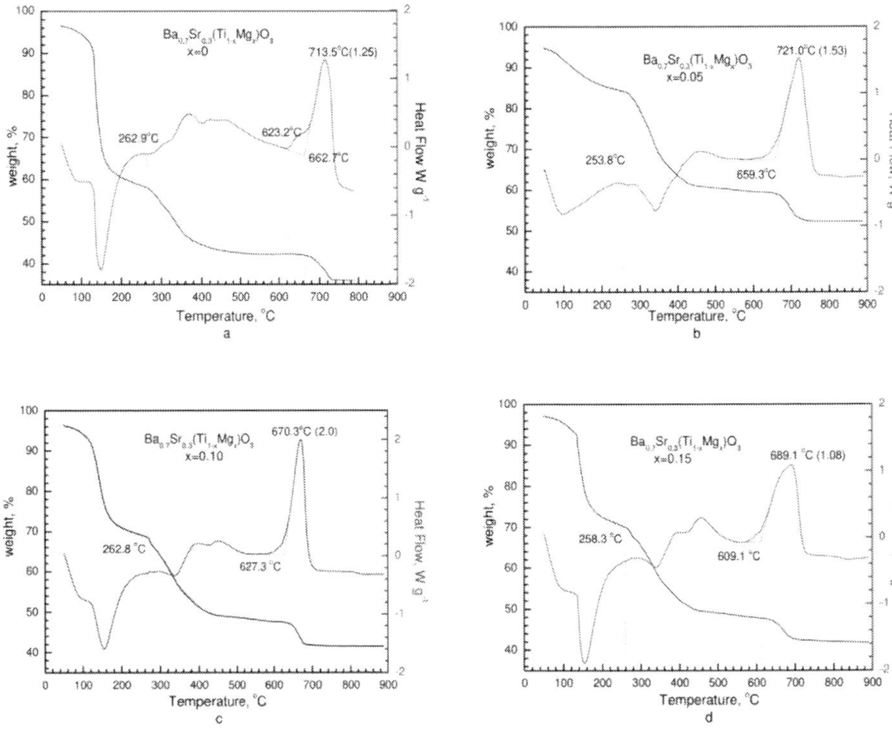

Figure 2. DSC-TG curves of $Ba_{0.7}Sr_{0.3}(Ti_{1-x}Mg_x)O_3$ gels a x=0, b x=0.05, c x=0.10, d x=0.15

3.2. XRD analysis

Figure 3 shows XRD patterns of $Ba_{0.7}Sr_{0.3}(Ti_{1-x}Mg_x)O_3$ powders after same thermal history described above. The diffraction peaks of all curves in Figure3 at 2θ value of 22.4°, 31.8°, 39.2°, 45.6°, 51.3°, and 56.7° are corresponding to the crystal planes of (100), (110), (111), (200), (210) and (211) of perovskite BST. However, the peak intensity of undoped sample is very small, as shown in Fig.3a. It indicated the crystallinity of BST crystal was low and the crystallizing process had not been completed. Besides the diffraction peaks of perovskite BST, the characteristic peaks of $(Ba,Sr)CO_3$ phase were detected at 2θ value of 24.2°, 34.7° and 42.7°; TiO_2 (rutile) phase was detected at 2θ value of 27.2° and $(Ba,Sr)_2TiO_4$ phase were detected at 2θ value of 27.8° and 47.7°, as marked on the figure. As described in the DSC-TG analysis, un-doped BST sample had two exothermal peaks during crystallizing process, indicating there were two kinds of crystallizing paths taken place. It has been reported [18] that after pyrolyzing the precursor of barium titanate didn't transform directly into perovskite phase but interim phase that finally form perovskite structure through solid reaction at high temperature. Gust et al [14] also found interim phase present in $Ba_{0.5}Sr_{0.5}TiO_3$ thin film. $(Ba,Sr)CO_3$, TiO_2 (rutile) and $(Ba,Sr)_2TiO_4$ phases found simultaneously in the un-doped sample, on the one hand, suggests the crystallizing reaction of BST system annealed at 750°C for 0.5 hour has not finished, needing more time or at higher temperature to obtain well-perovskited BST phase, and on the other hand, of more importance, confirms the point of view that the crystallizing process of un-doped BST system is undergone by a double phase-transformation, namely, a decomposing reaction of (Ba, $Sr)_2TiO_4$ and a solid reaction between $(Ba,Sr)CO_3$ and TiO_2.

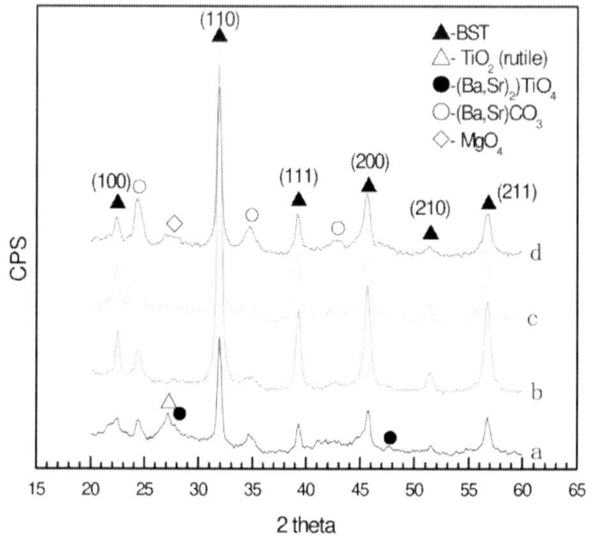

Figure 3. XRD patterns of of $Ba_{0.7}Sr_{0.3}(Ti_{1-x}Mg_x)O_3$ powders a x=0, b x=0.05, c x=0.10, d x=0.15

Compared the peak intensity of BST phase at curve-a with that at curve-b to -d in Figure 3, it was found that Mg-doping improved the crystallinity of BST crystal. When Mg were doped into the BST, neither TiO_2 nor (Ba, $Sr)_2TiO_4$ phase was detected, as shown at curves of b-d in Figure3. It suggested that the crystallizing reaction had finished in the Mg-doped samples through the same thermal history as the un-doped sample. Mg-doping quickens the crystallizing reaction of formation of BST perovskite, consistent with DSC analysis. The diffraction intensity of the BST crystal became lowered when Mg dopant concentration was x=0.15, as shown in Figure3d, and a flat peak appeared at $2\theta= 27.5^{\circ}$. Considering the Joint Committee of Powder Diffraction Standard (JCPDS) data, this peak may be attributed to the formation of the MgO_4 phase, indicating Mg-doping by this level was excessive. The redundant Mg did not enter into the lattice of ABO_3 structure of BST but was present as oxide of magnesium. Cole et al [15] also reported MgO present in the Mg-doped BST by sol-gel technique. Additionally, there were no diffraction peak of Mg and MgO in doped BSTs with Mg dopant concentration of x=0.05 and 0.10,shown in Fig.3b-c, indicating Mg element had entered into the lattice of ABO_3 structure of BST. So it could be concluded that the maximum Mg-doping level should be less than x= 0.15 in $Ba_{0.7}Sr_{0.3}(Ti_{1-x}Mg_x)O_3$ system.

However, there was a small quantity of $(Ba,Sr)CO_3$ phase still detected in Mg-doped BST system, of which the peak intensity increased with the Mg-doping level. This indicated that the amount of $(Ba,Sr)CO_3$ phase in the BST powder crystallized was increased with Mg-doping. It is known that the BST precursor solutions were confected by mol ratio of (Ba+Sr) and (Ti+Mg) of 1:1 supposing Mg atoms occupying Ti positions into ABO_3 structure [9,11-14]. $(Ba,Sr)CO_3$ phase detected means that Ti should exist as TiO_2 or other containing Ti. But there were no diffraction peak of TiO_2 or other containing Ti found in the Mg-doped samples. Therefore, a question emerges: who is with responsibility for the increased $(Ba,Sr)CO_3$ phase? Is the supposition of Mg atoms occupying Ti positions into ABO_3 structure wrong? If so, Mg atoms should occupy the A positions of ABO_3 structure, namely, substituting Ba or Sr atoms. And the mol of (Ba+Sr+Mg) should be more than that of Ti in the Mg-doped BST precursor solutions. As a result, excessive Ba and/or Sr would exist as $(Ba,Sr)CO_3$ phase, of which the amount would increase with the Mg-doping level as expected. So it is rational to think Mg atoms occupy the A positions of ABO_3 structure, instead of the B position. This supposition has been confirmed by XRD and XPS analysis on BSTs prepared by 1:1 mol ration of (Ba+Sr+Mg): Ti, which will be discussed in details in another paper.

4. Conclusion

(1) Mg-doping leads to a remarkable change of crystallizing process of BSTs from a double phase-transformation of non-doping sample into a single phase-transformation.
(2) With increase in the Mg-doping level, the onset temperature of crystallizing reaction of formation of perovskite BST first rose a little and then dropped down considerably, and the intensity of heat flow, representing the reaction rate, increases when the Mg-doping level was less than 0.15.
(3) The crystallinity of BST crystal by XRD analysis indicated Mg-doping quickens the crystallizing reaction of formation of BST perovskite, consistent with DSC analysis. The maximum Mg-doping level should be less than x= 0.15 in $Ba_{0.7}Sr_{0.3}(Ti_{1-x}Mg_x)O_3$ system.
(4) The amount of $(Ba,Sr)CO_3$ phase in the BST powders increases considerably with the Mg-doping level, indicating Mg atoms should occupy the A positions of ABO_3 structure of BST, instead of the B position.

References

[1] Flaviis F D, Alexopoulos N G and Stafsudd O 1997 *IEEE Trans Microwave TheoryTech* **45(6)** 963.
[2] Tombak A, Ayguavives F T and Maria J P 2000 *IEEE Trans Microwave TheoryTech* **48(3)** 1345.
[3] Padmini P, Taylor T R and Lefevre M J 1999 *Appl. Phys. Letter* **75** 3186.
[4] Chen C L, Shen J and Chen S Y 2001 *Appl. Phys. Letter* **78(5)** 652.
[5] Ding L, Liao H C and Jian Y F 2005 *Chemical Industry Times* **19(11)** 54.
[6] Rhim S M, Bak H J and Hong S M 2000 *J. Am. Ceram Soc.* **83(12)** 3009.
[7] Bao J B, Ren T L and Liu J S 2002 *Piezoelectrics & Acoustooptics* **24(5)** 389.
[8] Joshi P C, Ramanathan S and Desu S B 1998 *Integrated Ferroelectrics* **19(1-4)** 141.
[9] Yoon K H, Lee J C and Park J 2001 *Japn J. Appl. Phys.*, Part 1: Regular Papers and Short Notes and Review Paper **40(9 B)** 5497.
[10] Jeon Y A, Seo T S and Yoon S G 2002 *Japn J. Appl. Phys.*, Part 1: Regular Papers and Short Notes and Review Papers **40(11)** 6496.
[11] Zhang T J, Gu H S and Liu J H 2002 *J. Harbin Univ .Sci. Techn.* **7(6)** 51.
[12] Cole M W, Joshi P C and Ervin M H 2001 *J. Appl. Phys.* **89(11)** 6336.
[13] Chen Z H, Yin Z Q and Zhao X Z 2006 *Electronic Components and Materials* **(6)** 13.
[14] Gust M C, Momada L A and Evans N D 2001 *J. Am. Ceram Soc.* **84(5)** 1087.
[15] Cole M W, Hubbard C and Ngo E E 2002 *J. Appl. Phys.* **92(1)** 475.
[17] Cheng H R, Zhu J C and Jeon J H 2004 *J. Inorganic Mater.* **19(5)** 1023.
[16] Cole M W, Joshi P C and Ervin M H 2000 *Thin Solid Films* **374** 34.
[18] Gust M C and Evans N D 1997 *J. Am. Ceram Soc.* **80(11)** 2828.
[19] Kumar S, Messing G L and White W B 1993 *J. Am. Ceram Soc.* **76(3)** 617.
[20] Lange F F 1996 *Science* **273** 903.
[21] Chen X F, Lu W Q, Zhu W G and Lim S Y 2003 *Surf. Coat. Tech.* **167** 203.

Journal of Physics: Conference Series 188 (2009) 012010 doi:10.1088/1742-6596/188/1/012010

Synthesis and characterization of $In(OH)_3$ nanocubes

Xifeng Lu[1,2,3], Tao Wang[1], Xizhou Zhang[2,3], Aiguo Qiu[2,3], Deliang Cui[1,*]

[1] State Key Lab of Crystal Materials, Shandong University, Jinan 250100, P.R. China

[2] Lunan Institute of Coal Chemical Engineering, Jining 272000, P.R. China

[3] State Quality supervisor Center of Coal and Coal Chemical Product, Jining 272000, P.R. China

E-mail: cuidl@sdu.edu.cn

Abstract In this work, $In(OH)_3$ nanocubes were successfully synthesized in large quantities with $InCl_3{\cdot}4H_2O$ as indium source in the presence of sodium hydroxide by a hydrothermal process at 240 °C for 12 h. The structure and the morphology of the $In(OH)_3$ samples were characterized by X-ray powder diffraction (XRD), scanning electron microscopy (SEM), and transmission electron microscopy (TEM). The results revealed that the resultant $In(OH)_3$ nanocubes were nearly uniform in size. Photoluminescence (PL) measurement was carried out.

1. Introduction

As an important wide-gap semiconductor material, with a direct band gap of 5.15 eV, $In(OH)_3$ has recently received considerable attention due to its semiconductor and optical properties as well as photocatalytic properties [1–5]. To date, a variety of strategies have been applied to fabricate $In(OH)_3$ with well-defined shapes, sizes, and nanostructures, such as hydrolysis of indium nitrate [6], peptization of colloidal precipitates [1], double-jet precipitation [2], solvothermal method [7], hydrothermal method [8,9]. Among the as-synthesized samples, $In(OH)_3$ of various shapes have been controlled synthesis, such as nanosheets [10], spheres [9,11], nanorods [11,12], and nanocubes [10,11,12]. Herein, we reported the preparation of single-crystalline $In(OH)_3$ nanocubes via hydrothermal route at 240 °C without any catalysis and template based on sodium hydroxide and $InCl_3{\cdot}4H_2O$.

2. Experimental

2.1. Synthesis

All of the chemicals used in our experiment were of analytical-reagent (A.R.). In a typical synthesis, $InCl_3{\cdot}4H_2O$ (0.1 g) and sodium hydroxide (0.4 g) were dissolved in 16 mL of water and stirred at room temperature for 20 min. The resulting mixture was transferred into a Teflon-lined stainless steel autoclave (20 mL capacity) and heated at 240 °C for 12 h. After the autoclave was cooled to room temperature naturally, the precursors were separated by centrifugation, washing with ethanol and distilled water several times, and drying in an air atmosphere at 60 °C for 10 h.

2.2. Characterization

X-ray powder diffraction pattern was obtained on a Rigaku D/Max–γA X-ray powder diffractometer, with Ni filtered CuKα radiation (λ=1.54178 Å) at a voltage of 50 kV and a current of 100 mA, the scanning speed was 4°/min. The morphology of $In(OH)_3$ were observed with a Hitachi H-800 transmission electron microscope (TEM), using an accelerating voltage of 150 kV. HRTEM images were obtained by a JEM2100 high-resolution transmission electron microscope. The morphology of

* To whom correspondence should be addressed.

the final product was investigated with a scanning electron microscope (S-4800F, JEOL) at an accelerating voltage of 5 kV equipped with energy dispersive X-ray spectroscopy (EDS) after coating by gold sputtering. Fluorescence spectra were measured at room temperature on an Edinburgh FLS 920 fluorescent spectrometer.

3. Results and discussion

3.1. XRD pattern

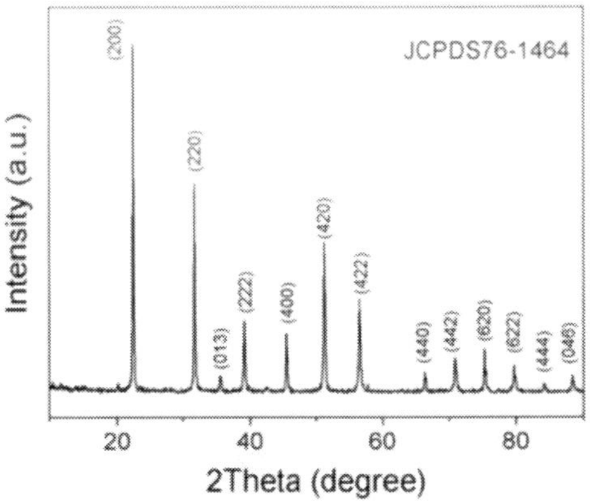

Fig. 1 XRD patterns of $In(OH)_3$ nanocubes.

Fig. 1 shows the XRD pattern of the as-synthesized products. The strong and sharp diffraction peaks can be perfectly indexed to the cubic phase $In(OH)_3$ [JCPDS No. 76-1464, space group: Im_3 (204)], The calculated lattice constant, a = 0.7965 nm, is in good agreement with that (a = 0.7974 nm) from the standard card (JCPDS No. 76-1464).

3.2. TEM image and HRTEM

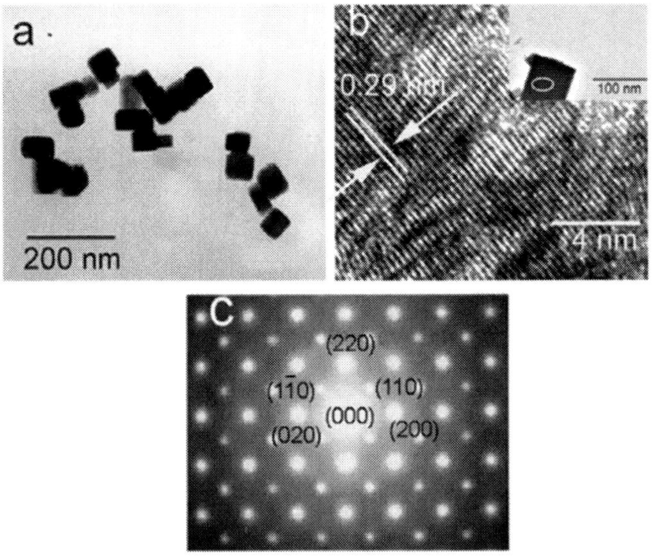

Fig. 2 TEM image of $In(OH)_3$ nanocubes (a), HRTEM image (b) and SEAD pattern corresponding to the inset in Fig.2 b.

Fig. 2a illustrates TEM images of as-prepared $In(OH)_3$. As shown in Fig. 2a, the particles are nearly uniform and monodisperse cubes with lengths of about 100 nm and widths of about 50 nm. The SAED pattern (Fig. 2c) taken from a single nanocube (the inset in Fig.2b) shows sharp diffraction spots that can be indexed to crystal faces of (200), (110), (220), $(1\bar{1}0)$ and (020) of the cubic $In(OH)_3$, indicating the single-crystalline nature of the $In(OH)_3$. The HRTEM pattern (Fig.2b) shows good crystallinity of the nanocubes. The interplane distance of 0.29 nm can be indexed to the (220) diffraction plane of the $In(OH)_3$.

3.3. SEM image and EDS spectrum

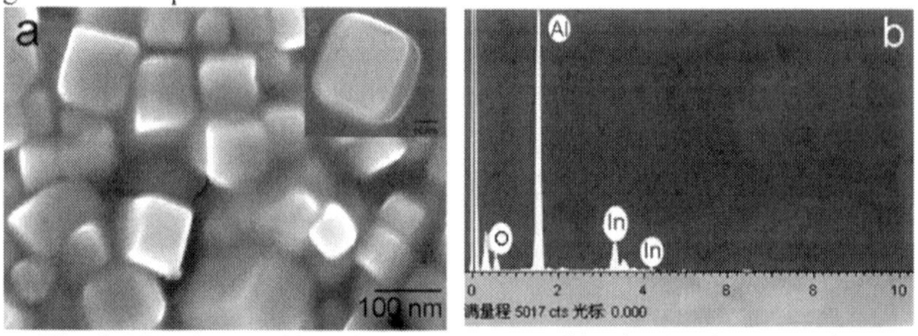

Fig. 3 SEM image (a) and EDS spectrum (b) of $In(OH)_3$ nanocubes.

The inset is a typical nanocubes. The overall morphology of the $In(OH)_3$ particles was also examined with SEM, as shown in Fig. 3a. The particles exhibit a cubic-like morphology, most of which are uniform in size, with the average edge length of about 50-100 nm. It is noted that some crystal planes are a little whirling, as shown in the inset of Fig. 3a. It may be attributed to the hydrothermal and crystallization process. The chemical composition of the product was characterized by energy-dispersive X-ray spectroscopy (EDS) (Fig. 3b). In addition to the Al element from sample table, the peaks of elements In and O are detected, confirming the element composition of $In(OH)_3$.

3.4. PL properties

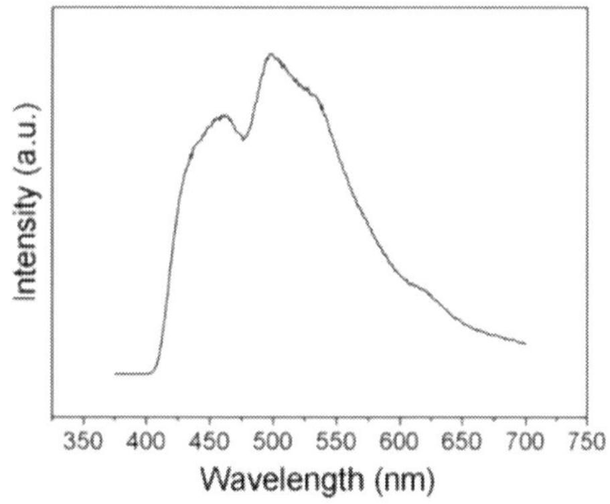

Fig. 4 PL emission spectrum of $In(OH)_3$ nanocube at room temperature

The photoluminescent emission spectrum of the as-prepared $In(OH)_3$ nanocubes is shown in Fig. 4. Upon excitated at 350 nm, $In(OH)_3$ nanocubes show intense photoluminescence centered at 460 nm and 498 nm. There are a few reports about the PL properties of $In(OH)_3$. Enlightened by the PL mechanism of ZnO and SnO_2 semiconductors [12,13], so the blue–green emission of $In(OH)_3$ nanocubes may be contributed to the luminescent centers such as oxygen vacancies, indium vacancies or interstitials, and so on [8,9]. Due to its regular morphologies, the recombination of an electron occupying luminescent centers and a photoexcited hole may cause the blue–green emission with maximum intensity at 498 nm [8,9].

4. Conclusion

In conclusion, single-crystalline $In(OH)_3$ nanocubes have been successfully synthesized by a simple hydrothermal method. XRD, TEM and SEM reveal that the $In(OH)_3$ cubes surface are very smooth, and are of a crystalline cubic structure. The as-synthesized $In(OH)_3$ samples exhibit blue emission luminescent properties.

Acknowledgements
The authors gratefully acknowledge the financial support from Shandong Science and Technology Council (04BS04001, 2005GG3203110).

References

[1] L. A. Perez-Maqueda, L. Wang, E. Matijevic, Langmuir 14 (1998) 4397.

[2] L. Wang, L. A. Perez-Maqueda, E. Matijevic, Colloid Polym. Sci. 276 (1998) 847.

[3] T. Ishida, K. Kuwabara, K. Koumoto, J. Ceram. Soc. Jpn. 106 (1998) 381.

[4] S. Avivi, Y. Mastai, A. Gedanken, Chem. Mater. 12 (2000) 1229.

[5] H. L. Zhu, K. H. Yao, H. Zhang, D. R. Yang, J. Phys. Chem. B, 109 (2005) 20676.

[6] X. H. Zhang, S. Y. Xie, Z. M. Ni, X. Zhang, Z. Y. Jiang, Z. X. Xie, R. B. Huang, L. Zheng, Inorg. Chem. Commun. 6 (2003) 1445.

[7] T. J. Yan, X. X. Wang, J. L. Long, H. X. Lin, R. S. Yuan, W. X. Dai, Z. H. Li, X. Z. Fu, New J. Chem. 32 (2008) 1843.

[8] T. J. Yan, X. X. Wang, J. L. Long, P. Liu, X. L. Fu, G. Y. Zhang, X. Z. Fu, J. Colloid Interface Sci. 325 (2008) 425.

[9] J. Yang, C. K. Lin, Z. L. Wang, J. Lin, Inorg. Chem. 45 (2006) 8973.

[10] J. H. Huang, L. Gao, Cryst. Growth Des. 6 (2006) 1528.

[11] B. X. Li, Y. Xie, M. Jing, G. X. Rong, Y. C. Tang, G. Z. Zhang, Langmuir 22 (2006) 9380.

[12] Q. Tang, W. J. Zhou, W. Zhang, S. M. Ou, K. Jiang, W. C. Yu, Y. T. Qian, Cryst. Growth Des. 5 (2005) 147.

[13] C. H. Liang, G. W. Meng, Y. Lei, Adv. Mater. 13 (2001) 1330.

Preparation and Characterization of Dysprosium-neodymium Modified Magnetic Fluid Coated by Agar

Yuan Xing-hai*, Liu Ru and Deng Yan-jun
Jiaying University, Meizhou, Guangdong 514015

Abstract: The experiment used agar as a coating agent to manufacture dysprosium-neodymium modified water-based magnetic fluid coated by agar with the method of chemical co-precipitation when the two rare-earth element dysprosium and neodymium were precipitated with Fe^{3+} and Fe^{2+} together, which has high stability and a strong magnetic field. After the experimental research on all aspects of factors influencing the magnetic fluid, the best conditions were summarized as follows: Under the premises those the mol ratio between Fe and (Nd^{3+}+Dy^{3+}) is 29:1and it between Fe^{3+} and Fe^{2+} is 1.70 to 1.75, the mol ratio between Nd^{3+} and Dy^{3+} should be 1:1; 25%$NH_3 \cdot H_2O$ is the precipitation reagent and the pH value regulator, which is used to control the pH value between 9 and 11; Under the above-mentioned conditions, dysprosium-neodymium modified water-based magnetic fluid is manufactured at the reaction temperature 35℃. And then the agar with a high degree of biological affinity is used as a coating agent to coat the above product, under the conditions of the temperature 55 ℃ and the pH value 2.5, whose best dosage is 0.0100g each 55mL liquid. The dysprosium-neodymium ferrite water-based magnetic fluid manufactured under the best conditions has black color, and its stability and magnetic properties are better than those of the ordinary one. It will appear a bright magnetic halo under the combined effect of the visible light irradiation and the role of magnetic fields. In addition, its characterization was tested in this experiment, such as the situation of surface coating, the stability, the viscosity, the magnetization intensity, etc.

Keywords: Agar, co-precipitation method, influencing factors

Introductions

Magnetic fluid is a novel functional nano-material. In the present study, highly magnetized material

*Corresponding author. E-mail address: mzysh@jyu.edu.cn

has been intensively studied. A number of researchers prepared the material with mixing Dy. However, the previous research has been limited to adding less than one element. In the work here, we describe a process to fabricate dysprosium-neodymium modified magnetic fluid coated by agar. The effect of various parameters like pH, temperature, viscosity, magnetization, coating and stability are investigated and data on viscosity and magnetization are presented. In addition, IR and spectrophotometry are performed for characterization of materials.

2. Materials and method

2.1 Apparatus

In this study, IR has been used for the determination of coating; viscosity and magnetic susceptibility were measured with NDJ-1 rotation viscometer and magnetic balance, respectively. Spectrophotometric measurements of Fe% were performed with UV-2401PC spectrophotometer.

2.2 Materials and reagents

$FeC1_2{\cdot}4H_2O$, $FeC1_3{\cdot}6H_2O$, $NH_3{\cdot}H_2O$ were all analytical reagent grade, Dy_2O_3 (≥99.5%) and Nd_2O_3(99.9%), agar was C.P.

2.3 Preparation

Magnetic fluid was prepared with co-precipitation method [1-4]; the factors affecting the temperature, pH and ratio of Dy and Nd, as well as the coating performance were investigated.

3. Results and discussion

3.1 Effect of Temperature

The optimal coating time of agar was at 55℃. It can be assumed that this reaction, on base of coating, was a chemical reaction, the higher temperature could be beneficial to it. However, the temperature over 55℃, particles tended to be oxidized or desorbed, at the same time, the coating was not complete at temperature lower than 55℃.

3.2 The dosage of agar

The quality of agar was taken 0.0010g, 0.0050g, 0.0100g, 0.0200g, 0.0300g respectively; the best quality is 0.0100g. If the amount of agar was low, the coating was not perfect, conversely, multilayer sorption could be formed and it leaded the magnetic properties reduced [5].

3.3 The effect of pH

The pH of co-precipitation: the $Dy_xNd_yFe_{(3-x-y)}O_4$ was prepared by co-precipitation method, the condition was controlled that the Fe^{2+}、Fe^{3+}、Nd^{3+}、Dy^{3+} which were simultaneously precipitated. The pH ranged 9 from11, due to the Fe^{3+} was more easily precipitated, but Fe^{2+}、Nd^{3+}、Dy^{3+} precipitated could be high concentration of OH^-. The pH of coating: the pH from 2 to 3 was best. Keeping pH 2, the magnetic liquid was received oxidation resistance. However, Zeta Potential could be produced by the lower pH and it had negative effects to liquid [7].

3.4 Character of products

3.4.1 IR studies

Using the IR spectrometer to detect surface active agent was adsorbed on materials. From the spectrogram, it demonstrated the magnetofluid was loaded by agar.

3.4.2 Settleability

Let it stand for one year or stirring 3000r/min for 30 min, the colloid had not laminating and precipitation, the result indicated that it was stability and good coating; Of all the reasons, the grain diameter was less than 16nm, it due to the more stability.

3.4.3 Thermal stability

The magnetofluid was not stability, it lost magnetism and became the brick red fluid, when the temperature was over 75℃.

3.4.4 Calculating of Fe%

The percentage of Fe was analyzed by colorimetry of Sulphosalicylic acid, at the 450nm. The computing formula as shown:

$$Fe\% = \frac{C\times100\times10^{-6}}{m\times\frac{5}{250}\times\frac{25}{100}}\times100\% \qquad 1.1$$

(c: concentration of Fe; m: quality of sample)

The ratio of ideal magnetofluid was $n(Fe):[n(Dy^{3+})+n(Nd^{3+})]=29:1$, total mol of Dy and Nd had 1/30; the molecular formula was $(DyNd)_{0.04839}Fe_{2.9033}O_4$. The true value of Fe% was 61.36% using the 1.1computing formula to calculate. Comparing with theoretical value the 67.34%, the result could be acceptable.

3.4.5 Viscosity and percentage

The viscosities were monitored by using a rotation viscometer at 30℃, it was 2.07 mPa·s. In the continuous mechanics opinion, the viscosity increases with the increase of internal friction of solid particle as fluid flow. Using Einstein formula [1] to calculate Φ, it was 0.43. Einstein formula:

$$\eta = \eta_0(1+5\varphi/2) \qquad 1.2$$

(η viscosity of magnetic liquid; η_0 viscosity of non-magnetic liquid; Φ percent by volume)

3.4.6 Magnetic property

The magnetofluid prepared by different mixing ratios of Dy and Nd was determined by magnetic balance at the H=100mT, the results were showed in table 1.

Table 1 magnetization of magnetofluid with different ration of Dy and Nd

Mol ratio of Dy and Nd	0:10	2:8	4:6	1:1（5：5）
Dosage of Dy/mL	0.0	2.0	4.0	5.0
Dosage of Nd/mL	10.0	8.0	6.0	5.0
d（g/mL）	1.0010	0.9986	1.0005	1.0002
Mass susceptibility χ	13.1068	16.7488	14.0460	20.4100

continuing

Mol ratio of Dy and Nd	6:4	8:2	10:0	0:0
Dosage of Dy/mL	6.0	8.0	10.0	0.0
Dosage of Nd/mL	4.0	2.0	0.0	0.0
d（g/mL）	1.0040	1.0088	1.0102	0.9937
Mass susceptibility χ	17.2700	15.6474	13.2724	7.8544

Table 1 reveals that the products prepared by adding Dy and Nd has the 2.6 times magnetism than not adding Dy and Nd, the best condition is that the mol ratio of 1:1 and the dosage of Dy and Nd is 5.0mL. The structure of Fe_3O_4 with spinel-type: $(Fe^{3+})[Fe^{3+}Fe^{2+}]O_4$, (Fe^{3+}) as the A and $[Fe^{3+}Fe^{2+}]$ as B. The metal can be padded and replaced Fe of $(Fe^{3+})[Fe^{3+}Fe^{2+}]O_4$, because of the A and B had many gaps[8]. The magnetic moment can be improved due to Dy^{3+} and Nd^{3+} padded and replaced the B. Reaction principle was suggested as follow:

$$(1-x-y)Fe^{2+}+2Fe^{3+}+8OH^-+(y)Nd^{3+}+(x)Dy^{3+} == Dy_xNd_yFe_{(3-x-y)}O_4+4H_2O$$

4. Conclusions

Finding showed that the dysprosium-neodymium modified magnetic fluid was obtained by coprecipitation method at the n(Fe):[n(Nd^{3+})+n(Dy^{3+})]=29:1,n(Fe^{3+})： n(Fe^{2+})=1.70~1.75, n(Nd^{3+})：n(Dy^{3+})=1:1, T=35 ℃，pH rang 9 from 11，stirring at 3000r/min, ammonia as precipitator and regulator. The products coating by agar was stability when agar as surface active agent was 0.0100g/55mL, pH 2.5, T=55℃, stirring at 3000r/min.

Acknowledge

The work was supported by plan of Science and Technology Department of Guangdong Province (Grand No. 2005B16001065)

References

[1] Li C D Magnetic fluid theory and application [M]. Bei jing: Scientific Press, 2003:3-8.

[2] Yuan X H, Li Yong, Zhang L T 2008 J. *Journal of Jiaying University*. **26** 42-45.

[3] Yuan X H, Yang D H, Lin S Y 2008 J. *Nanoscience & Nanotechnology*.**5** 31-34.

[4] Yuan X H, Lin S Y, Huang J W 2007 J. *Journal of Jiaying University*. **25** 39-40.

[5] Ren H Y, Liu Lei, Liu T J 2003 J. *New Chemical Materials*. **31** 15-17.

[6] Jiang Y L; Liu Y J, Liu Y C, Yin W X. 2003 J. *Journal of China University of Mining & Technology*, **11** 98-101.

[7] Zhang J S ,Yin Y S ,Zhang Y Y 2003 J. *Chemical Industry and Engineering Progress*. **22** 494-498.

[8] Jiang R L, Liu Y C, YIN W X 2004 J. *Journal of Sichuan University (Engineering Science Edition)*. **36** 32-37.

Journal of Physics: Conference Series **188** (2009) 012012 doi:10.1088/1742-6596/188/1/012012

Synthesis of $LiFePO_4/C$ cathode materials by carbothermal reduction method using two kinds of Fe^{3+} precursors

M E Zhong and Z T Zhou[1]

College of Materials Science and Engineering, South China University of Technology, Guangzhou 510640, China

E-mail: mcztzhou@scut.edu.cn

Abstract: $LiFePO_4/C$ cathode materials were synthesized by carbothermal reduction method using $FePO_4$ and citrate ferric ($FeC_6H_5O_7{\cdot}5H_2O$) as Fe^{3+} precursors. The samples were characterized by X-ray diffraction (XRD), scanning electron microscope (SEM), X-ray photoelectron spectrometry (XPS) and charge-discharge test. The results showed that both the crystallization degree and particle size increased markedly with the increase of synthesis temperature. Among the synthesized cathode materials, $LiFePO_4/C$ prepared at 650 ℃ showed the best electrochemical properties with an initial discharge specific capacity of 138 $mAh{\cdot}g^{-1}$ at 0.2 C, 128 $mAh{\cdot}g^{-1}$ at 0.5 C and 116 $mAh{\cdot}g^{-1}$ at 1.0 C, respectively.

1. Introduction

$LiFePO_4$, proposed by Padhi et al. in 1997, has attracted much attention as a promising cathode material for rechargeable Li-ion batteries because of its low cost, nontoxicity, good thermal stability and superior safety [1-3]. However, there are two main obstacles preventing $LiFePO_4$ to be a commercial material. One is the inherent low electronic conductivity, which limits the electrochemical properties of this material. To enhance the electronic conductivity, three approaches have been widely in progress, including minimizing the particle size [4], introducing conductive additives [5-7], and doping with supervalent metal ions [8]. Especially, coating the active particles with conductive carbon was very effective in the improvement of electronic conductivity. The other problem of $LiFePO_4$ is that it is difficult to prevent Fe^{2+} from oxidation. The conversion of Fe^{2+} to Fe^{3+} has been usually avoided by the solid-state reactions with a reductive or inert gas atmosphere [9-11]. Nevertheless, expensive divalent iron compounds, typically iron (II) oxalate ($FeC_2O_4{\cdot}2H_2O$), or iron (II) acetate ($Fe(OOCCH_3)_2$), were always selected as a starting material to synthesize $LiFePO_4$, which increased the cost. In 2003, Barker et al. [12] first reported that $LiFePO_4/C$ composite materials were synthesized by a carbothermal reduction (CTR) method, which was relied on the C to CO transition, starting with cheap Fe_2O_3 as the iron precursor. The CTR approach seems ideally suited to favouring the stabilization of iron as Fe^{2+}, as well as offering a degree of control of the material morphology and conductivity. Therefore, some recent studies have highlighted the effectiveness of the CTR method for the synthesis of low-cost, phase-pure and high electronic conductivity of $LiFePO_4$. In this work, we report a simple and novel CTR method to synthesize the $LiFePO_4/C$ composite by employing cheap citrate ferric ($FeC_6H_5O_7{\cdot}5H_2O$) and $FePO_4$ as Fe^{3+} precursors in which citrate ferric acts as reducing

[1] Corresponding author

agent, carbon source and part of iron resource. The effect of sintering temperature on the phase-purity, particle size, morphology and electrochemical performance of the materials are investigated.

2. Experimental

The $LiFePO_4/C$ cathode materials were prepared with citrate ferric ($FeC_6H_5O_7{\cdot}5H_2O$), $FePO_4$, Li_2CO_3, and $(NH_4)_2HPO_4$ as raw materials. Citrate radical containing in citrate ferric acted as both carbon resource and reductive reagent. The amount of citrate radical used was 35 wt. % of calculated $LiFePO_4$ product. A stoichiometric mixture of the raw materials was thoroughly mixed in an alcohol medium by ballmilling in a planetary mill for 6 h. The resulting product was decomposed at 300 ℃ for 10 h, further sintered at temperatures between 600 and 750 ℃ for 24 h, finally cooled to room temperature. All these processes were carried out in a nitrogen atmosphere.

The phase structure of the $LiFePO_4/C$ material was analyzed by X-ray diffraction (XRD) with Cu Kα radiation (Siemens D-500, Germany). The material morphology was investigated by a scanning electron microscopy (SEM, Hitachi S-550, Japan). The X-ray photoelectron spectroscopy (XPS) was performed with an Axis Ultra instrument (Kratos Analytical Ltd., UK). The amount of carbon was determined by C-S 800 Determinator (Eltar, Germany).

The performances of the cathode materials were evaluated using a coin type cell with lithium as the anode. The cathode film of $LiFePO_4/C$ was composed of active material, acetylene black and polytetrafluoroethylene (PTFE) with a weight ratio of 85:10:5. The separator used was a celgard 2300 microporous membrane. The electrolyte was 1M $LiPF_6$ in ethylene carbonate/dimethyl carbonate (1:1, v/v). The cells were assembled in an argon-filled glove box. The charge-discharge test was performed between 2.0 and 4.2 V versus Li/Li^+ on BS-9300 cell test system. Cyclic voltammograms were run on a CHI660A electrochemical workstation at a scanning rate of 0.1 $mV{\cdot}s^{-1}$ between 2.8 and 4.2 V. All electrochemical measurements were carried out at room temperature.

3. Results and discussion

Figure 1 shows the XRD results of $LiFePO_4/C$ composites sintered at different temperatures. It can be seen that the calcining temperature has obvious influence on the structure of $LiFePO_4$ materials. For the samples heat-treated at 600-700 ℃, all of the peaks in the XRD patterns are assigned to the triphylite $LiFePO_4$. As the annealing temperature is increased to 750 ℃, an additional peak corresponding to the Fe_2P phase appears. The generation of Fe_2P can be ascribed to the reduction of phosphate and iron during the sintering process by the residue coating carbon originated from the citrate ferric. From figure 1 we can also see that the peak intensity of $LiFePO_4$ increases slightly with increasing the synthesis temperature from 600 to 750 ℃, indicating that the crystallization degree improves with the enhancement of synthesis temperature. On the other hand, residual carbon can not be found in the patterns, suggesting their amorphous state and/or low content. The results obtained from the C-S determinator demonstrate that the amount of carbon in the samples calcined at 600, 650, 700 and 750 ℃ are 2.38 wt. %, 2.19 wt. %, 1.80 wt. % and 1.23 wt. %, respectively.

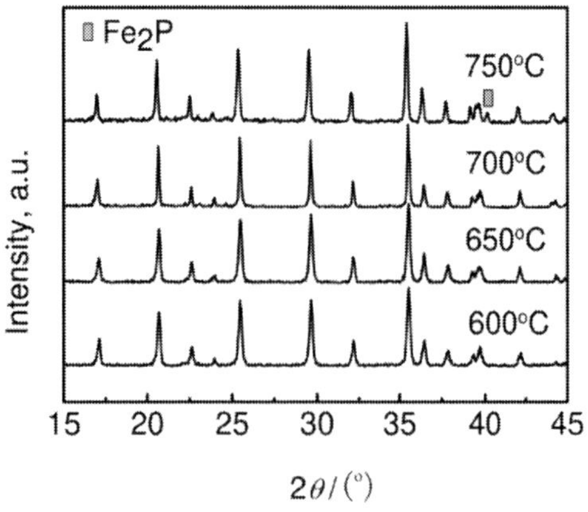

Figure.1 XRD patterns of the $LiFePO_4/C$ composites synthesized at different temperatures.

The chemical composition and valent state of the as-prepared $LiFePO_4/C$ composites was confirmed by XPS analysis and results for the sample synthesized at 650 ℃ are shown in figure 2. Figure 2(a) shows the wide range scanning spectrum and it can be seen that this sample consists of Li, Fe, P, O and C elements. Figure 2(b) illustrates the XPS narrow spectra of the Fe 2p and it is evident that the Fe 2p signal derives from the contribution of two peaks. The peaks of the binding energy (BE) at 710.9 and 724.2 eV represents the Fe^{2+} valence state, suggesting that the product consists only of one valance state of iron atoms [13,14]. The results of both XRD and XPS analyses demonstrate that this material preserves a crystal structure of $LiFePO_4$ and that it is comprised of all the constituent elements of $LiFePO_4/C$.

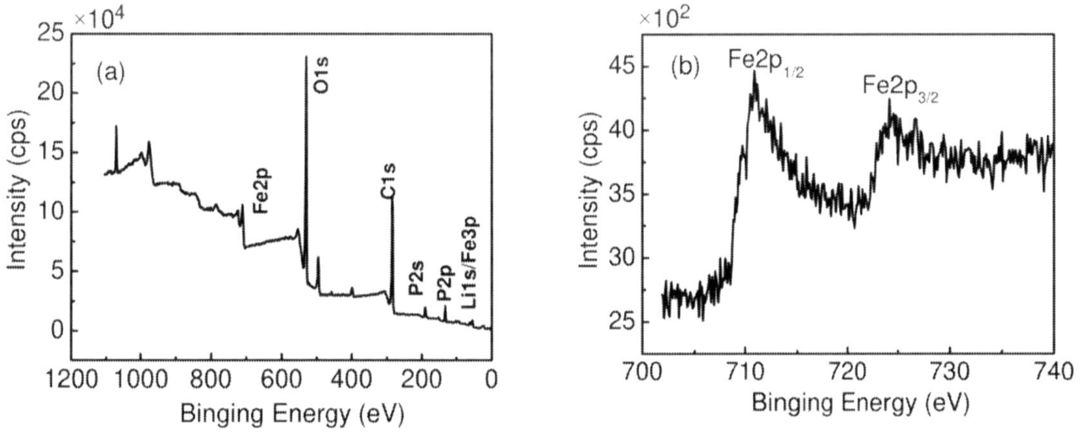

Figure.2 The XPS spectra of the $LiFePO_4/C$ powders calcined at 650 ℃: (a) wide range scanning spectrum, (b) the XPS narrow spectra of the Fe 2p.

Figure 3 shows the SEM images of $LiFePO_4/C$ composites synthesized at 650 and 750 ℃. With the increase of the temperature, a significant increase in particle size is observed. The sample prepared at 650 ℃ has a small particle size of about 500 nm although some micrometer-sized particles exist. However, the sample prepared at 750 ℃ shows a particle-agglomerated morphology, whose particle size is about 1-5 μm. The increase in particle size can be attributed to the higher crystallization degree of the samples prepared at high temperature.

Figure.3 SEM photographs of $LiFePO_4/C$ composites synthesized at different temperatures.

The initial discharge curves of $LiFePO_4/C$ samples synthesized at different temperatures at 0.2 C rates are illustrated in figure 4. It can be seen that the initial discharge capacities of $LiFePO_4/C$ samples synthesized at 600, 650, 700 and 750 ℃ are 125, 138, 120 and 117 $mAh{\cdot}g^{-1}$, respectively.

From the XRD and SEM measurements of the samples, we can explain why the $LiFePO_4/C$ powders sintered at 650 ℃ delivers the highest discharge capacity. It is known that a small particle size and perfect crystal structure are two key factors to enhance the electrochemical performance of $LiFePO_4$ [15,16]. As described in the previous section, particles of $LiFePO_4/C$ powder sintered at lower temperatures show smaller particle sizes but lower crystallinity. Moreover, the $LiFePO_4/C$ powder sintered at higher temperatures has higher crystallinity but larger particle sizes. As a result, in the range of 600-750 ℃, 650 ℃ is the optimum synthetic temperature for the $LiFePO_4/C$ with both small particle size and perfect crystal, which results in the best electrochemical performance.

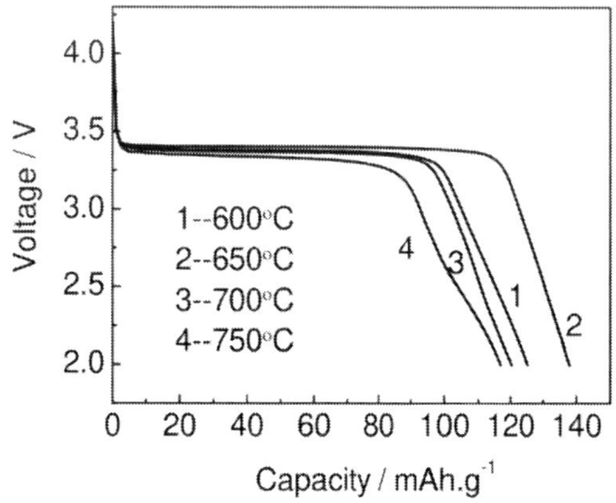

Figure.4 Initial discharge curves of $LiFePO_4/C$ synthesized at different temperature under 0.2 C.

Figure 5 presents the first charge-discharge profiles of the $LiFePO_4/C$ powder prepared at 650 ℃. Charge-discharge rate is varied from 0.2-1.0 C. The first discharge capacities are 138 $mAh{\cdot}g^{-1}$ at 0.2 C, 128 $mAh{\cdot}g^{-1}$ at 0.5 C and 116 $mAh{\cdot}g^{-1}$ at 1.0 C, respectively. It is noted that the potential plateau remains flat even for the 1.0 C curve except for a slight decrease, which indicates the excellent kinetics of $LiFePO_4$ cathode materials. This sample also shows relatively stable charge-discharge cycle property as illustrated in figure 6. The initial discharge capacities of this sample at 0.2 and 1.0 C are 138 and 116 $mAh{\cdot}g^{-1}$, respectively. After 20 cycles, its discharge capacities at 0.2 and 1.0 C are 137 and 115 $mAh{\cdot}g^{-1}$, respectively. Above results demonstrate that the powder prepared at 650 ℃ possesses good rate performance and cycling stability. It could be attributed to the fact that the high phase purity, small particle size and improved conductivity through surface carbon coating.

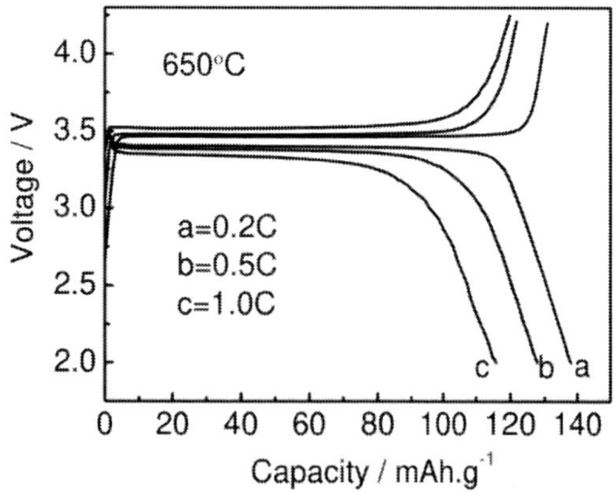

Figure.5 The first charge-discharge curves of $LiFePO_4/C$ synthesized at 650 ℃ at various rates.

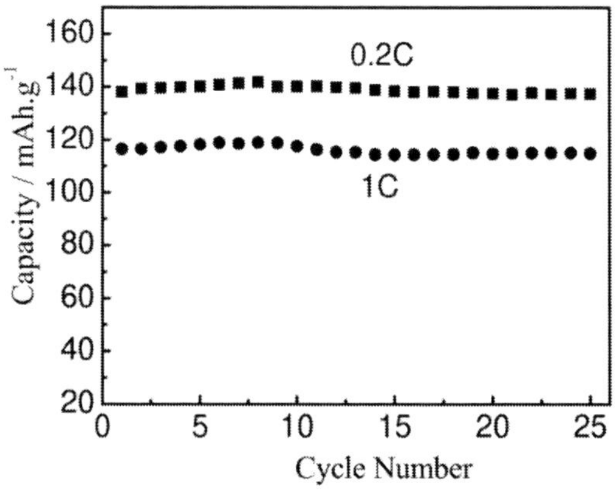

Figure.6 Cycle properties of $LiFePO_4/C$ synthesized at 650 ℃.

4. Conclusion

A simple and low-cost carbothermal reduction (CTR) method was developed to prepare pure-phase $LiFePO_4/C$ by using $FePO_4$ and citrate ferric ($FeC_6H_5O_7 \cdot 5H_2O$) as Fe^{3+} precursors. It is demonstrated that the particle size and crystallization degree increased with the increase of synthesis temperature. The largest discharge capacities of $LiFePO_4/C$ are 138 $mAh \cdot g^{-1}$ at 0.2 C and 116 $mAh \cdot g^{-1}$ at 1.0 C after sintering at 650 ℃, and these values indicate a good electrochemical performance probably due to the small particle size and perfect crystal structure.

References

[1] Wu S H, Hsiao K M and Liu W R 2005 *J. Power Sources* **146** 550
[2] Choi D and Kunta P N 2007 *J. Power Sources* **163** 1064
[3] Jin H F, Liu Z, Teng Y M, Gao J K and Zhao Y 2009 *J. Power Sources* **189** 445
[4] Koleva V, Zhecheva E and Stoyanova R 2009 *J. Alloys Compd.* **476** 950
[5] Bewlay S L, Konstantinov K, Wang G X, Dou S X and Liu H K 2004 *Mater. Lett.* **58** 1788
[6] Belharousk I, Johnson C and Amine K 2005 *Electrochem. Commun.* **7** 983
[7] Cho Y D, Fey G T K and Kao H M 2009 *J. Power Sources* **189** 256
[8] Wang G X, Bewlay S L, Konstantinov K, Liu H K, Dou S X and Ahn J H 2004 *Electrochim. Acta* **50** 443
[9] Kim J K, Cheruvally G, Choi J W, Kim J U, Ahn J H, Cho G B, Kim K W and Ahn H J 2007 *J. Power Sources* **166** 211
[10] Zhang M, Jiao L F, Yuan H T, Wang Y M, Guo J, Zhao M, Wang W and Zhou X D 2006 *Solid State Ionics* **177** 3309
[11] Mi C H, Cao G S and Zhao X B 2005 *Mater. Lett.* **59** 127
[12] Bark J, Saidi M Y and Swoyer J L 2003 *Electrochem. Solid-State Lett.* **6** A53
[13] Wang Z L, Su S R, Yu C Y, Chen Y and Xia D G 2008 *J. Power Sources* **184** 633
[14] Liu H W and Tang D G 2008 *Solid State Ionics* **179** 1897
[15] Zhi X K, Liang G C, Wang L, Ou X Q, Zhang J P and Cui J Y 2009 *Solid State Ionics* **189** 779
[16] Mi C H 2005 *Study on Olivine-type $LiFePO_4/C$ Composite Cathode Materials* (Zhejiang: zhejiang university) p 18

Journal of Physics: Conference Series **188** (2009) 012013 doi:10.1088/1742-6596/188/1/012013

A Study on the Influence of Surface Active Agent Molecules on the AFM Scanning and Imaging of SWCNTs

Ke Xu [1,2], Cheng-dong Wu [3], Hai-bo Yu [1,2], Xiao-jun Tian [1], Zai-li Dong [1]

(1. State Key Laboratory of Robotics, Shenyang Institute of Automation, CAS, Shenyang 110016, China；

2. Graduate University of the Chinese Academy of Sciences, Beijing 100049, China；

3. Northeastern University, Shenyang 110004, China)

E-mail: xuke08@sia.cn

Abstract: The paper proposed a preprocessing method for scanning samples based on rinsing technology, and solved the effective fabrication problem of AFM scanning samples based on facilitated dispersing SWCNTs with SDS surface active agents. First, the ultrasonication oscillation method was applied, and the uniform alignment of SWCNTs in SDS was realized. Then, SDS solution of different concentrations was scanned and imaged with AFM, the influence of SDS solution on the imaging quality of SWCNTs was analyzed, and the method to effectively fabricate scanning samples after dispersing SWCNTs was found. The results of the experiments showed that the preprocessing of the SWCNTs solution scanning samples was the decisive factor to influence SWCNTs imaging quality, that the result of rinsing SWCNTs scanning samples for 5s at 0.1ml/s with di-ionized water was the best, and that with the same rinsing rate and angle, the rinsing di-ionized water quantity would influence the alignment degree of SWCNTs on the substrates and SDS macromolecules' residual quantity on SWCNTs scanning samples.

1. Introduction

Since Iijima, a Japanese scientist, found carbon nanotubes (CNTs) in 1991, CNTs have received extensive attention due to the special physical structure and electrical characteristics[1]. CNTs have extreme potential in such areas of nano electrical devices and new functional materials as field-effect transistors, photoelectric near infrared spectral emitters, sensors and polymer compound materials[2]. As one-dimensional nano materials, due to the attraction of Van der Waals force, single-walled carbon nanotubes (SWCNTs) tend to twine with one another so as to gather easily and decrease the total surface energy of the system[3]. The gathering form usually destroys the excellent characteristics of SWCNTs in mechanics, optics, electricity and calorifics, which limits the use of CNTs in the fabrication of nano devices. For example, when SWCNTs are used in the assembly of carbon nanotube field effect transistor (CNT-FET), SWCNTs are required to accurately align and singly disperse between source and drain[4]. Thus, it's of great necessity to study SWCNTs dispersion property. SWCNTs are lack of active genes, difficult to dissolve in organic solution and water, which makes the research on the chemical qualities hard[5]. Therefore, the research on how to improve SWCNTs dispersion property has become the precondition in the actual application of SWCNTs[6].

The key point in obtaining ideal dispersed SWCNTs is to solve the problem of uniform dispersion and dispersion stability of SWCNTs on substrates[7]. The surface active agent has special structure, which is formed by both hydrophilic groups and oleophilic groups, and thereby has good adsorption and chemical reaction activity[8]. The SWCNTs surface activity can be changed by decreasing the surface tension of SWCNTs. The surface active agents adsorb on the surface of SWCNTs, forming colloids covering SWCNTs, and due to ionization or adsorption, the surface of colloids have charges[9]. Because of electrostatic repelling force, colloids can gather together, which makes nano particles in stable situation. Moreover, the long tail ends of surface active agents form space steric hindrance on the surface of nano particles, preventing SWCNTs from re-gathering. Due to adsorption, the surface active agents cover nano materials, forming a space barrier, keeping nano materials from colliding and combining with one another, which makes SWCNTs disperse in solution stably. Therefore, the dispersion, reaction and the surface structure of nano particles of SWCNTs in different dielectrics are improved. However, when the surface active agents reach a certain concentration, the surface active agent molecules would combine into colloid polymers from single ion or molecule, forming colloid groups, and the solution quality will mutate and form unimolecule film and make the adsorption ability decrease sharply. Meanwhile, the dispersion of SWCNTs will be influenced greatly[10].

Sodium Dodecyl Sulfate (SDS) is a comparatively long negative ion surface active agent. It can adsorb on the surface of solution and form diffusion layer, preventing SWCNTs from re-gathering. Thus, SWCNTs have good dispersion property. Whereas the macromolecules of SDS will influence the scanning and imaging of SWCNTs, even make it hard to obtain clear AFM images of SWCNTs. Against the influence of SDS on the scanning and imaging of SWCNTs, the paper proposed a post treatment method based on rinsing technology, and solved the effective fabrication problem of atomic force microscope (AFM) scanning samples based on facilitated dispersing SWCNTs with SDS surface active agents[11].

2. Experiment materials and methods

2.1 Experiment reagents and instruments

The nano materials used in this experiment are SWCNTs produced by Chengdu Institute of Organic Chemistry (model SH071011, diameter of SWCNTs 0.8-2nm, length 2-5μm, purity >90%, surface area >380m^2/g). SDS surface active agents were purchased from Beyotime Institute of Biotechnology Co., Ltd. Ultrasonic oscillator were purchased from Shanghai KeDao Limited Company of Ultrasonic Instrument (type: SK5210LHC). Atomic force microscope(AFM) was purchased from Veeco Company of United States (type: Dimension 3100).

2.2 The experiment flow

First, four kinds of 10ml SDS solution were prepared, whose concentrations were respectively 1%, 0.1%, 0.01%, and 0.001%. They were put in four test tubes and were sonicated for 10 minutes. Then 2μl solution of each was obtained with pipette to be scanned with AFM. The accurately matched SWCNTs solution (The quality percentages of the matched SWCNTs solution and SDS solution were respectively 0.1% and 1%.) was put in test tubes and was sonicated for 4 hours. 2μl solution was obtained to be scanned with AFM, watched and compared. The experiment flow can be shown in Fig.1.

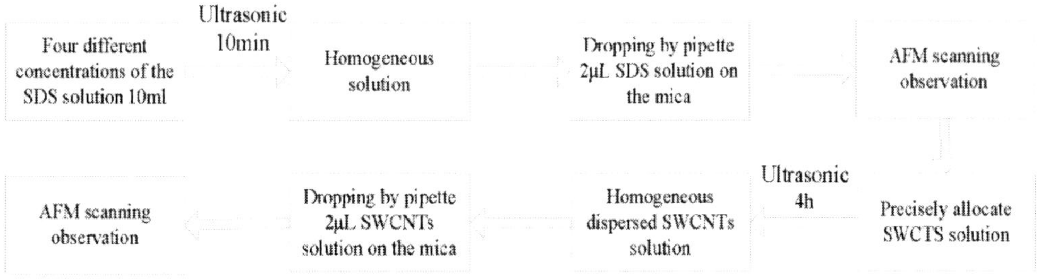

Fig. 1 The experiment flow

Journal of Physics: Conference Series **188** (2009) 012013 doi:10.1088/1742-6596/188/1/012013

2.3 Observation of SDS solution samples of different concentrations scanned with AFM

The gathering activity of SDS surface active agents on the water solution interface was helpful to improve the dispersion stability of SWCNTs, whose concentration change played a key role in colloid system and the directed alignment of molecules. Meanwhile, when SDS facilitated the dispersion of SWCNTs, the coverage of SWCNTs prevented AFM scanning from imaging clearly, and part of SWCNT was covered and couldn't show the superior electrical characteristics. So the study on the influence of SDS solution of different concentrations on the imaging of SWCNTs is also a key step in SWCNTs preprocessing.

10mg SDS solid powder was put in the test tube with 9.99ml di-ionized water, made into SDS solution whose concentration was 1%, sonicated for 10min to solve completely in the ultrasonication oscillator. Then 2μl reagent was dropped on a mica sheet with pipette, laid uniformly on the sheet with capillary quickly, weathered with N_2 immediately and scanned with AFM. 5ml SDS solution whose concentration was 1% was diluted 10 times into 0.1% solution with di-ionized water. All the same, SDS solution whose concentration was 0.01% and 0.001% was obtained and scanned with AFM. During imaging, the AFM system was working in tapping mode with frequency of about 300kHz, and the AFM probe used here was model MPP-11100 with rectangle Si cantilever, tip radius less than 10nm. The experiment process is showed in Fig. 2.

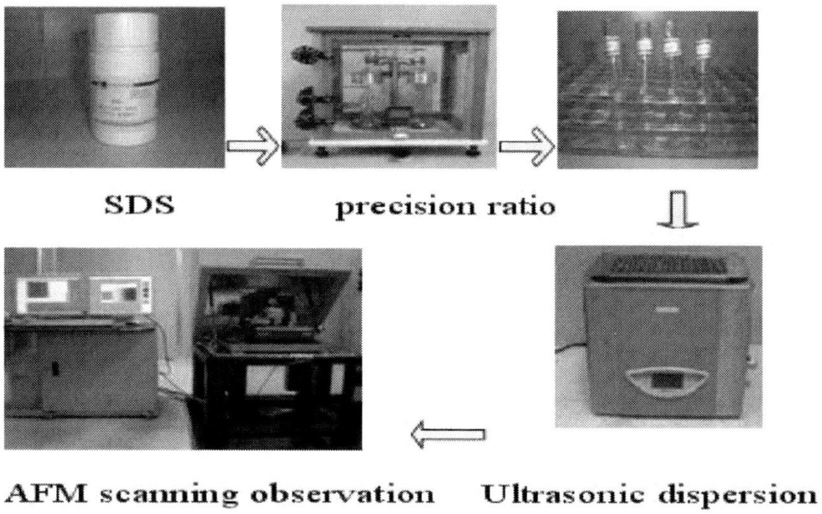

Fig. 2 The experiment process under different SDS concentrations

2.4 Observation of the scanning of SWCNTs with AFM

2.4.1 The SWCNTs dispersion in SDS surface active agents

The accurately matched SWCNTs solution (The quality percentages of the matched SWCNTs solution and SDS solution were respectively 0.1% and 1%.) was put in a test tube, sealed with plastic film, sonicated for 2h with frequency of 59kHz in the ultrasonication oscillator, rested for 1h, sonicated again for 2h, rested for over 3h and disposed by precicpitating. During ultrasonication, the location and the inclination angle of the test tube in the ultrasonication oscillator were regulated in the process of sonication so that SWCNTs solution in the test tube was "boiling" to mix completely uniformly. At the same time, the test tube was heated to 50°C to facilitate the dispersion.

2.4.2 SWCNTs rinsing disposition and AFM observation

2μl SWCNTs solution was dropped on a mica sheet with pipette, laid uniformly on the sheet with capillary quickly, rested for 1m and weathered with N_2. Sample A was obtained. All the same, 2μl SWCNTs solution was dropped on a mica sheet twice, laid uniformly and rested for 1m. The mica sheet was down dipped 30 degree angle. Then the samples were rinsed with 0.5ml and 1ml di-ionized water at the rate of 0.1ml/s, and blew in very short time with nitrogen impulse gas flow of a certain pressure (about 0.15MPa). Sample B and C were obtained. The post treatment process is showed in Fig. 3.

Samples A, B and C and the SWCNTs solution sample that was not dispersed were scanned and compared with AFM in tapping mode.

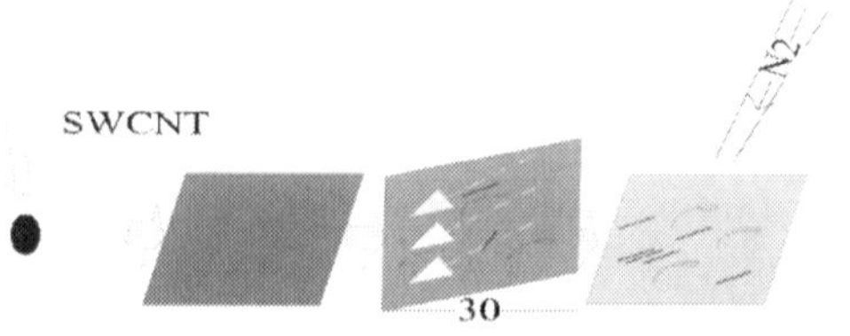

Fig. 3 SWCNTs post treatment principle

3 The experiment results and analysis

Fig. 4 is the AFM characterization observation figure of SDS of four different concentrations: 1%, 0.1%, 0.01% and 0.001% in micron scope. From fig. 4(a), we can see that SDS surface active agent molecules form colloid, so that the contact of water and air reduces and the surface tension of the solution decreases. In Fig. 4(b), some SDS molecules gather and the hydrophobic groups of these SDS surface active agents rely on one another, forming small colloid groups, so the surface tension of the solution increases. SDS colloid groups in fig. 4(c) are much less than those in (b) and (c), and the surface tension is bigger. The surface energy of SWCNTs decreases with it. In fig. 4(d), the concentration of the SDS surface active agents is so low that air and water contact directly, there are a few surface active agent molecules in water, SDS has no colloid groups, and the electrostatic repelling force between colloids becomes smaller and the adsorbency decreases with it.

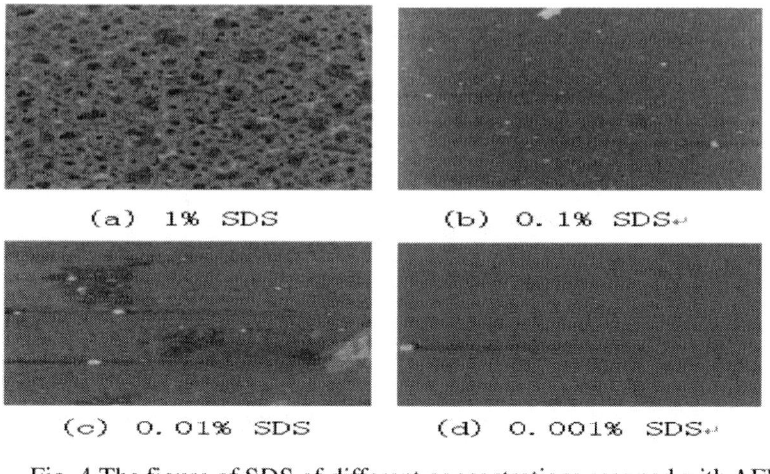

Fig. 4 The figure of SDS of different concentrations scanned with AFM

SWCNTs were dispersed by the method in 1.4 and the dispersed reagents were light black uniform

Journal of Physics: Conference Series **188** (2009) 012013 doi:10.1088/1742-6596/188/1/012013

transparent bodies. There were no graininess suspended substances in the reagents through naked eye observation. Such SWCNTs sample was characterized with AFM. From fig. 5(a), it can be seen that SWCNTs which was dispersed only by ultrasonication oscillation and not added into SDS surface active agents gather seriously. Fig. 5(b) is SWCNTs solution that has not been rinsed. Some SWCNTs exist singly and don't twine with one another, but seen from the AFM height characterization value, the height of single SWCNTs is less than 2nm and the dispersion situation is not ideal. Compared with fig. 4(a), because SDS colloid groups are too many and the length and the diameter of SDS molecules are far larger than SWCNTs, SDS surface active agents form saturant adsorbency on SWCNTs and cover most SWCNTs, so that more SWCNTs can't be characterized with AFM. And just because of the influence of macro molecules of SDS active agents on the scanning and imaging of SWCNTs, SWCNTs solution can be post treated by rinsing. Fig. 5(c) is the figure of SWCNTs solution scanned with AFM that is washed with di-ionized water at 0.1ml/s for 5s. From the AFM height characterization value, it can be seen that the height of single SWCNTs is less than 2nm. And compared with fig. 5(b), some SDS molecules adsorbed on SWCNTs are flushed out with water, thus the adsorbency quantity decreases, but the single SWCNTs are comparatively more. Therefore, SDS is good for decreasing the surface energy of SWCNTs and increasing the mutual repelling force, and the macro molecules of SDS surface active agents also have less influence on the scanning and imaging of SWCNTs. Fig. 5(d) is the figure of SWCNTs solution scanned with AFM that is washed with di-ionized water at 0.1ml/s for 10s. It is showed in the figure that the SDS concentration in the solution is low, but because of rinsing, most SWCNTs lose and the SWCNTs stability isn't good.

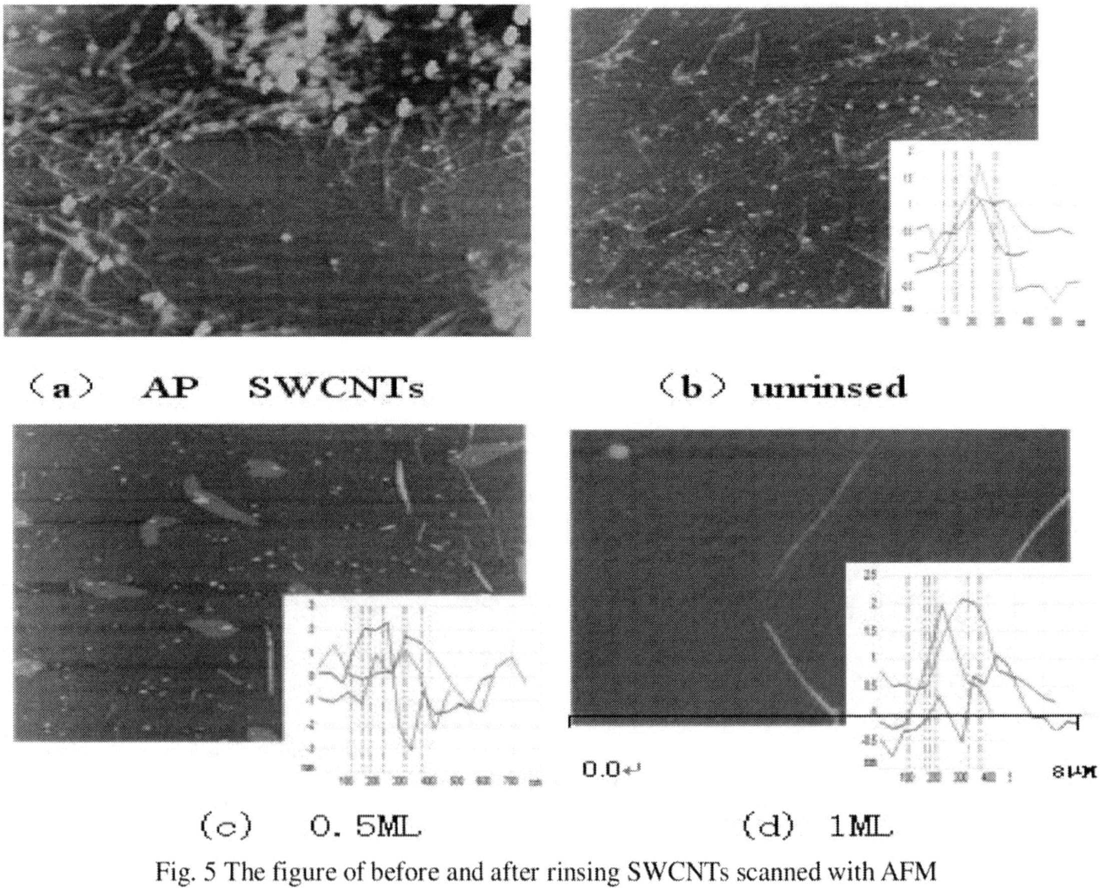

Fig. 5 The figure of before and after rinsing SWCNTs scanned with AFM

4. Conclusion

The results of the research show that the preprocessing for scanning samples method based on rinsing technology can solve the effective fabrication problem of AFM scanning samples based on facilitated dispersing SWCNTs with SDS surface active agents. Before centrifugating purification, the research on the influence of the macro molecules of SDS surface active agents on the scanning and imaging of SWCNTs is a meaningful step in the preprocessing of SWCNTs. In the fabricated SWCNTs solution, besides foreign substances brought by self growth of SWCNTs, such as carbon nano particles, amorphous carbon , fullerene particles and metal catalysts. SDS surface active agents can block the imaging of pure SWCNTs. From the SDS characterization figure obtained by observing AFM scanning, the SWCNTs characterization figures before and after rinsing, the influence degree of SDS molecules on the scanning and imaging of SWCNTs can be clearly shown. What's more, the factors influencing the imaging of SWCNTs are properly considered, which will provide important foundation for how to obtain ideal dispersion results of SWCNTs.

Acknowledgments

The Authors thank the support from the CAS FEA International Partnership Program for Creative Research Teams.

References

[1] Hu Y F, Yao K, Wang S 2007 Appl Phys Lett **90** 223116
[2] Krupke R, Hennrich F, Lohneysen H V 2003 Science **301** 344
[3] Larsson P, Larsson J A, Ahuja R 2007 Phys Rev B **75** 115419
[4] Michael S, Alexander A 2006 Nature **138** 60
[5] Jeffrey A, Matthew L, Jaehun C 2008 Langmuir **24** 13880
[6] Yuta S, Kazuhiro Y, Yasumitsu M 2008 Nano Letters **8** 3151
[7] Richard M 2008 American Chemical Society Nano **2** 2195
[8] Blackburn J, Barnes T, Beard M 2008 American Chemical Society Nano2 1266
[9] Biswas C, Kim K K, Geng H Z 2009 AMER CHEMICAL SOC **6** 10044
[10] Zarrin T, Ribeiro R, Banda S 2009 PROCEEDINGS OF THE STLE/ASME INTERNATIONAL JOINT TRIBOLOGY CONFERENCE 2008
[11] M. C. Hersam 2008 Nature Nanotechnology **6** 387

8th China International Nanoscience and Technology Symposium (CINSTS09) IOP Publishing
Journal of Physics: Conference Series **188** (2009) 012014 doi:10.1088/1742-6596/188/1/012014

Degradation of tetrachloromethane and tetrachloroethene by Ni/Fe bimetallic nanoparticles

Y Y HUANG[1], F LIU *[2] and H D LI[3]

[1] National Research Center for GeoAnalysis, Beijing 100037, China

[2]China University of Geosciences(Beijing)/Beijing Key Laboratory of Water Resources & Environmental Engineering 100083, China

[3]Laboratory of Crustal Deformation and Process, Institute of Geomechanics, Chinese Academy of Geological Sciences, 100081, Beijing, China

Abstract. The study investigated the potential of nanoscale Ni/Fe bimetallic particles reduction for carbon tetrachloride (CT) and tetrachloroethene (PCE). BET specific surface areas of the laboratory synthesized Ni/Fe (2% wt.) particle, with diameter on the order of 20-60nm, was approximately 52.61m^2/g. Batch studies demonstrated that rapid transformations of PCE and CT were achieved with nanoscale Ni/Fe particles. The degradation process appeared to be pseudo-first-order. Values of the surface area normalized rate coefficients (K_{SA}) of PCE and CT for the reaction with nano Ni/Fe were 2.068mL/(m^2·h), 10.08mL/(m^2·h), respectively. This indicated that the degradation rate of CT was about 5 times larger than that of PCE under comparable environmental condition. Significant amounts of DCM were detected for the reaction with CT unlike the PCE transformation where ethane was the only end-product, amount to 103% of the initial PCE carbon. Both DCM (~15%) and methane (~27%) were the major end products for CT reaction. Based on the rapid rate of degradation and no or less chlorinated byproducts, the nanoscale particles technology offered great opportunities for both fundamental research and technological application for remediation of contaminated ground water.

Keywords: nanoscale Ni/Fe bimetallic particles; reaction rate constant K_{SA}; PCE; CT; dechlorination

1. Introduction

Because of many chlorinated solvents, such as tetrachloroethene (PCE), trichloroethene (TCE) and carbon tetrachloride (CT), were known or potential threats to public health and the environment, so

*Corresponding author. Tel:+86-10-82320920; E-mail: feiliu@cugb.edu.cn

there was an urgent need to develop effective treatment methods. A survey of groundwater quality from east-northern area, China showed that TCE, followed by PCE, was detected. Approximately 44% of the total samples contained TCE, concentration of that above the drinking water standards issued by the US EPA, and the max concentration of TCE was almost 1000μg/L, 200 times greater than the standard [1].

Catalytic hydrodehalogenation with noble metal and direct reduction with zero-valent metal represent the two most widely studied processes. In the last two decades, Fe^0 permeable reactive barriers technology had received much interest as an alternative to pump and treat systems [2-5]. However, many challenges still exist for implementation of the zero-valent metal technology, such as slow reaction rate, accumulation of persistent byproducts, and progressive deactivation of the ZVI by the precipitated corrosion products [6-8]. To overcome these, various nanoscale bimetallic particles have been synthesized and used to improve the performance of the ZVI technology [5] [7-12]. Nanoscale metal particles, with diameter in the range 1~100 nm, were characterized by high surface area to volume ratio, high level of stepped surface, and high surface energy. There were also publications on the application of nano-iron to removal of heavy metals [13], transformation of nitrogen[14] or remediation of PCBs contaminated soils [15]. Nanoscale metal particles could be injected directly to contaminated soils, sediments, and aquifers for in situ remediation chlorinated hydrocarbons, instead of building permeable reactive barriers in the subsurface. The nanoscale particles could be anchored onto activated carbon and zeolite for ex situ treatment of contaminated waters and industrial effluents [7]. In the aqueous solution, the nanoscale iron particles could remain suspended under very gentle agitation. Bimetallic Pd/Fe particles have been shown to exhibit a high efficacy for the transformation of many chlorinated compounds. Among all the noble metal catalysts, Pd has been proved to be the best catalyst with high selectivity of C–Cl bond cleavage and resistance against catalyst deactivation. Hydrogen gas, metal hydride, and other small molecules such as formate, methanol and ethanol, can act as hydrogen donors in this reaction. It is well known that its cost for application in engineering is high because Pd is a noble metal. Less is known about the reaction kinetics and products formed during PCE or CT dechlorination by nano Ni/Fe particles. We presented low-cost nickle substitute for noble metal Pd, which removed PCE or CT using laboratory-synthesized nanoscale Ni/Fe. We present applications of the lab-synthesized nano-Ni/Fe for transformation of PCE and CT. The purpose of this study was aimed to measure the rate and extent of dechlorination and to characterize and quantify reaction intermediates and final products. Kinetic analysis was conducted by using pseudo-first-order rate equation and activity was expressed as both observed and surface-area-normalized rate constants.

2. Material and methods

2.1. Synthesis of nanoscale iron particles

$FeCl_3{\cdot}6H_2O$ was dissolved into 1L of 30% anhydrous ethanol, 70% deionized water (v/v). An appropriate amount of 1.0mol/L $FeCl_3{\cdot}6H_2O$ was placed in a conical flask, and purged with pure nitrogen for two hours. Synthesis of nanoscale iron particles was achieved by adding 1:1 volume ratio of $NaBH_4$ (1.6mol/L) dropwise to $FeCl_3 \cdot 6H_2O$ (1.0 mol/L) .The solution was mixed vigorously on the magnetic agitator under 22±1℃ for 5 minutes [16]. Fe^{3+} was reduced by borohydride according to the following reaction:

$$4Fe^{3+} + 3BH_4^- + 9H_2O \rightarrow Fe\downarrow + 3H_2BO_3^- + 12H^+ + 6H_2 \qquad (1)$$

The nano-iron formed from the above reaction was rinsed with large volume of 0.5mol/L hydrochloric acid, de-oxygen water and ethanol for at least three times, respectively. Bimetals could be prepared by soaking freshly prepared nanoscale iron particles with ethanol solution containing 2% (wt.) $NiCl_2$. The mixture was stirred for 20 min and then filtered through a 0.2μm filter. The amount of Ni deposited on the nano-Fe surface can be got by analyzing the Ni^{2+} concentration in the solution before and after reaction:

$$Ni^{2+} + Fe \rightarrow Ni + Fe^{2+} \qquad (2)$$

Ni/Fe particles were washed three times with ethanol and dried at 105℃ for 4 h under a flow of N_2. Synthesizing nanoscale iron particles had average BET specific surface area of $52.61 m^2/g$ by nitrogen adsorption method (Autosorb-1 model SSA and porosity, USA). BET data were supplied by the laboratory of Analytical Chemistry in China University of Geosciences (Beijing). The micrographs (figure 1) showed that the synthetic particles were in the range of 20~60nm by S-4300F scan electron microscopy (SEM) at 30,000 magnification.

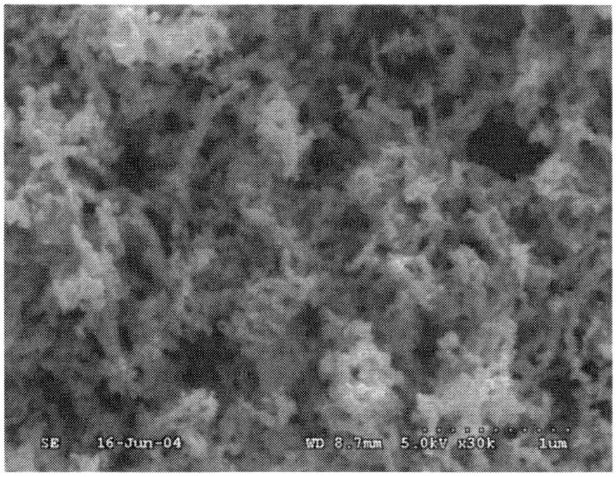

Figure 1. SEM image of nano-Ni/Fe(2%)

2.2. *Batch experiments*

Batch experiments were conducted to test reactivity of the laboratory synthesized nanoscale particles for dechlorination of PCE. Take 0.4g freshly-made nano-Ni/Fe (2%) particles into120-mL reaction bottle capped with a Teflon valve. The initial concentration (C_0) of PCE was 15.83mg/L and volume was 50mL. The bottle was mixed on a rotary shaker at a speed of 170rpm at 20℃. Parallel experiments were also performed without metal particles (control sample). Analyses of organic mass in the control samples indicated that the total mass of parent chlorinated carbon in the batch bottles varied from 95 to 110% of the initial input when the experiment were completed. About 10% error of organic concentrations in the control samples was largely caused by volatilization of PCE or CT, and the gas chromatograph (GC) analysis. Analytical working standards were prepared daily prior to analysis.

2.3. Methods of analysis

2.3.1. Chlorinated hydrocarbons

Periodically, 50μL of the aqueous solution was withdrawn by a 100-μL gas-tight syringe, and spiked quickly into a 10mL vial charged with 0.45mL purified water. Concentrations of chlorinated ethenes were measured by a HP 6890 GC equipped with a HP-5 capillary column (30m×0.25mm×0.25μm) and an electron capture detector (ECD). The carrier gas was ultra-pure nitrogen.

GC HP-6890: inlet 160℃, column flow 1.0 mL/min, oven 70℃, oven run time 10min, ECD 300℃.

Headspace sampler: vial 50℃, loop 60℃, transline 70℃, vial equation time 10min, injection time 1min, shake time 5 min, carrier gas flow 30 mL/min [17].

2.3.2. Hydrocarbons products

Hydrocarbon products in the headspace ware qualitatively identified and quantified with GC analysis by comparing retention times and peak areas with standard gas samples(ethane, ethylene, acetylene and methane).The GC was equipped with a flame ionization detector and a HP-PLOT/Al_2O_3 capillary column (50m×0.53mm×15.0μm. Oven temperature was set at 100℃, injection port temperature at 150℃, and detector temperature at 160℃. The carrier gas was ultrapure nitrogen (99.999%).

2.3.3. Method detection limit (MDL)

The MDLs were got by experiments and calculation. Firstly prepared 7 samples with signal to noise ratio of 3, and analyzed these 7 samples to get their experimental concentration by calculation. The MDLs equal to 3 times the standard deviation value [18].The results of MDLs were listed in Table 1.

Table 1 Detection limit of chlorinated hydrocarbon and hydrocarbon products (μg/L)

Compounds	MDL
DCM	0.083
TCM	0.076
CT	0.022
TCE	0.041
PCE	0.021
DCE	0.083
VC	0.096
C_2H_6	0.48
C_2H_4	0.48
C_2H_2	2.4
CH_4	0.85

Journal of Physics: Conference Series **188** (2009) 012014 doi:10.1088/1742-6596/188/1/012014

2.4. Chemicals

Five halogenated hydrocarbons dissolved in the methanol were purchased from the Institute of Reference Materials belonging to the Environmental Monitoring General Station of China, including 200μg/mL CF, 50μg/mL CT, 200μg/mL TCE, 100μg/mL PCE and 200μg/mL bromoform (BF). DCEs were from National Center for Reference Materials with a concentration of 0.93μg/mL. VC was from Chemistry Service, PA, USA. Standard gas samples were acquired from Beijing Hua Yuan gas Chemical Industry Co., Ltd. $FeCl_3{\cdot}6H_2O$, $NaBH_4$, $NiCl_2$, HCl and anhydrous ethanol were analytical grade reagents.

3. Results and discussion

3.1. Degradation of PCE with nanoscale Ni/Fe bimetallic particles

Figure 2(a) showed the results of degradation of PCE by using nanoscale Ni/Fe bimetallic particles. The reaction kinetics of the degradation of PCE was modeled with a pseudo-first-order rate equation. Plots of the natural logarithm of concentrations of PCE versus time through linear regression analysis gave straight line results. Linear regression analyses were used to obtain observed first-order rate constants (k_{obs}).Disappearance of PCE reacting with nano-Ni/Fe was plotted in the form of $\ln(C/C_0)$ as a function of time in figure 2(a) where C_0 was the initial concentration of PCE (15.83mg/L).Data presented in figure 2(a) indicated that the disappearance of PCE exhibited pseudo-first-order behavior (R^2=0.9311, n=6).PCE was rapidly degraded by nano-Ni/Fe. The observed first-order rate constant (k_{obs}) was 0.8487 h^{-1}. Half-life time ($t_{1/2} = ln2/k_{obs}$) was 0.82h.

To better compare the reaction rates observed with different metal loading and particle size, it is useful to normalize the reactivity per unit metal surface area. The rate of dechlorination in a batch system can be described by the following equation (Johnson et al.1996):

$$dC/dt = -k_{SA}\, P_a\, C. \qquad (3)$$

Where C=concentration of organic compound in the aqueous phase (mg/L).t=time (h). Specific surface area constant $P_a = SSA \times W/V$ (m^2/mL) where W was the weight of reactive materials and V the reaction solution volume. For a specific batch system, k_{SA} and P_a were constants. The above equation therefore represents a pseudo-first-order kinetics. The reaction rate constant and half-life time can be standardized by specific surface area constant P_a. In this experiment, P_a was 0.4 m^2/mL, so k_{obs} and $t_{1/2}$ was normalized per unit metal surface area. The surface-area-normalized rate coefficient ($K_{SA} = k_{obs}/P_a$) and the surface-area-normalized half-life time ($t_{50\text{-}N} = P_a \times t_{1/2}$) was 2.122mL/($m^2{\cdot}$h) and 0.33h, respectively.

Reaction of PCE with the nano Ni/Fe bimetallic particles were shown in figure 2(b). Concentrations in the figure were expressed as the molar ratio to initial organic concentrations. As shown in figure 2(b), 99% of PCE was reduced using nanoscale Ni/Fe particles within 6h. We observed the immediate appearance of ethane corresponding to the disappearance of PCE. Ethane concentrations increased steadily during the entire reaction time. No chlorinated intermediates or products such as TCE, DCEs and VC were detected above the detection limit. PCE was essentially reduced to ethane. Average yield of ethane from PCE was ~ 103% after 6h. Only trace levels of ethene (~6%) appeared briefly in the headspace. It maybe much ethane or hydrogen adhering to the surface of nanoscale Ni/Fe particles, prevented PCE from arriving at the reactive situation, so that reactive rate was reduced rapidly. On the other hand, nanoscale Ni/Fe reactivity was decreased over time, probably due to the precipitation on the surface of Ni/Fe.

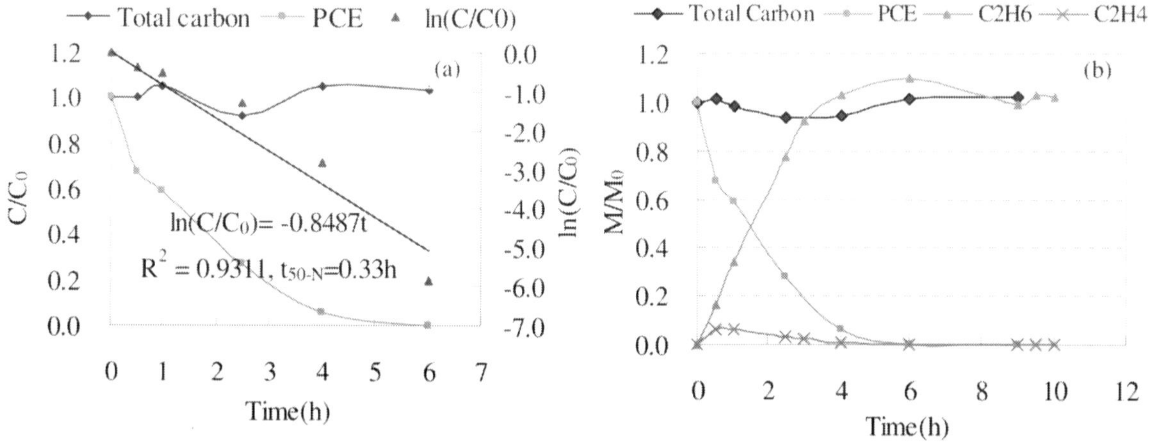

Figure 2. Reactions of nanoscale Ni/Fe bimetallic particles with PCE. Initial organic concentration was 15.83mg/L, and metal to solution ratio was 0.4g/50mL.

The concentration of Ni^{2+} ion in solution, determined by ICP-MS, was less than 10μg/L when the reaction completed, which suggested that Ni likely served as catalyst within a bimetallic system. At present, there are different opinions on mechanisms of the dechlorination in a bimetallic system. Zhang et al. [5] and Mallat et al.[19]) thought that presence of the noble metal on the surface of iron certainly creates many galvanic cells thus promoting the electron release reactions of iron corrosion. Quan et al.[20]) studied the characteristics on the degradation of TCM, CT and TCE in water using different bimetallic systems (Cu/Fe, Zn/Fe, Pd/Fe, Ni/Fe). The results showed hydrogenation catalyst such as Pd and Ni acted as important function during the transfer of H_2. In addition, Pd and Ni as transition metal had vacant orbits, which could form transition compounds with the *p* electron pair or π bond of chlorine atom in chlorinated hydrocarbons, and decreased the reactive energy.

3.2. Degradation of CT with nanoscale Ni/Fe bimetallic particles

Figure 3(a) showed the results of degradation of CT by using nanoscale Ni/Fe bimetallic particles. Rapid transformations of CT were achieved with nanoscale Ni/Fe bimetallic particles. CT was reduced below detection limits within 60 min. Disappearance of CT reacting with nano-Ni/Fe was plotted in the form of $\ln(C/C_0)$ as a function of time in figure 3(a) where C_0 was the initial concentration of CT (26.28mg/L). Plots of the natural logarithm of concentrations of CT versus time through linear regression analysis gave straight line results. Data presented in figure 3(a) indicated that the disappearance of CT exhibited pseudo-first-order behavior (R^2=0.996, n=5). The observed first-order rate constant (k_{obs}) was 4.242h^{-1}. Half-life time ($t_{1/2} = ln2/k_{obs}$) was 0.16h. The surface-area-normalized rate coefficient ($K_{SA}=k_{obs}/P_a$) and the surface-area-normalized half-life time ($t_{50\text{-}N}=P_a \times t_{1/2}$) was 10.60 mL/(m^2·h) and 0.07h, respectively.

The rate of CT dechlorination observed in this experiment was comparable to the results for PCE.

Observed rate constants of CT and PCE were 4.242 and 0.8487 (h^{-1}), respectively. This indicated that the degradation rate of CT was 5 times larger than PCE.

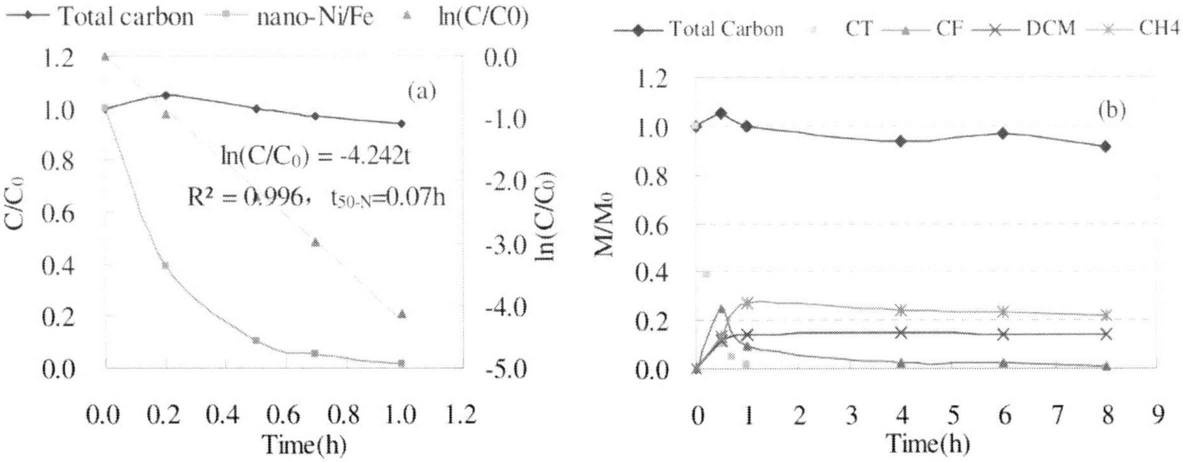

Figure 3. Reactions of nanoscale Ni/Fe bimetallic particles with CT. Initial organic concentration was 26.28mg/L, and metal to solution ratio was 0.4g/50mL.

Figure 3(b) showed the time course of the product formation when degrading CT by nanoscale Ni/Fe particles. CT in the amount of 0.426mM (26.28mg/L) was completely reduced within 1h while CF, DCM and methane were produced. CF emerged very quickly and peaked (25%) around 0.5h.Complete reduction of CF was achieved around 4h.On the other hand, concentrations of DCM increased continuously to about 15% of the initial CT carbon after 1h and remained essentially constant after that. Methane accumulated with a final yield around 27%. Both DCM (~15%) and methane (~27%) were the major end products for CT reaction.

4. Conclusions

In summary, nanoscale Ni/Fe particles had indicated distinct dechlorination for PCE and CT. The degradation process appeared to be pseudo-first-order. Many advantages of nanoscale Ni/Fe bimetallic particles for treatment of PCE or CT include: 1) due to high specific surface area and high surface reactivity, the nanoscale Ni/Fe particles were observed to suspend in the aqueous solution, which favored to increase the dechlorination rate. Freshly synthesized nanoscale Ni/Fe bimetal for PCE exhibited higher activity. 2) Due to the presence of catalyst Ni on the surface of iron, increased reactivity per unit surface area, and reduced the yield of chlorinated by-product. The reduction of PCE with nanoscale Ni/Fe particles would then completely dechlorinated to non-chlorinated products. 3) cost-effectiveness. However, before this technology can be fully optimized for environmental applications, further study is needed. Considered the low cost of nickel (relative to palladium) and its high activity, Ni/Fe could be a promising reactive reagent for remediation of groundwater contaminated with halogenated carbon. Nevertheless, a better understanding of the kinetics and mechanisms of the rate-limiting process is required to take full advantage of this new treatment technology. The dechlorination of CT with Fe as electron donor was not sequential. Reductive dechlorination of alkyl chlorides is probably initiated by a one-electron transfer, producing an alkyl radical and liberating a chloride ion [21]. The alkyl radical can rapidly uptake a second electron and a proton from solution [22].

Acknowledgement

This work is supported by the National Natural Science Foundation of China (No.40372109 and 40572146) and the Special Fund of Basic Fundamental Scientific Research from National Research Center for GeoAnalysis (200607CSJ14). We wish to acknowledge all partners of the projects introduced in this paper. The analytical data were supplied by the Lab of Water Resources and Environmental Engineering in China University of Geosciences (Beijing), which is also appreciated.

References

[1] Liu F, Huang Y Y and Cui W H 2006. Impact factors on removal of perchloroethylene with nano-Ni/Fe methodology. *Acta Geologica Sinica (English edition)* **80**, 262-266.

[2] Matheson L J and Tratnyek P G 1994. Reductive dehalogenation of chlorinated methanes by iron metal. *Environ. Sci. Technol.* **28**, 2045-2053.

[3] Agrawal A and Tratnyek P G 1996. Reduction of nitroaromatic compounds by zero-valent iron metal. *Environ. Sci. Technol.* **30**, 153-160.

[4] Quan X and Yang F L 1997. Dechlorination of trichloroethylene in water by Pd/Fe^0 catalytic reduction. *Journal of Dalian University of Technology* **37**, 46-48.

[5] Zhang W X, Wang C B and Lien H L 1998. Treatment of chlorinated organic contaminants with nanoscale bimetallic particles. *Catalysis Today* **40**, 387-395.

[6] Vidic R D and Pohland F G. 1996. in: Technology Evaluation Report TE-96-01:T*reatment Walls, Ground-water Remediation Technologies Analysis Center*, Pittsburgh, PA.

[7] Wang C B and Zhang WX 1997. Synthesizing nanoscale iron particles for rapid and complete dechlorination of TCE and PCBs. *Environ. Sci. Technol.* **31**, 2154-2156.

[8] Cheng S F and Wu S C 2000. The enhancement methods for the degradation of TCE by zero-valent metals. *Chemosphere* **41**, 1263–1270.

[9] Li W and Klabunde K J 1998. Ultrafine zinc and nickel, palladium, silver coated zinc particles used for reductive dehalogenation of chlorinated ethylenes in aqueous solution. *Croat. Chem. Acta* **71**, 853–872.

[10] Scherer M M, Balko B A and Gallagher D A 1998. Correlation analysis of rate constants for dechlorination by zero-valent iron. *Environ. Sci. Technol.* **32**, 3026.

[11] Chung Y M and Rhee H K 2003. Dendrimer-templated Ag–Pd bimetallic nanparticles. *J. Colloid Interf. Sci.* **271**, 131–135.

[12] Liu Y, Majetich S A, Tilton R D, Sholl D S and Lowry G V 2005. TCE dechlorination rates, pathways and efficiency of nano-scale iron particles with different properties. *Environ. Sci. Technol.* **39**, 1338–1345.

[13] Ponder S M, Darad J G and Mallouk T E 2000. Remediation of Cr(VI) and Pb(II) aqueous solution using supported nanoscale zero-valent. *Environ.Sci.Technol.* **34**, 2564-2569.

[14] Cheo S, Chang Y Y, Hwang K Y and Khim J 2000. Kinetics of reductive denitrification by

nanoscale zero-valent iron. *Chemosphere* **41**, 1307-1311.

[15] Varanasi P, Fullana A and Sidhu S 2007. Remediation of PCBs contaminated soils using iron nano-particles. *Chemosphere* **66**, 1031-1038.

[16] Lien H L and Zhang W X 2001. Nanoscale iron particles for complete reduction of chlorinated ethenes. *Colloids and Surfaces* **191**, 97-105.

[17] Liu F 2002. Study on volatile chlorinated hydrocarbons in groundwater using the permeable reactive barrier of zero valence iron. *A dissertation for doctor's degree of China University of Geosciences* (Beijing).

[18] Liu F, Tang M G, He X J, Huang Y Y and Zhao J 2002. Laboratory study of chlorinated hydrocarbon in drinking water using Fe^0, *Earth Science-Journal of China university of Geosciences* **27**, 182-186.

[19] Mallat T, Bodnar Z and Petro J, 1991. Reduction by dissolving bimetals. *Tetrahedron* **47**, 441-446.

[20] Quan X, Liu H J, Yang F L, Xue D M and Zhao Y Z 1998. Dechlorination of three polychlorinated hydrocarbons in water using bemetallic systems. *China Environmental Science* **18**, 333-336.

[21] Vogel T M, Criddle C S and Mccarty P L 1987. Transformations of halogenated aliphatic compounds. *Environ. Sci. Technol.* **21**,722-736.

[22] Macalady D L, Tratnyek P G and Grundl T M 1986.Abiotic reduction reactions of anthropogenic organic chemicals in anaerobic system: a critical review, *J.Cont Hydrol,* 1-28.

Polyelectrolyte multilayer film-assisted formation of zero-valent iron nanoparticles onto polymer nanofibrous mats

Shili Xiao[1,2], Siqi Wu[3], Mingwu Shen[3], Rui Guo[3], Shanyuan Wang[2], and Xiangyang Shi[1,3]

[1]State Key Laboratory for Modification of Chemical Fibers and Polymer Materials,
[2]College of Textiles, and
[3]College of Chemistry, Chemical Engineering, and Biotechnology, Donghua University, Shanghai 201620, China

E-mail: xshi@dhu.edu.cn

Abstract. A facile approach that combines the electrospinning technique and layer-by-layer (LbL) assembly method has been developed to synthesize and immobilize zero-valent iron nanoparticles (ZVI NPs) onto the surface of nanofibers for potential environmental applications. In this approach, negatively charged cellulose acetate (CA) nanofibers fabricated by electrospinning CA solution were modified with bilayers composed of positively charged poly(diallyl-dimethyl-ammoniumchloride) (PDADMAC) and negatively charged poly(acrylic acid) (PAA) through electrostatic LbL assembly approach to form composite nanofibrous mats. The composite nanofibrous mats were immersed into the ferrous iron solution to allow Fe(II) ions to complex with the free carboxyl groups of PAA, and then ZVI NPs were immobilized onto the composite nanofibrous mats instantly by reducing the ferrous cations. Combined scanning electron microscopy (SEM), transmission electron microscopy (TEM), energy dispersive spectroscopy (EDS), and thermogravimetry analysis demonstrated that the ZVI NPs are successfully synthesized and uniformly distributed into the polyelectrolyte (PE) multilayer films assembled onto the CA nanofibers. The present approach to synthesis ZVI NPs opens a new avenue to fabricating various materials with high surface area for environmental, catalytic, and sensing applications.

1. Introduction

LbL self-assembly technique has drawn much attention as a powerful and highly versatile approach to prepare nanostructured materials with tailored properties.[1, 2] In a typical process, the oppositely charged polyelectryolyte solutions sequentially deposited onto a substrate, through controlling the dipping cycles and composition of polyelectrolyte (PE), the obtained composites with desired thickness, morphology and composition can be easily realized. The versatility of LbL assembly technique makes a broad range of materials including polymers, nanoparticles, and biomolecules etc. be able to be assembled onto various substrates (e.g. planar, solid spheres) to form functional nanomaterials.[3-7] In our previous work,[8] we immobilized ZVI NPs into the PE multilayer films assembled onto microsized glass beads via LbL assembly approach to mitigate their aggregation.[9] The ZVI NPs thus obtained had a narrowly size distribution from 1 to 4 nm and displayed an excellent reactivity for the dechlorination of trichloroethylene, a kind of typical organic contaminants.

In this study, we attempted to extend our previous work to the synthesis and immobilization of ZVI NPs onto a continuous, porous medium with high surface-to-volume ratio by combining electrospinning and LbL assembly techniques, and thereby developing a functional filtration material for environmental applications. To our best knowledge, immobilization of ZVI NPs onto surface of electrospun nanofibers using LbL assembly approach is not reported in literature.

2. Materials and methods

Poly(acrylic acid) (PAA, Mw = 240,000, 25 wt% in water) and poly(diallyl-dimethyl-ammoniumchloride) (PDADMAC, Mw = 100,000-200,000, 20 wt% in water) were obtained from Aldrich. Sodium borohydride was from J&K Chemical ®. $FeCl_2.4H_2O$ (ACS reagent grade), cellulose acetate (powder, viscosity = 300.0 ~ 500.0 mPa.s), acetone, and N, N-dimethylformanide (DMF) were all in reagent grade and used as received. Water used in all experiments was purified using a Milli-Q Plus 185 water purification system (Millipore, Bedford, MA) with resistivity higher than 18 MΩ cm.

CA nanofibers were fabricated by electrospinning a 12 mg/mL CA solution with an applied voltage of 20 kV and a collection distance of 20 cm. The homogenous CA solution was prepared by dissolving a measured amount of CA powder into a mixed solvent of acetone and DMF (v/v = 2:1). The negatively charged electrospun CA nanofibrous mats were then assembled with PDADMAC and PAA alternatively using the LbL assembly approach to form six bilayers of PAA/PDADMAC (a polymer pair of PAA/PDADMAC was considered as a bilayer). Subsequently, the nanofibrous mats were immersed into ferrous iron solution to form a Fe(II)-nanofiber complex, and then ZVI NPs were simultaneously synthesized and immobilized into the PE multilayer-modified CA nanofibers by in situ reducing the Fe (II) ions.

Morphologies of the prepared samples were observed using scanning electron microscopy (SEM) and transmission electron microscopy (TEM). X-ray energy dispersive spectroscopy (EDS) was used to confirm the presence of ZVI NPs. Thermogravimetry analysis (TGA) was used to determine the iron loading capacity on the nanofibrous mats.

3. Results and discussion

The SEM images and diameter distribution histograms of CA nanofibers and ZVI NP-immobilized CA nanofibrous mats are shown in Figure 1. The morphology of CA nanofibers does not significantly change after the immobilization of ZVI NPs. Some membrane-like structures are observed between the adjacent nanofibers for ZVI NP-immobilized CA nanofibers, which could be caused by the assembly of multilayer films in the limited space in between the adjacent fibers.[10] The average diameter of pure CA nanofibers (Figure 1b) was estimated to be 295 ±145 nm. However, after the immobilization of ZVI NPs, the fiber diameter was increased to 461±157 nm (Figure 1c).

In order to observe the distribution of ZVI NPs in the composite nanofibrous mats, the ZVI NP-immobilized composite nanofibrous mats were embedded in epoxy resin and cut into ultrathin sections with ultramicrotome equipped with a diamond knife. The cross-sectional images of the sample were observed by TEM, which was shown in Figure 2. It was clear that the PE multilayers were successfully coated onto the surface of CA nanofibers (Figure 2a). But, it was difficult to identify the layered structure of PE multilayers due to the fuzzy characteristics of the LbL structure associated with the interpenetration between the layers.[2] And, it seems that the thickness of ZVI NPs/PE multilayer films is irregular and cannot be estimated, which might be closely related to the deposition conditions (e.g. pH value, depositon time and the number of bilayers) and the overlapping of samples during the sample preparation. Figure 2b displays the magnified cross-sectional TEM image of ZVI NPs immobilized composite nanofibrous mats. ZVI NPs were dispersed into the PE multilayers with dense and uniform distribution. The iron particle size was narrowly ranged from 0.8 to 2.2 nm, and the average size of ZVI NPs was measured to be 1.4 ±0.29 nm.

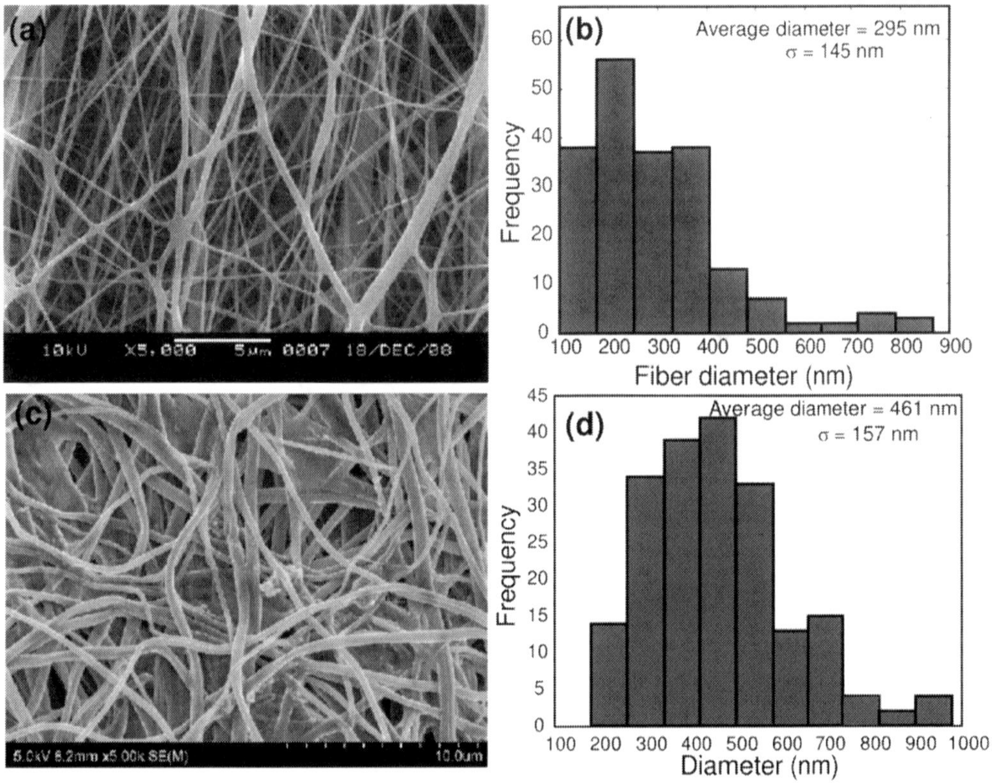

Figure 1. SEM images and diameter distribution histograms of pure CA nanofibers (a and b) and ZVI NP-immobilized CA nanofibers (c and d).

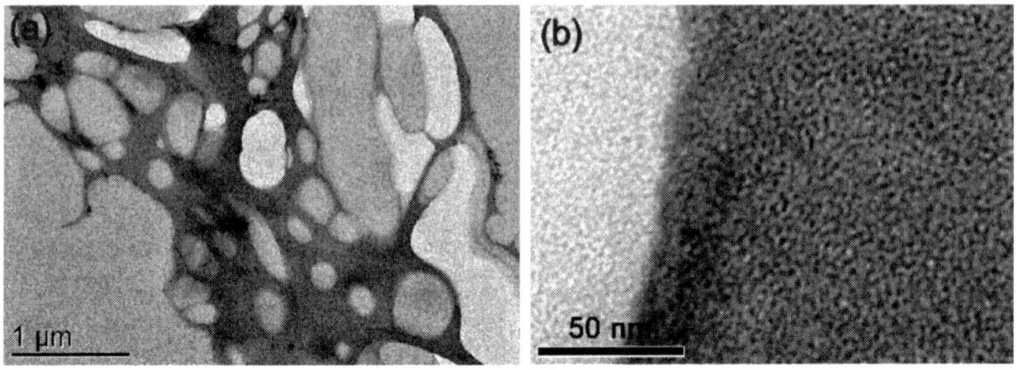

Figure 2. A typical TEM image (a) and a magnified TEM image (b) of cross-sectional ZVI NP-immobilized composite nanofibers embedded in epoxy resin.

Figure 3 shows the EDS spectrum of ZVI NP-immobilized composite nanofibrous mats. The presence of iron siginals from EDS spectrum suggests that ZVI NPs are successfully synthesized and immobilized in the PE multilayer films assembled on CA nanofibers. The presence of oxygen signal may stem from the iron/iron oxide core-shell structure, which is consistent with literature.[8] And, the oxygen is also partly from the carboxyl groups of the PAA polymer layers and the CA nanofiber templates. The element carbon was from the polymer components and the carbon thin film coating during the sample preparation. Element sodium was likely from sodium borohydride residues used for ferrous reduction.

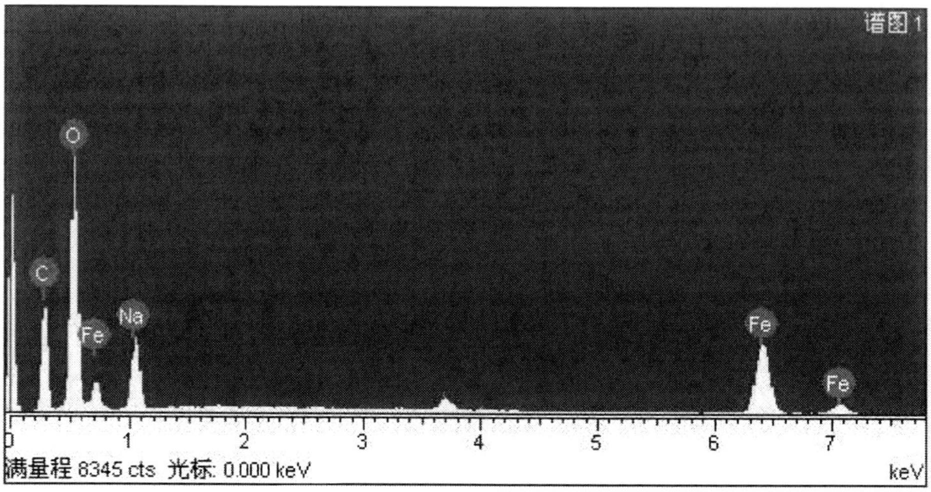

Figure 3. EDS spectrum of ZVI NP-immobilized composite nanofibrous mats.

Iron loading of six PE bilayers assembled onto the CA nanofibrous mats with 2 cycles of binding/reduction process was measured using thermogravimetric analysis (TGA) at a heating rate of 10 ℃/min under air (Figure 4). It is clear that at 600 ℃, all the polymer component was nearly burned out and only the iron oxide was left (Fe2O3). Thus, we could calculate that the iron loading of composite nanofibrous mats was 19.7 %.

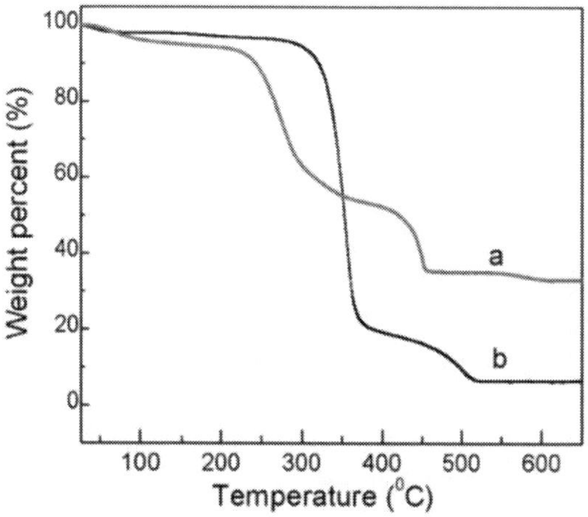

Figure 4. TGA results of composite nanofibrous mats before (b) and after (a) immobilization of ZVI NPs.

4. Conclusion

In summary, ZVI NP-immobilized composite nanofibrous mats were successfully fabricated by a unique combination of electrospinning technology and electrostatic layer-by-layer self-assembly techniques. The synthesis of ZVI NPs in the composite nanofibrous mats were confirmed by TEM and EDS. We show that ZVI NPs are uniformly dispersed into the PE multilayer films with a dense and even distribution. The average particle size of ZVI NPs was 1.4 nm. Such ZVI NP-immobilized fibrous mats with high surface-area-to-volume ratio and porous structure should have great potential for environmental applications. Furthermore, the developed approach to immobilizing ZVI NPs onto electrospun polymer nanofibers with LbL approach may open a new avenue in the fabrication of 3-dimentional porous nanostructured materials with tunable size, composition, and surface properties for various applications.

5. References

[1] Decher G and Hong JD. Ber. Bunsen-Ges. Phys. Chem. 1991;95:1430-1434.
[2] Decher G. Science 1997;277:1232-1237.
[3] Caruso F, Caruso RA, and Mohwald H. Science 1998;282:1111-1114.
[4] Donath E, Sukhorukov GB, Caruso F, Davis SA, and Mohwald H. Angew. Chem. Int. Ed. 1998;37:2201-2205.
[5] Shi X, Sanedrin RJ, and Zhou F. J. Phy. Chem. B 2002;106:1173-1180.
[6] Shi X, Shen M, and Moehwald H. Prog. Polym. Sci. 2004;29:987-1019.
[7] Shi X, Wang S, Chen X, Meshinchi S, and Baker JR, Jr. Mol. Pharmaceutics 2006;3:144-151.
[8] Huang G, Shi X, Pinto RA, Petersen E, and Weber WJ, JR. Environ. Sci. Technol. 2008;42:8884-8889.
[9] Wang C-B and Zhang W-X. Environ. Sci. Technol. 1997;31(7):2154-2156.
[10] Ding B, Fujimoto K, and Shiratori S. Thin Solid Films 2005;491:23-28.

Journal of Physics: Conference Series **188** (2009) 012016 doi:10.1088/1742-6596/188/1/012016

A Study on AFM Manipulation of Single-Wall Carbon Nanotube

Xiaojun Tian, Zaili Dong, Peng Yu, Zhu Liu
State Key Lab. of Robotics, Shenyang Institute of Automation, Chinese Academy of Sciences, Shenyang 110016, P. R. China

E-mail: xjtian@sia.cn

Abstract. As single-wall carbon nanotube (SWCNT) has special electrical and physical property, it can be used as excellent material to construct various nano electronic device. However, in the fabrication process, the modification of size, shape and even the electronic property, especially to the metallic SWCNT, is a key problem to be overcome. Here a modified nanomanipulation technology based on atomic force microscope (AFM) is utilized to perform various kinds of SWCNT manipulation, such as SWCNT separation, catalyst remove, continual nano buckles fabrication and even stretch to break, thus to modify the size, shape and eventually the electrical property of the SWCNT. In addition, the manipulation results are analyzed based on the mechanical mechanism.

1. Introduction

Since single-wall carbon nanotube (SWCNT) was successfully synthesized in 1993 [1], much research has been done on it for its special electrical and physical property, and it has been used to fabricate various nano electronic device, such as transistor, sensor and even simple logic circuit [2][3]. With the development of SWCNT based nano electronic device, the modification of SWCNT's size, shape and electrical property has been one of a excited research fields. And limited by the SWCNTs synthesizing technology, the raw SWCNTs material consists of both semi-conducting SWCNTs and metallic SWCNTs, while the metallic SWCNTs are unnecessary in nano electronic device. For that, di-electrophoresis technology is utilized to separate the metallic from semi-conducting SWCNTs [4], and also the manipulation through a probe of an atomic force microscope (AFM) is a feasible way to modify the electrical property of the metallic SWCNT.

However, limited by dispersion technology of SWCNT bundles, most of the manipulation objects are MWCNT [5] or SWCNT bundles [6], little has been reported about the manipulation of single SWCNT. Aimed to overcome the barrier described above, here a modified AFM manipulation technology is utilized to perform various kinds of manipulation to modify the size,

Journal of Physics: Conference Series **188** (2009) 012016 doi:10.1088/1742-6596/188/1/012016

shape and electrical property of the SWCNT, thus to eventually develop a method to modify the electrical property of the nano device.

2. Materials and methods

Before the alignment process, SWCNTs raw materials produced by arc discharge, is dispersed according to the following steps. Di-ionized water with sodium dodecyl sulfate (SDS) as additives is used as solvent[7], after accurate mass weighing with 0.1% SWCNT and 1% SDS proportion, the mixture solution in test tube is sonicated (40kHz) for about two hours. During sonication, the position and tilt angle of test tube in ultra-sonicator is adjusted to keep SWCNT solution 'boiling' in order to mix them uniformly and sufficiently, and at the same time the solution is heated to 40~60 centigrade degree for improving the dispersion effect. After the sonication, the solution keeps unmoved for sedimentation for about 6 hours.

After the dispersion of SWCNTs, a modified AFM based nanomanipulation system (AFM system, model Dimension3100 of Veeco Co., is modified according to the same method previously presented in[8]) is used for imaging and manipulating the SWCNT at room temperature and atmosphere condition. During imaging, the AFM system is working in tapping mode with frequency 312.5kHz, and the AFM probe used here is model MPP-11100 with rectangle Si cantilever, tip radius less than 10nm and force constant 40N/m. After imaging, the manipulation with force feedback is performed without changing the probe, and as force applied on the tip can be online controlled, wear or destroy of the AFM probe is significantly reduced, which can be verified by the fact that the resolution of the image keeps unchanged even after manipulation.

3 Results and Discussion

After the dispersion, the modified AFM based manipulation was performed to the SWCNT in the array with several kinds of results shown in the following.

3.1 Stretch to break of SWCNT

After the dispersion, scan area is reduced step by step and two SWCNTs are clearly imaged as shown in Fig. 1. Note here that as the scan size of the AFM system is 90um*90um*7um, which is relatively large at the nanometer scale, the effective signal is at the same scale of background noise when object less than 1 nm is imaged, thus parameters, such as setting point, vibration frequency, proportional gain and integer gain, have to be continually adjusted for obtaining image with acceptable resolution, and even thus the resolution of the scanned image is still deteriorated.

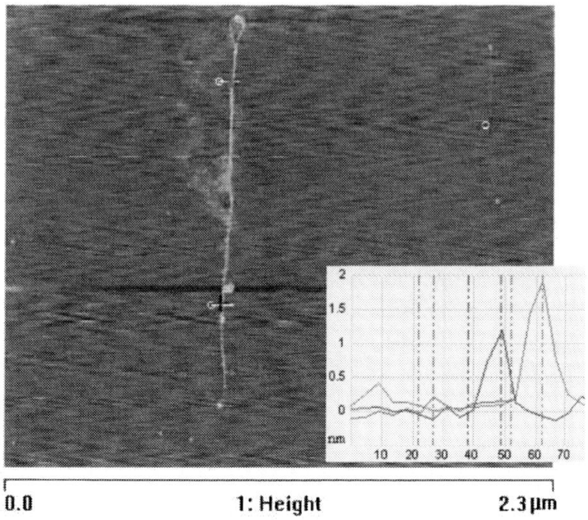

(a)

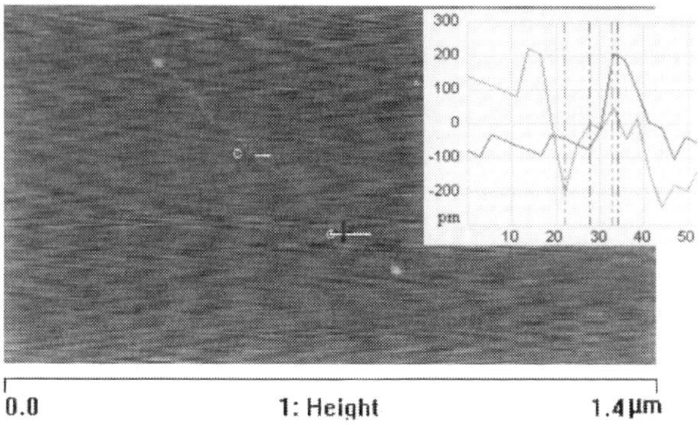

(b)

Fig. 1 AFM scan image of SWCNT (a) scan image of two SWCNTs, (b) image of the right slim SWCNT after rotation 45 degree

As shown in Fig. 1(a), two SWCNTs are parallel aligned and also straightened after the alignment process, and the left SWCNT with larger diameter actually consists of two single SWCNTs grown from the same catalyst. And we can also see that the height of the SWCNT consisted of two single SWCNTs is 1.8nm, which is close to the sum of the heights of two single SWCNTs with diameter of 1.2nm shown in the section curve shown in Fig. 15(a). In Fig. 5 it is noted that the height of the slim SWCNT on the right side is only 0.26nm as shown in the section curve of Fig. 1(a), and for verifying its repeatability the sample is rotated 45 degree and then scanned, after that the characterized height value of the slim SWCNT are still 0.24nm and 0.26nm as shown in Fig. 1(b). With the consideration of the push-down effect by probe during scanning by tapping mode, which means that the measured height of SWCNT is a little less than the original diameter due to the push force of the probe and the height value changes a little with the setting point changing.

After the imaging, the SWCNT is manipulated to bend and then break by controlling the probe trajectory of the AFM based nanomanipulation system, and the manipulation result and its stretch stress analysis is shown in Fig. 2.

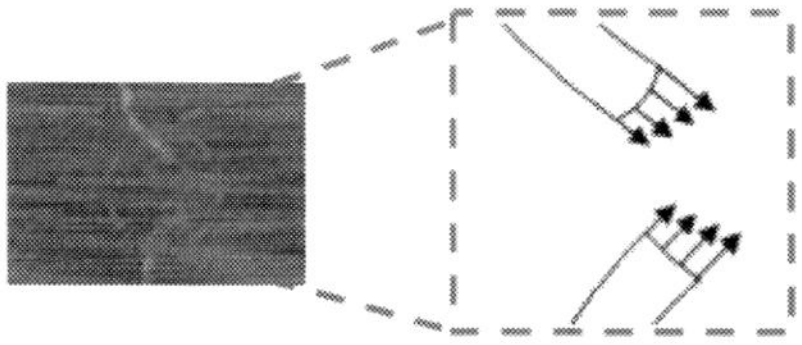

Fig.2 bend and break of slim SWCNT (red arrow shows the probe trajectory, and the right image circled by the dashed line shows the stretch stress of the SWCNT)

From the experiments, it can be seen that the SWCNT bend and then break under the effect of the probe, here the break is mainly caused by stretch stress with the plausible explanation as follows. With strong adhesion force acted on SWCNT by the substrate, such as Van der Waals force and etc [9], the SWCNT is strongly adhered on the substrate and then stretched along its axis after bend, which results in much larger stretch stress on the extrusive side of the SWCNT as shown in Fig. 2. And when the stretch stress exceeds the extreme value, the covalent bond of carbon atoms breaks and thus the SWCNT breaks.

3.2 Other manipulation results of SWCNT

By the proposed alignment method, another SWCNT is aligned and various kinds of manipulation are

Journal of Physics: Conference Series 188 (2009) 012016 doi:10.1088/1742-6596/188/1/012016

performed with results shown in Fig. 3.

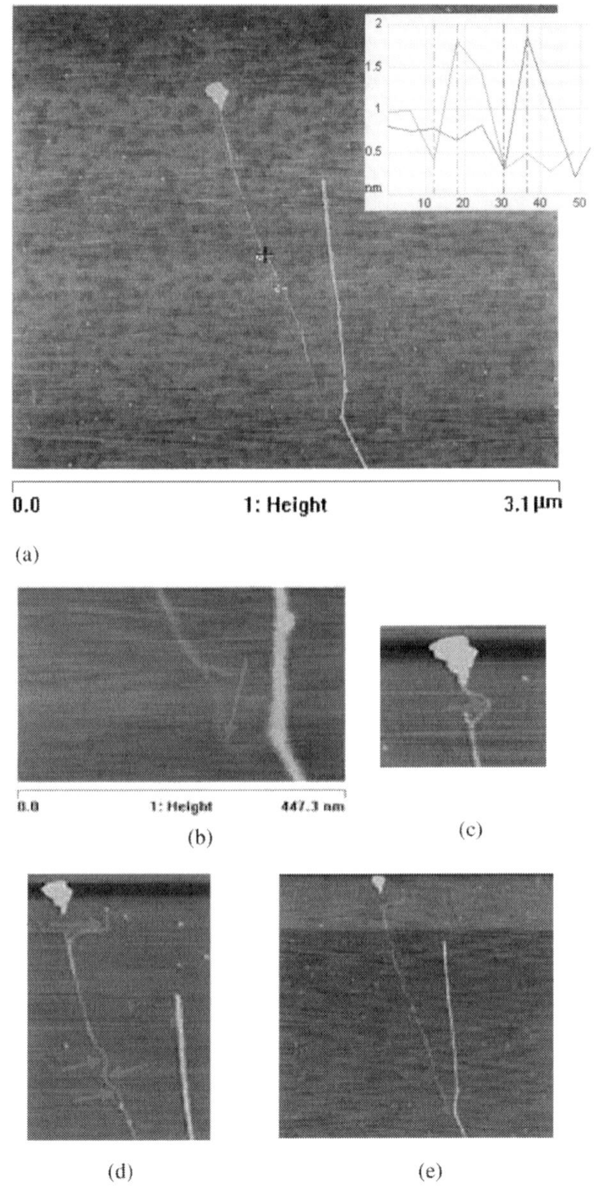

Fig. 3 Various AFM manipulation of SWCNT
(a) SWCNT image before manipulation, (b) separation of SWCNT from 'Y' CNTs, (c) bend of SWCNT, (d) removal of catalyst at one end of SWCNT and form of three nano bucklings, (e) SWCNT image after manipulation

In the image of SWCNT shown in Fig. 3(a), the height of the slim SWCNT is 1.5nm, and after the local scan the manipulation is performed by controlling the probe trajectory to successfully separate the SWCNT from the 'Y' CNTs as shown in Fig. 3(b), where the trace at the original position of the SWCNT is the left amorphous carbon particle. And the manipulation results in Fig. 3(c) and (d) show that although the catalyst strongly bonds to SWCNT [10], the catalyst particle is still removed by several steps of manipulation. And by three continual manipulations, three nano buckles are also formed on the SWCNT to change its electron transport property. Besides, small nano particles in Fig. 3 (a) and (e) are mainly nano particles of amorphous carbon without exclusion of possible nano bubbles [11].

In addition, compared with the motion mode of multi-wall carbon nanotube (MWCNT) during manipulation [22][27], the experimental results shown in Fig.6 and 7 demonstrate that the motion mode of the SWCNT is no longer the rotation around a rotation center under the manipulation of the AFM probe. On one side, due to the stronger quantum effect with size at the nanometer scale, SWCNT will suffer relatively stronger adhesion force with the consideration of its nanometer size. On the other side, different from polycarbonate or silicon substrate with relatively rougher surface, here SWCNT aligns on mica substrate with atomic flatness and thus the contact area between the SWCNT and mica is relatively larger. Dependent on both adhesion force and contact area [9], the nano frictional force is relatively much larger and the push force applied by the probe is not enough to move the whole SWCNT, which results in the local deform and local move of the small SWCNT section in essence, or the break of the SWCNT when the deform caused by the push force is overlarge.

4. Conclusions

In summary, a modified AFM based nanomanipulation technology is utilized to perform various kinds of manipulation on the SWCNT after the SWCNT raw materials is dispersed, such as SWCNT separation, catalyst remove, continual nano buckles fabrication and even stretching to break, which are used to control the size or shape and the electrical property of the SWCNT eventually. This manipulation technology shows a feasible way to perform the modification of the SWCNT's size, shape and eventually the electrical property.

Acknowledgments

The Authors thank the support from the CAS FEA International Partnership Program for Creative Research Teams.

Reference:

[1] Iijima S and Ichihashi T 1993 *Nature* **363** 603
[2] Avouris Ph, appenzeller J, Martel R 2003 *J. Proc. of the IEEE* **91** 1172
[3] Dai H J, Javey A, Pop A, Mann D 2006 *NANO: Brief Reports and Reviews* **1** 1
[4] Krupke R, Hennrich F, Lohneysen H V 2003 *Science* **301** 344
[5] Tian X J, Wang Y C, Xi N 2006 *Journal of Chinese Electron Microscopy Society* **25** 490
[6] Liu S J, Shen Z Y, Hou S M 2003 *Acta Physico-Chimica Sinica*, **3** 233
[7] Vigolo B, Penicaud A, Coulon C 2000 *Science* **290** 1331
[8] Tian X J, Wang Y C, Liu L Q 2006 *Chinese Journal of Scientific Instrument*, **27** 661
[9] Bhushan B 2004 in Springer Handbook of Nanotechnology, Springer Berlin Heidelberg, Berlin
[10] Larsson P, Larsson J A, Ahuja R 2007 *Physical Review B* **75** 115419
[11] Zhang X H, Lou P T, Zhang Z Q 2003 *Journal of Chinese Electron Microscopy Society* **22** 136

Effects of substrate temperature on the properties of heavy Ga-doped ZnO transparent conductive film by RF magnetron sputtering

Shufang Wang[1, 2], Xifeng Li[1], Jianhua Zhang[1, 2*]

[1] Key Laboratory of Advanced Display and System Applications, Ministry of Education, Shanghai University, Shanghai 200072, China

[2] School of Mechanical & Electronic Engineering and Automation, Shanghai University, Shanghai 200072, China

E-mail address: jhzhang@staff.shu.edu.cn

Abstract. Gallium-doped zinc oxide (GZO) transparent conductive thin films were deposited on glass substrate by RF magnetron sputtering using a ceramic ZnO target with 10 wt. % Ga_2O_3 doped. The effects of substrate temperature on structural, electrical and optical properties of the films were investigated. X-ray diffraction (XRD) measurements indicated that GZO films were polycrystalline and strongly c-axis oriented. The surface roughness evaluated in terms of R_{RMS} measured by atomic force microscope (AFM) was 1.8 nm for the smoothest surface obtained at 250 ℃. The electrical resistivity decreases with increasing substrate temperature and the lowest resistivity obtained was 5.8×10^{-4} Ω·cm at 350 ℃. The average transmittance in the visible region was over 90%. The carrier concentration increased with substrate temperature increasing, and the highest mobility value was 22.1 $cm^2v^{-1}\cdot s^{-1}$.

1. Introduction

TCOs with transmittance and resistivity values close to that of ITO have been widely researched in order to substitute ITO [1-3]. Among these TCOs, ZnO is the most favourable material for its wide band energy (3.37 eV), abundant resources and non-toxic [4, 5]. Otherwise, ZnO not only can be grown at low temperature but also can be grown by diverse simple methods. These advantages are of considerable interests for practical applications such as transparent electrode, window material in display, solar cell, and various optoelectronic devices.

The electrical behaviour of ZnO thin films could be significantly improved by replacing Zn atoms with higher valence elements, such as In, Al and Ga [6, 7]. From many reports of the doping effect of impurities on ZnO, it is widely demonstrated that Al and Ga are the best dopants for transparent conductive ZnO films [8]. However, most of the works related to ZnO use Al as dopant. Ga-doped zinc oxide films (GZO) have some advantages compared with Al-doped zinc oxide films (AZO). First of all, they are very close in atom radius: the radius of Ga^{3+} (0.062 nm) is closer to that of Zn^{2+} (0.074 nm) than that of Al^{3+} (0.053 nm) [9, 10]. Secondly, Ga is more resistive to oxidation compared to Al. The third is that it makes less diffusion-related problems since the diffusivity of Ga is lower than that of Al at the same temperature [11]. In spite of these advantages, GZO has been relativity less studied than AZO. Consequently, there is comparatively less information on the influence of the deposition conditions on the characteristics GZO thin films [12-14]. For the deposition of GZO films, the radio frequency magnetron sputtering method is outstanding among kinds of deposition methods due to its low cost of the source materials, and the simplicity of the deposition system required [15, 16].

Substrate temperature is one of the most important parameters in film deposition. It can enormously affect the crystallinity of the film, which can change the optical, electrical properties and microstructures of the film [12]. However, there are few studies of RF magnetron sputtering deposition substrate temperature conditions. In this respect, in order to improve and control the electrical properties of transparent conductive ZnO films, different substrate temperatures have been introduce to investigate the substrate temperature effects on the film properties.

* Corresponding author: Tel.: 86-21-56334778; Fax: 86-21-56334778; E-mail address: jhzhang@staff.shu.edu.cn

In this paper, high-quality ZnO: Ga transparent conductive films were prepared by RF magnetron sputtering using a 10 wt. % Ga_2O_3 doped ZnO target. The substrate temperature was ranging form 50 ℃ to 350 ℃. The dependence of the structural, electrical and optical properties of the Ga doped ZnO films on substrate temperature was investigated.

2. Experimental details

GZO films were deposited on glass substrate by JPGF-700B radio frequency magnetron sputtering system with a frequency of 13.56 MHz. The target with a mixture of ZnO (99.95% purity) and Ga_2O_3 (99.99 purity) was employed as source material. The amount of Ga_2O_3 added to the target was 10 wt. % [10]. The size of the target was 250 mm and 80 mm in length and width separately.

The substrate was rinsed ultrasonically in alcohol, acetone, and deionized water for 15 minutes respectively and baking-dried in the oven. The RF sputtering power was 200 W, the sputtering substrate temperatures were varied form 50 ℃ to 350 ℃. The sputtering time was 40 minutes and films with a thickness between 600-900 nm were obtained. The distance between target and substrate was 6 cm and the target was water-cooled. The sputtering gas argon (Ar) with 99.99% purity was controlled via a gas mass flow controller.

The thickness of the films was measured by surface profiler (alpha-step 500). The sheet resistance was measured with 5601-Y four-point probe instrument. The surface profile of the films was observed with Nanomavi SPA-400 SPM atomic force microscopy (AFM) system. The structural property of the films was determined with D/Max-2200 X-ray diffractometer (XRD) which uses Cu Ka radiation. The carrier concentration and electron mobility were measured, at room temperature in a normal atmosphere, by Accent HL5500PC Hall Effect measurement system. The optical transmittance measurements were performed with U-3000 spectrophotometer.

3. Results and discussion

3.1. Structure properties

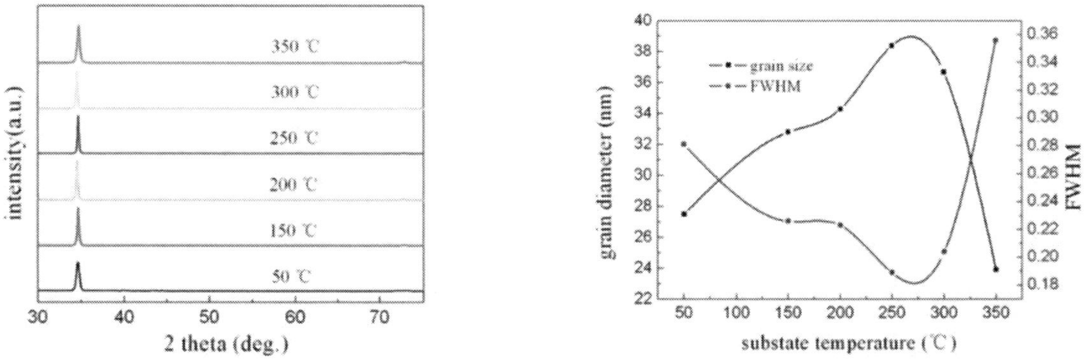

(a) (b)

Figure 1. Structural properties of GZO films as the function of substrate temperature: (a) X-ray diffraction pattern of GZO films; (b) FWHM and grain size of the GZO films.

Figure 1 (a) shows the XRD pattern of GZO films deposited at different substrate temperatures. From figure 1 (a), it can be seen that all of the films have shown a strong (002) c-axis growth orientation. The X-ray diffraction angle of GZO films deposited at different substrate temperatures was between 34.45° and 34.68°, which is a little bigger than the diffraction angle of standard ZnO powder (34.45°) [4]. This manifest there has been tensile stress in the plane of the films as compared with unstressed powder. This may be mainly due to the radius of Ga^{3+} (0.062 nm) is smaller than that of Zn^{2+} (0.074 nm)[9, 10], The replacement of Zn^{2+} with Ga^{3+} caused the decrease of the c-axis lattice parameter, which may result in the increase of the diffraction angle. What's more, there is a fluctuation of diffraction angle as substrate temperature increases. In general, the diffraction angle is getting closer to 34.45° until 300 ℃. This indicates higher temperature can improve the crystal quality of the films.

Figure 1 (b) shows the full width at half maximum (FWHM) and grain size of the GZO films. The grain size was calculated by Scherer formula: $C=0.9\lambda/B\cos\theta$ [19]. It depicts that with increasing substrate temperature, there has a distinct decrease in the FWHM and reaches the minimum at medial temperature of 250 ℃. However, the grain size goes reversely with the FWHM. This indicates that high growth temperature can improve the crystallinity of the GZO film. At lower substrate temperature, migration and diffusion of the sputtering species is inhibited and this definitely suppresses the growth of the nuclei. With increasing substrate temperature, the diffusion of sputtering species is enhanced and the grains become bigger. However, there is a significant increase in FWHM and grain size becomes smaller when the substrate temperature exceeds 250 ℃; the result is similar to those of literatures [11, 12, 18]. This fact implies the substrate temperature have a remarkable effect on the structure of ZnO, therefore, high crystalline ZnO films required optimum substrate temperature.

Figure 2 shows the film thickness of GZO films. Below 250 ℃, the film thickness is growing with substrate temperature increasing and the thickness became stable when the temperature is over 250 ℃. There is probably reorganization of certain chemical bonds and rupture process happen at the surface of the substrate. Therefore, with elevating substrate temperature the chemical absorption is speed up, thus the deposition rate is increased. However, the vapour atomic deposition is an equilibrium process of chemical adsorption and physical desorption. When the temperature is over 250 ℃, the rate of chemical adsorption and physical desorption reaches a balance, and results in the stable of growth rate.

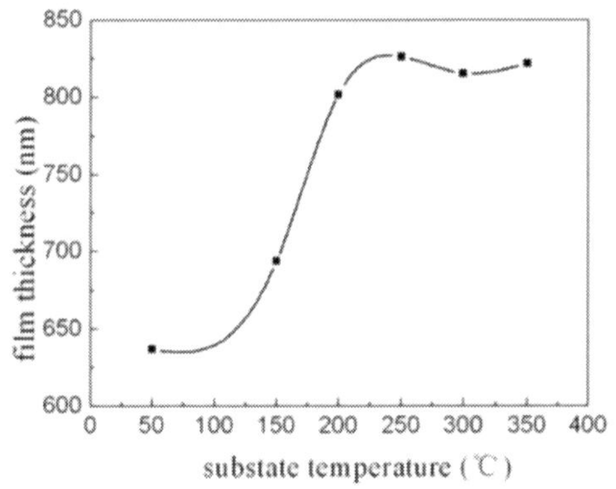

Figure 2. Substrate temperature effects on the film deposition rate.

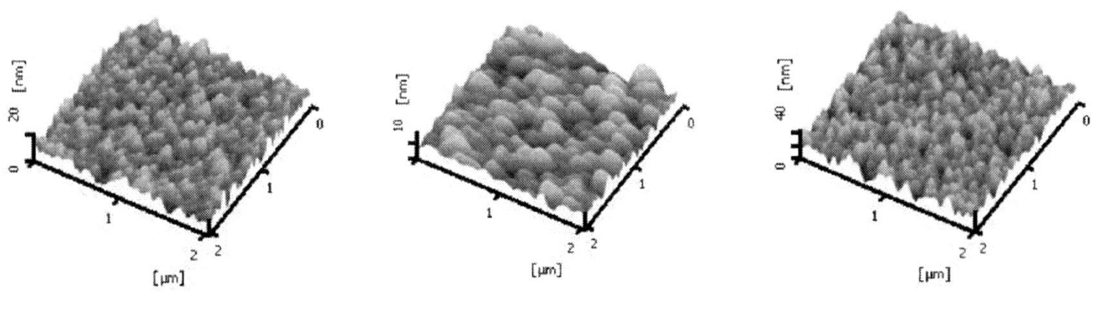

(a) (b) (c)

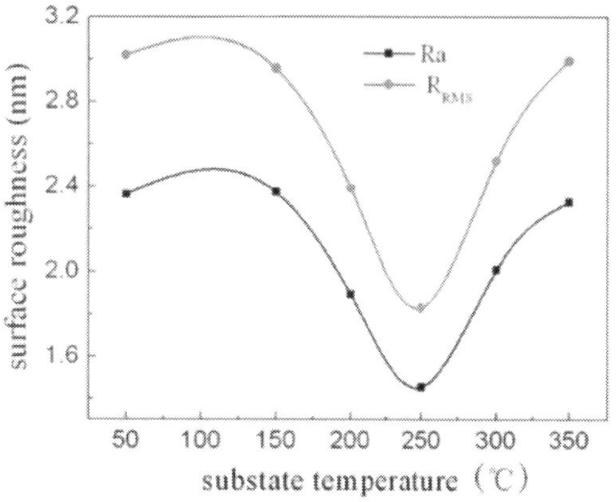

(d)

Figure 3. The AFM surface morphology of film deposited as the function of substrate temperature: (a) 50 ℃; (b) 250 ℃; (c) 350 ℃; (d) Surface roughness of films.

Figure 3 (a) - (c) shows the AFM surface morphology of GZO films obtained at the substrate temperature of 50, 250 and 350 ℃. Figure 3 (d) shows the roughness of the surface which was calculated according to the roughness tested by AFM. It shows that at the lower temperature the surface is comparatively rougher. As substrate temperature increases, the surface getting smother and reaching the minimum R_{RMS} value of 1.8 nm at 250 ℃. With further increasing temperature, the film became rougher again, which is comparatively identical to the XRD result.

3.2. *Electrical properties*

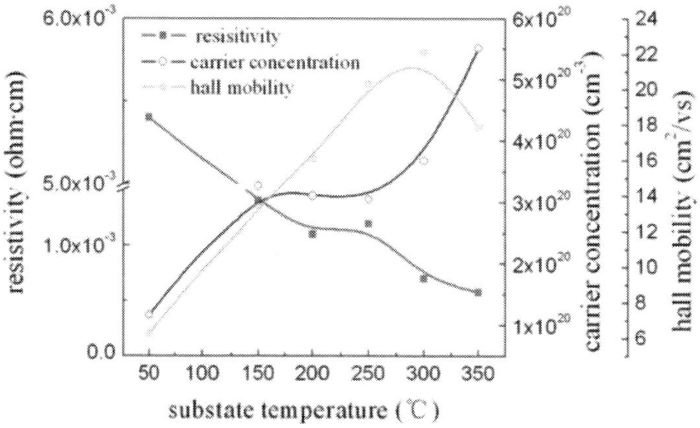

Figure 4. Electrical properties of the films as function of substrate temperature.

Figure 4 shows the electrical behaviour of the GZO films as a function of substrate temperature. The film resistivity decreases dramatically as the substrate temperature is increased from 50 to 250 ℃, and shows a slow downward tendency with further increasing substrate temperature. The as-deposited GZO films with the lowest resistivity of 5.8×10^{-4} ohm·cm is obtained at substrate temperature of 350 ℃. This result can be explained on the basis of an electrical activation of the impurities contributing to the free carrier concentration in films [13]. The initial decrease in resistivity can be ascribed to the increase in both carrier concentration and carrier mobility, then the slow decrease in resistivity may be due to the balance between the increase of carrier concentration and decrease of carrier mobility. The phenomena have been confirmed by the carrier concentration and mobility curve in the figure.

As shown in figure 4, the carrier concentration monotonously increases with the substrate temperature increasing. However, the hall mobility increases linearly, reaching a maximum value of 22.1 cm^2 $v^{-1}s^{-1}$ at 300 ℃. After that it goes down with further increasing substrate temperature. With increasing substrate temperature, more Ga atoms are activated into the ZnO lattice as active donor, which leads to the monotonous increase in carrier concentration. The improvement of mobility with the increasing of substrate temperature is probably attributed to the grain boundary scattering. It can be explained that the grain size goes up and surface roughness is improved with the increasing of temperature as shown in figure1 (b) and figure 3 (d), which reduce the electronics' diffraction of the grain boundary and improve the mobility.

However, the drop of the mobility when the substrate temperature is over 300 ℃ can be associated with the scattering mechanism of the film. According to the above discussions, the increase of carrier concentration results in the increasing of ionized impurity center which enhance the ionized impurity scattering. At the same time, the reduction of grain size and the deterioration of the surface roughness lead to strong grain boundary scattering. Therefore, the electron mobility is deteriorated with further increasing of substrate temperature. The relation of mobility with substrate temperature implies that the dominant scattering mechanism of as-deposited GZO film may be determined by both grain boundary scattering and ionized impurity scattering.

3.3. Optical properties

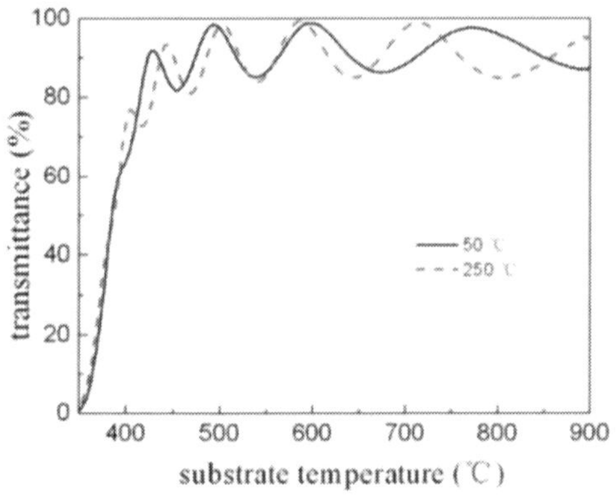

Figure 5. Transmittance of the films.

The optical transmittance of GZO films deposited at the substrate temperature of 50 and 250℃ are shown in figure 5. The optical transmittance measurements between 400 and 900 nm show that the films are highly transparent in the visible region; the average transmittance reaches up to 90% in the visible region. The high transparency is associated with a good structural homogeneity and crystallinity. With increasing substrate temperature, the absorption edge of the film shifts to the shorter wavelength due to the Burstein-Moss effect [20, 21]. However, the transmittance of the films does not change with increasing substrate temperature. This may be the reason that Ga in the films has basically become a full oxidation state. Thus the temperature increase can no longer significantly improve the transmission.

4. Conclusion

High transparent conductive gallium doped zinc oxide (GZO) thin films were deposited on glass substrate by RF magnetron sputtering. The effect of substrate temperature on structural, electrical and optical properties of the films was studied. The films are polycrystalline and have strong c-axis preferred orientation. FWHM lower than 0.3 shows good crystalline of the films. The electrical resistivity decreases with substrate temperature increasing and the lowest resistivity is $5.8\times10^{-4}\ \Omega\cdot$cm deposited at 350 ℃. The hall mobility increases with the substrate temperature increasing and the highest mobility is 22.1 $cm^2v^{-1}\cdot s^{-1}$. The deposition rate is increasing with the substrate temperature increasing and gets stable after 250 ℃. The average transmittance is over 90% in the visible region and it does not change much as the substrate temperature increasing. The results show that the high quality ZnO: Ga thin films can be obtained at the optimum growing temperature.

References

[1] V. Bhosle, A. Tiwari, J. Narayan, *J. Appl. Phys.* 100, 033713 (2006).
[2] T. Minami, *Thin Solid Films* 516 (2008) 5822-5828.
[3] T. Minami, *Thin Solid Films* 516 (2008) 1314-1321.
[4] D.-H. Kim, H. Jeon, G. Kim, S. Hwangboe, V. P. Verma, W. Choi, M. Jeon, *Opt. Commun.* 281 (2008) 2120-2125.
[5] N. A. Suvorova, O. Usov, L. Stan, R. F. DePaula, A. M. Dattelbaum, Q. X. Jia, A. A. Suvorova, *Appl. Phys. Lett.* 92, 141911 (2008).
[6] C.-S. Hong, H.-H. Park, J. Moon, H.-H. Park, *Thin Solid Films* 515 (2006) 957-960.
[7] X. H. Yu, J. Ma, F. Ji, Y. H. Wang, X. J. Zhang, H. L. Ma, *Thin Solid Films* 483 (2005) 296-300.
[8] Q.-B. Ma, Z.-Z. Ye, H.-P. He, J.-R. Wang, L.-P. Zhu, B.-H. Zhao, *Vacuum* 82 (2008) 9-14.
[9] K. Yim, Ch. Lee, *J. Mater. Sci. - Mater. Electron.* 18 (2007) 385-390.
[10] V. Khranovskyy, U. Grossner, O. Nilsen, V. Lazorenko, G.V. Lashkarev, B. G. Svensson, R. Yakimova, *Thin Solid Films* 515 (2006) 472-476.
[11] S. Kim, J. Jeon, H. W. Kim, J. G. Lee, Ch. Lee, *Cryst. Res. Technol.* 41 (2006) 1194-1197.
[12] S. Kim, W. I. Lee, E.-H. Lee, S. K. Hwang, Ch. Lee, J. Mater. Sci. 42 (2007) 4845-4849.
[13] H. G′omez, M. L. Olvera, *Mater. Sci. Eng. B* 134 (2006) 20-26.
[14] B. D. Ahn, S. H. Oh, H. J. Kim, M. H. Jung, Y. G. Ko, *Appl. Phys. Lett.* 91, 252109 (2007).
[15] X. H. Yu, J. Ma, F. Ji, Y. H. Wang, X. J. Zhang, C. F. Cheng, H. L. Ma, *Appl. Surf. Sci.* 239 (2005) 222-226.
[16] B. D. Ahn, S. H. Oh, C. H. Lee, G. H. Kim, H. J. Kim, S. Y. Lee, *J. Cryst. Growth* 309 (2007) 128-133.
[17] S.-M. Park, T. Ikegami, K. Ebihara, *Thin Solid Films* 513 (2006) 90-94.
[18] K. Ellmer, A. Klein, B. Rech, *Transparent Conductive Zinc Oxide*, Springer Press, Berlin, 2008, pp. 258.
[19] J. K. Yao, J. D. Shao, H. B. He, *Appl. Surf. Sci.* 253 (2007) 8911-8914.
[20] E. Burstein, *Phys Rev* 93 (1954) 632.
[21] I. Hamberg, C. G. Granqvist, *J. Appl. Phys* 60 (1986) R123.

Acknowledgments
This work was supported by the National Nature Science Foundation of China (NSFC) under the grant number 50675130, Program for New Century Excellent Talents in University (NCET) under the grant number NCET-07-0535, and the Science and Technology Committee of Shanghai under the grant number 08DZ1140603 and 08QH14007.

Synthesis, characterization and photocatalytic activity of Co^{2+}-doped titania

DUAN Su Hua[1], ZHANG Kunfeng[1], LI Daorong[1] ,YIN Qiya[1], HU Changwen[2]

1. School of Chemistry and Chemical Engineering , Henan University of Technology, Zhengzhou, 450001, China
2. China Department of Chemistry, Beijing Institute of Technology, Beijing, 100081, China;)

E-mail:ldrgroup@126.com

Abstract. Photocatalyst of Co^{2+}-doped titania was simply prepared, and the photocatalytic activity was tested under UV irradiation for methyl orange degradation. The influence of doping quantity, calcination temperature, calcination time and quantity of catalyst was investigated for photocatalytic efficiency of the photocatalyst. The sample was characterized by UV diffuse reflectance spectra (UV/DRS), X-ray diffraction (XRD), X-ray photoelectron spectroscopy (XPS), transmission electron microscopy (TEM) and IR. The results showed that the photocatalytic activity Co^{2+}-doped TiO_2 which calcined at 200 °C for 5 h is high, and which diameter is about 4-7 nm presents anatase phase. The sample absorption band of UV/DRS moves to visible region and reaches 580 nm. This kind of catalyst is also effective for degradation other organic dyes and beer wastewater.

Keywords: sol-gel; titanium oxide; photocatalytic degradation; visible light.

1. Introduction

In 1972, Fujishima and Honder reported the first photoelectrochemical cell water splitting, by using a rutile TiO_2 photoanode and Pt counter electrode [1]. The interest in developing artificial systems capable of converting solar to chemical or electrical energy is increasing[2-5]. One of the most active photocatalytic areas is that semiconductor particles as photocatalysts for the oxidation of organic pollutants in the water or air[6-8]. In spite of this, the active phase-anatase, do not absorb significantly over 387 nm, and the activity decrease when only visible light is used. The efforts to achieve a more efficient photocatalytic process, are in direction to improve the TiO_2 response, people look for new materials with ion doped, etc.

The biggest challenge is getting high activity catalysis. The method is doping ions to increase the efficiency of TiO_2. The doping ions include transitions metal ions, rare earth metal ions, inorganic function ions, etc. Metal ions doped to TiO_2 have been frequently reported[9-12]. Photocatalytic activity of metal doped TiO_2 depends on its preparation method [13-14], which affected the physicochemical properties of the photocatalyst. Therefore it is necessary to develop a simple method of catalyst synthesis, which doped transition metal ions. In this work, we reported the synthesis and characterization of Co^{2+} doped TiO_2 by simple preparation, which use cheap raw material $TiCl_4$. We studied the photocatalytic activity of the catalyst by degradation of methyl orange in aqueous dissolution. The result showed that the photocatalytic activity of samples was high when it calcined at lower temperature.

2. Experimental

2.1 Synthesis

Take 2 mL $TiCl_4$ in beaker, water and solution of $CoCl_2$ were added under stirring. Then an aqueous solution of NH_4OH was added under continuous stirring to adjusted pH value of the solution to ca. 7. The reaction mixture was maintained at room temperature until a gel was formed, and then colloid was stirred gently for about 1.5 h. It was dried in the oven at 85 °C. Finally the sample was calcined at different temperature in muffle.

2.2 Photocatalysis

Photocatalystic activity was evaluated by degradation of methyl orange solution. A quantity of 50 mg of the catalyst was added to a vessel containing 20 mL of methyl orange aqueous solution ($C_0 = 0.02$ g/L). The solution was irradiated with mercury lamp (250 W). During the process of the photocatalysis tests, the concentration of methyl orange was detected by a visible spectrophotometer (Shanghai 722s) at λ_{max} =464 nm.

The decontamination of wastewater was evaluated by COD degradation in beer wastewater.

2.3 Characterization

TEM images were recorded on a H-800 transmission electron microscope. DR-UV-Vis spectra was recorded on a Cary 500 UV-VIS-NIR spectrometer. XRD patterns were recorded on a Shimadzu XP-7000 with Cu-Kα radiation (λ = 0.154178 nm) and XPS patterns were recorded on a PHI 5300 ESCA X-ray photoelectron spectroscopy. IR images were obtained on a HITACHI 260-10 IR spectrometer. COD were detected on an AQUA LYTIC AL-32 COD vario.

3 Results and discussion

3.1 Optimization of photocatalyst

3.1.1 Influence of calcination temperature

In this paper, we discussed the influence of the calcination temperature of Co^{2+} doped titania. The samples were calcined at 150, 200, 300, 400, 500, 600 and 700 °C for 5 h. The degradation results were shown in Fig. 1. From Fig. 1, it can be seen that from 150 °C to 700 °C, the activity of Co^{2+}-TiO_2 is higher than TiO_2. The sample which calcined at 200 °C preserved high degradation. As the calcination temperature is above 400 °C, the activity of Co^{2+}-doped TiO_2 decreased quickly.

3.12 Influence of doping quantity

$CoCl_2$ solution (0.03 mol/L) was added to titania gel, the quantity of $CoCl_2$ is 0.05, 0.075, 0.1, 0.125, 0.15, 0.2, 0.3 mL respectively. Then the samples were calcined at 200 °C for 5 h. The methyl orange degradation experiments were shown in Fig.2. The degradation gradually increased along with the increment of the doping ion quantity ($Co^{2+} \leq 0.003$ mmol), and the photocatalytic activity of sample containing 0.003 mmol Co^{2+} is high.

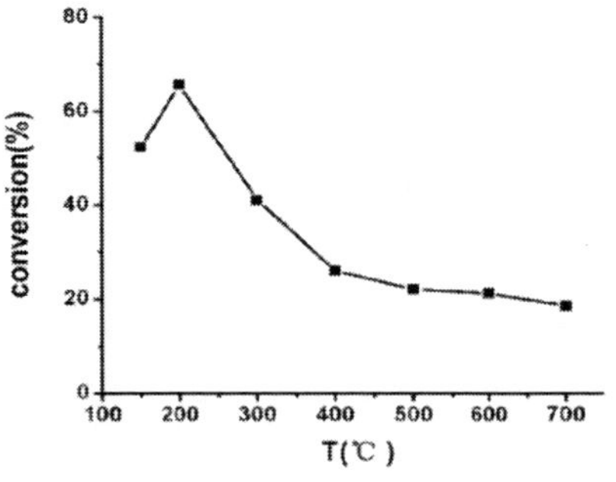

Fig. 1 Photocatalytic activity of catalyst calcined at different teperature

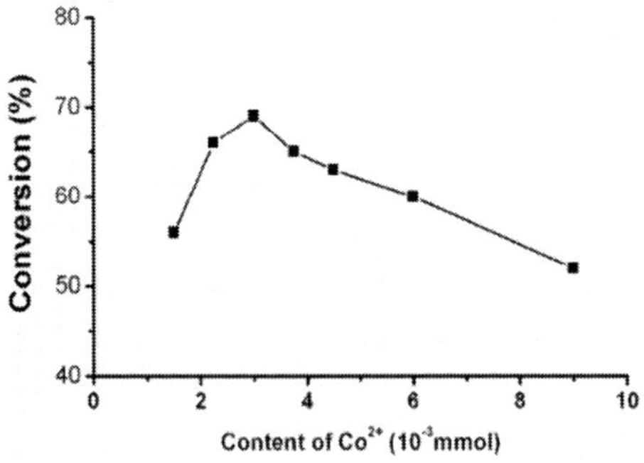

Fig. 2 Photocatalytic activity of different contents Co^{2+}-doped titania.

3.1.3 Influence of calcination time

A quantity of 0.1 mL $CoCl_2$ solution (0.03 mol/L) was added to titania gel. Then the samples were calcined at 200 °C for 3h, 4h, 5h, 6h, 7 h respectively. The methyl orange degradation experiments result was shown in Fig.3. From Fig.3, it can be seen that with the calcination time increasing, the photocatalytic activity of sample was high. However, the optimum time was 5h by considering energy resource.

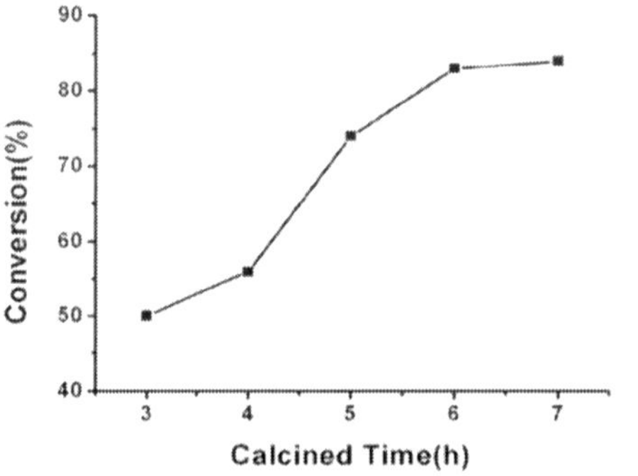

Fig. 3 Photocatalytic activity of Co^{2+}-doped titania calcined for different time

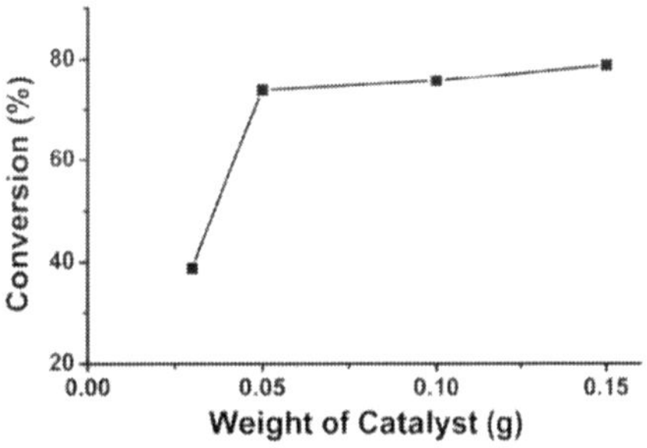

Fig. 4 Photocatalytic activity of Co^{2+}-doped titania with different quantity.

3.14 Influence of catalyst quantity
Methyl orange solution (0.02 g/L) was degraded by catalyst of 0.03, 0.05, 0.10 and 0.15 g, respectively. The degradation result was shown in Fig. 4. The photocatalytic activity was increased with a rise in dosage of catalyst. However, more than 0.05 g, the photocatalytic activity of catalyst increases slowly. Hence, the quantity of catalyst we chose was 0.05 g.

3.2 Activity of Co2+-doped TiO_2 for other dyes and beer wastewater
In this work, we choose Rosaniline, Bromophenol blue, Methylene blue to study the activity of Co^{2+}-doped TiO_2. The results were listed in Tab. 1. The photocatalytic activity of P25 and TiO_2 is lower when irradiated under visible light.

In addition, beer wastewater was chose to study the activity of Co^{2+}-doped TiO_2. A quantity of 50mg of catalyst was added to a vessel containing 20 mL wastewater solution. Under stirring, the solution was irradiated with a Hg lamp for 2 h. The COD was decreased from 1431 to 376, and the COD degradation rate was 73.7%, compared with degradation rate of 69.4% of P25 .

Table 1 Photocatalytic activity of Co^{2+}-doped TiO_2 composite for organic dyes degradation.

Dye	C(mg/L)	V/(mL)	UV light		Visible light	
			Time/h	Degradation/%	Time/h	Degradation/%
Rosaniline	20	20	1	95.7	5	75.2
Bromophenol blue	20	20	2.5	97.9	5	67.9
Methylene blue	20	20	2.5	89.6	5	63.8
Methyl orange	20	20	2.5	95.6	5	79.6

3.3 Characterization

3.3.1 UV diffuse reflectance spectra
The UV/DRS spectra of Co^{2+} doped TiO_2 showed that Co^{2+}-doped TiO_2 exhibited broad and strong absorption in the range from 200 to 400 nm, and the red shift was observed compared with pure TiO_2, the absorption peak shift to 580 nm. The result was explained degradation of methyl orange and beer wastewater test.

3.3.2 X-ray diffraction
From the XRD pattern in Fig. 5, it was shown that the effect of calcination temperature to the crystal of TiO_2. As shown in Fig. 5, the peak shapes of the samples calcined at 150 and 200 °C were obviously different from those of 400, 500 and 700 °C. The XRD patterns of 150 and 200 °C had characteristic peak of NH_4Cl phase at $2\theta = 23°$, 32.7°and had characteristic peak of anatase phase at 2θ

= 25.4°.It illustrated that amorphism of TiO_2 had changed into anatase at 150 °C. From 150 °C to 700 °C, it can be seen that the breadth of peak is minished, with the calcination temperature increased, and the crystal size was enlarged. When temperature higher than 700 °C, the rutile characteristic peaks at 2θ = 27.4°, 36.0° and 54.3° appeared in the XRD pattern. Through the calculation by Scherrer the particle size calcined at 150, 200, 400, 500, 700 °C for 5 h was roughly 6, 7, 11, 15, 33 nm, respectively. The temperature of crystallize transform is 150 °C, which was shower than the other report [15-16].

3.3.2 TEM and X-ray photoelectron spectroscopy

The TEM images of Co^{2+}-doped TiO_2 calcined at 200, 500 and 700 °C are showed in Fig. 6 From Fig. 6, it can be seen that the average diameter of the sample calcined at 200 °C was about 4-7 nm and the selected areaelectron diffraction (SAED) was showed that the composite has a uniform polycrystalline structure (Inset in Fig. 6A). When the calcination temperature increased, the average diameter of the sample enlarged gradually. The diameter of the sample calcined at 500 °C was roughly 15-20 nm and the diameter calcined at 700 °C was roughly 30-40 nm. It was also in accordance with the results of XRD.

3.3.3 XPS and Infrared spectroscopy

The XPS images of Co^{2+}-doped TiO_2 calcined at 200 °C showed that the bands at 780 and 795.9 eV were corresponding to Co 2p3/2 and Co 2p1/2 of CoO[17], respectively. Because XPS can only detect the photoelectron for several nanometer on the surface of the samples (for inorganic compound as 2-4 nm), Therefore Co^{2+} exists on the surface of the samples.

The IR images of Co^{2+}-doped TiO_2 calcined at 200 °C for 5 h showed the NH_4^+ IR characteristic peaks (3130 cm^{-1} and 1400 cm^{-1}), and anatase phase TiO_2 peaks (1650 cm^{-1} and 700 cm^{-1}). It can be deduced that the samples contain NH_4Cl and anatase TiO_2. Otherwise, it illustrated that TiO_2 was crystallized under 200 °C.

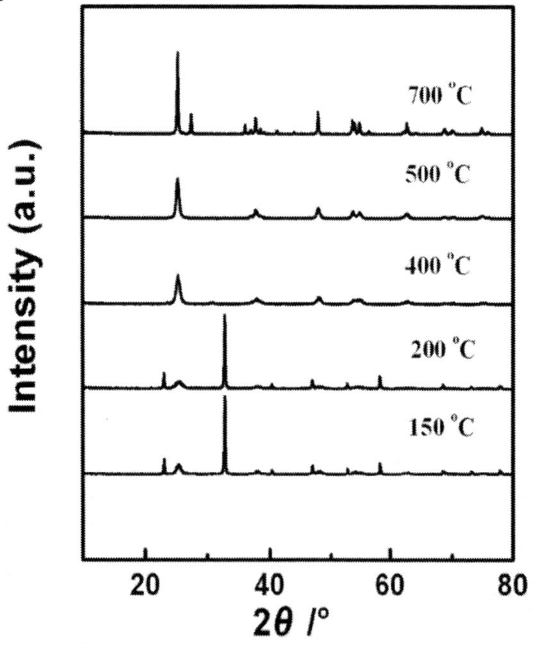

Fig. 5 XRD patterns of Co^{2+}-doped TiO_2 calcined at different temperature.

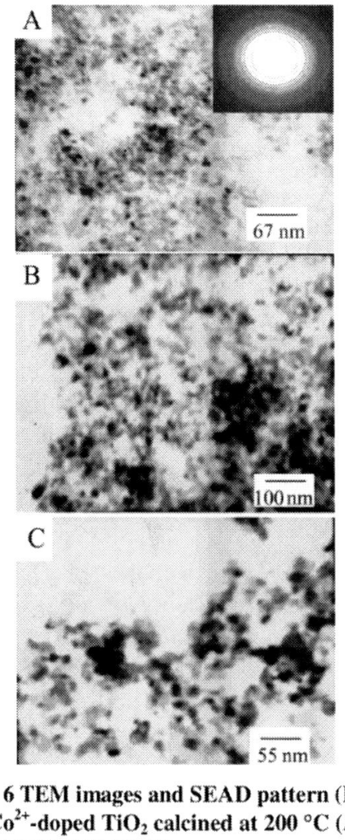

Fig. 6 TEM images and SEAD pattern (Inset A) of Co^{2+}-doped TiO_2 calcined at 200 °C (A), 500 °C (B), 700 °C (C).

4. Conclusion

In this paper, we used $TiCl_4$ as the raw material to prepare Co^{2+}-doped TiO_2, and discussed the factors effect on the activity. The result showed that activity of Co^{2+}-doped TiO_2 calcined at 200 °C is high. This crystal type of Co^{2+}-doped TiO_2 is anatase and absorption band moves to the visible region 580 nm, and has the utilization of visible-light. The organic dyes and beer wastewater were effectively degraded by the catalyst, and the graduation rate is high.

Acknowledgements

The Natural Science Foundation Council of China (20271007 and 20331010), and Specialized Research Fund for the Doctoral Program of Higher Education (20030007014) are acknowledged for financial support.

References

[1] Fujishima A, Honda K. Nature, 1972,238:37.
[2] Schnadt J, Bruhwiler P A, Patthey, L, et al. Nature, 2002, 418: 620.
[3] Ohko Y, Tatsuma T, Fujii T, et al, Nature Materials, 2003, 2: 29.
[4] Khan U S M,.Al-Shahry M, Ingler Jr B, W Science, 2002, 297: 2243.
[5] Eric W, Farland M, Jing T. Nature, 2003, 421:616.
[6] Shi J E, Yan J C, Shang S X, et al. Chemical Journal of Chinese Universities, 2007, 28(7): 1325.
[7] Ding Z, Feng X G, Chen X D, et al. Environmental Science, 2006, 27(9): 1814.
[8] Peng S Q,Jiang F Y, Li Y X. Journal of Functional Material, 2005,36(8):1207.
[9] Lu W Q, Zhao L M, Liu G L, et al. Chinese Rare Earths, 2006,1: 30.
[10] K. J. Zhang, W. Xu, X. J. Li, et al. CEJC, 2006, 4(2): 234-245.
[11] Shi J W, Zheng J T, Hu Y, et al. Journal of the Chinese Ceramic Society. 2007,35 (2):182.
[12] Wu S X, Ma Z, Qin Y N, et al. Acta Phys. Chim. Sin. ,2004,20(2):138-143.
[13] X. Yu, G. Wang, Y. Luo, et al, Chemical Research and Appl., 2000, 12(1): 14.
[14] M. Shui, L. Yue, Z. Xu, Acta Physico-chimica Sinica, 2001, 17(3): 282.
[15] S. J. Kim, S. D. Park, Y. H. Jeong, et al, J. Am. Ceram. Soc., 1999, 82(4) :927.
[16] J. Ovenstone, K. Yanagisawa, Chem. Mater., 1999, 11: 2770.
[17] N.S. McIntyre, Analytical Chemistry, 1977, 49:1521.

Journal of Physics: Conference Series **188** (2009) 012019 doi:10.1088/1742-6596/188/1/012019

Preparation, characterization and photocatalytic performance of rutile TiO_2 nanowire arrays via a novel hydrothermal process

Lei Ge Jing Liu
email: gelei08@eyou.com
(China University of petroleum Beijing, Beijing, 102249)

Abstract The single crystal rutile TiO_2 nanowire arrays with flat tetragonal crystallographic plane were successfully synthesized via a novel hydrothermal method. The prepared samples were characterized by X-ray diffraction (XRD) and scanning electron microscopy (SEM). The photocatalytic activities of the TiO_2 nanowires were determined by degradation of methyl orange molecules in water under UV-light irradiation. The results indicated that the TiO_2 nanowires were rutile phase. The reaction temperatures, pH values and time did not change the crystal structures, while the morphologies of nanowire arrays changed. The photocatalytic experiments revealed that different synthesis conditions can affect the photocatalytic activity of the sample under UV-light irradiation.

Key words TiO_2 nanowire arrays, rutile, hydrothermal, photocatalytic

Introduction

Photocatalyst is a kind of semiconductor materials[1]. With multiple chemical properties such as photocatalytic performance[2], degradation of various types of chemicals[3] and killing bacteria[4] under UV even visible light irradiation. This kind of material includes nano-oxides[5], sulphides[6] and multiple oxides[7] as well as its modified and multiple doping materials[8] etc. All the environmental pollutant can be decomposed by the photocatalysis of semiconductor materials, with a spectrum nature. Compared with the other photocatalytic materials, TiO_2 is to be the most commonly used catalyst, because TiO_2 has the advantages of nontoxic, stable property of photochemical, effective photocatalysis, killing bacteria and low cost etc. Therefore, to get the technology , controlling the structure of TiO_2 to nano-size , is very important.

Nano-materials of TiO_2 include nano-particles, nanotubes and nanowires. Contrasting to the general nanomaterials, the identical orientation nanowire array structure has more excellent performance. The published preparation methods of titanium dioxide nanowires are sol-gel[9], hydrolysis[10], hydro-thermal[11] etc. This paper presents a new way of hydrothermal to compound single crystal rutile TiO_2 nanowire arrays.

1. Experiment

1.1 Preparation

TiO_2 nanowire arrays were synthesized on slide glass which initially cleaned by ethanol, then cleaned by ultrasonic for five minutes, and finally rinsed with deionized water. The slide glass substrates were placed within a teflon-lined sealed stainless steel reactor which contains hydrochloric acid (37%), toluene, tetrabuty titanate (1 M in toluene), and titanium tetrachloride (1 M in toluene). The reaction temperature of 150 ℃ and 180 ℃ were used, with reaction times lasting from 12h to 36h. After the reaction period, the nanowires were obtained on the slide galss.

1.2 Characterization

The X-ray diffraction (XRD) measurements were carried out on a XD-2(Beijing Purkinjie) X-ray diffractometer using CuKa radiation (λ=0.15405nm). The morphologies and microstructures of as-prepared samples were inspected with scanning electro microscopy (SEM, FEI Quanta-600).

1.3 Photocatalytic Reaction

Visible-light photocatalytic activities of the obtained samples were measured by the decomposition of methyl orange in an aqueous solution at ambient temperature. A 250W high-pressure Hg lamp was used as a light source. In each experiment, a 0.1g amount of photocatalyst was added into 50mL methyl orange solution with a concentration of 10 mg L^{-1} . Prior to irradiation, the mixed liquor was placed in the dark for 30 min to obtain the saturated absorption of methyl orange onto the catalysts. At irradiation time intervals of every lh, the concentrations of the methyl orange was monitored using a UV-vis spectrophotometer (Japan Shimadzu UV-Vis 1700) by checking the absorbance at 510 nm .

2. Results and Discussion

2.1 XRD patterns analysis

Fig.1

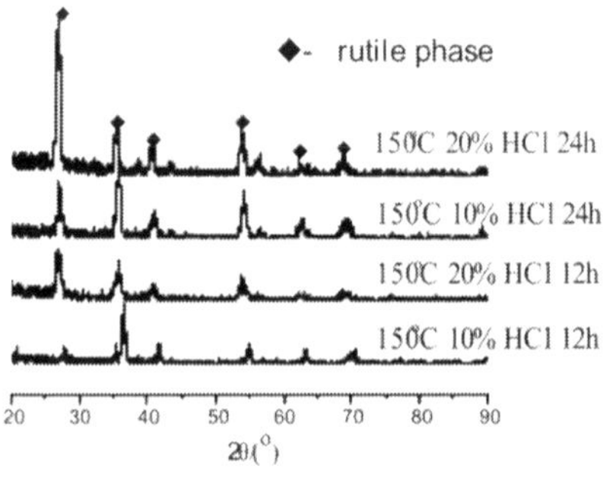

Fig.2

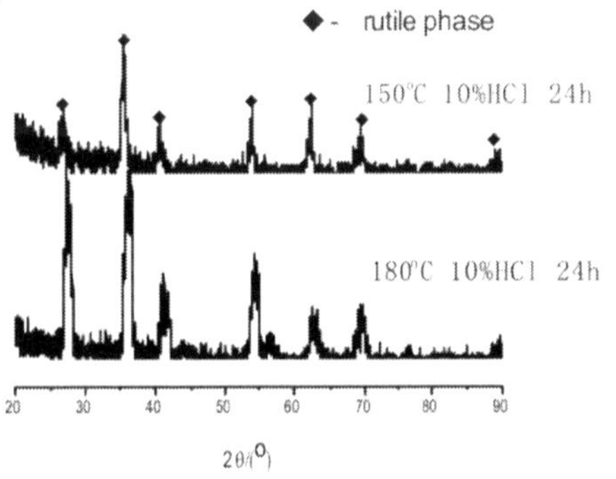

Fig.1 XRD patterns of TiO_2 nanowire array samples synthesized at 150℃

Fig.2 XRD patterns of TiO_2 nanowire array samples synthesized at 150℃ and 180℃

Fig.1 and 2 show the XRD patterns of TiO_2 nanowires with different reaction time, pH value and reaction temperature. The XRD patterns reveal that nanowires can be indexed to rutile phase. The enhanced peaks indicate the nanowire is well crystallized. From Fig.1, as other conditions are equal, the lower pH value leads to the higher the sharp peak that means the rutile nanowire crystal better. Alkaline environment tends to generate anatase, while acidic environment is in favor of the synthesis of rutile [12].

2.2 Morphologies of the sample

Fig.3 top-view of vertically oriented TiO_2 nanowire arrays (a)180℃,12h,10%HCl (b)150℃,12h,10 % HCl

Fig.3 is SEM images of nanowire arrays top-surface. Fig.3 a shows uniform and densely packed array of nanowires with flat tetragonal crystallographic plane. Fig.3 b is not as packed as Fig.3 a. The reason is that reaction temperatures affect the crystal nucleation. The higher temperature will obtain more crystal nucleus, which makes the nanowires more packed.

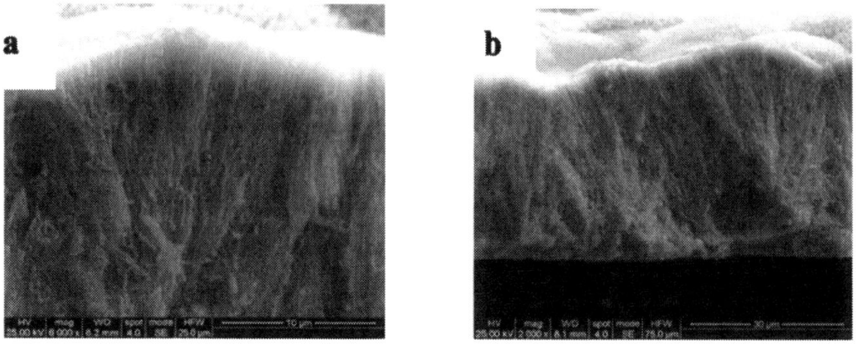

Fig.4 cross-sectional of vertically oriented TiO_2 nanowire arrays under the condition of 180℃, 20%HCl, 24h

Fig.4 is SEM cross-sectional images of nanowire arrays, showing that the nanowires grow almost perpendicularly from the substrate. It is found that nanowires grow slow with time, while no increase in nanowire length is achieved for reaction times beyond 24h. The experimental results show that the reaction time mainly affects the length of the nanowires. If other conditions are equal, the nanowires will be longer with increasing time.

2.3 Photocatalytic degradation

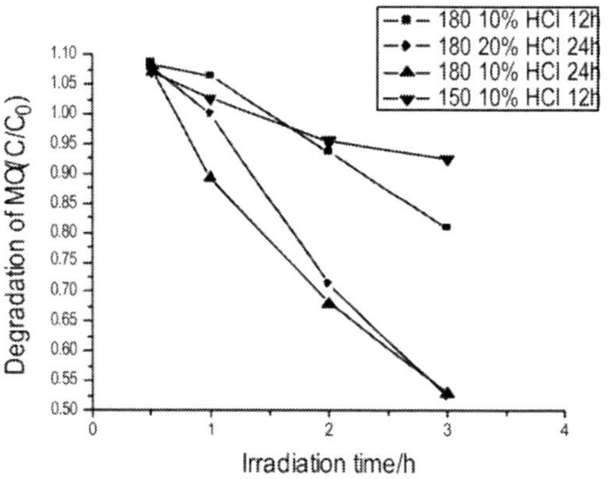

Fig.5 Photocatalytic degradation of methyl orange over TiO_2 photocatalysts synthesized with different conditions under UV-light irradiation

Fig.5 shows that the photocatalytic activity of nanowires synthesized under the condition of 180℃, 20%HCl, 24h were better. The reason is that nanowires have good crystalline state under this condition. The better cryatalline state has less imperfection which makes electron-hole pairs have less chance to recombine. Therefore more electron-hole pairs will participate in oxidation-reduction reaction, which makes better photocatalytic performance.

3. Conclusion

A new type of hydrothermal synthesis is used to fabricate ordered rutile TiO_2 nanowire arrays. The XRD patterns reveal that samples obtained are rutile phase. The SEM images show samples are nanowires and the reaction time affect the growth of nanowires. The photocatalytic experiments indicate the photocatalytic performance of the single rutile TiO_2 nanowire arrays obtained at 180℃ with 20% HCl for 24h is the best.

Acknowledgment

The authors acknowledge the financial support of the Key Project of Chinese Ministry of Education (Grant No. 108024), and Doctor Foundation of Chinese Ministry of Education (Grant No. 200804251014), and Beijing Nova Program (Grant No. 2008B76).

Reference

[1] Liu LN, Song HW. Inorganic-organic hybrid semiconductor nanomaterials: $(ZnSe)(N_2H_4)(x)(C_5H_5N)(y)$[J]. Materials Research Bulletin,2009, 44:1385-1391.

[2] Alem A, Sarpoolaky H, Keshmiri M. Sol-gel preparation of titania multilayer membrane for photocatalytic application[J]. Ceramics International, 2009 , 35 :1837-1843.

[3] Huang HJ, Lid Z, Lin Q. Efficient degradation of benzene over $LaVO_2/TiO_2$ nanocrystalline heterojunction photocatalyst under visible light irradiation[J]. Environaental Science & Technology, 2009, 43 :4164-4168.

[4] Wu PG, Xie RC, Imlay JA, Shang JK. Visible-light induced photocatalytic inactivation of bacteria by composite photocatalysts of palladium oxide and nitrogen-doped titanium oxide[J]. Applied Catalysis B-environmental, 2009,88:576-581.

[5] Geng Wang, Da Chen, Hao Zhang Tunable photocurrent spectrum in well-oriented zinc oxide nanorod arrays with enhanced photocatalytic activity[J]. Phys.chem 2008, 112 :8850-8855.

[6] Save F, Popescu M. Structural modeling of nano-carbons and nano-chalcogenides[J]. Proceedings of the romanina academy series a-nathematics physics technical sciences information science,2009,10:19-25.

[7] Hua Gui Yang, Hua Chun Zeng. Synthetic Architectures of $TiO2/H_2Ti_5O_{11}\cdot H_2O$, $ZnO/H_2Ti_5O_{11}\cdot H_2O$, $ZnO/TiO_2/H_2Ti_5O_{11}\cdot H_2O$, and ZnO/TiO_2 nanocomposites[J]. Articles, 2005, 127:270-278.

[8] Liu Xiang-zhi, Xu Ming-xia. Preparation of transition-metal ion doped titania nanowires and the wisible photocatalytic activity[J]. Journal of wuhan university of technology,2007,29:173-177.

[9] Liu FF, Li XY, Zhao QD. Structural and photowoltaic properties of highly ordered ZnFe2O4 nanotube arrays fabricated by a facile sol-gek template method[J]. 2009, 57:2684-2690.

[10] Hu XL, Masuda Y, Ohji T. Micropatterning of ZnO Nanoarrays by forced hydrolysis of anhydrous zinc acetate[J]. Langmuir,2008, 24 :7614-7617.

[11] Masaaki kitano, Ryo mitsui. Synthesis of nanowire TiO_2 thin films by hydrothermal treatment and their photoelectrochemical properies [J]. Catal Lett, 2007,119:217-221.

[12] Xiao yi-fan, Liu song. The progresses in research survey of photocatalytic properties of nano-titania preparation by hydrothermal method[J]. bulletin of the Chinese ceramic society, 2007,26(3):523-528.

About the Authors:

Lei Ge (1979/08) ,male (Han), Hebei Hengshui, Ph. D, Lecturer (Professor jobs), Novel Photocatalytic Synthesis and Reaction Mechanism

Jing Liu (1985/02), female (Han), Hubei Jinzhou, Master in reading, Student, Novel Photocatalytic Synthesis and Reaction Mechanism

Contacts:

First Author: Lei Ge

Address: Material Department, China University of Petroleum (Beijing), No.18 Fuxue Road, Changping, Beijing,102249

Journal of Physics: Conference Series 188 (2009) 012020 doi:10.1088/1742-6596/188/1/012020

Preparation and reduction behavior of Mo-Cu powders by Sol-Gel

Ming Zhao, Jinshu Wang*, Wei Liu and Meiling Zhou

Key Laboratory of Advanced Functional Materials Ministry of Education, College of Material Science and Engineering, Beijing University of Technology, Beijing, 100124, China

wangjsh@bjut.edu.cn

Abstract. Mo-Cu composite powders were prepared by the Sol-Gel method using citric acid as chelating reagent. The composition, morphology and particle size of the powders were studied by the XRD and TEM. TEM observation result shows the particle size of Mo-Cu powders is about 150nm after reduction and the Mo grains are covered homogeneously by Cu on the surface. The reduction behavior and mechanism were analyzed systematically. Temperature programmed reduction (TPR) results indicate that the appropriate two-step reduction temperatures are 500°C and 700°C, respectively. The results from FT-IR show that the bond of Mo=O is weakened due to the bond of Cu-O-Mo, which leads to the increase of the reduction capability of the Mo-Cu composite oxide powders.

1. Introduction

Due to the excellent properties of high electrical and thermal conductivities, arc erosion resistance and low thermal expansion coefficient, Mo-Cu alloy is extensively used as heavy-duty electronic contacts, welding electrodes, and other parts used in electronic packaging and military fields [1-4].

In general, full densification for Mo-Cu compact is difficult to achieve by liquid phase sintering. The low solubility of Mo in Cu makes the contribution of grain-shape accommodation by solution-re-precipitation negligible, so densification can be attributed to either particle rearrangement due to capillary forces or sintering of the Mo skeletal structure by solid state diffusion. Consequently, the particle size and mixing uniformity of the Mo and Cu raw powders make a great impact on the densification behavior and final microstructure owing to the lack of homogenization during sintering [5,6]. Recently, mechanical alloying [7-9] which has been successfully used to fabricate Mo-Cu alloy possessing high mechanical property aroused great attention. Homogenization of the ingredients occurs by a process involving repeated extrusion, cold welding and crush of the powder particles during milling, accompanied by the gradual refinement of the mixed powder particles. At the same time, the formation of a large amount of dislocation, lattice distortion and the nano-grain boundaries increase the activity of powders, which is beneficial to the sintering process [10]. Unfortunately, high energy ball milling includes long time for achieving ultra-fine particle size and involves introduction of impurities. Therefore, Sol-Gel as a chemical procedure [11-13] has been explored to prepare homogeneously dispersed ultra-fine Mo-Cu powders, possessing the advantages of good uniformity of

* Corresponding author: Jingshu Wang

different substances, small particle size and high purity. However, little attention has been focused on the reduction process. Reduction temperature is very important since it should be higher than that for the reduction reaction taken place but as low as possible to prevent grain growth during the reduction period. In this study, Mo-Cu composite powders were successfully produced by the Sol-Gel method followed by reduction. In order to get the optimum reduction technique for the preparation of Mo-Cu powders with small particle size, the reduction behavior was analyzed using temperature programmed reduction (TPR) and infrared spectrum (IR).

2. Experimental

To produce the Mo-30wt.%Cu composite powders, $Cu(NO_3)_2 \cdot 3H_2O$ and $(NH_4)_6Mo_7O_{24} \cdot 4H_2O$ as starting materials were dissolved in distilled water, and citric acid was used as chelating reagent in the precursor solution. The mixed solution was adjusted to pH=1 by adding ammonia or dilute nitric acid solution. Afterwards the solution was evaporated to composite sol by heating at 80°C in water bath with continuous stirring. Then the xerogel was formed at 120°C and was calcined at 500°C in air. Finally, the obtained metal oxide powders were reduced by a two-step reduction at temperatures of 500°C and 700°C for 90min, respectively.

The phase analysis was conducted by X-ray diffraction (XRD) using Cu Kα radiation (SHIMADZU XRD-7000). The Microstructure was characterized by TEM (JEM-2010) and SEM (S4800). The particle size was evaluated by laser particle size analyzer (LA-920). The fourier transform infrared (FT-IR) spectrum was measured by BRUKER VERTEX 70 infrared spectrometer (squashed with KBr). TPR was performed in $Ar+5\%H_2$ of flow rate $35mL \cdot min^{-1}$. Heating rate was $10K \cdot min^{-1}$ and the bridge current was 120mA.

3. Result and discussion

3.1. Analysis of calcined product

In order to obtain composite oxide powders, the xerogel was calcined under air to remove impurities such as C and N elements. The samples were calcined at different temperatures and the XRD results are shown in Figure 1. Figure 1(a) presents XRD pattern of xerogel calcined at 400°C. Both amorphous peaks caused by the residue of organic substance and XRD peaks identified as $CuMoO_4$, CuO and MoO_3 phases, were observed in the XRD patterns. Figure 1(b) is XRD patterns of calcined products at 500°C. No amorphous peak is found, indicating that organic substance has been decomposed completely. Therefore, 500°C was chosen as the appropriate calcination temperature.

Figure 2 shows the TEM micrograph of the samples calcined at 500°C. Direct TEM observation shows that Mo-Cu oxides agglomerate and single particle size is about 200nm.

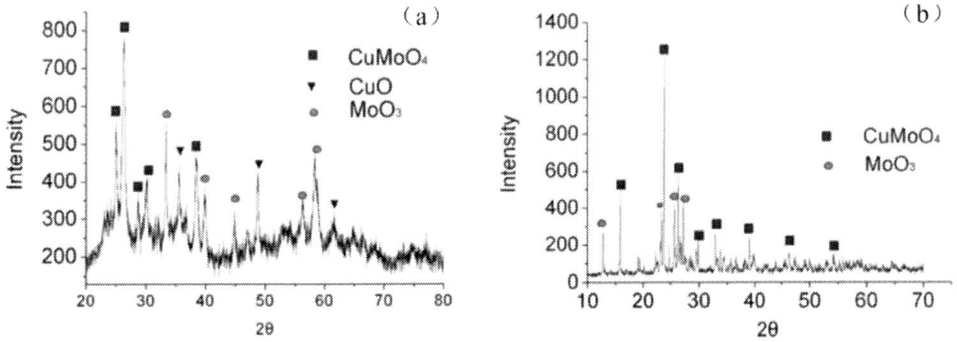

Figure 1. XRD patterns of xerogel calcined by different temperatures :

(a) 400°C; (b) 500°C.

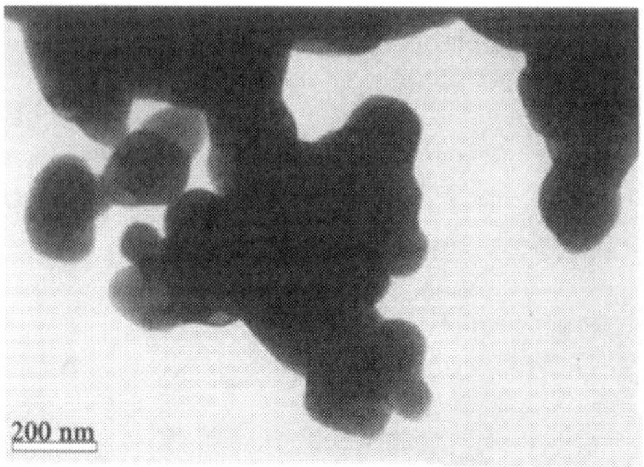

Figure 2. TEM image of calcined powders.

3.2. Reduction process

TPR was applied to determine the appropriate reduction process of calcined product. Figure 3 shows the TPR spectra of pure MoO_3 and Mo-Cu composite oxide powders prepared by Sol-Gel methods. The reduction of pure MoO_3 to Mo consists of two reduction stages, namely, $M_O^{6+} \rightarrow M_O^{4+} \rightarrow M_O^{0}$. However, only one peak of hydrogen consumption appears in the TPR spectra of pure MoO_3 at a temperature of 655°C (see Figure 3(a)), indicating that the hydrogen consumption corresponding to the two-step reduction of $M_O^{6+} \rightarrow M_O^{4+} \rightarrow M_O^{0}$ has been merged into one peak. On the other hand, Figure 3(b) indicates clearly that the Mo-Cu composite oxide powders show different reduction behavior. The reduction process of the oxides is divided into three stages. The hydrogen consumption peak at 250°C corresponds to the oxygen active site of the Cu-O bond for hydrogen consumption. The other two peaks appearing at 380°C and 620°C might be caused by reduction of two stages for $MoO_3 \rightarrow MoO_2$ and $MoO_2 \rightarrow Mo$. The relevant chemical equations are listed as follows:

$$CuMoO_4(s)+H_2(g)=Cu(s)+MoO_3(s)+H_2O(g) \quad (1)$$

$$MoO_3(s)+H_2(g)=MoO_2(s)+H_2O(g) \quad (2)$$

$$MoO_2(s)+2H_2(g)=Mo(s)+2H_2O(g) \quad (3)$$

Each peak of hydrogen consumption for the composite oxides is associated with a different oxygen active site. The peaks move towards lower temperatures side compared with pure MoO_3. From Fig.3, it can be seen that additive Cu is favorable for the formation of active oxygen centers with high activation energy, leading to a low temperature of the hydrogen consumption peak. It may be assumed the interaction between Cu and Mo atoms have significant influence on coordination environment for Mo and bond of Mo-O, which can form the different oxygen active sites to separate continuous reduction into different processes.

Based on the TPR result, 500°C and 700°C are determined as the optimum two-step reduction temperatures for Mo-Cu powders in this work in order to get fully reduced metallic Mo-Cu powder.

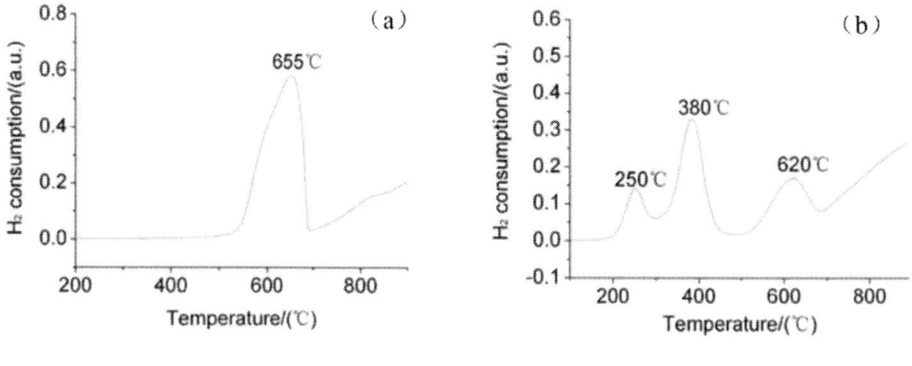

Figure 3. TPR spectra of MoO_3 and Mo-Cu composite oxides: (a) MoO_3; (b) Mo-Cu.

3.3. Characterization of reduced powders

Figure 4 shows the XRD patterns of the samples reduced by a two-step of 500°C and 700°. The patterns exhibit well-defined crystalline peaks completely identified as Mo and Cu phase, indicating that MoO_3 and $CuMoO_4$ composite powders have been reduced completely above 700°C.

TEM micrographs of the Mo-Cu powders after reduction at 700°C are shown in Figure 5. From Figure 5(a), it can be seen that the powders has uniform particle size with the mean of around 150nm. In addition, the single particle observation (Figure 5(b)) indicates that the Mo grains are homogeneously covered by copper phase since a thin layer is found on the particle surface. This kind of particle is the key point to obtain Mo-Cu alloy with homogeneous structure and good properties.

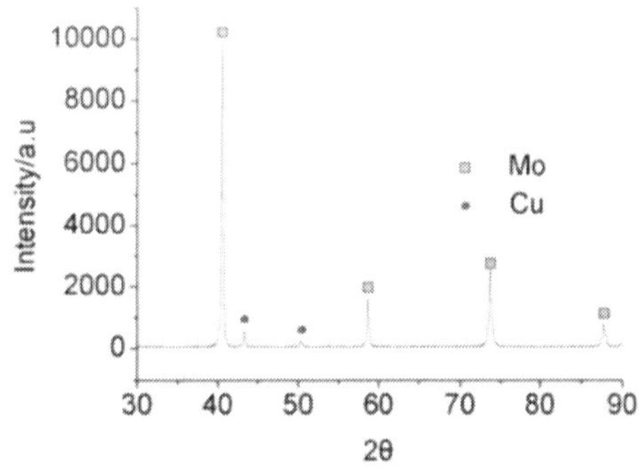

Figure 4. XRD patterns of Mo-Cu composite powders reduced at 700°C.

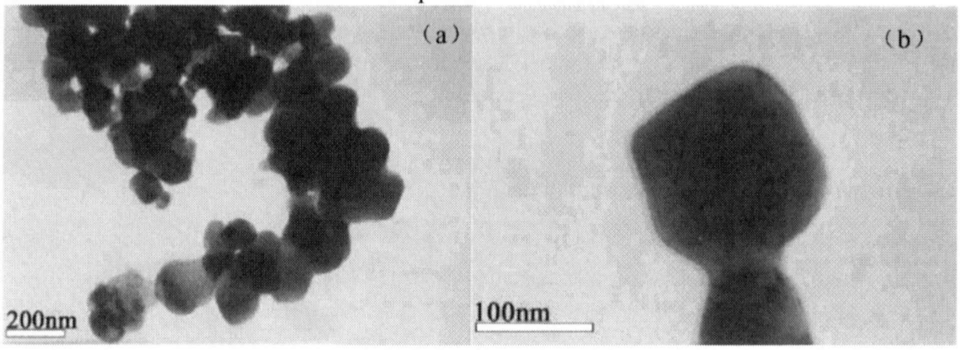

Figure 5. TEM images of Mo-Cu composite powders.

In order to further characterize the particle size, particle size analysis for the powder mixtures was employed. The particle size distribution of Mo-Cu powders reduced at 700°C is presented in Figure 6. The result shows that the average powder particle size D(n,50) is 407nm, much larger than the result of TEM observation. This difference results from the particle agglomeration because ultra-fine particles with high specific surface area and specific surface energy caused a relatively unstable thermodynamic state. The aggregation between particles could reduce free energy of the system. However, this aggregation is a kind of soft agglomeration caused by physical interaction, which is generally believed that this will not affect either the uniformity of powders or the densification during sintering process.

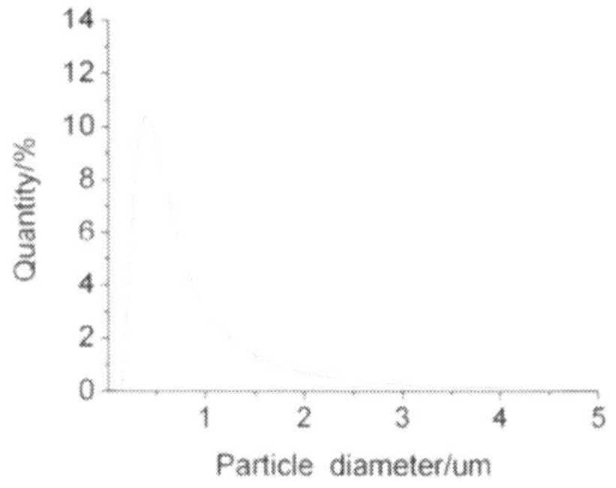

Figure 6. Particle distribution of powders after reduction at 700 °C.

3.4. Reduction behavior

TPR results illustrate the reducibility of Mo-Cu composite oxides is enhanced compared with pure MoO_3. Copper is easy to be reduced in Mo-Cu composite oxides, and the reduction behavior of Mo is closely related to the nature of Mo-O bond, which is dependent on the coordination environment and interaction with other components for Mo atom. Figure 7 shows IR spectra of MoO_3 and Mo-Cu composite oxides. There are distinct absorption peaks in both MoO_3 and Mo-Cu oxides located in 800～1000cm^{-1} range. It is reported that the absorption peak in 800～900cm^{-1} corresponds to expansion vibration of metal-oxide-metal band (M-O-M), and the absorption peak in 900～1000cm^{-1} is attributed to the terminal metal=oxide band (M=O) [14]. In this study, The absorption peak of Mo-Cu oxides defined as Mo-O-Mo and Cu-O-Mo bands is located at 863cm^{-1}, which is nearly the same with the result of MoO_3 (862 cm^{-1}). Moreover, the Mo=O band of Mo-Cu oxides at 954cm^{-1} is lower than that of MoO_3 samples (993cm^{-1}). This red shift indicates that the bond of Cu-O-Mo makes the Mo=O bond weakened and the activity for terminal oxygen atoms increased, which leads to the increase of the reduction capability of the Mo-Cu composite oxide powders. Therefore the reduction temperature of Mo-Cu composite oxide powders is lower than that of pure molybdenum oxides.

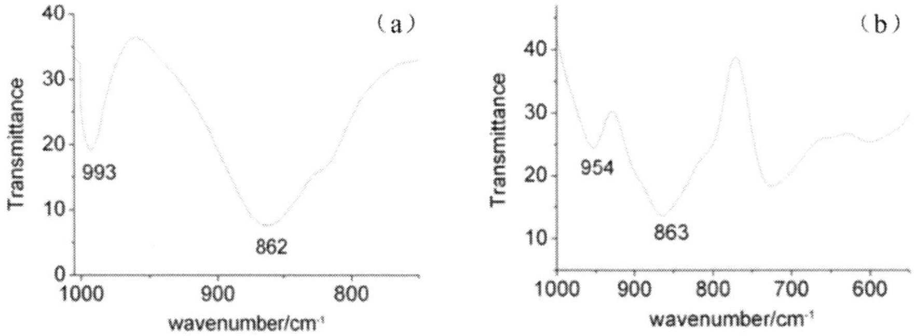

Figure 7. IR spectra of MoO_3 and Mo-Cu composite oxides: (a) MoO_3; (b) Mo-Cu.

Based on the crackling core model for the reaction of solid particles [15], a simple model is proposed to describe the reduction process of Mo-Cu particles under hydrogen gas. Figure 8 shows the schematic diagram of reduction process for Mo-Cu oxides. A lot of cracks and micropores form on the Mo-Cu oxide particle from the surface to the interior in initial reduction stage. Cu phase possessing low melting point, is reduced firstly, then it will be soften and precipitated onto the surface of reduced Mo particles by capillary action with the increase of reduction temperature. After cooling, a copper layer coated on Mo particles will be formed. Since the composition of Mo-Cu oxides is well-mixed by Sol-Gel, Mo grains are covered homogeneously by Cu phase after reduction.

Metallic Mo-Cu particles are smaller than Mo-Cu oxides in size, which is also caused by the reduction process. During the first reduction period of MoO_3, cracks and micropores are extend to the center of particle with the decrease of oxygen content and the increase of vacancies. A great deal of MoO_2 nuclei and an intermediate gaseous transport phase are formed. This transport phase is deposited on MoO_2 nuclei, which leads to grain growth. The grain size of the final product is determined by the growth of MoO_2 grain.

In addition, the addition of Cu element also plays a significant role in obtaining ultra-fine Mo-Cu powders. On the one hand, precipitated copper phase are favorable for the formation of micropores and cracks. On the other hand, during the reduction period, many defects such as dislocations and crystal deformation are formed by the enlargement of either phase interface between reduced Mo and MoO_2 phases or contact surface between the crystallites formed in incubation period and MoO_2. The reduced Cu deposits around MoO_2 grain could pin these dislocations and prevent the grain growth.

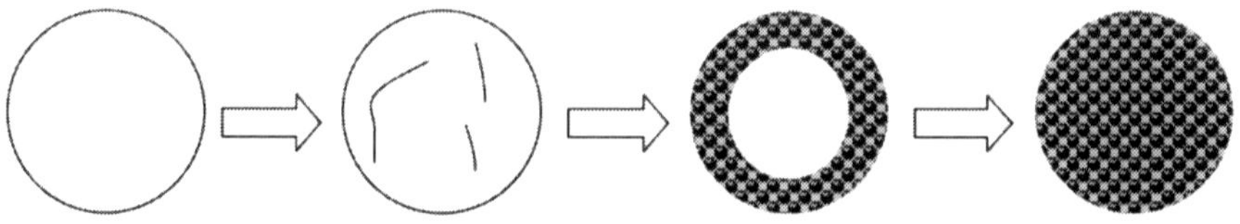

Figure 8. The reduction process of Mo-Cu composite oxides.

4. Conclusion

Mo-Cu composite powders have been successfully produced by the Sol-Gel method followed by two-step hydrogen reduction. Obtained Mo-Cu powders consist of metallic Mo-Cu composite particles about 150nm in size in which the Mo grains are homogeneously dispersed in copper phase. The optimum two-step reduction temperatures are 500°C and 700°C.The bond of Mo=O is weakened due to the bond of Cu-O-Mo, which leads to the increase of the reduction capability of the Mo-Cu composite oxide powders.

Acknowledgement

The authors would like to acknowledge gratefully the financial support of National High-tech R&D Program of China (Grant No. 2006ZZ03Z524 and 2008AA031001) and the National Natural Science Foundation of China (Grant No. 50801001).

References

[1] Johnson JL, German RM 1999 *Inter. J. Powder Metall.* **35** 39
[2] Martinez VD, Aguilar C, Marin J, Ordonez S and Castro F 2007 *Mater. Lett.* **61** 929
[3] Wang GJ, Wang DZ 2007 *Trans. Nonferrous. Met. Soc.China* **17** s803
[4] Zweben C 1998 *J. Miner. Met. Mater. Soc.* **50** 47
[5] Johnson JL, German RM 2001 *Metall. Mater. Trans. A* **32** 605
[6] Johnson JL, Brezovsky JJ and German RM 2005 *Metall. Mater. Trans. A* **36A** 2807
[7] Tikhii GA, Kachalin NI, Belova VP, Nikitin VI 2007 *Met. Sci. Heat. Treat.* **49** 448
[8] Aguilar C, Ordonez S, Marın J, Castro F and Martınez V 2007 *Mat. Sci. Eng. A* **464** 288
[9] Jia CC, He Z, Su XK, Yang ZQ, Zhao J and Xie ZZ 2001 *Rare Metals* **20** 112
[10] Benjamin JS, Volin TE 1974 *Metall. Mater. Trans. B* **5** 1929
[11] Srikanth R, David L 1999 *P/M Sci. Technol.* **1** 9
[12] Song P, Cheng JG, Wan L, Zhao JS, Wang YF and Cai YB 2009 *J. Alloy. Compd.* **476** 226
[13] Kang ZY, Chen WG and Ding BJ 2005 *Rare Metal Mat. Eng.* **34** 990
[14] Liu W, Zhou ML, Wang JS, Chen YX and Li N 2007 *Chin. J. Nonfer. Metal.* **17** 820
[15] Park JY, Levenspiel O 1975 *Chem. Eng. Sc.* **30** 1207

The Study of Optimizing Growth Conditions for Improving Field Emission Property of $W_{18}O_{49}$ Nanorod Arrays

L. F. Chi[1, 2], S. Z. Deng[1], N. S. Xu[1, a)], J. Chen[1], J. C. She[1], X. H. Liang[1]

1 State Key Lab of Optoelectronic Materials and Technologies, Guangdong Province Key Laboratory of Display Material and Technology, Sun Yat-Sen (Zhongshan) University, Guangzhou 510275, People's Republic of China

2 Department of Physics, Shantou University, Shantou, 515063, People's Republic of China

a) The corresponding author's E-mail address: stsxns@mail.sysu.edu.cn

Abstract Large-area tungsten oxide nanorod arrays were synthesized by the catalyst-free tungsten hot-filament thermal vapor method. The diameter, height, interval and alignment of the nanorods were all found to be sensitive to the distance between hot filament and substrates. And the field emission property of the nanorod arrays prepared at various distances was shown to be significantly different. This demonstrates that optimal conditions can be found for preparing $W_{18}O_{49}$ nanorod arrays of better field emission property.

1. Introduction

Lately, one-dimensional (1D) nanostructure materials were found to be the potential candidate as the cold cathode in field-emission devices [1] and vacuum nanoelectronics [2]. Tungsten oxide (WO_x) is an important functional material, its nanowires, nanorods and nanotips can be synthesized by thermal vapor deposition method [3-8], and were shown to have excellent field emission property [3, 4]. But it is still a challenge to control the field emission performance of WO_x nanorod array. It was reported that the field emission of WO_x nanostructures is related to their phase structure, and can be adjusted by changing the vapor source temperature [7]. In this paper, we used tungsten hot-filament array as vapor source for preparation of large-area tungsten oxide nanorod arrays. Then the $W_{18}O_{49}$ nanorods array growth condition, i.e. the gap between hot filament array and substrates, was researched to tune their field emission properties. Since this growth technique can be scaled up, the results reported here may be important to the development of large-area cold cathode field electron sources.

2. Experimental and Discuss

2.1 The synthesis of samples and characterization

The synthesis approach is similar to that described in our previous report elsewhere [8]. Briefly, an array of clean tungsten filaments was installed in the center of a vacuum chamber as the heating vapor source. The clean silicon substrates were placed beneath the tungsten filament array. In a series of experiments, the distance (d) between filament array and substrates was set at 2, 3, 4 and 5 mm respectively, for studying the effect of the d value. Base pressure and reactive pressure were 4.0×10^{-2}

Torr and 0.8 Torr, respectively. A tunable AC current was allowed to pass the filaments and heat them. The high purity argon (99.99%) with 200 standard cubic centimeters per minute (sccm) was introduced to the vacuum chamber during the synthesis process, which lasted for 30 minutes.

Figure 1(a)-(d) show the cross-section views of nanorod arrays synthesized at d = 2, 3, 4 and 5 mm, respectively. The typical height of nanorods prepared at d = 2 mm is about 4.5 μm, which is longer than that of the nanorods grown at d = 3 (~3 μm), 4 (~3 μm) and 5 mm (~1 μm). The typical diameters of the nanorods synthesized at different d values are diverse. The nanorods prepared at d = 2 mm is the largest, with a diameter of about 150 nm. The typical diameter of the nanorods changes with increasing d, with diameter of about 100, 80 and 100 nm, corresponding to the d = 3, 4 and 5 mm, respectively. The nanorod arrays prepared at d = 4 mm have the largest aspect ratio value (~38). In addition, the nanorod arrays synthesized at d = 3 mm is seen to have best alignment, while the alignment of the nanorod arrays grown at d = 5 mm is the worst.

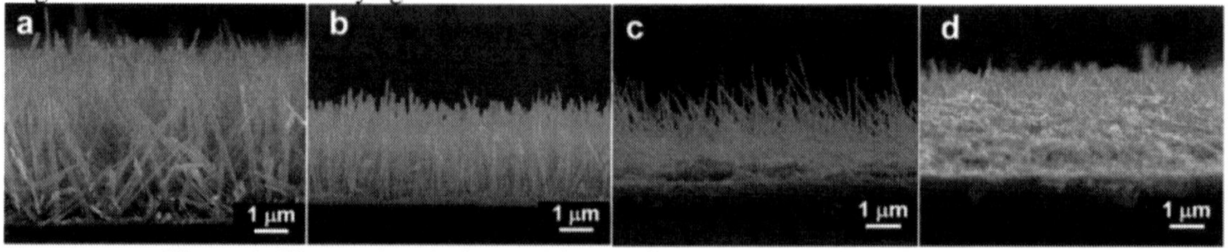

Figure 1. The sectional views of four typical tungsten oxide nanorods array.

Figure 2(a)-(c) show the typical high resolution TEM (HRTEM) image of a single nanorod, the corresponding selected area electron diffraction (SAED) pattern and the EDS result of the nanorods, respectively. Figure 2 (c) indicates that the nanorods consist of W and O elements, while the C and Cu signals were from the TEM copper grid. From Figure 2(a), the crystallographic planes with interplanar spacing of 0.38 and 0.46 nm can be observed, which correspond to the (010) and (103) planes of monoclinic $W_{18}O_{49}$ (JCPDS 05-0392). The SAED pattern shown in Figure 2(b), indicates that the nanorod is a single crystal with monoclinic structure and that the growth direction is [010]. For there was not any catalyst used in the whole synthesis process, the purity of $W_{18}O_{49}$ nanorods is high. The growth mechanism of $W_{18}O_{49}$ nanorods might be attributed to the vapor-solid (V-S) process [8].

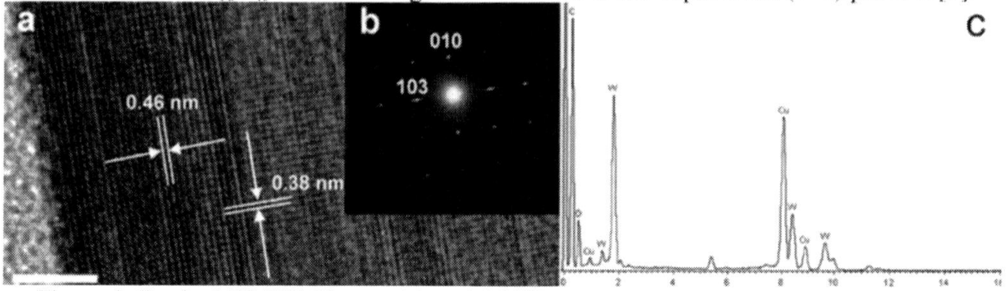

Figure 2. Typical HRTEM image (a) and SAED pattern (b) of tungsten oxide nanorods, scale bar is 5 nm. (c) is the EDS result of the nanorod.

2.2 The field emission property and discuss

The field emission characteristics of the $W_{18}O_{49}$ nanorod arrays were measured in an ultrahigh vacuum system at room temperature. The transparent-anode and a Mo anode were applied to obtain the emission site distributions and measure turn-on / threshold fields, respectively. The vacuum gap between the anode and the surface of sample was set to 200 μm.

The characteristics of emission current density versus applied electric field (J-E curves) of the $W_{18}O_{49}$ nanorod arrays are shown in Figure 3(a). The sample of d = 3 mm has the lowest turn-on field (E_{to}) and threshold field (E_{th}), which are defined as the macroscopic fields required to produce current density of 10 μA/cm^2 and 10 mA/cm^2 respectively. The typical E_{to} and E_{th} values were 2.5 and 5.5 MV/m respectively for the $W_{18}O_{49}$ nanorod array prepared at d = 3 mm, which are compatible to the best of those of tungsten oxide nanotips [9] and nanowires [10]. For the sample synthesized at d = 5 mm, its typical E_{to} (8.3 MV/m) and E_{th} (12.2 MV/m) were the highest among the four different arrays. The E_{to} of nanorod array prepared at d = 2 and 4 mm were 3.8 and 8.3 MV/m, respectively, and E_{th} of them were 6.6 and 8.9 MV/m in sequence. The difference in the field emission characteristics among the four samples is significant. The results indicated that the nanorod array should have a combination of both good alignment and medium interval (see figure 1) to give rise to better field emission property. This is similar to the findings from the early studies of the field emission property of the carbon nanotubes film [9, 10].

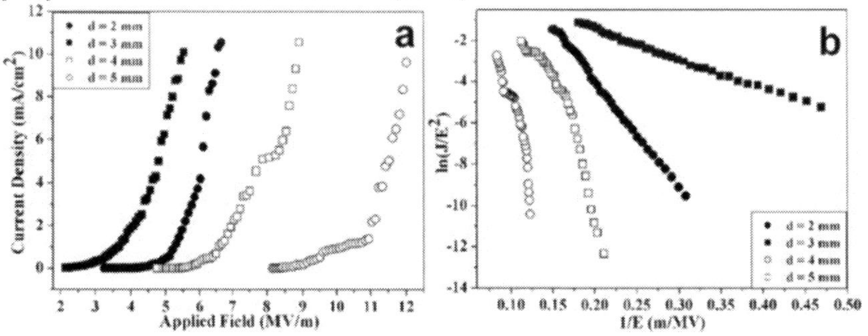

Figure 3. (a) shows the J-E curves of the four type tungsten oxide nanorod arrays, and (b) the corresponding F-N plots. The symbols ●, ■, □ and ○ denote the field emission property of tungsten oxide nanorods array prepared at d = 2, 3, 4 and 5 mm, respectively.

Figure 3(b) shows the corresponding Flower-Nordheim (F-N) plots of four typical nanorod arrays of different d values, which exhibited distinctly different features from each other. For the nanorods of the d values of 2 and 3 mm, the F-N curves exhibits approximately linearity; however, for those nanorods of the d values of 4 and 5 mm, their F-N curves showed a significantly non-linear behavior, and can be approximately divided into several nearly-straight lines. This was similar to the results reported in the study of non-metal nanostructures field emission property [11-13].

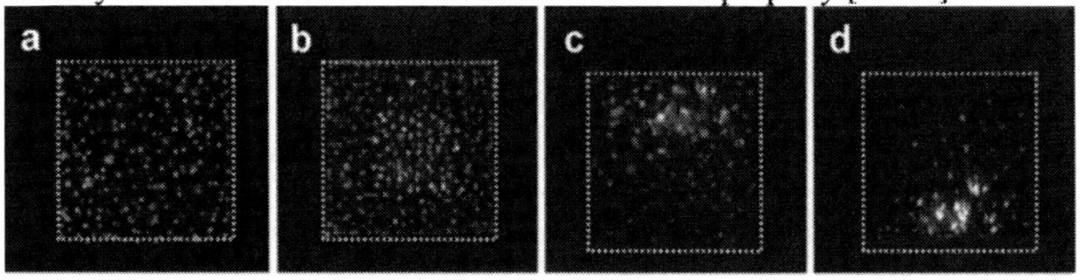

Figure 4. (a), (b), (c) and (d) show the emission site distributions of nanorods array prepared at d = 2, 3, 4 and 5 mm, respectively. The white dashed rims enclose the areas of the samples.

The variation of another important field emission property, i.e. the field emission site distribution is shown in Figure 4. The emission site distributions were recorded at the emission current of 1 mA/cm^2. It is noticed that the field emission site distributions of the nanorod arrays prepared at d = 2 and 3 mm were quiet uniform, as compared to those samples prepared at d = 4 and 5 mm. The effective emission site density of four typical nanorod arrays prepared at d = 2, 3, 4 and 5 mm was calculated to be 3.0×10^3 cm^{-2}, 4.2×10^3 cm^{-2}, 6.5×10^2 cm^{-2} and 4.1×10^2 cm^{-2}, respectively. So the emission uniformity of sample of d = 3 mm is the best, this may be attributed to the good alignment and medium interval of the nanorods array.

From the above experiment results, one can see that the $W_{18}O_{49}$ nanorod array synthesized at d = 3 mm had the best field emission properties. The explanation to why its corresponding turn-on and threshold fields are lower may be given: a typical single nanorod of the array of d = 3 mm has larger aspect ratio, i.e. 30. However, it is smaller than the value of nanorod prepared at d = 4 mm (~38). So the relative high aspect ratio is not a key factor for the observed low turn-on and threshold fields. The values of field enhancement factor (β-value) were calculated based on the low field section of the F-N curves with the assumption of the value of work function of 5.27 eV. The four β-values of the samples prepared at four d values (d =2, 3, 4 and 5 mm) are calculated to be 1620, 5700, 560 and 270 respectively, which are a lot larger than their corresponding aspect ratios, i.e. 30, 30, 38 and 10. Thus, the field emission characteristics of $W_{18}O_{49}$ nanorod arrays are not determined only by the aspect ratio of nanorods, and the other factors should be also considered. We believe that the large low field β-values are due to the fact that the semiconducting nature of $W_{18}O_{49}$ nanorods is not properly included into our calculation. The F-N theory assumes that the nanorods are metallic; however, $W_{18}O_{49}$ is an n-type semiconductor that may have poor conductivity. So the following two facts should be considered: field penetration into the tip end of the nanorod and the band bending at large emission current. If these happen, the surface barrier height may not be much smaller. This may lead to low field β-values, and it needs a lot more theoretical study.

2.3 Conclusion

The tungsten oxide nanorods arrays were synthesized by the catalyst-free hot-filament thermal vapor approach. The aspect ratio, interval and alignment of nanorods can be tuned by changing the distance between hot-filament and substrate set for a growth experiment. The field emission property of tungsten oxide nanorod arrays is strongly dependent on the aspect ratio, the interval of nanorods and the alignment of nanorods, so by optimizing the distance between hot-filament and substrate and nanorod arrays with better field emission property can be obtained.

Acknowledgements

N S Xu, S Z Deng and Jun Chen gratefully acknowledge the financial support of the project from the National Natural Science Foundation of China (Grant No. U0634002, 50725206, 60571035, 50672135, 50021202), Science and Technology Ministry of China (Grant No. 2003CB314701, 2007CB935501), Education Ministry of China, the Science and Technology Department of Guangdong Province, the Education Department of Guangdong Province, and the Science and Technology Department of Guangzhou City.

References

[1] Xu N S and Huq S E 2005 *Mat. Sci. Eng. R.* **R48** 47
[2] Huang X M H, Zorman C A, Mehregany M and Roukes M L 2003 *Nature* **421** 496
[3] Zhou J, Gong L, Deng S Z, Chen Jun, She J. C, Xu N S, Yang R S and Wang Z L 2005 *Appl. Phys. Lett.* **87** 223108.
[4] Li Y B, Bando Y and Golberg D 2003 *Adv. Mater. (Weinheim, Ger.)* **15** 1294
[5] Liu Z W, Bando Y and Tang C C 2003 *Chem. Phys. Lett.* **372** 179
[6] Mahan A H, Parilla P A, Jones K M and Dillon A C 2005 *Chem. Phys. Lett.* **413** 88
[7] Liu J G, Zhang Z J, Zhao Y, Su X, Liu S, Wang E G and Wang E G 2005 *Small* **1** 310
[8] Chi L F, Xu N S, Deng S Z, Chen Jun and She J C 2006 *Nanotechnology* **17** 5590
[9] Nilsson L, Groening O, Emmenegger C, Kuettel O, Schaller E, Schlapbach L, Kind H, Bonard J-M, 2000 *Appl. Phys. Lett.* **76** 2071
[10] Suh J S, Jeong K S, Lee J S and Han I T 2002 *Appl. Phys. Lett.* **80** 2392
[11] Chen Jun, Deng S Z, Xu N S, Wang S H, Wen X G, Yang S H, Yang C L, Wang J N and Ge W K 2002 *Appl. Phys. Lett.* **80** 3620
[12] Chen Y, Deng S Z, Xu N S, Chen J, Ma X C and Wang E G. 2002 *Mat. Sci. Eng.* **A327** 16
[13] Li S Q, Liang Y X and Wang T H 2006 *Appl. Phys. Lett.* **88** 053107.

Journal of Physics: Conference Series **188** (2009) 012022 doi:10.1088/1742-6596/188/1/012022

Effect of nano-segregation phases on electrochemical property of high active Al alloy anode

S Q Liang, Y Zhang, Z W Mao, Y Tang and D K Guan
School of Material Science and Engineering, Central South University, Changsha 410083, China

E-mail: cw04zhangyong@126.com (Y Zhang)

Abstract. The effect of nano-segregation phases formed during rolling process on the electrochemical property of Al-Mg-Sn-Bi-Ga-In alloy anode in alkaline solution（80℃, Na_2SnO_3 + 5mol/L NaOH）was analyzed according to the chronopotentiometry (E-T curves), hydrogen collection tests and modern microstructure analysis. The results show that when controlling the rolling temperature and pass deformation at 370℃ and 40% respectively, the Al alloy anode undergoes the dynamic recrystallization, which benefits to the uniform distribution of nano-segregation phases and improvement of electrochemical property of Al alloy anode. The optimum Al alloy anode has the more negative electrode potential of about -1.48V（vs.Hg/HgO）and the lower hydrogen evolution rate of 0.1889mL/ (min·cm^2).

Key words. Al alloy anode; nano-segregation phases; electrochemical property; rolling process

1. Introduction

The aluminum / air (Al/O_2) fuel cells or aluminum / silver oxide (Al/AgO) sea-water cells, with alkaline sea-water as electrolyte, oxygen in the air or silver oxide as cathode, and the aluminum alloy with high electrochemical equivalent and negative electrode potential as anode, have hold the leading position among the cells with excellent performance due to their high specific energy, high specific capacity, free contamination and no noise [1-3]. In addition, the other advantages of aluminum alloy, such as rich in storage, easy to manufacture and so on, will result in the low cost of this kind of cells [4]. Since 1970s, the research on active aluminum alloy anode has been paid great attention by the energy, transportation, telecom and defense departments in many countries [5-7]. However, during the working process, pure aluminum reacts intensively with electrolyte and produces a large amount of hydrogen because of the solution of oxide film on aluminum surface. Especially, when the cell works under large current density, the stable working potential will become fairly low and the anode polarization is very serious. In recent years, these situations had been greatly improved by optimizing alloying of the Al anode [8]. It has been indicated that different heat treatment technics have significant effect on microstructure and properties of Al anode [9-11], which will be mainly attributed to the size, shape and distribution of segregation phases formed during heat treating. Especially when the size of segregation phase reduces into the nanometer range, the Al anode exhibits interesting electrochemical property compared to other anode materials. However, the effect of nano-segregation phases formed during rolling process on electrochemical properties of Al anode has been little investigated. This research mainly focuses on investigating the effect of nano-segregation phases on the microstructure and properties of a high active Al alloy anode. The main purpose is to find an optimum rolling technology.

2.Experimental procedure

2.1. Sample preparation

Conventional melting and casting technique had been applied to produce Al-Mg-Sn-Bi-Ga-In alloy anode. The surfaces of the alloy ingots were scraped by machining. The alloy ingots were then heat treated for 5 h at 510℃ before quenching them with cold water, which was aimed to reduce the crystal defects in alloy and maximize the solid solubility of alloying elements. After that, the ingots were hot-rolled at different pass deformation and rolling temperature to obtain plates with 0.5mm thickness.

2.2. Microstructure analysis

The microstructure of the Al alloy was studied by a Tecnai$G^2$20 TEM after being polished by machine and then submitted to a conventional twin jet electropolishing in 30%HNO_3+70%CH_3OH (vt%) solution. The surface topography and alloy elements distribution of Al anode were examined by Sirion-200 SEM after the surface was burnished and polished, eroded in 0.5%HF solution for 30 s and then washed with distilled water before it was dried. The composition of the segregation phases was also analyzed by means of EDAX.

2.3. Performance measurements

Electrochemical measurements were operated in a classical three-electrode glass cell. The working electrode was made of Al alloy chip with 1.0cm^2 in area. The auxiliary electrode and reference electrode was graphite flake and mercury/mercuric oxide (Hg/HgO) electrode respectively. The working electrolyte was 5mol/L NaOH solution with Na_2SnO_3 as inhibitor. The chronopotentiometry with current density of 700mA/cm^2 was performed at CHI660C/CHI680 electrochemical working station. All the experiments were conducted at a constant temperature of 80℃. The hydrogen corrosion rates of the Al anode in alkaline solution （80℃, Na_2SnO_3 + 5mol/L NaOH） were determined by hydrogen collection method, which could be used to estimate the erosion degree of the Al anode under different conditions.

3.Results and discussions

3.1. Effect of pass deformation on the nano-segregation phases

The TEM photographs of Al alloy anode with different pass deformation under 370℃ are shown in figure 1.

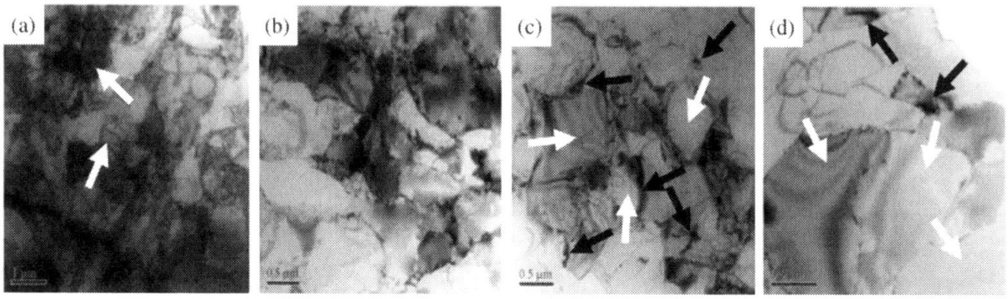

Figure 1. TEM photographs of Al alloy anode under different pass deformation: (a)20%, (b)30%, (c) 40% and (d) 50%.

With the increase of pass deformation, the microstructure of Al alloy anode undergoes a transformation from the typical dislocation tangle and cellular structure to the dynamic recrystallized structure, finally to the secondary dynamic recrystallized structure. When the pass deformation is 20%, the dislocation tangles and cellular structures , as the arrows show in figure 1 (a), are the main structure in Al alloy. When the pass deformation is 40%, a lot of fine recrystallized grains with clear and straight boundary, as the white arrows show in figure 1 (c), can be observed in alloys. At the same time, lots of nano-segregation phases with the size of 100-200nm, as the black arrows show, can also be found among the grain boundary. When the pass deformation is up to 50%, some abnormal big grains have been produced due to the second recrystallization, which promotes the gather of nano-segregation phases, as the black arrows show in figure 1 (d).

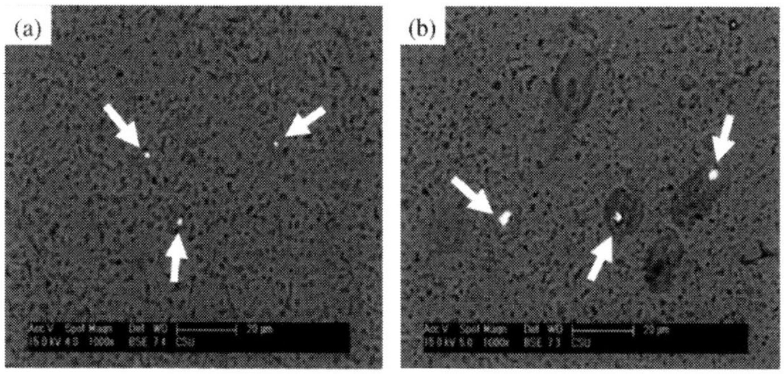

Figure 2. SEM photographs of Al alloy anode under different pass deformation: (a)40% and (b)50%.

Figure 2 is the SEM photographs of above samples under different pass deformation. It can be found that there are lots of nano-segregation phases, as the arrows show in figure 2 (a), which distributes uniformly due to the dynamic recrystallized under pass deformation of 40%. And the corrosion pits on Al surface also seem to be finer and uniform. While under the pass deformation of 50%, lots of large size segregation phases, as the arrows show in figure 2 (b), begin to form due to the gather of nano-segregation phases. The corrosion degree of Al alloy becomes much more severe correspondingly. The EDAX microanalysis on the nano-segregation phases according to figure 3 and table 1 shows that the active elements Sn, In are the main components, which demonstrates that the main body of the nano-segregation phases is Sn-rich phase. As active sites, the distribution of nano-segregation phases plays an important role in improving the properties of Al alloy anode.

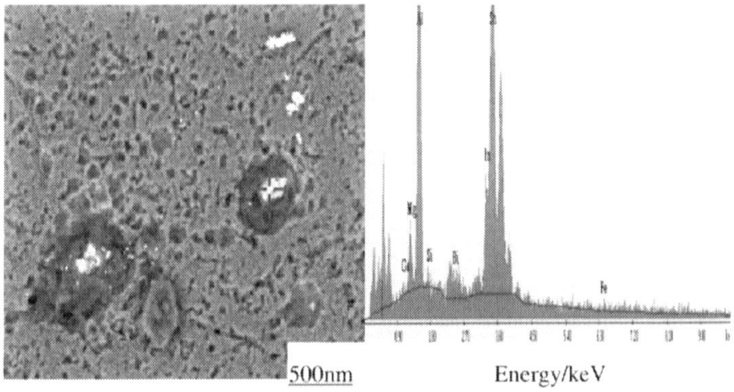

Figure 3. EDAX analysis of nano-segregation phases of Al alloy anode.

Table 1. Quantity results of nano-segregation phases of Al alloy anode.

element	Sn	In	Bi	Ga	Fe	Si	Mg	Al
mass fraction, %	40.08	11.19	3.36	0.71	1.08	0.50	2.54	40.55
mole fraction, %	15.99	4.62	0.76	0.48	0.92	0.85	5.03	71.34

Hence, the pass deformation, which could result in dynamic recrystallization [12,13], is a key factor for the formation and uniform distribution of nano-segregation phases in Al anode. A lot of new fine grains could be produced by the dynamic recrystallized process, which will reduce the average size of grains and affect the precipitation and distribution of nano-segregation phases in Al alloy. The optimum pass deformation which causes dynamic recrystallization is 40% in our study.

3.2. Effect of nano-segregation phases on the properties of Al anode

Figure 4 is the constant-current discharge curves and hydrogen evolution curves of Al alloy anodes prepared under different pass deformations in 80℃, Na_2SnO_3 + 5mol/L NaOH electrolyte with current density of 700mA/cm^2. The correlative parameters are listed in table 2. It can be found from figure 4(a) that under the pass deformation of 40% which benefits to the uniform distribution of nano-segregation phases, the Al alloy anodes own the more negative electrode potential of about -1.48V（vs.Hg/HgO）. But when the pass deformation increases to 50%, the electrochemical activity of Al anode becomes much lower than other anodes, and its discharge curve also becomes much unstable. The electrochemical activity of Al anode is mainly controlled by the distribution of nano-segregation phases in Al matrix. For the anodes prepared at the pass deformation of 40%, the nano-segregation phases containing active elements Sn, In, is of the much smaller size, and distribute homogeneously in Al matrix, which leads to the increase of the quantity of active points in Al matrix and improves electrochemical activity of Al anode. With the increase of the pass deformation, the quantity of active point decreases with the gather of nano-segregation phase, the electrochemical activity of Al anode drops accordingly.

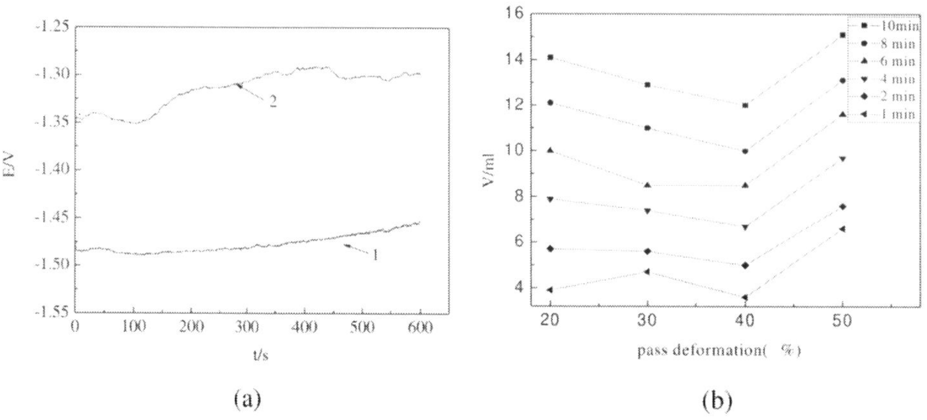

Figure 4. Constant-current discharge curves (a) and hydrogen evolution curves (b) of Al anode under different pass deformation: (1) 40% and (2) 50%.

Table 2. H_2 evolution parameters of Al alloy anode (6.35cm^2) under different pass deformation.

Rolling Temperature/ (℃)	Pass deformation/ (%)	Working time/ (min)	hydrogen volume / (mL)	corrosion rate / (mL·min^{-1}·cm^{-2})
370	20	10	14.1	0.2220
	30		12.9	0.2031
	40		12.0	0.1889
	50		15.1	0.2378

From figure 4 (b) and table 2, it could be found that the Al anode achieves its lowest hydrogen corrosion rate of 0.1889mL/cm^2·min due to the uniform distribution of nano-segregation phases under pass deformation of 40%. That means, the best anti-corrosion property has been obtained. While the corrosion degree for other samples are still severe. The reason of these changes is that under pass deformation of 40%, the uniform distribution of nano-segregation phases is improved, and a more homogeneous microstructure is developed. All of these are helpful for the improvement of the anti-corrosion property of Al anode.

4. Conclusions

The pass deformation had great effects on the formation and distribution of nano-segregation phases in Al anode. With the increase of pass deformation, the microstructure of the Al alloy anode underwent a transformation from the typical dislocation tangle and cellular structure to the dynamic recrystallized structure, finally to the secondary dynamic recrystallized structure. The optimum size and distribution of nano-segregation phases could be achieved by the dynamic recrystallization under pass deformation of 40%. The uniform distribution of nano-segregation phases promoted the Al alloy anode to achieve well electrochemical activation as well as the anti-corrosion property. The optimum aluminum alloy anode had the more negative electrode potential with the value of about -1.48V（vs.Hg/HgO）and the lower hydrogen evolution rate of 0.1889mL/ (min·cm^2).

Reference

[1] Reboul M C, Gimenez P H and Rameau J J 1984 *Corrosion* **40** 366–371
[2] Wang Z W, Li Y X, Li Q F, Gao B L and Qiu Z X 2002 *Nonferrous Metals* **54** 19–22
[3] Tang Y G, Huang B Y, Lu L B and Liu D R 2004 *The Special Subject of Physics and New Energy Materials* **33** 85–89
[4] Ma Z Q, Li W X, Tan D Q and Xiao Y D 2001 *Ordnance Material Science and Engineering* **24** 48–50
[5] Tuck C D S, Hunter J A and Scamans G M 1987 *J Electrochem Soc.* **123** 2970–2981
[6] Ei S H A, Abd E W F M and Zein E A S 2001 *Corrosion Sicience* **43** 643–654
[7] Venugopal A, Veluchamy P, Selvam P, Minoura H and Raja V S 1997 *Corrosion* **53** 808–812
[8] Li Q F, Hjuler H A, Berg R W and Bjerrum N J 1991 *J Electrochem Soc.***138** 763–766
[9] Gurrappa I 1997 *Corrosion Prevention and Control* **44** 69–80
[10] Lin J C and Shih H C 1987 *J Electrochem Soc.* **134** 817–823
[11] Qi H F, Liang G C, Li G L, Liang J S and Meng J P 2005 *Materials Engineering* **10** 27–30
[12] Kassner M E, Myshlyaev M M and McQueen H J 1989 *Material Science and Engineering* **A108** 45–61
[13] Zhu Q, McQueen H J and Blum W 1998 *High Temperature Materials and Processes* **17** 289–297

Journal of Physics: Conference Series **188** (2009) 012023 doi:10.1088/1742-6596/188/1/012023

Nano-structure fabrication for HgCdTe ultra-fast infrared sensors

Shao-Wei Wang[1], Jianrong Yang[2], Yifang Chen[3]*, Honglou Zhen[1], Xiaoshuang Chen[1], Wei Lu[1]

1 National Laboratory for Infrared Physics, Shanghai Institute of Technical Physics, Chinese Academy of Sciences, Shanghai 200083, China

2 Shanghai Institute of Technical Physics, Chinese Academy of Sciences, Shanghai 200083, China

3 Micro and Nanotechnology Centre, Rutherford Appleton Laboratory, Chilton, Didcot, OX11 0QX, UK

*E-mail: yifangchen@stfc.ac.uk

Abstract. High-speed infrared sensors are necessary for broad applications in optic fibre communications with high data capacity and high security and distance sensing by laser. In this paper, an ultra-fast infrared sensor with 100 nm ballistic channels has been proposed and successfully fabricated by electron beam lithography. The area of photosensitive surface is as large as 100 μm × 100 μm to ensure its high sensitivity. The principle is based on the ballistic transport of electrons in the nano-scale channels. The frequency is expected to be in the range of 10 GHz, one or two orders higher than conventional photovoltaic sensors.

1. Introduction

Infrared sensors are widely applied for domestic applications for environmental monitoring, weather forecast, safety and health protection, agriculture production and optic-fibre communications. In all these applications, ultra fast sensors beyond hundreds of megahertz are essential. For example, the study on chemical changes needs very high speed, especially when infrared output is very low. Conventional high-speed infrared sensors are made using photovoltaic II-V compounds with p-i-n structure junction for low capacity. The typical operational frequency is about a few hundreds MHz. For such kind of sensors, there is a trade off between the operation speed and the sensing area. One can improve the frequency by reducing the p-i-n junction area, while with the cost of lowering its sensitivity. This trade-off limits the operation speed with high sensitivity.

This paper proposed a novel optoelectronic device, using short channel as the sensing area in a photoconductive configuration. It is well known that there is a significant overshoot in the electron drift velocity under an electric field in most of semiconductors[1,2]. Especially in III-V or II-V compounds, such kind of overshoot become most pronounced due to the existence of heavy mass side valley. When the channel length is within ballistic regime (the channel length is shorter than the mean free path of the carriers), the photoconductive electrons will drift at the transient velocity two to four times higher than stationery one before the electric field is applied. Therefore, using nanofabrication technique to fabricate short channel photoconductive sensors in the ballistic regime, the working frequency of the device is expected to increase by a factor of 2 to 4. In this work, we report our technical development for the fabrication of nanosize photoconductive sensors in mercury cadmium and telluride (MCT) compounds grown by liquid phase epitaxy (LPE) method. The characterisation of the fabricated devices will be present elsewhere.

2. Design of the device layout

Photoconductive sensors work under a constant bias between source and drain, which creates an electric field to accelerate electrons drifting in the conducting channels. In the electric field, electrons gain its transient velocity and momentum over their stationery values due to the overshoot effect. The distance between the source and the drain should be so short that the time for the electrons travelling in the channel is shorter than the momentum relaxation time when electrons transit to the heavy-mass valleys. In this situation, the photoconductive electrons will always drift in the active region at much higher velocity and the devices shall operate at high frequency. Ideally, the device design should try to achieve the maximum speed of the electrons. This becomes possible when the electrons are travelling in the ballistic regime. In the ballistic regime, there are no inelastic scattering such as between electron-electron or electron-phonons. As the result, the overshoot effect is further enhanced.

For the n-type $Hg_{1-x}Cd_xTe$ layer material with the alloy composition of x=0.16, the electron mobility typically ranges from 2.0×10^4 to 2×10^5 cm/V corresponding to the temperature from 300K down to 77K [3]. The mean free path is from 300 nm (300 K) to 3 μm (77 K). Therefore, the conducting channel with the length of 100 nm from the source to the drain was designed. Fig.1 shows the schematic diagram of the ultra-fast infrared sensor we proposed. Fig.1(a) depicts the overall device layout on the MCT thin film grown by liquid phase epitaxy (LPE). The central part is comb-shape electrodes as shown in Fig. 1(b). Both the line width and gaps are 100 nm. The gaps are transparent to light, which act as both the sensitive regions for receiving lights and conducting channels. The carriers will transport without scattering when the channels between two electrodes are much smaller than the mean free path. The frequency is estimated to be elevated by the factor of 2 to 4. To achieve such short channel, nanofabrication by electron beam lithography has been carried out and this paper concentrates on reporting our recent progress in fabricating nanoscale ballistic photoconductive sensors in n-type HgCdTe thin film.

(a) (b)

Figure 1. (a) The schematic diagram of the ultra-fast infrared sensor (not to scale). The grey central part is the 100 × 100 μm photosensitive area and two sides are the electrodes. (b) is the magnified image of the central part from (a). Both the line width and gap are 100 nm. The gaps form the ballistic channels.

3. Nanofabrication of ballistic infrared sensors

The photosensitive area of the sensor is designed as 100 × 100 μm with 100 nm lines/spaces metallic graings. A mix and match strategy in lithogrpahy is applied in the fabrication of sensors. For the critical part, i.e. the comb shaped electrodes with nanometre scale was replicated by electron beam lithography. Its flow chart for the fabrication procedure is shown in Fig.2. For the large electrodes, photolithography is used.

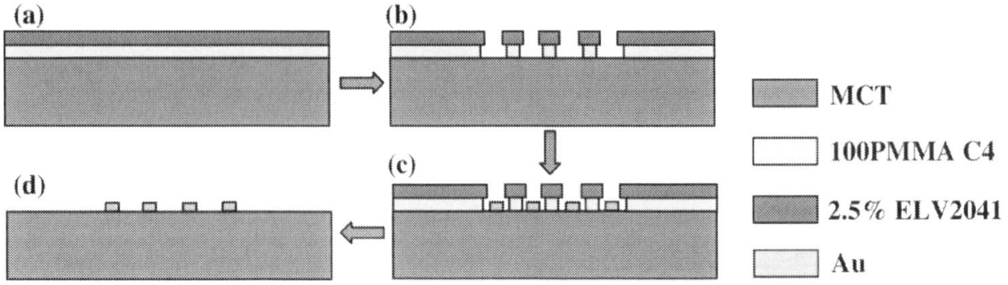

Figure 2. Flow chart for the fabrication procedure of nano-scale ballistic channels.

In the electron beam lithography for the central part with 100 nm ballistic channels, bi-layer resists stack has been used to ease the lift-off process. The resists of 100PMMA C4 (100k MW) (as the bottom layer) and PMMA ELV2041 (350k MW) are chosen for the sensitivity difference to electron beam irradiation to create a clear undercut [4,5], as shown in Fig.2. In Fig. 3(a), contrast curves of the

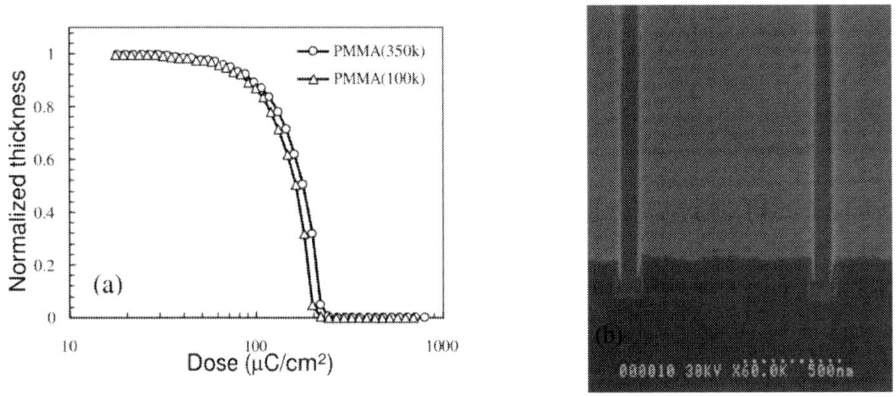

Figure 3. The contrast curves of the two resists used in this work and the resist profile in the bilayer after EBL. In (a), the PMMA (100k) is symbolized by the open triangles and PMMA(350k) by the open circles. In (b), the scanning electron microscope of the resist profile. As can be seen that there is a clear difference in the sensitivity which was used for the form of undercut profile.

two resists show the difference in sensitivities, that is about 20 $\mu C/cm^2$ from 218 $\mu C/cm^2$ (PMMA 350K) and 198 $\mu C/cm^2$ (PMMA 100K). Resist coating was carried out by spinning at the speed of 3000 rpm for PMMA (100K) and 2000 rpm for PMMA (350K), respectively, followed by soft bake in oven for one hour at 180 ^oC. Electron beam exposures of the resists were carried out by the high-resolution vector beam writer, VB6 delivered by Vistec Lithography Ltd at 100 keV for the electron acceleration and 500 pA for the beam current. Conventional development in MIBK:IPA (1:3) for one minute was undertaken. Such a difference in sensitivity creates a 20 nm undercut between the top and bottom layer of PMMA resists as shown in Fig.3(b), which is exactly desired because larger undercut is not applicable in dense patterns like gratings. A 30 nm Au was deposited onto the pattern with a 15 nm thin layer of Cr on the bottom for a good adhesion (Fig.2 (c)). The deposition processes were carried out by SVS V2000 E-beam Evaporator. The central 100 nm ballistic channels were formed after lifting off in acetone. After the nanofabrication of the central area of nanometre scale conducting channels, photolithography was subsequently carried out to form the big leads and bonding pads for photo-electronic characterisation, using negative tone resist, UVN-30. Metalisation was carried out by depositing 15 nm Cr and 100 nm Au, ended by a lift off process in acetone.

4. Results and discussion

In this work, various geometric designs for the device configuration were made to optimise the optical electronic device structure. Figure 4 presents the SEM photos for one of the patterned devices in PMMA resist layer (a) and the close-up view of the line width (Fig.4 (b)). The result shows clearly 100 nm line-width in resist. Figure 5 shows the device after metalisation and lift-off. From Fig.5(b), it is seen that all the metal lines are separated without connection. The width of ballistic channels is ~100 nm agreeing with the designed result very well.

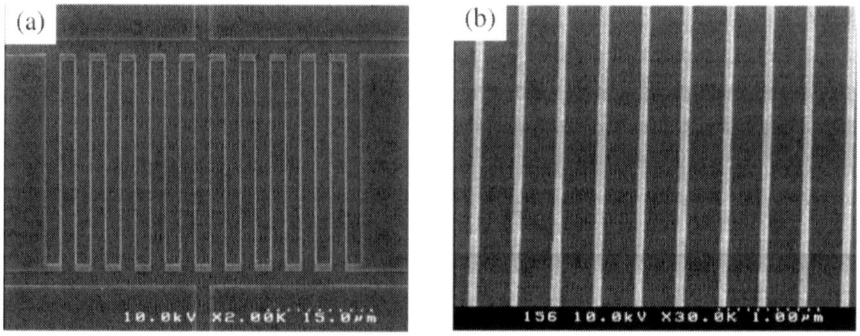

Figure 4. SEM image of the photosensitive area of the ultra-fast infrared sensor in resist (a) and its magnified image for a close-up view (b). As can be seen, resist lines with the width shorter than 100 nm was produced.

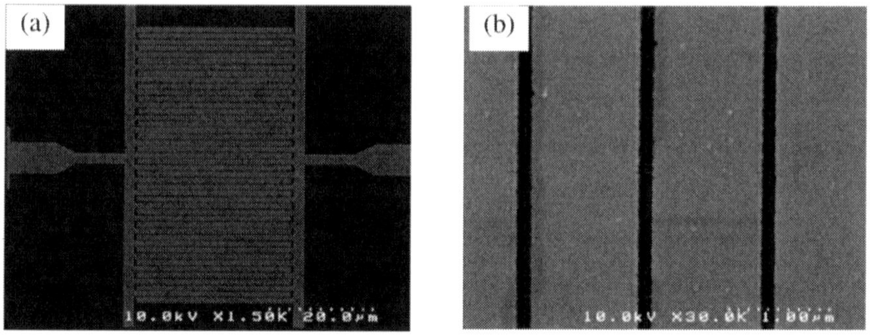

Figure 5. SEM image of the photosensitive area of the ultra-fast infrared sensor in metallised film (a) and its magnified image for a close-up view (b). As can be seen, 100 nm gaps in the metal layer are the sensitive area for receiving and conducting photoconductive electrons.

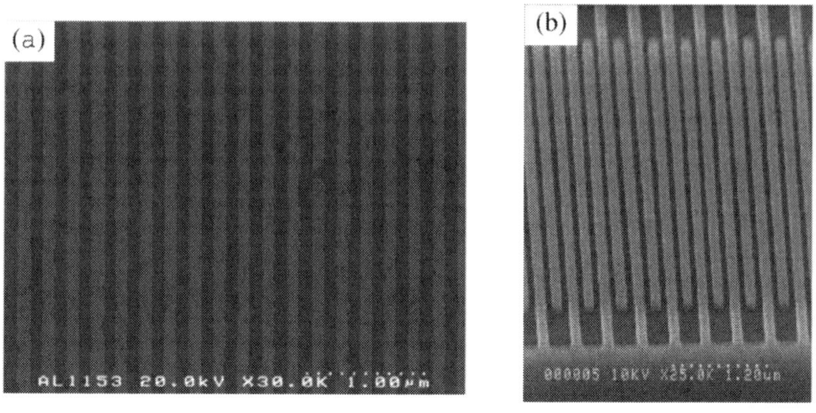

Figure 6. The fabricated ballistic sensor with 100 nm channels (a) and 50 nm channels (b).

However, the weakness of such a geometry is that the active area is too small, comparing with the large electrical leads. We have further improved the design by greatly reducing the metallic electrodes. Fig.6 shows the ballistic sensors with inter-digitated electrodes of equal lines/spaces electrodes. The electrode has the same line-width as the gap (conducting channel). Fig.6(a) is the top view of the fabricated electrodes with very clear gaps of 100 nm gaps. Fig.6(b) is the tilted view of a ballistic sensor with 50 nm gap. The study on the device properties will be reported in other place.

5. Summary

An ultra-fast infrared sensor structure has been proposed and successfully fabricated by a mix and match process using electron beam lithography and photolithography. The photosensitive area is composed of 100 nm ballistic channels with the over all area of 100 μm × 100 μm. Various designs of the device layout have been successfully tried. This ballistic sensor has the prospect of higher operation frequency than conventional ones by a factor of 2 to 4, owing to the ballistic transport of photoconductive electrons in the nano-scale channels. It opens up a new way for achieving ultra-fast infrared sensors.

Acknowledgments

This work was partly sponsored by the National Natural Science Foundation of China (10874196, 60508018), the Shanghai Rising-Star Program (08QH14025, 05QMX1459), and Shanghai Science and Technology Foundation (08DZ2201000).

References

[1] Ali Javey, Pengfei Qi, Qian Wang, and Hongjie Dai 2004 *PNAS* **14** 13408.
[2] Kees M. Schep, Paul J. Kelly, and Gerrit E. W. Bauer 1998 *PHYSICAL REVIEW B* **57** 8907.
[3] S. Sone, N. Oda, T. Sasaki and M. Kawano 1992 *Journal of Crystal Growth* **117** 218.
[4] Yifang Chen, Kaiwu Peng, Zheng Cui 2004 *Microelectronic Engineering* **73–74** 662.
[5] Yifang Chen, Kaiwu Peng, Zheng Cui 2004 *Microelectronic Engineering* **73–74** 278.

Thermoelectric Effect on Electroosmotic Flow in Microchannel

Zheng Xu, Xiao-yu Yu, Li-qun Du, Lun-lei Yang, Chong Liu, Li-ding Wang

Key Laboratory for Micro/Nano Technology and System of Liaoning Province, Dalian University of Technology, 116024 Dalian, Liaoning, China

xuzheng@dlut.edu.cn

Abstract. The present work reports the thermoelectric effect on electroosmotic flow (EOF) in microchannel. The thermoelectric (TE) effect on EOF was analyzed and the switch process of fluids controlled by TE unit was numerically simulated. The mathematical model was established that is composed with electrokinetics fluid, liquid-solid phase transition, and heat transfer. The microfluidic chip with the integrated TE unit was fabricated that includes the microchannel layer, heat conducting layer and refrigerating layer. As the transition from the liquid phases to the solid phase can significantly decrease electrical conductivity of electrolyte, the electrical current is available to make sure the status of phase transition. From the results, the controlling integrated TE unit is effective to close/open the EOF in microchannel. The freezing phenomena were described by results from numerical simulation and experiment.

1. Introduction

The development of lab-on-a-chip requires efficient fluids controlling for preventing contamination, adjusting flow velocity and direction, tunable sample supplying. The tunable flux and open-close switch of fluid are often required. EOF has been the most useful method to drive the fluid, due to the simplicity and easy-control advantages. But there are few reports about microvalves for EOF.

Various fluid switching methodologies have been developed for microfluidic devices. An interesting microvalve is the ice valve or phase changing (PC) valve. The PC valve is opened and closed by thermally-actuated phase changing (freezing or melting) of the local working fluid (ice plug). The thermal actuation is commonly realized by using the Peltier effect. Comparing to other valves, it is relatively noninvasive, reusable, and leakage-free. The principle of PC valve was firstly reported by Bevan etc (1995) [1]. Later Gui and Liu (2004) [2] tested cooling & heating the liquid in a cylindrical tube *via* a TE unit. More recently Welle (2005) [3] integrated some mini TE units with a more complicated microfluidic chip.

The present work mainly deals with the thermoelectric effect on EOF in microchannel. In the following sections, the developed methodology is described, including the mathematical formulation, the numerical simulation, the prototype fabrication, and the evaluation of results.

2. Mathematical formulation

Consider an ideal 2D rectangular microchannel in which is filled with incompressible Newtonian electrolyte [4] [5].

2.1. Electrical field
In the bulk liquid, thin electrical double layers (EDL) and zero net charge density are assumed. The external electric potential ϕ in the liquid can be described by the Laplace equation for electrostatics

$$\nabla.\sigma\nabla\phi = 0 \tag{1}$$

Here σ is the electrical conductivity.

For the thin EDL, Smoluchowski equation $\vec{u}_{eo} = -\varepsilon\varepsilon_0\zeta\vec{E}/\mu$ can be safely used as the slip boundary of velocity. where $\vec{u}_{eo}$ is the slip velocity vector, ε is the relative permittivity, ε_0 is the free space permittivity, ζ is the zeta potential, μ is the dynamic viscosity of the electrolyte solution, $\vec{E}$ is the electric field strength defined as $\vec{E} = -\nabla\phi$.

2.2. Flow field, Heat Transfer and Phase Change
The EOF equations in microchannels are the modified incompressible NS equation and the continuity equation

$$\frac{\partial\vec{u}}{\partial t} + (\vec{u}.\nabla)\vec{u} = -\frac{1}{\rho}\nabla p + \mu\nabla^2\vec{u} + \rho_e\vec{E} \tag{2}$$

$$\nabla.\vec{u} = 0 \tag{3}$$

As the thermal capacity and thermal conductivity often change evidently following phase change, the different variables are induced. The energy equations for heat transfer are given respectively by

$$\rho.C_p^L\left(\frac{\partial T}{\partial t} + \vec{u}.\nabla T\right) = k^L\nabla^2 T + \sigma^L.\vec{E}^2 \tag{4}$$

$$\rho.C_p^S\frac{\partial T}{\partial t} = k^S\nabla^2 T + \sigma^S.\vec{E}^2 \tag{5}$$

where superscript L and S are the abbreviate of liquid and solid respectively, k is the thermal conductivity of electrolyte, C_p is the thermal capacity of electrolyte, T is the temperature, and the source term of $\sigma.\vec{E}^2$ expresses the influence of Joule heating in EOF.

The theory of phase change can be expressed as

$$H = h + \Delta H \tag{6}$$

$$\Delta H(T) = \begin{cases} L & T \geq T_{PC} \\ 0 & T < T_{PC} \end{cases} \tag{7}$$

$$h = T.C_p \tag{8}$$

where H is the enthalpy, h is the sensible heat, L is the latent heat, T_{PC} is the phase changing (PC) temperature. The domain can be regarded as a porous medium and the porosity λ can be expressed as

$$\lambda = 1 - f_s = \begin{cases} 0 & \text{Solid} \\ 0\sim 1 & \text{Hybrid} \\ 1 & \text{Liquid} \end{cases} \tag{9}$$

where f_s is the local solid fraction. The fluid velocity $\vec{u}$ in *NS* equation should be modified as $\lambda.\vec{u}_f$, $\vec{u}_f$ is the actual fluid velocity from *NS* equation.

3. Numerical Simulation
The numerical simulations are performed using a commercial solver of the NS equations. The 2D microchannel is rectangular ($80\mu m \times 1200\mu m$). The 10mM KCL solution is selected as the work buffer. In order to simplify the process, we only assigned the independent parameters in solid phase and liquid phase. The summary parameters are shown in table 1.

Table 1. Summary Parameters [6] [7]

	Liquid	Solid		Liquid	Solid
Density (kg/m^3)	1000	1000	**Dynamic viscosity (kg/m.s)**	0.000855	
Zeta potential (V)	-0.1	-0.1	**Latent heat (J/kg)**	324000	324000
Deby thickness(nm)	10	10	**Thermal capacity (J/Kg.K)**	4179	1960
σ (S/m)	0.13	0.0013	**Thermal conductivity (W/m.K)**	0.613	2.400
Relative permittivity	78.5	78.5	**PC temperature (K)**	273	273

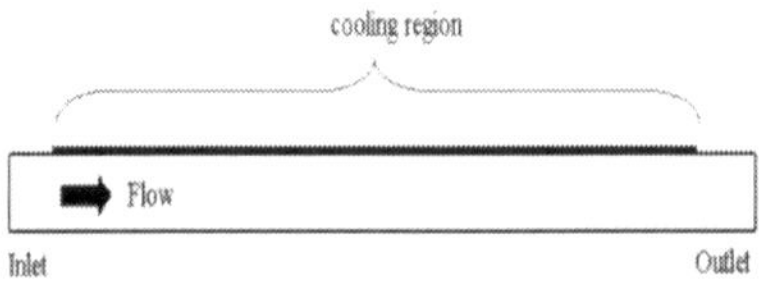

(a). Geometry of microchannel model

(b). Phase changing of liquid

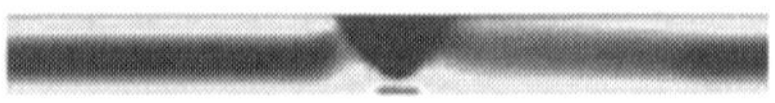

(c). Flow transition from plug flow to saddle

Figure 1. Simulation process of thermo-electric effect on EOF

As shown in figure 1.a, the boundary condition of electrical insulation is applied on all walls. On top of microchannel, the negative heat flux is applied for the endothermic&exothermic process (1000 μm length), except two adiabatic boundaries adjacent to inlet and outlet (100 μm length). On the bottom of microchannel, all boundaries are assumed adiabatic (1200 μm length). A potential drop and zero pressure drop condition are applied between inlet and outlet. The steady simulation was done. In all simulation, the ice cores always appears near the middle of cooling area firstly and then extends until heat balance(figure 1.b). In the freezing process, the distribution of EOF velocity transits from plug flow to saddle flow, due to the pressure gradient induced by phase change (figure 1.c).

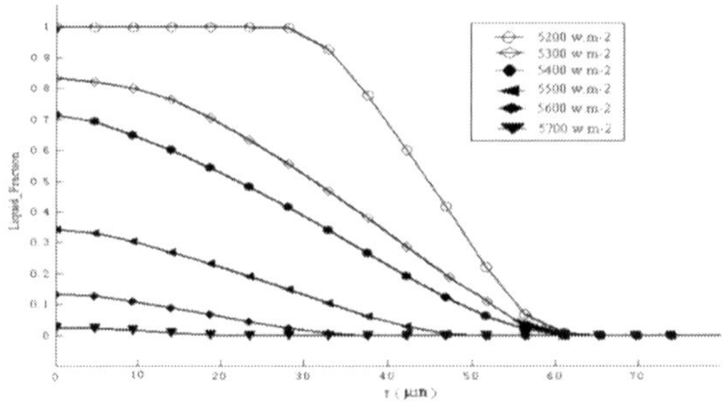

(a). 1V/cm external electrical field

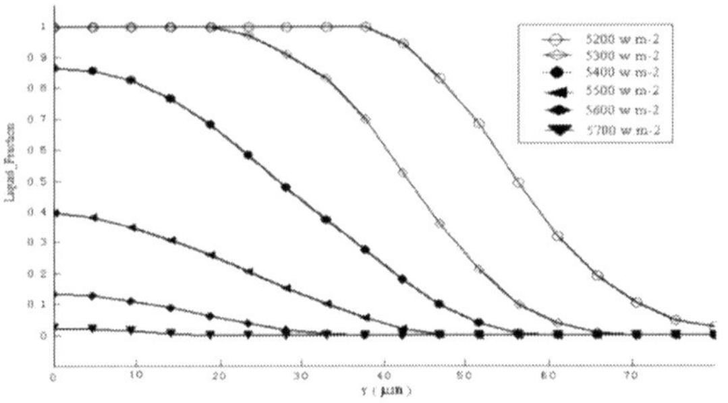

(b).5V/cm external electrical field

Figure 2. PC ratio with various cooling power density in the section of middle microchannel

The performance of PC valve is evaluated by the maximum PC ratio in microchannel section. The figure 2 shows PC ratio in the section of middle microchannel with different parameters. With the 1V/cm external electrical field of EOF, when the cooling power density (CPD) increases to $5200 w.m^{-2}$, the ice cores with 18 μm widths appears, and almost full clogging the channel with $5700 w.m^{-2}$ CPD. As the EOF external electrical field increases to 5V/cm, the width of ice cores almost decreases to zero with $5200 w.m^{-2}$ CPD. But following the increasing of CPD, the PC ratio curves approximately equal to the result of 1V/cm external electrical field and the CPD of full clogging is almost $5700 w.m^{-2}$ too. We have calculated the process with the condition of 10V/cm external electrical field too and the CPD of full clogging is still approximately $5700 w.m^{-2}$.

4. Experimental Setup

As shown in figure 3, the designed microfluidic chip is composed of microchannel layer, heat conducting layer, and refrigerating layer from top to down: The microchannel layer is formed from

PDMS (DowCorning Corp) by casting on the SU-8 mold. The size of microchannel is 60 μm ×80 μm ×80mm (W×H×L). The conductive layer is a glass slide with 120 μm thickness that is irreversible bonded with microchannel layer. The refrigerating layer includes one TE unit and a frame for fixing the TE units. The TE unit is a Peltier cooler (4mm×4mm×2.4mm). The maximum heat flux density is 8.0e4W/m^2. The frame is formed by a laser cutter from a PMMA piece (2mm thickness). The refrigerating layer is linked to heat conducting layer by glue.

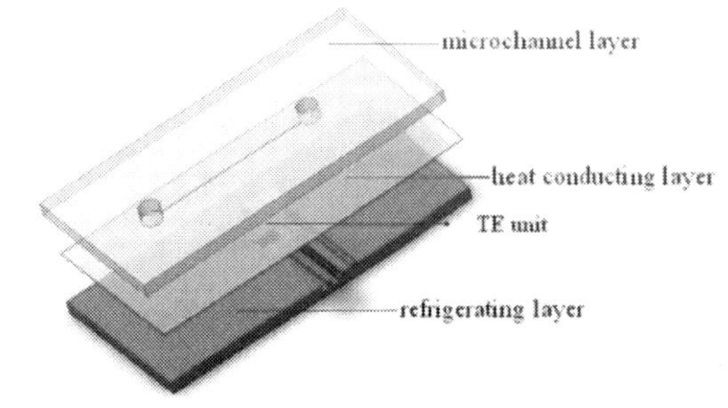

Figure 3. The microfluidic chip

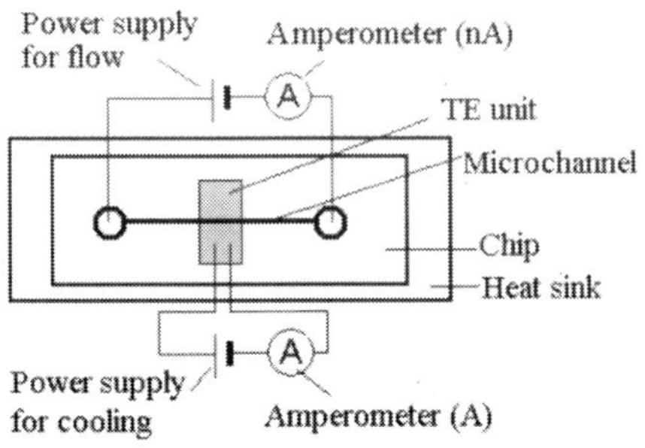

Figure 4. The setup of EOF driving and PC measurement

A 10mM.L^{-1} KCL electrolyte is selected as buffer solution, which electrical conductivity at 23°C is about 0.13S/m. Although there are still absent of exact information about electrical conductivity of KCL electrolyte in solid phase, mostly it will fall to one orders of magnitude at least [6] [7]. So the phase transition is ensured indirectly through the electrical current decreasing. The setup of EOF driving and PC measurement are shown in figure 4. A power supply is used to drive fluid flowing and another is used to drive the TE unit. A precise amperemeter (resolution 1nA) is used to measure the electrical current in microchannel. A steel heatsink is applied to advance the heat dissipation from TE heat tips.

5. Test Results

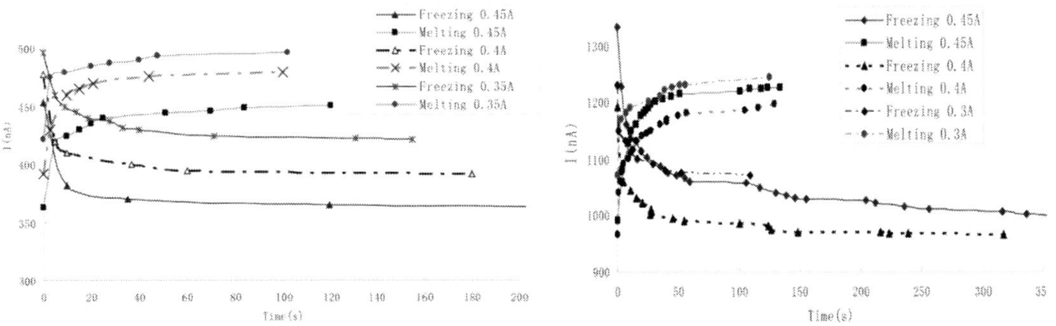

(a). 11.5v/cm external electrical field condition (b). 19.2 v/cm external electrical field condition

Figure 5. The fluctuation of electrical current in EOF

Previously the electrical conductivities have been measured *via* the precise amperemeter by freezing or melting all electrolytes in microchannel. The electrical conductivity in solid phase is about twentieth to that in liquid phase from the results. The heat flux of TE units is measured *via* a thermographic camera (A40 FLIR) and the typical TE heat flux values with electrical current (in brackets) were ensured such as 1.38e5w/m^2 (0.45A), 1.21e5w/m^2 (0.40A), 1.07e5w/m^2 (0.35A), 0.85e5w/m^2 (0.3A). Then the fluctuation of electrical current in EOF is dynamically monitored with different heat flux. Figure 5.a shows that the process of fluctuation with 11.5v/cm for EOF in freezing and melting. Figure 5.b shows that the process of fluctuation with 19.2 v/cm for EOF in freezing and melting.

We use the following equation $\frac{I_{Start}}{I_{End}} = \frac{(L-x)/\sigma_L + x/\sigma_S}{L/\sigma_L}$ to evaluate the length of ice plug x.

With 11.5v/cm external electric field for EOF, the lengths of ice plug are 340 μm (0.45A), 300 μm (0.4A), and 236 μm (0.35A). With 19.2v/cm external electric field for EOF, the lengths of ice plug are 474 μm (0.45A), 250 μm (0.4A), and 201 μm (0.3A). When the TE heat flux deceases to 0.85e5w/m^2, no obvious fluctuation of electrical current is observed. When the external electric field for EOF increases to 38.0 V/cm, obvious fluctuation of electrical current isn't observed too. From results, the freezing periods of are 35s (0.45A), 60s (0.4A), 75s (0.35A) with 11.5V/cm external electric field for EOF and 408s (0.45A), 220s (0.4A), 109s (0.35A). Due to the difference of latent heat between liquid phase and solid phase, the melting processes often increases a little faster than freezing process.

6. Conclusions

The thermoelectric control method by cooling the liquid plug for EOF was presented, and it does not require any moving part. A mathematical formulation was given, in which electrical field, flow field, heat transfer, and phase change were considered. Furthermore, we simulated the freezing phenomena in EOF. The flow transition from plug flow to saddle flow was found in the phase change process. PC valve performance was predicted. For 1~10V/cm external electrical field condition of EOF, the cooling power density of full clogging is about 5700w.m^{-2}.

According to the fluctuation of electrical current, we experimentally found that its response time ranges from 35s to about 408s, and the length of ice plug ranges from 474 μm to 200 μm, depending on the external electric field of EOF and heat flux of TE. Issues of the measurement of the flow velocity distribution will require further research.

Acknowledgments

Authors wishing to acknowledge assistance or encouragement from the National High Technology Research and Development Program of China (863 Program, No: 2007AA04Z302), National Natural Science Foundation of China (No: 50705012, No: 20890024), Major State Basic Research Development Program of China (No: 2007CB714502).

References

[1] Bevan C D, Mutton I M 1995 *J.Chromatogr.* A **697** 541–548.
[2] Gui L, Liu J 2004 *J. Micromech. Microeng.* **14** 242–246.
[3] Welle R P, Hardy B S etc. 2005 *International Conference on Thermoelectrics.*
[4] Barz D, Ehrhard P 2005 *J. PAMM.* **5** 535 - 536.
[5] Kler P A, Ezequiel J L etc. 2009 *J. Comput. Methods Appl. Mech. Engrg.* **198** 2360 - 2367.
[6] Rusiniak L 2004 *J. Acta Geophysica Polonica.* **52** 364-380.
[7] Wolff E W, Miners W D etc. 1997 *J. Phys. Chem.* B **101** 6090-6094.

In-situ Polymerization and Characterization of Poly (ε -caprolactone) Urethane/ SiO_2 Nanocomposites

Xiao-dong Chen[a,b*], Nan-qiao Zhou[a], Hai Zhang[b]

[a] National Engineering Research Center of Novel Equipment for Polymer Processing, South China University of Technology, Guangzhou, Guangdong, 510640, China

[b] Guangzhou SCUT Bestry Technology Joint-stock Co. Ltd, Guangzhou, Guangdong, 510640, China

E-mail: cxdzlgzhnlg2003@163.com

Abstract. Nanocomposites of poly(ε -caprolactone) (PCL) castable polyurethane elastomer (CPUE) with nano-SiO_2 particles of different surface properties were prepared via in-situ polymerization. An electronmechanical universal testing machine, a durometer, a rubber resilience experimental machine, a dynamic-mechanical analyzer, a thermogravimetric analysis and a scanning electron microscope were used to investigate the macro -static/dynamic mechanical properties, micro-dispersed state and thermostability. The results showed that the modulus at 100% and 300%, elongation at break, tensile strength and tear strength of poly(ε -caprolactone) urethane nanocomposites were increased by introducing a certain amount of nano-SiO_2. Especially the tensile strength and tear strength at 100℃ of the PCL CPUE with 5% nano-SiO_2 pretreated by γ-glycidochloropropyl methyl trimethoxy silane (trade name A-187) were 1.50 and 1.94 times than those of the pure PCL CPUE, respectively. The addition of the nano-SiO_2 had little effect on the hardness, but the impact resilience decreased slightly. The loss factor peaks of two nano-SiO_2 polyurethane composites were higher obviously than the pure PCL CPUE and the glass transition temperature (T_g) of the two nano-SiO_2 polyurethane composites increased to higher temperature region. The surface treatment by the optimum silane coupling agent influenced the dispersibility of nano-SiO_2 in the PCL CPUE distinctly. The agglomerating phenomenon, and even some nano-agglomerates with more than 1 μ m diameter can be observed in the PCL CPUE with 5% untreated nano-SiO_2, but the nano-SiO_2 pretreated by A-187 was dispersed in the PCL CPUE at nano-scale.

Introduction

The preparation methods of polymer/inorganic nano-particle composites are sol-gel method, intercalation method, melt-compounding method and *in-situ* polymerization process[1-5]. Here the in-situ polymerization process is defined that the monomers dispersed with nano-scale inorganic powders can bulk polymerize and a certain nano composites are obtained. It is benefit for the dispersion of nano-scale inorganic powders in polymer matrix by using this method, and the obtained composites usually feature outstanding processing properties. At the same time, silane coupling agents with active groups are employed widely in the surface treatment of nano-scale inorganic powders, especially for those powders with hydroxyl groups on their surface. The alkoxy groups of silane coupling agents can be hydrolyzed and form chemical bonding with the surface of inorganic powders and the active groups can generate chemical bonds with polymer matrix. Then the interfacial bonding of inorganic powders and polymer matrix can be improved, also a high performance composite can be obtained[1].

Castable polyurethane elastomers (CPUEs) have been applied widely in national defense, metallurgical industry, mining, electro-machinery vocation, textile industry, shoemaking and automobile fields. But the application of the traditional CPUEs is still unsatisfied in some tough circumstance because there are some disadvantages such as high heat generation and poor heat resistance[6]. Many engineers are working hard to find a solution for this problem and promoting new applications for the CPUEs.

Polymer nanocomposites exhibit many outstanding properties when compared with the universal materials with micrometre fillers[7-10], this point attracts attention of the world. The most prominent feature is that the introduction of a small amount of nano-scale fillers can modify the mechanical properties of the nanocomposites significantly[7,11]. So many studies have been carried out on the polymer nanocomposites[7,12-14]. As an increasingly advancing material, polyurethane is a focus what has been developed for several decades. According to the search data, more than one hundred papers had been adopted by SCI after the first thesis of polyurethane nanocomposites was published in 1998.

The research field is mainly concentrated on the structure and properties of polyurethane/surface modified nano clay composites, and different research teams obtained different conclusions. Tien and Wei[15] found that the young's modulus and the elongation at break increased substantially, while Zilg[16] et al. reported an improvement of elongation at break, but a reduction in the young's modulus. Han[17] et al obtained a result that the dynamic tensile modulus decreased, however, the elongation at break decreased. Xia[18] et al. found that the tensile strength and modulus increased, whereas the elongation decreased with the increasing of the exfoliation degree. Zheng[19] et al. not only carried out the study on the polyurethane nano clay and nano Al_2O_3 composites, but also discussed the polyurethane/ZnO nanocomposites. The experiments presented some surprising results: the introduction of nano-scale fillers resulted in a decrease both in the modules and the elongation at break, also the decline amplitude was variable with the different kinds of nano-scale fillers. The modification of the nano SiO_2 on the polyurethane has become the hot topic in recent years. Park[20] prepared the polyurethane acrylate anionomer /silica composites by the sol-gel reaction. The mechanical properties of the composites were enhanced drastically. Other researchers[1,21,22] worked hard on the preparation of polyurethane/SiO_2 nanocomposites used as adhesives & coatings via the solvent method to improve the hardness, abrasion resistance, surface free energy and tensile properties of the polymer films. It is interesting that few studies are concerned with the preparation of castable polyurethane/SiO_2 nanocomposite elastomers via in-situ bulk polymerization and the investigation of the micro-dispersibility of nano-scale SiO_2 with different surface characteristics in prepolymers & CPUEs and the influence mechanism of nano-scale SiO_2 on the mechanical properties.

Nano-scale SiO_2 fillers are ultrafine SiO_2 powder obtained by the pyrohydrolysis of silicon/organosilicon chloride and within hydroxyl groups/adsorption water on the surface. They exhibit the characteristics such as small particle size (<100nm), huge specific surface area (>100m^2/g), high chemical purity and good dispersibility. They have been employed extensively in the rubber, coatings, medicine and papermaking because of their excellent stability, reinforcement, thickening and thixotropic properties. The tensile/ tear strength, wear resistance, aging characteristic and thermostability of the modified rubber with nano-scale SiO_2 fillers are all improved[23-26].

Poly (ε -caprolactone) polyurethane/SiO_2 nanocomposites were prepared via in-situ polymerization and characterized by a scanning electronic microscope, an electronic universal testing machine, a thermogravimetric analyzer and a dynamic thermomechanical analyzer. The surface treatment effect of coupling silicone agent on the nano-scale SiO_2 and its effect on the properties of CPUE composites were investigated. It is believed that this work can provide some thoughts for the design of novel materials.

1. Experimental

1.1. Raw materials

ε -caprolactone monomer, purchased from Chunshin Chemical Co., Ltd, was stored below 5℃ in nitrogen atmosphere and purified by distillation before usage. 1, 5-pentanediol, supplied by Nanjing Oriental Pearl Industry & Trade Co., Ltd, was distilled under reduced pressure before use. Tetraisopropyl orthotitanate, the catalyst for ring-opening polymerization, was used as received. Nano-scale SiO_2 (trade mark SS1) spherical particles of Zsmrnmclgs production, with a mean particle size from 20 to 50 nm and a purity of more than 99.5%, was stored in a desiccator. Its specific surface area was160±20m2/g and hydroxyl group content was more than 19%. γ-glycidochloropropyl methyl trimethoxy silane, with a trade name of A-187, was produced by American Union Carbide Company. 2, 4-toluene diisocyanate (TDI), manufactured by Bayer, was used as reviewed. The production of Albemarle Corporation, 3, 5-dimethythiotoluene diamine, was dehydrated before use.

1.2. Synthesis of poly(ε -caprolactone) glycol

Nitrogen was imported into a 500ml glass reactor equipped with a stirrer, a thermometer and a reflux condenser. Stoichiometric quantities of ε-caprolactone monomer and tetraisopropyl orthotitanate were added, then 1, 5-pentanediol, used as the initiator, was introduced and the mixture was heated with an intensive stirring. The polymerization was carried out under nitrogen atmosphere for 48h at 140 to 160℃. After reaction, the mixture was vacuum-evaporated at 120 to 140, and the unreacted monomer was removed. A milk-white solid was obtained and stored in a desiccator after cooled. The polymerization process is presented in Formula 1 and the mean molecular weight was determined before the preparation of prepolymer.

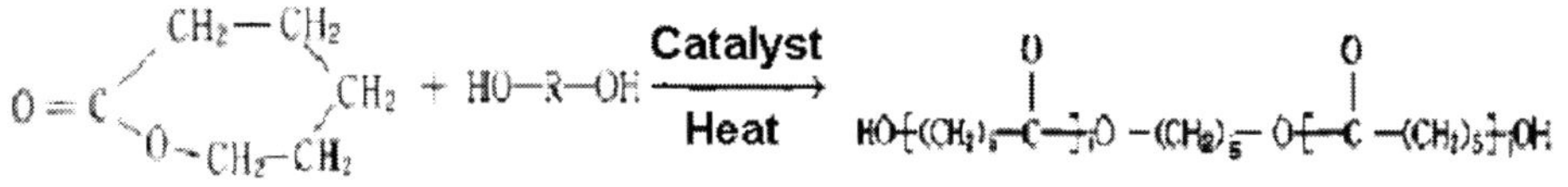

(1)

1.3. Preparation of CPUE

1.3.1. Surface treatment of nano-scale SiO_2 particle

γ-glycidochloropropyl methyl trimethoxy silane (A-187) was added into acetone solvent and dispersed via ultrasonic wave for 10 minutes to form a solution with 20% content. Nano-scale SiO_2 particles were added into the solution and mixed uniformly. Then the mixture was treated via ultrasonic wave for 30 minutes and coupling-reacted for 2 hours, and the solvent was distillated and recycled. The obtained nano-scale SiO_2 particles were desiccated and ready for next operation.

1.3.2. Synthesis of NCO terminated prepolymer

Poly (ε -caprolactone) glycol was melted and the nano-scale SiO_2 particles were added with high speed stirring. Nano SiO_2/polylol dispersion was formed after 15 to 30 minutes stirring with the speed of 4000 to 5000 rpm. The dispersion was heated to 100 to 120℃ and dehydrated at vacuum for 2 to 3 hours. After cooled to 60℃, the dispersion was mixed with a stoichiometric quality of diisocyanates and maintained the temperature at 80℃ for 3 hours. The reactant was degassed after the concentration of isocyanate groups was determined by the amine equivalent method.

1.3.3. Preparation of polyurethane elastomer

In this stage, the precisely dehydrated 3, 5-dimethythiotoluene diamine was poured into the NCO terminated prepolymer at 80℃ at the quantity which was obtained from the designed molar ratio (NH/NCO group=0.88). After 1 min of intensive stirring, the mixture was poured into a mold which had been placed on a vulcanizing press and preheated up to 100℃. The mold was locked while the mixture initially formed a gel. A piece of 2mm thick test plate was obtained after 30 min curing period and demolded from the mold. The elastomer samples were subjected to the tests after vulcanized in an oven at 120℃ for 24 hours and stored at room temperature for a week.

2. Test and characterization

2.1. Determination of the mean molecular weight of poly(ε -caprolactone) glycol

The mean molecular weight of poly (ε -caprolactone) glycol was determined via a Knauer vapour pressure osmometer (model: K-7000) by using toluene as solvent. The test temperature was 37± 0.01℃. The test method was based on the Ragout's law. At the same time, it was also measured by testing the terminated hydroxyl groups via the phthalic anhydride- pyridine acylation method.

2.2. Characterization of the dispersity and storage stability of nano-scale SiO_2 in the poly(ε -caprolactone) prepolymer

The melting prepolymer was casted on the 0.2mm thick glass slide and formed a film, then the film was observed by a Nikon YS100 optical microscope after stained by $KMnO_4$ solution. And the storage stability can be judged by the appearance of the prepared prepolymer and the state after long time storage.

2.3. Characterization of the dispersity of nano-scale SiO_2 in CPUE

The CPUE samples were fractured in liquefied nitrogen. The brittle fractured surface was sputtered with gold. The morphology of sample surface and the dispersity of nano-scale SiO_2 in CPUE were observed via a Philip XL30 scanning electron microscope (SEM).

2.4. Properties determination

2.4.1. mechanical properties

The mechanical properties of CPUE samples were tested via an electronmechanical universal testing machine (INSTRON Co. LTD, Model 5566,USA). The test of tensile strength was based on ASTM D 412-1998A named "Standard Test Methods for Vulcanized Rubber and Thermoplastic Elastomers Tension". And the tear strength was tested based on ASTM D624-00e1 named "Standard Test Method for Tear Strength of Conventional Vulcanized Rubber and Thermoplastic Elastomers".

2.4.2. Hardness

A LX-A durometer made by ShangHai HongSheng Instrument Co., Ltd. was utilized to determine the sample hardness according to ASTM D 2240 named "Standard Test Method for Rubber Property—Durometer Hardness".

2.4.3. Resilience

A rubber resilience experimental machine made by ShangHai Chemical Machinery Company was used for testing the impact resilience of samples and ASTM D 1054-2002 Standard titled "Test Method for Rubber Property - Resilience Using a Goodyear-Healey Rebound Pendulum" was the basic method.

2.4.4. Characterization of the thermostability of samples

Thermogravimetric analysis (TGA) was carried out on a NETZSCH Instruments analyser TG209 in nitrogen atmosphere. All samples were heated from 25 to 500℃ at a heating rate of 10℃/min.

2.4.5. Characterization of the dynamic mechanical properties of samples

A NETZSCH Instrument, Dynamic-Mechanical Analyzer DMA242, was employed to determine the dynamic properties of the sample which was cut as following size: 2.0×5.8×10.0mm. The test was carried out at 10 Hz frequencies, ±2N maximum dynamic stress, ±40μm maximum deformation amplitude. The temperature range was -100～150℃ and the heating rate was 5℃/min.

3. Results and discussion

3.1. The dispersity and storage stability of two sorts of nano-scale SiO_2 in PCL prepolymers

A key technique of nano-scale particle modification on CPUE is how to disperse the nano-scale particle into material matrix with nanometer grade successfully. At the same time, the storage stability is a great factor on the manufacturability of prepolymers. First of all, it is necessary to investigate the dispersing state and storage stability of nano-scale SiO_2 in the prepolymer.

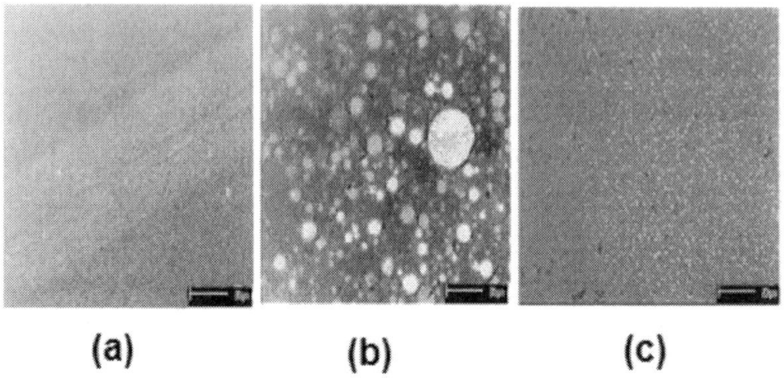

Figure 1. Optical micrographs of (a) pure prepolymer, (b) PCL based prepolymer with 5% untreated nano-SiO_2 and (d) PCL based prepolymer with 5% nano-SiO_2 pretreated by A-187(1000×)

The films of two types of prepolymers, the PCL based prepolymer with 5% untreated nano-SiO_2 and the PCL based prepolymer with 5% nano-SiO_2 pretreated by A-187, were observed and different images were obtained. In order to get a comparison, an image of the pure prepolymer was also captured and listed in figure 1. It was obviously that the nano-SiO_2 pretreated by A-187 formed a homogenizing-dispersion, while the untreated nano-SiO_2 aggregated and even formed some agglomerates within 1 μ m diameter because of the incompatibility caused by the polarity difference between two components.

Table 1. Storage stability of two sorts of nano-scale SiO_2 in PCL prepolymers

Prepolymer sample	Storage stability
PCL based prepolymer with 5% untreated nano-SiO_2	White subtransparent liquid with some deposition after 48 hours storage at room temperature
PCL based prepolymer with 5% nano-SiO_2 pretreated by A-187	Colourless transparent dispersion without any change during storage life (30d)

As shown in table 1, the untreated nano-SiO_2 with 5% content was dispersed in PCL based prepolymer, and a white subtransparent liquid was obtained. After stored for 48 hours in room temperature, there was some deposition in this system because of the poor interface compatibility between the untreated nano-SiO_2 and the prepolymer matrix. While the nano-SiO_2 pretreated by A-187 with 5% content was dispersed uniformly and a colourless transparent homogeneous system was obtained. It was so stable that there was no change during 30 days of storage. This phenomenon also confirmed the micro-dispersing difference between two systems.

3.2. The dispersity difference of two sorts of nano-scale SiO_2 in PCL CPUEs

The SEM fractographs in figure 2 present the surface morphology of three samples. After compared the fractographs, a conclusion can be made that there are obvious differences in the dispersibility of two sorts of nano-scale SiO_2 in the PCL CPUEs. The surface of the Pure PCL CPUE is smooth and clear, whereas both the polyurethane/nano-scale SiO_2 composites have rough surface. It could be contributed that the difference in surface recovery capability what is depended on the soft segment mobility. The soft segments within low glass transition temperature （Tg） and low surface energy could move and minimize the surface area. In the nanocomposites, the soft segment could not effectively migrate and was unable to recover surface because it was restricted by the interfacial interactions between the polymer matrix and the nano-scale particles[19]. This explanation also confirmed by the comparison of the DMA curves of the samples in figure 5. On the other hand, the polarity difference between two components resulted in different dispersion states of polyurethane/nano-scale SiO_2 composites. In the PCL CPUE with 5% untreated nano-SiO_2, there existed agglomerating phenomenon, and even formed some nano-agglomerates having more than 1 μm diameter. But the stable and homogeneous nano-dispersion of nano-scale SiO_2 in the PCL CPUE was achieved because of the surface modification on nano-scale SiO_2. This phenomenon corresponded well with the dispersion states of two nano-scale SiO_2 in the PCL prepolymers.

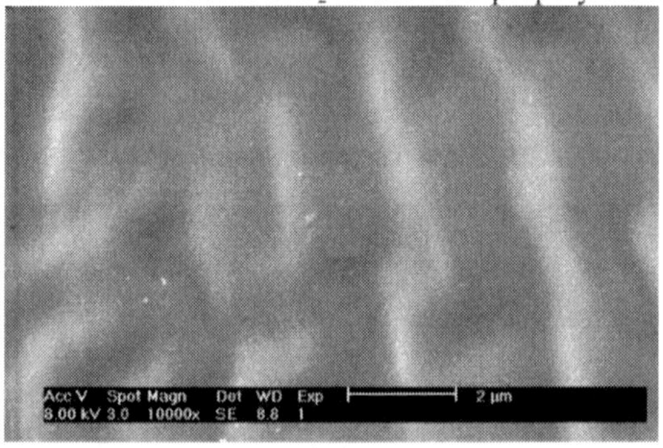

(a)

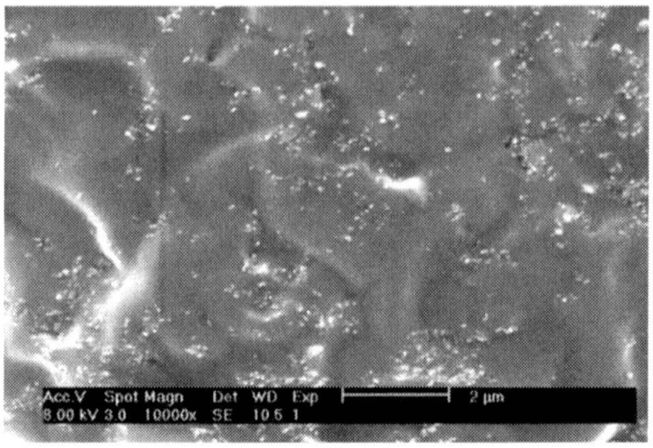

(b)

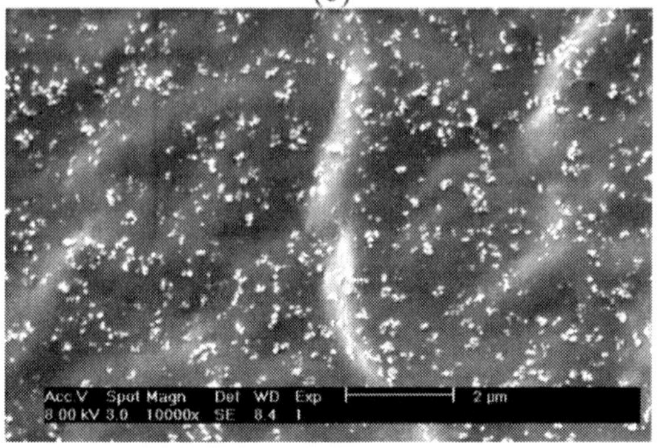

(c)

Figure 2. SEM fractographs of (a) Pure PCL CPUE, (b) PCL CPUE with 5% untreated nano-SiO_2 and (c) PCL CPUE with 5% nano-SiO_2 pretreated by A-187 (10000×)

3.3. The thermostability difference of two sorts of nano-scale SiO_2 in PCL CPUEs

In figure 3, the comparison of three TGA curves can indicate remarkably that the introduction of nano-scale SiO_2 in the PCL CPUE via in-situ polymerization increased the initial thermal degradation temperature, moreover, the PCL CPUE with 5% nano- SiO_2 pretreated by A-187 exhibited better thermostability than the PCL CPUE with 5% untreated nano-SiO_2 because more efficient chemical bonds could be found due to the reactive groups on the surface of nano-SiO_2 pretreated by A-187.

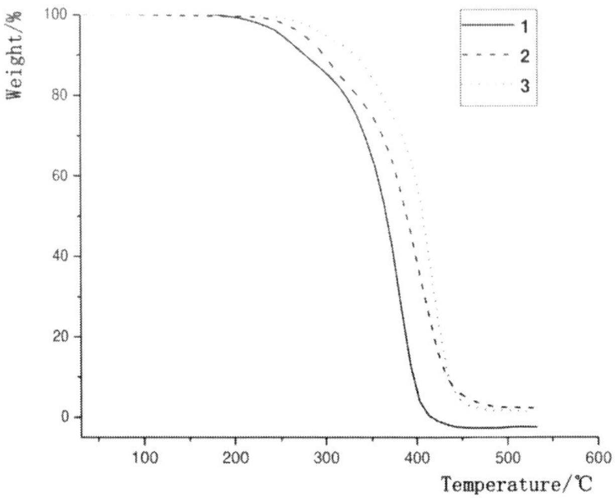

Figure 3. TGA thermograms for (1) Pure PCL CPUE, (2) PCL CPUE with 5% untreated nano-SiO_2 and (3) PCL CPUE with 5% nano-SiO_2 pretreated by A-187

3.4. Effect of nano-scale SiO_2 on the mechanical properties of PCL CPUEs

Table 2. Mechanical properties of three CPUE samples

Sample	Hardness (Shore A)	Modulus 100%, MPa	Modulus 300%, MPa	Elongation at break, %	Tensile strength, Mpa		Tear strength, KN/m		Resilience,%
					Ambient temperature	100℃	Ambient temperature	100℃	
Pure PCL CPUE	83	4.33	8.23	582.0	39.24	8.48	79.5	41.0	51
PCL CPUE with 5% untreated nano-SiO_2	83	4.58	8.51	597.1	40.70	9.43	85.7	68.7	50
PCL CPUE with 5% nano-SiO_2 pretreated by A-187	84	5.22	10.46	622.5	46.39	12.72	99.7	79.5	48

Note : All the CPUE samples were prepared by PCL and TDI, the content of NCO group was 3.4%, the chainextender was DADMT, and crosslinking ratio was 0.88.

From the data in table 2, a conclusion could be made that the introduction of two sorts of nano-SiO_2 increased the modulus at 100% and 300%, elongation at break, tensile strength and tear strength. What should be specially noted was that the improvement of the tensile strength and tear strength at 100℃ of the PCL CPUE with 5% nano-SiO_2 pretreated by A-187 was the best. When compared with the data of the pure PCL CPUE, they were increased by 1.50 and 1.94 times, respectively. At the same time, the introduction of nano-SiO_2 via in-situ polymerization had little influence on the hardness, but the impact resilience of nano composites decreased slightly.

These results are understandable according to the systematical analysis on the nature of the nano composites. The dispersion state of nano-SiO_2 in the polyurethane matrix played an important role when the elastomer materials were impacted by an external force. The nano fillers could improve the tension resistence effectively because they acted as crosslinking points. It was remarkable that the optimum coupling agent, γ-glycidochloropropyl methyl trimethoxy silane (A-187), not only formed hydrogen bonds with polar groups contained by soft/hard segments of polyurethane, but also ring-opening reacted with isocyanate groups & hydroxyl/amine groups during the preparation process by the terminated epoxy groups and achieved stable chemical bonds. This was favorable for the modification of the mechanical properties, especially for the performance at the elevated temperature of the CPUEs.

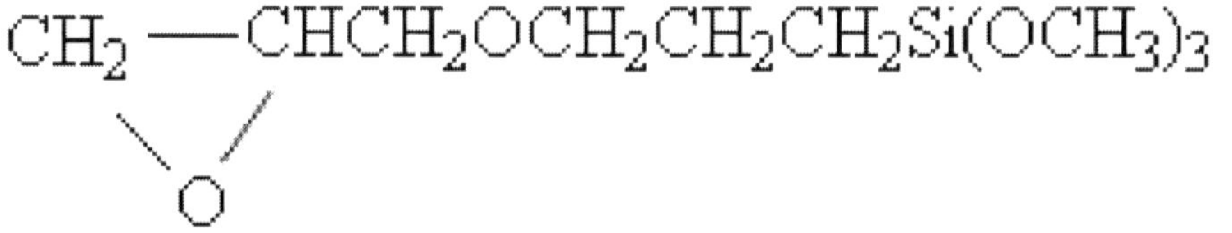

Figure 4. Molecular structure of γ-glycidochloropropyl methyl trimethoxy silane (A-187)

3.5. Effect of nano-scale SiO2 on the dynamic mechanical properties of PCL CPUEs

As shown in figure 5, T_g of two nano-SiO_2 polyurethane composites increased to higher temperature region than the pure PCL CPUE. T_g of the PCL CPUE with 5% untreated nano-SiO_2 increased about 2.5℃ and that of the PCL CPUE with 5% nano-SiO_2 pretreated by A-187 increased about 7.5℃. And the loss factor peaks of the two nano-SiO_2 polyurethane composites were higher obviously than that of the pure PCL CPUE. In addition, the loss factor (Tanδ) of the PCL CPUE with 5% nano-SiO_2 pretreated by A-187 was the biggest one among the three samples. This may be explained by the differences in the microphase separation degree caused by the modification of nano-SiO_2 on the CPUEs. More hard segments were dispersed in the soft matrix, which can restrict the movement of soft chain and result in the migration to higher temperature zone of T_g. At the same time, the material hysteresis became bigger. This phenomenon was correlated to the decrease in the resilience of the two nano-SiO_2 polyurethane composites.

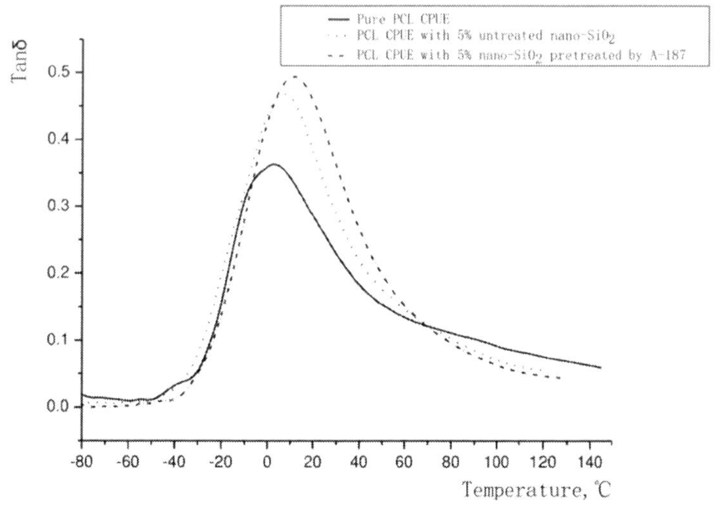

Figure 5. DMA curves for (1) Pure PCL CPUE, (2) PCL CPUE with 5% untreated nano-SiO_2 and (3) PCL CPUE with 5% nano-SiO_2 pretreated by A-187

Conclusion

The following conclusions can be made according to the investigation on the two nano-SiO_2 polyurethane composites.

(1)The surface treatment by the optimum silane coupling agent influenced the dispersibility of nano-SiO_2 in the PCL CPUE distinctly. The agglomerating phenomenon, and even some nano-agglomerates with more than 1 μ m diameter can be observed in the PCL CPUE with 5% untreated nano-SiO_2, but the nano-SiO_2 pretreated by A-187 dispersed in the PCL CPUE in the nano-scale degree.

(2)The introduction of a certain amount of the nano-SiO_2 in the PCL CPUE via in-situ polymerization improved the initial thermal degradation temperature. And for the PCL CPUE with 5% nano-SiO2 pretreated by A-187, it was boosted close to 30℃ compared with the date of the pure PCL CPUE.

(3)The introduction of a certain amount of the nano-SiO_2 via in-situ polymerization intensified the modulus at 100% and 300%, elongation at break, tensile strength and tear strength. The most remarkable fact is that the tensile strength and tear strength at 100℃ of the PCL CPUE with 5% nano-SiO_2 pretreated by A-187 were 1.50 and 1.94 times than those of the pure PCL CPUE, respectively.

(4) Hardness of the nanocomposites had nothing to do with the addition of nano-SiO_2, but the impact resilience decreased slightly.

(5) T_g of the two nano-SiO_2 polyurethane composites increased to higher temperature region compared to the pure PCL CPUE. Here, T_g of the PCL CPUE with 5% untreated nano-SiO_2 increased about 2.5℃ and that of the PCL CPUE with 5% nano-SiO_2 pretreated by A-187 increased about 7.5℃.

(6)The loss factor peaks of two nano-SiO_2 polyurethane composites were higher obviously than that of the pure PCL CPUE. In addition, the loss factor (Tanδ) of the PCL CPUE with 5% nano-SiO_2 pretreated by A-187 was the biggest one among the three samples.

Acknowledgements: This research is funded by the Important Projects from National Natural Science Foundation of China (10590351), and supported by the Polyurethane Department of GuangZhou SCUT Bestry Technology Joint-stock Co. Ltd. The authors are grateful to the teacher of National Engineering Research Center of Novel Equipment for Polymer Processing for their helpful advice

References:

[1] Widmaier, J.-M.; Bonilla, G. Polymers for Advanced Technologies 2006, 17, 634-640.
[2] Hajji, P.; David, L.; Gerard, J. F.; Pascault, J. P.; Vigier, G. Journal of Polymer Science Part B: Polymer Physics 1999, 37, 3172-3187.
[3] Hsiue, G.-H.; Chen, J.-K.; Liu, Y.-L. Journal of Applied Polymer Science 2000, 76, 1609-1618.
[4] Kojima, Y.; Usuki, A.; Kawasumi, M.; Okada, A.; Kurauchi, T.; Kamigaito, O. Journal of Polymer Science Part A: Polymer Chemistry 1993, 31, 983-986.
[5] Wu, L.; Luo, X.; Wang, X. Journal of Applied Polymer Science 2006, 102, 5472-5482.
[6] Industry, S. P. I. o. C. Polyurethane Elastomer Handbook; Chemical Industry Press: Beijing, 2001.
[7] Yano, K.; Usuki, A.; Okada, A.; Kurauchi, T.; Kamigaito, O. Journal of Polymer Science Part A: Polymer Chemistry 1993, 31, 2493-2498.
[8] Singh, C.; Balazs, A. C. Polymer International 2000, 49, 469-471.
[9] Ginzburg, V. V.; Balazs, A. C. Advanced Materials 2000, 12, 1805-1809.
[10] Giannelis, E. P. Applied Organometallic Chemistry 1998, 12, 675-680.
[11] Kornmann, X.; Berglund, L. A.; Sterte, J.; Giannelis, E. P. Polymer Engineering & Science 1998, 38, 1351-1358.
[12] Zhang, J.; Jiang, D. D.; Wang, D.; Wilkie, C. A. Polymers for Advanced Technologies 2005, 16, 800-806.

[13] Messersmith, P. B.; Giannelis, E. P. Journal of Polymer Science Part A: Polymer Chemistry 1995, 33, 1047-1057.
[14] Anit, K. G. Advanced Materials 1997, 9, 163-166.
[15] Tien, Y. I.; Wei, K. H. Journal of Applied Polymer Science 2002, 86, 1741-1748.
[16] Zilg, C.; Thomann, R.; Mülhaupt, R.; Finter, J. Advanced Materials 1999, 11, 49-52.
[17] Han, B.; Cheng, A.; Ji, G.; Wu, S.; Shen, J. Journal of Applied Polymer Science 2004, 91, 2536-2542.
[18] Xia, H.; Shaw, S. J.; Song, M. Polymer International 2005, 54, 1392-1400.
[19] Zheng, J.; Ozisik, R.; Siegel, R. W. Polymer 2006, 47, 1-9.
[20] Park, N.-H.; Lee, J.-W.; Suh, K.-D. Journal of Applied Polymer Science 2002, 84, 2327-2334.
[21] Chen, Y.; Zhou, S.; Yang, H.; Wu, L. Macromolecular Materials and Engineering 2005, 290, 1001-1008.
[22] Chen, X.; Wu, L.; Zhou, S.; You, B. Polymer International 2003, 52, 993-998.
[23] Zhu, Y.; Sun, D.-X. Journal of Applied Polymer Science 2004, 92, 2013-2016.
[24] Lee, S.-I.; Hahn, Y. B.; Nahm, K. S.; Lee, Y.-S. Polymers for Advanced Technologies 2005, 16, 328-331.
[25] Zhou, S.-X.; Wu, L.-M.; Sun, J.; Shen, W.-D. Journal of Applied Polymer Science 2003, 88, 189-193.
[26] Petrovi, Z. S.; Javni, I.; Waddon, A.; Bánhegyi, G. Journal of Applied Polymer Science 2000, 76, 133-151.

Analysis of the meniscus and viscous effects in digital micromirror device

Sisi Liu, Chaohui Zhang

School of Mechanical, Electronic and Control Engineering, Beijing Jiaotong University, Beijing, 100044

E-mail: 08121840@bjtu.edu.cn

Abstract. Digital micromirror device (DMD) is the core of digital light processing projected display technology wherein the spring tip-landing site adhesion and wear are the substantial clause leading to failure due to frequent contacts. In an attempt to analyze the adhesive force between the spring tip and the landing site, two main aspects are considered: i.e. the meniscus force and the viscous resistance during separation. The simulation results show that the influences of the surface energy, liquid volume and the separate velocity are prominent. The work will provide some hints for control of adhesive force in DMD components.

1. Introduction

Since researchers in University of California Berkeley exhibited micro electro-mechanical system (MEMS) products [1] in the late 80's of the last century, it has brought enormous changes to our life and will continue to do in the same way. This is similar to the impacts which brought to the information technology by the appearance of integrated circuit (IC). The development of MEMS technology enables it to become the important area in nanotribology research [2, 3].

The ratio of area to volume increases continuously along with dimension shrinkage, so the properties related to material surface nature (such as adhesive force, friction, meniscus force, viscosity traction, surface tension and so on) play the decisive role. Moreover, the admissible tolerance of these equipments is so small that physical direct contact is nearly inevitable, which will lead to stronger ratio of adhesive force from adjacent components [4]. To find an anti-sticking technology becomes important. Generally, the origin is adhesive force of intervenient liquids or the capillary condensation which forms from the condensed vapor in ambient environment [5]. So the hydrophobic/hydrophilic performances of material surface in micro/nano apparatus are very essential.

In order to reduce the adherence in MEMS, it is extremely important to let the material surface at de-wetting or super-hydrophobic state. Reducing the humidity is better for limiting meniscus's formation; therefore, it can reduce adherence, friction and wear. Super-hydrophobic surface has large contact angle (>150°) but small contact angle hysteresis (<10°). In order to improve hydrophobic performance of materials, one way is to reduce surface energy of materials to deteriorate the interaction between the material surface and the water [6-8], another one is to increase the materials surface roughness (Ra), by which forming a multilayered roughness structure in micro/nano scale, specially forming compound surfaces [9, 10]. However, some researches pointed out that this kind of multilayered structure of roughness distribution is unnecessary [11]. Moreover, there is also a need to increase adhesion in certain situation in MEMS. It's a suitable choice to form micro/nano roughness as "spatula" on gecko's bristle [12]. Hiratsuka et al. experimentally measured average friction coefficient down to 8.3×10^{-4} when water drop intervened in hydrophobic surfaces [13].

Normal load is generally small and mutual contact surfaces are relatively smooth in micro/nano device in MEMS. At this time, the coherence which causes by the meniscus force and the viscous force is possibly quite large and becomes the primary cause to affect the equipment works reliably. Adhesive force depends on area, separate distance, surface tension and liquid adhesion of formed meniscus. In the separation process the meniscus force decreases along with the distance increase, while the viscous resistance reacts in opposite [14, 15].

As a successful commercial exemplification of MEMS, the digital micromirror device (DMD) is one kind of large scale integrated three-dimensional light beam regulator. It mainly applies in the projected display equipment and the electrostatic printer [16]. DMD is composed of rotatable aluminum mirror arrays (figure 1). These aluminum mirrors are processed with one kind of semiconductor processing technology. Processed surface array has hundreds of thousands to 1.3 million micromirrors which can control reflection independently [5, 14, 17, 18]. The size of each micromirror is around 12um × 12um. The adhesion and wear between tip which locates at the end of cantilever spring and landing site below may lead to reset failure.

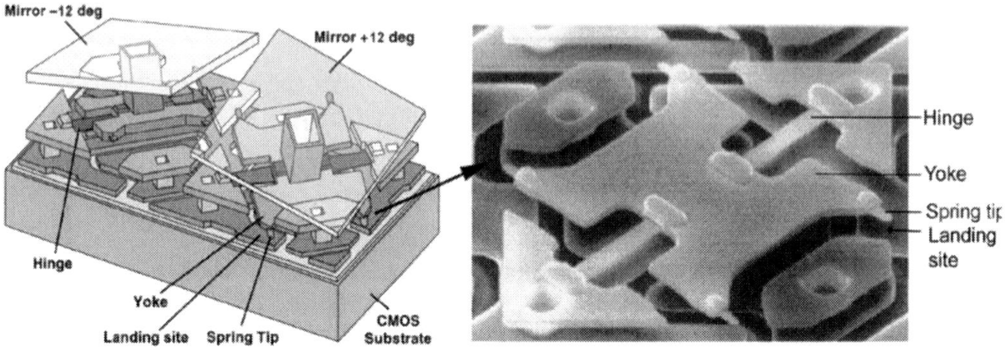

Figure 1. Schematic view of DMD device [17].

This paper aims at the contact problem of spring tip- landing site in DMD. The meniscus force and viscous resistance during separation are analyzed, which will shed some lights to the mechanism of the adherence property and lay a feasible foundation for the future application.

2. Summary of lubrication technology in DMD

DMD is the display technology which developed by Texas Instrument (TI) in 1987 [19]. The DMD beam control function of light switch (figure 1) provided the best compromise of contrast gradient ratio and the entire system's brightness efficiency [20]. In a DMD system, the micromirror connects tightly on the following yoke piece. Because of the electrostatic attraction, the micromirror can turn in two directions (micromirror and yoke piece connect on deflection/resetting voltage). Each micromirror slewing angle is approximately ±12°, stops slewing until meet with mechanical resist. The yoke piece and electrode surface contact guarantees the real binary digital operation. Each picture element causes to electrode surface impacting energy is approximately 9×10^{15}J [17]. Thus, the yoke piece tip rubs electrode surface back and forth during operating process.

In order to land at another direction, the tip must be released from original position after removing the deflecting voltage, leaves landing position when rebound. The commercial micromirror life request is also getting higher and higher, it requests running surpasses 100,000 hours not to have the picture quality drop. The micromirror modulation frequency is 7 kHz. Thus, each micromirror rotates as many as 2,500 billion times in the average life span [21]. Meniscus force or van der Waals force between contact elements might be too big to overcome by reset pulse, thereby cause flaw such as macula or lightspot and so on. In order to overcome the influence which the contact adherence brings about, a low surface energy self-assembly layer is usually deposited on the tip and the landing site surface. However, high adhesion still possibly exists in certain position. Therefore to raise the manufacture output, it's necessary to research adhesive problem in DMD, in an attempt to find out its mechanism and to propose a probable plan to reduce adhesion. It's a significant work for either the industrial community or the academic circles.

3. Analysis of contact between spring tip and landing site

3.1 Analysis of contact angle

In DMD system, the meniscus and the viscous resistance are the main sources to dominant adhesive force performance.

Material wetting ability is related to contact angle θ. If the contact angle θ between solid and liquid varies from 0° to 90°, ($0°\le\theta\le90°$), liquid can soak the solid, then the solid surface is called a hydrophilic surface. If the contact angle θ varies from 90° to 180°, ($90°\le\theta\le180°$), liquid cannot soak the solid, then the solid surface is called a hydrophobic surface. A lot of factors influence contact angle, for example, surface roughness, surface pretreatment ways and cleaning degree and so on. Surface with low surface energy usually is hydrophobic. So we can by means of sputtering a layer of low surface energy self-assembly film on high energy surface to improve hydrophobic performance. In DMD, usually deposit a low surface energy layer of perfluorodecanoic acid on high-energy aluminum mirror surface to make it became hydrophobic surface, (whose contact angle is about 113° [22]). The influence caused by surface roughness to contact angle is given by the Wenzel equation [23]:

$$\cos\theta = f_r \cos\theta_0 \tag{1}$$

where θ is the contact angle of rough surface, θ_0 the contact angle of smooth surface, and $f_r\,(\ge1)$ the roughness coefficient, respectively. Because of increasing actual contact areas, equation (1) indicates: to the hydrophobic surface, roughness existence increases the surface hydrophobic capability, but to hydrophilic surface, roughness existence increases the surface hydrophilic capability. To design a surface with appropriate roughness, an ideal hydrophobic or hydrophilic surface is crucial [24, 25]. If there is bubble between the rough peak gaps, composing a solid-liquid-gas compound interface, it is better for dewetting performance of hydrophobic surface [26], but this state usually unstable.

In DMD system, the spring's surface roughness (Rq) is about 0.8 nm, while the value of landing site's surface roughness (Rq) is usually about 3.2 nm (figure 2), the interface is a rectangle size of 0.41 um×0.65um. Therefore two surfaces are treated as smooth in the following simplified analysis.

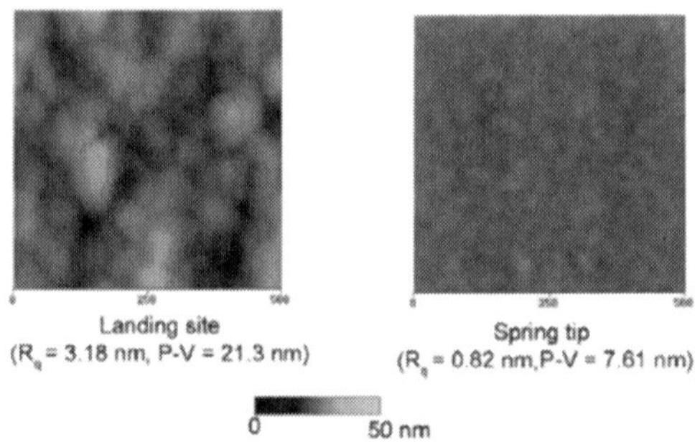

Figure 2. SEM image of contact surface [17].

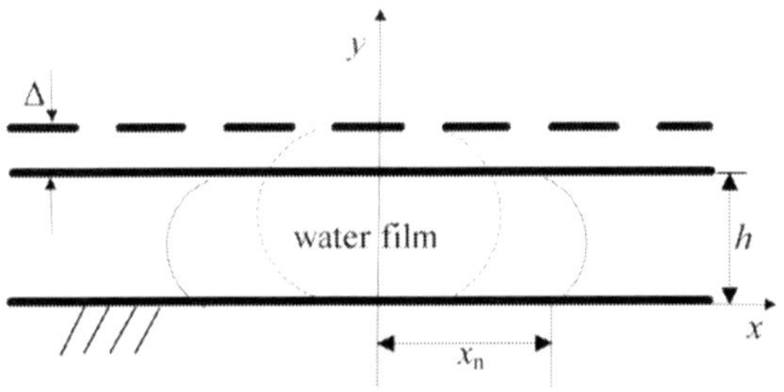

Figure 3. Meniscus shape between smooth surfaces.

When a thin water layer enters into two smooth surfaces, it will form a meniscus due to surface energy. Generally speaking, the hydrophilic surface will form concave surface but the hydrophobic surface will lead to convex surface.

The pressure difference (i.e. the Laplace pressure) caused by the surface tension γ, is given by the equation below:

$$\Delta p = \gamma\left(\frac{1}{r_1}+\frac{1}{r_2}\right) \tag{2}$$

In the above equation, r_1, r_2 are the radius of meniscus in two orthogonal planes respectively, they are related with Kelvin radius as follows:

$$\frac{1}{r_k} = \frac{1}{r_1}+\frac{1}{r_2} \tag{3}$$

The Kelvin radius r_k can be educed as follows

$$r_k = \frac{\gamma V}{\Re T \log\left(p/p_0\right)} \tag{4}$$

where V is the Moll volume, $\Re$ the gas constant, T the absolute temperature, p the ambient pressure and p_0 the saturation vapor pressure when temperature is T, respectively.

Meniscus force is calculated on the basis of the equation listed below (figure 3):

$$F_{\mathrm{m}} = \iint_{\Omega} \frac{\gamma}{r_{\mathrm{k}}} d\Omega + 2\pi\gamma x_{\mathrm{n}} \sin\theta_2 \tag{5}$$

where Ω is the meniscus area. The expression of meniscus height is:

$$h = r_{\mathrm{k}} \left(\cos\theta_1 + \cos\theta_2 \right) \tag{6}$$

The meniscus force equation is:

$$F_{\mathrm{m}} = \frac{\pi x_{\mathrm{n}}^2 \gamma \left(\cos\theta_1 + \cos\theta_2 \right)}{h} + 2\pi\gamma x_{\mathrm{n}} \sin\theta_2 \tag{7}$$

3.2 The calculation of adhesive force

Liquid flow between two surfaces during separating is described by Reynolds equation. Incompressible Reynolds equation reads:

$$\frac{\partial}{\partial r}\left(r h^3 \frac{\partial p}{\partial r} \right) = 12\eta r \frac{\partial h}{\partial t} \tag{8}$$

where η is the fluid viscosity. We can get:

$$\Delta p = 3\eta \frac{r^2 - x_{\mathrm{n}}^2}{h^3} \frac{\partial h}{\partial t} \tag{9}$$

The maximal pressure difference and the mean pressure difference are:

$$\Delta p_{\max} = -3\eta \frac{x_{\mathrm{n}}^2}{h^3} \frac{\partial h}{\partial t} \quad \text{and} \quad \Delta p_{\mathrm{avg}} = -\frac{3}{2} \frac{x_{\mathrm{n}}^2}{h^3} \eta \frac{\partial h}{\partial t} \tag{10}$$

And then, using the above equation the viscous resistance reads

$$F_{\mathrm{v}} = \int_0^{x_{\mathrm{n}}} 2\pi \Delta p_{\mathrm{m}} r dr = \frac{3\pi\eta x_{\mathrm{n}}^4}{4t_s} \left(\frac{1}{h_s^2} - \frac{1}{h_0^2} \right) \tag{12}$$

where h_s is the liquid height when meniscus fracture during separation, and h_s and x_n are determined real-timely in the computational process, which is expressed in reference [27]. Parameter h_0 is the initial liquid height.

The meniscus force and viscous force during separation will give rise to the following relation:

$$F_{\mathrm{a}} = \frac{\pi x_{\mathrm{n}}^2 \gamma \left(\cos\theta_1 + \cos\theta_2 \right)}{h} + 2\pi\gamma x_{\mathrm{n}} \sin\theta_2 + \frac{3\pi\eta x_{\mathrm{n}}^4}{4t_s} \left(\frac{1}{h_s^2} - \frac{1}{h_0^2} \right) \tag{13}$$

3.3 Calculation results

Table 1 and 2 show the adhesive force (the sum of meniscus force and viscous force here) variation features with different conditions, where the separation velocity is set to 4 m/s. From table 1, it is shown that the increase in the initial liquid height h_0 will lead to a decrease in F_a (absolute value). Table 2 tells that a larger contact angle (which means a more strong hydrophobicity) will give rise to an increase in F_a (also related to absolute value).

Table 1. Adhesive force of different liquid volume on hydrophobic surface.

($\theta_1 = \theta_2 = 120^\circ$)

x_{n0} (nm)	h_0 (nm)	F_a (nN)
100	4	-981.11
100	6	-743.61

Table 2. Adhesive force of different contact angle on hydrophobic surface.

($x_n = 100\text{nm}, h_0 = 4\text{nm}$)

θ_1 (°)	θ_2 (°)	F_a (nN)
120	120	-981.11
120	150	-1156.80

The separate velocity effects are shown in table 3. Clearly, the larger of the velocity, the stronger the viscous force is. The variation of parameter F_a during separation is plotted in figure 4, by which it is abruptly decrease to zero during the separation.

Table 3. Adhesive force of different separate velocity on hydrophobic surface.

($\theta_1 = \theta_2 = 120^\circ, x_n = 100\text{nm}, h_0 = 4\text{nm}$)

V(ms^{-1})	F_a (nN)
4	-981.11
6	-1169.33
8	-1357.56

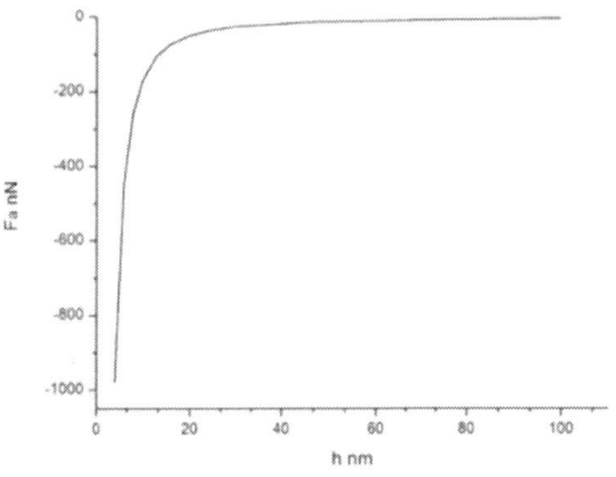

Figure 4. Adhesive force on hydrophobic surface: $x_n = 100\text{nm}, h_0 = 4\text{nm}, \theta_1 = \theta_2 = 120^\circ, V = 4\text{m}/s$.

4. Conclusions

In digital micromirror device (DMD) system, there are many factors affect the adhesive force between the spring tip and the landing site, among which the meniscus and viscous effects usually take a dominant role. The contributions of meniscus force and viscous resistances are analyzed, and the parameters of different liquid volume, contact angle as well as separate velocity are considered. It will provide some feasible candidate solutions to reduce adhesion so as to withstand a longer operation life of DMD components.

Acknowledgements

The financial supports of the National Natural Science Foundation of China (50705006) and Beijing Natural Science Foundation (3082015) are gratefully acknowledged.

References

[1] Kim S H, Asy D B and Dugger M T 2007 *Nanotoday*. **2**(5): 22-29
[2] Braun O M and Naumovets A G 2006 *Surface Science Reports*. **60**: 79-158

[3] Chaohui Zhang 2005 *Tribol. Int.* **38**(4): 443-448
[4] Zhao Y P 2003 *Acta Mechanica Sinica*. **19**(1): 1-10
[5] Bhushan B 2007 *Microelectronic Engineering*. **84**: 387-412
[6] Wang S T, Song Y L and Jiang L 2007 *Nanotechnology*. **18**(015103): 1-5
[7] Guo Z G, Fang J, Wang L and Liu W M 2007 *Thin Solid Films*. **515**: 7190-7194
[8] Sun M H, Luo C X, Xu L P, Ji H, Ouyang Q, Yu D P and Chen Y 2005 *Langmuir*. **21**: 8978-8981
[9] Nosonovsky M and Bhushan B 2005 *Microsyst Technol*. **11**: 535-549
[10] Nosonovsky M and Bhushan B 2007 *Ultramicroscopy*. **107**: 969-979
[11] Guo Z G and Liu W M 2007 *Plant Science*. **172**: 1103-1112
[12] Autumn K, Liang Y A, Hsieh S T, Zesch W, Chan W P, Kenny T W, Fearing R and Full R J 2000 *Nature*. **405**: 681-685
[13] Hiratsuka K, Bohno A and Kurosawa M 2001 *Mathematics, Physics and Chemistry*. **10**: 345-348
[14] Cai S B and Bhushan B 2007 *Nanotechnology*. **18**(465704): 1-14
[15] Cai S B and Bhushan B 2008 *Mat. Sci. Eng. R*. **61**: 78-106
[16] Henck S A 1997 *Tribol. Lett.* **3**: 239-247
[17] Liu H W and Bhushan B 2004 *Ultramicroscopy*.**100**: 391-412
[18] Douglass M R 2003 *Proc SPIE Int. Soc. Opt. Eng.* **4980**: 1-11
[19] Hornbeck L J 1996 Proceedings of the 1996 IEEE/LEOS Summer Topical Meeting Keystone, CO, USA.
[20] Zhou J X and Wu R Z 2003 *Digital micromirror device*. Chinese J. **18**(6): 445-449 (in Chinese)
[21] Douglass M R 1998 Proceedings of the 1998 36th IEEE International Reliability Physics Symposium. Reno, NV, USA. 9-16
[22] Lee K K, Cha N G, Kim J S, Park J G and Shin H J 2000 *Thin Solid Films*. 377-378: 727-732
[23] Wenzel R N 1936 *Industrial Engineering Chemistry*. **28**: 988-994
[24] Sun M H, Luo C X, Xu L P, Ji H, Ouyang Q, Yu D P and Chen Y 2005 *Langmuir*. **21**: 8978-8981
[25] Bhushan B and Sayer R A 2007 *Microsyst Technol*. **13**: 71-78
[26] Cassie A and Baxter S 1944 *Faraday Soc*. **40**: 546-55
[27] Zhang C H, Liu S S, Xu X and Ying L. Calculation of meniscus force during separation of two flat micro-surfaces. *Chinese Science Bulletin* (submitted)

Journal of Physics: Conference Series **188** (2009) 012027 doi:10.1088/1742-6596/188/1/012027

Preparation and photocatalytic activity of La and Y co-doped nano TiO_2

Tianyong Zhang*, Huixian Shi, Quisheng Yang, Xu Liu, Cong Han
(School of Chemical Engineering and Technology, Tianjin University, Tianjin 300072, China)

Abstract: La and Y co-doped TiO_2 nano photocatalyst (La-Y/TiO_2) was prepared by sol-gel method, and characterized by X-ray diffraction (XRD), specific surface area (BET), scanning electron microscope (SEM) and high resolution transmission electron microscopy (HRTEM). The photocatalytic activity of the La-Y/TiO_2 was evaluated by the degradation of methylene blue (MB) under UV light. Parameters effecting photocatalytic process such as La and Y doping concentration, calcination temperature, the dosage of catalyst and the pH of MB solution were investigated. The co-doped TiO_2 catalyst showed obviously higher photocatalytic activity for the degradation of methylene blue (MB). The presence of La and Y ions in the TiO_2 particles would prevent the recombination of the electron-hole on the surface of TiO_2 effectively.

Keywords: rare earth; photocatalyst; nanometer; TiO_2; methylene blue

1. Introduction

Photocatalytic degradation and complete mineralization of toxic organic compounds in wastewater and air by nanometer semiconductor catalysis has received much attention in the recent years. TiO_2 is considered the most promising photocatalyst due to its high efficiency, excellent physical, chemical stability, nontoxicity, and relative cost. However, the use of TiO_2 is impaired by its wide band gap, which requires ultraviolet irradiation for photocatalysis. Modification of TiO_2 is necessary to enhance the efficiency to enhance the efficiency of TiO_2 photocatalysis and to improve the photocatalytic activity. Using dopants in TiO_2 lattice is one of the methods to reach this goal [1-7]. Some studies indicated that the photocatalytic activity of TiO_2 can be significantly enhanced by doping with lanthanide ions/oxides with 4f electron configurations and the interaction of the functional groups with their f-orbital. Numerous studies of rare earth ions in recently years have been focused on the photocatalytic activities of rare earth elements hosted in crystalline matrices[8-11]. However, two kinds of rare earth metal co-doped TiO_2 and their catalytic properties have seldom been presented so far. In this paper, La and Y co-doped TiO_2 nano photocatalyat (La-Y/TiO_2) was prepared by sol-gel process. The photocatalytic activity of the prepared catalyst was evaluated via the degradation of methylene blue (MB) in the aqueous solution under UV irradiation. Parameters effecting photodegradation process such as variant doping amount, catalyst dosage, calcination temperature, and pH were studied.

2. Experimental

2.1 Materials

Tetra-*n*-butyloxy titanium [$Ti(OBu)_4$], $La(NO_3)_3$, $Y(NO_3)_3$, acetic acid, absolute ethanol, nitric acid, MB are all analytically pure.

Tianyong Zhang, born in 1966, male, Prof. Tel: +8622-27406610, E-mail: tyzhang@tju.edu.cn

2.2 Preparations of catalysts

Catalysts were prepared by sol-gel method with following procedure: 34mL $Ti(OBu)_4$ was dissolved in 44mL absolute ethanol with stirring for several minutes, then 3mL acetic acid was added as suppressant to given solution A. Solution B containing 44mL absolute ethanol, 7.2mL H_2O, and rare earth metal salt in the required stoichiometry was slowly added to solution A. The mixture was hydrolyzed at room temperature for 30min under agitation and a transparent sol was obtained. The gelation was prepared by aging the sol for 24h at room temperature. The gelation was dried at 60℃, then ground to power and calcined at different temperture.

2.3 Characterization of the catalyst

The XRD (X-ray diffraction) patterns were recorded on a PANAalytical X'pert Por X-ray diffractometer using Cu Kα radiation as the X-ray source. The diffractograms were recorded in the 2θ range 20~100° in steps of 0.017°.

The specific surface area (BET method), specific pore volume and average pore diameter (BJH method) of the sample were determined by nitrogen adsorption-desorption isotherms using Quantachrome NOVA-2000 sorption analyzer. The samples were analyzed at 77K by nitrogen adsorption-desorption.

The particle size and morphology of La-Y/TiO_2 was observed using high resolution scanning electron microscope (HRSEM) using VEGA TS-5130SB produced by TESCAN Company in Czech. Light source of electron-beam was tungsten lamp, the voltage was 10~30 kV.

2.4 Photocatalytic degradation

The photocatalytic activity of La-Y/TiO_2 was evaluated by degradation of MB. The initial concentration of MB was 20mg·L^{-1}, put a fixed weight of the catalyst. Prior to the irradiation, the reaction mixture was kept in the dark for 20min to ensure sufficient adsorption of the MB. The mixture was then irradiated with UV light of 125W Hg lamp in a pyrex cylindrical photo reactor. The samples were collected at regular intervals of time and concentration changes of MB solution were measured using a UV-vis spectrometer at 665nm (λ_{max}). The activity of catalyst was evaluated by the decolor rate (D) of the samples. The equation of the decolor rate followed

$$D = (A_o - A_t) / A_o \times 100\%$$

Where D is the decolor rate, A_o is the initial absorbance of MB and A_t is the absorbance of MB after '*t*' minutes.

3. Results and discussion

3.1 Characterization of the photocatalysts

XRD patterns of nano TiO_2 (prepared by sol-gel method), La-TiO_2 (dopant of La is 0.005mol), La-Y/TiO_2 (the dopant of La is 0.005mol and the dopant of Y is 0.008mol) calcined at 600℃ are shown figure 1. The pattern of TiO_2 corresponds to both anatase and rutile phase but the former is the predominant one. The XRD patterns of La-TiO_2 and La-Y/TiO_2 indicate the anatase phase. It can be seen from figure 2 that the peak intensity of anatase increases with the increase of calcination temperature from 400℃ to 900℃, and the width of (101) plane becomes narrow. The rutile phase appears at 900℃ for La-Y/TiO_2, the phase transformation to rutile but the anatase phase was dominant up to 900℃. Thus the dopant is expected to play a significant control in the selective crystallization of anatase phase during sol-gel process [12]. Further, among the two main kinds crystalline of TiO_2, anatase is proved to exhibit higher photocatalytic activity. The anatase phase also exhibits low rate of recombination in comparision to rutile due to its ten fold greater rate of hole traping[13].

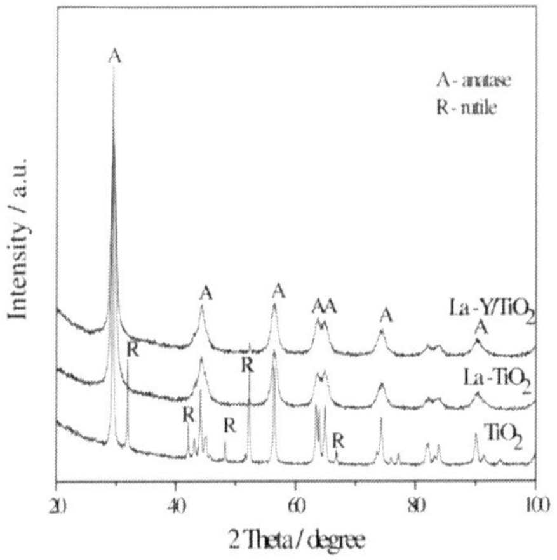

Figure 1 XRD patterns of TiO_2, La-TiO_2, La-Y/TiO_2 calcined at 600℃

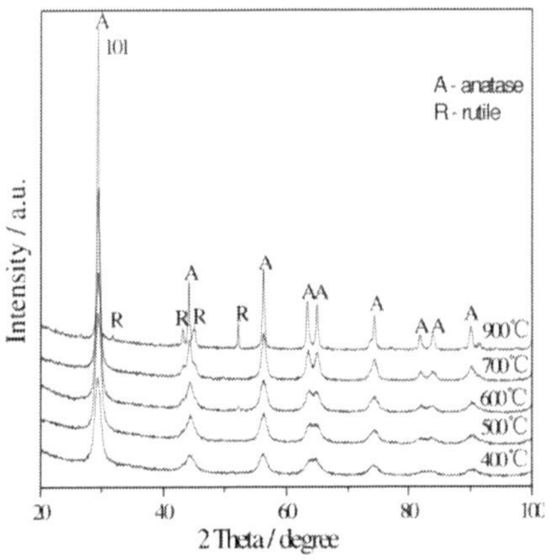

Figure 2 XRD patterns of La-Y/TiO_2 calcined at different temperature

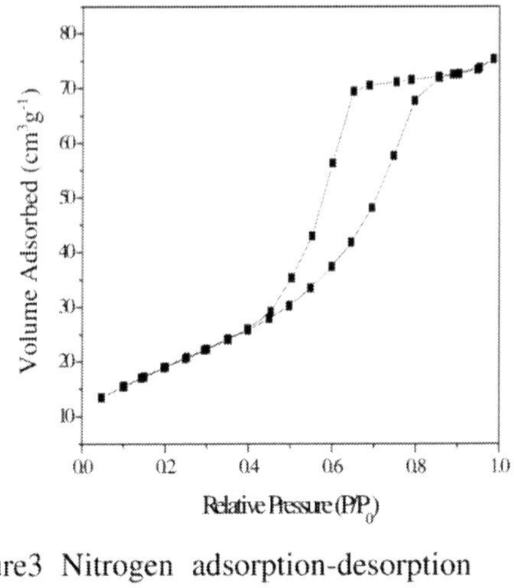

Figure3 Nitrogen adsorption-desorption isotherm La-Y/TiO_2

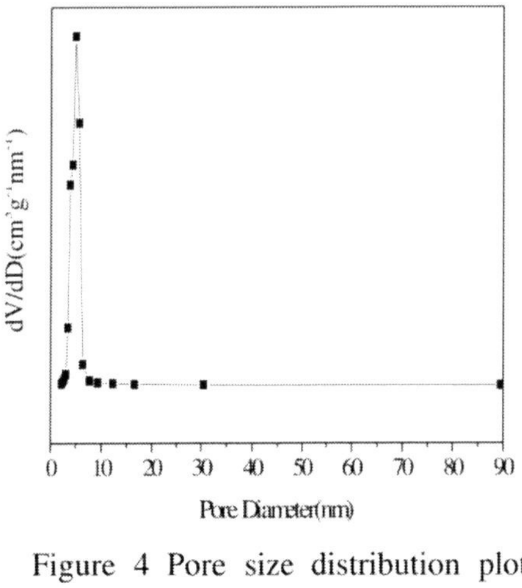

Figure 4 Pore size distribution plot of of La-Y/TiO_2

The nitrogen adsorption-desorption isotherm and the pore size distribution plot of the La-Y/TiO_2 can be seen in the figure 3, figure 4 respectively. The nitrogen adsorption-desorption isotherm was characteristic of a type Langmuir IV isotherm with a H2 hysteresis loop. The pore size distribution calculated from the desorption branch of nitrogen isotherm by BJH method, which indicated that the pore diameter of the sample was about

3-7nm, the pore volume was $0.127 cm^3 g^{-1}$ and the BET specific surface area of the sample was $102 m^2 g^{-1}$. These pores may allow rapid diffusion of MB molecules during photocatalytic reaction, and also enhance the adsorption of MB and its intermediates on nano La-Y/TiO_2 surface.

Figure 5 SEM of La-Y/TiO_2 calcined at 500℃

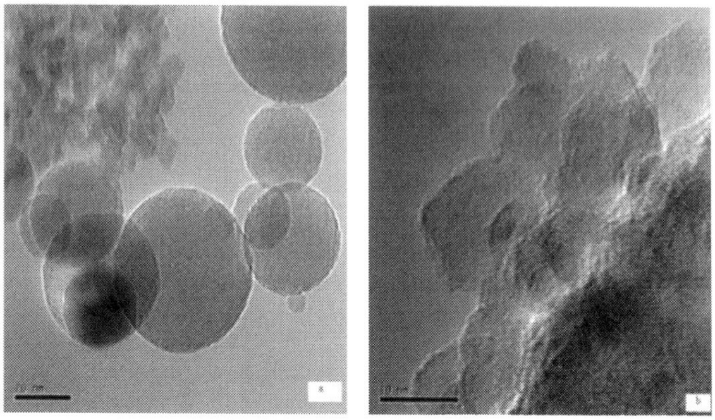

Figure 6 HRTEM of La-Y/TiO_2 calcined at 500℃ (a) and 900℃ (b)

Morphology of the La-Y/TiO_2 catalyst was determined by SEM (figure 5) and HRTEM (figure 6) pictures. SEM and HRTEM pictures conformed that the La and Y co-doped nano TiO_2 particles are spherical with an average grain size of 50-70nm. Further, the HRTEM micrographs in figure 6 characterized the surface morphology of the sample. It is observed that the sample is consist of good crystallized area of nanosized La-Y/TiO_2.The HRTEM picture of La-Y/TiO_2 calcined at 900℃ showed that many small particles deposit over nano TiO_2 surface, and these tiny particles may be La^{3+} and Y^{3+}, This is quite reasonable as the ionic radius of La^{3+} and Y^{3+} is larger than Ti^{4+}. So it can prevent the recombination of the electron-hole on the surface of TiO_2 effectively.

3.2 Photocatalyic degradation

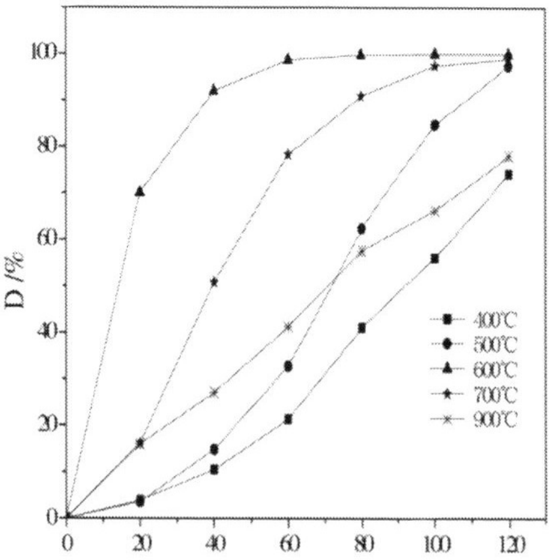

Figure 7 Effect of calcination temperature

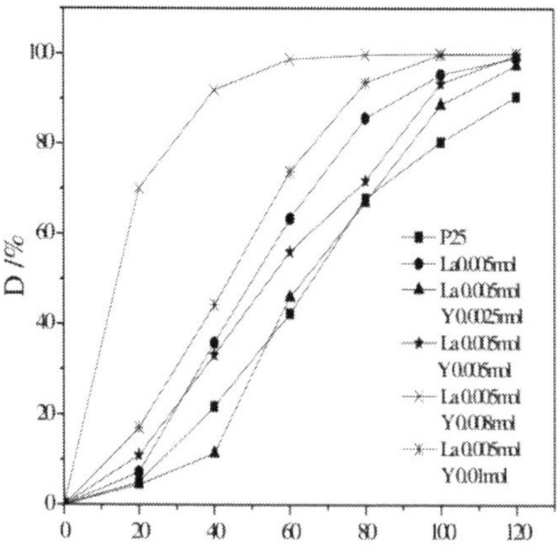

Figure 8 Effect of La and Y dopant amount

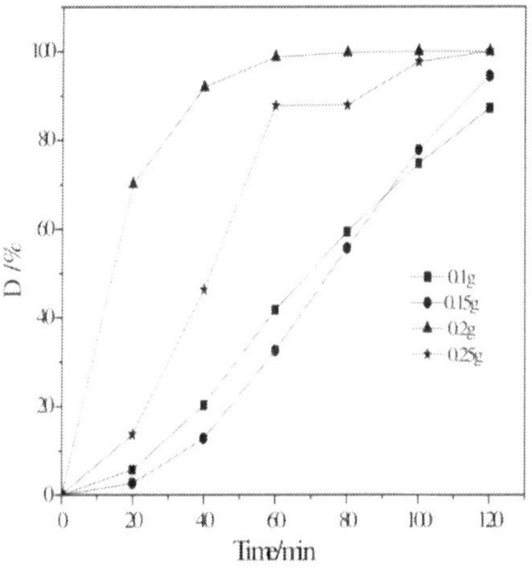

Figure 9 Effect of the catalyst dosage

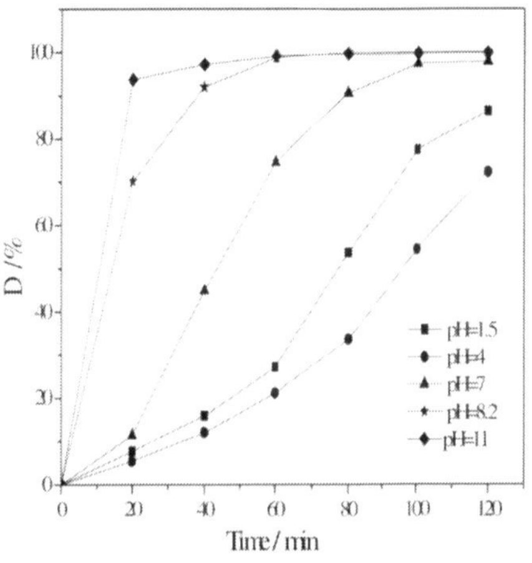

Figure 10 Effect of pH

The efficiency of MB photodegradation as a function of operating parameters is shown in figure 7-10. To study the effect of catalyst structure on photodegradation, the sample should be calcined at different temperature that shows different crystal structures. It can be shown from figure 7 that the highest degradation is achieved with the sample calcined at 600℃, and the peak intensity of anatase increases and the width of (101) plane becomes narrow. The anatase structure has a better photocatalytic activity [14]. The catalytic activity of La-Y/TiO_2 declined when calaining temperature rise 900℃ where the rutile structure formed, but the decolor rate of MB still can reach 78% after irradiation 120 min with 125W Hg lamp. So the doping La and Y can enhance the stability of the catalyst.

Figure 8 shows the effect of La^{3+} and Y^{3+} dopant concentration on MB photodegradation. For a solution containing initially 20mg/L MB after 120 min UV irradiation, the photocalystic degradation efficiency of La-Y/TiO_2 is higher than P-25 and La-TiO_2. The degradation efficiency increase with increasing Y concentration and a maximum of 100% is obtained with sample containing 0.005mol La^{3+} and 0.008mol Y^{3+}, a further increase in Y^{3+} concentration to 0.01mol leads to a slight decrease in degradation efficiency to 99%. As the content of doping ions increases, the surface barrier becomes higher, and the space charge region becomes narrower. The electron-hole pairs within the region are efficiently separated by the large electric field before recombination which led to the higher photocatalytic activity. However, when the content of doping ions is excessively high, the space region becomes very narrow and the penetration depth of the light into TiO_2 greatly exceeds the space charge layer[15], therefore the recombination of the photo-generated electron-holes pairs becomes easier, which leads to lower photocatalytic activity for MB degradation. Otherwise, the La^{3+} an Y^{3+} co-existence leads to a synergistic effect.

The effect of the dosage of La-Y/TiO_2 on MB degradation is illustrated in figure 9. It is expected that MB degradation increases with increasing the dosage of La-Y/TiO_2 (the dopant of La is 0.005mol; the dopant of Y is 0.008mol), and reaches a maximum of 100% at a dosage of 0.2g/50mL. However, a further increase in catalyst dosage slightly reduces the photodegradation efficiency. It is known that the photodegradation rate of the solution is affected by not only the active sites but also the photo-absorption of the catalyst. Adequate dosage of the photocatalyst increases the generation rate of electron-hole pairs for enhancing photodegradation, but a high dosage of the photocatalysts will decrease the light penetration by the suspension and reduce the degradation rate [16].

Solution pH is an important variable in the evaluation of aqueous phase mediated photocatalytic degradation reactions. It influences adsorption and dissociation of the substrate, catalyst surface charge, oxidation potential of the valence band and other physico-chemical properties [17]. The effect of solution pH on MB degradation is illustrated in figure 10. It can be seen that the highest photocatalytic activity can be reach when the pH of MB solution is 11, the high degradation rate in the alkaline pH range due to the MB is cationic dye, the catalyst surface has negative charge in the alkaline condition, which benefit to absorption of MB.

4. Conclusions

La-Y/TiO_2 nano photocatalyst was prepared via an acid-peptized sol-gel method. Particles that were annealed at 600℃ revealed the highest MB photodegradation efficiency, primarily because of the anatase structure. When La^{3+} and Y^{3+} doped 0.005 mol and 0.008 mol respectively in 1 mol TiO_2, the photocatalyst has the highest activity, the decoloration rate of MB can reach 100% in irradiation time 120 min with 20 mg catalyst and 125W Hg lamp. The photodegradation efficiency is higher than La-YiO_2 and P-25 in the same conditions.

References

[1] Gandhe A R and Fernandes J B 2005 *J. Solid Stat. Chem.* **178** 2953
[2] Chatterjee D and Dasguple S 2005 *J. Photochem. Photobiol.* C **6** 185
[3] Masih D, Yoshitake H and Izumi Y 2007 *Appl. Catal.* A **325** 276
[4] Mohamed M M and Al-Esaimi M M 2006 *J. Mol. Catal.* A. **255** 53
[5] Zhou M, Yu J and Cheng B 2006 *J. Hazard. Mater.* B **137** 1838
[6] Wu C S and Chen C H 2004 *J. Photochem. Photobiol.* A. **163** 509
[7] Kubacka A, Fuerte A, M-Arias A and Garcia M F 2007 *Appl. Catal.* B **74** 26
[8] Zhou J, Takeuchi M, Ray A K, Anpo M and Zhao X S 2007 *J. Colloid Interface Sci.* **311** 497
[9] Zhu J, Zhu K and Chen L 2006 *J Non-Cryst Solids* **352** 150-154
[10] Gacon J C, Horchani K, Jouini A and Dujardin C 2006 *Opt. Mater.* **28** 14-20
[11] Koepke C, Wisniewski K, Sikorski L, Piatkowski D, Kowalska K and Naftaly M 2006 *Opt. Mater.* **28** 129-136
[12] Hirano M, Nakahara C, Ota K and Inagaki M 2002 *J. Am. Ceram. Soc.* **85** 1333
[13] Sakthivel S, Hidalgo M C, Bahnemann D W, Geissen S U, Murugesan V and Vogelpohl A 2005 *Appl. Catal.* B **63** 31
[14] Barakat M A, Chen Y T and Huang C P 2004 *Appl. Catal.* B **53** 13-20
[15] Chun-Hua liang, Fang-Bai Li and Cheng-Hui Liu 2008 *Dyes and Pigments* **76** 477-484
[16] Chiou C H and Juang R S 2007 *J. Hazard. Mater.* **149** 1-7
[17] Venkatachalam N, Palanichamy M and Murugesan V 2007 *J. Mol.Catal.* A **273** 177-185

Journal of Physics: Conference Series **188** (2009) 012028 doi:10.1088/1742-6596/188/1/012028

Growth of copper-indium nanorods on Si substrate using porous anodic alumina as template

Z W Zhang[1], H Lam[2], J Li[1], M Wang[1], G S Jiang[1], C Surya[2] and C F Zhu[1,3]

[1] Department of Materials Science and Engineering
University of Science and Technology of China, Hefei, Anhui 230026, Peoples Republic of China
[2] Department of Electronic and Information Engineering
The Hong Kong Polytechnic University, Hong Kong, China

E-mail: cfzhu@ustc.edu.cn

Abstract. Copper-indium nanorod arrays have been synthesized by electrodeposition using porous anodic alumina nano template supported on a silicon substrate. Porous anodic alumina template was fabricated by evaporating aluminium film onto silicon substrate and then anodizing the aluminium film in diluted phosphoric acid. The approach employs a plasma etching to penetrate the alumina pore barrier layer before electrodeposition, which enables direct electrical and chemical contact with the silicon substrate electrode. The resulting template and copper-indium nonorods obtained were characterized using scanning electron microscopy (SEM) and energy-dispersived X-ray spectroscopic analyzer (EDS), the pores of alumina are found to have dimensions of 150–250 nm pore diameters and 330–510 nm pore spacings, partial filling of the pores of the alumina template by Cu-In is achieved. The results of this work reveal that the alumina nano template is particularly well suited to the etching mask and template-assisted growth of nanostructures to be integrated into rigid substrate.

1. Introduction

Ordered one-dimensional nanorods have attracted considerable interests recently for their fundamental importance and potential applications in many fields, such as data storage, field emission display, light-emission diodes and energy conversion [1-3]. To fabricate the nanorod arrays, it has been considered that the template-directed synthesis method is the simplest and convenient pathway as an inexpensive alternative to lithographic patterning method. Porous anodic alumina (PAA) has been frequently used for the templated synthesis of nanorods, since this porous template material is consisted of a self-assembled honeycomb array of uniform size parallel channels [4,5], pores with different diameters and pore interval distance can be created by choosing the proper anodization condition of the aluminium layer. The majority of previous studies used Al foils to prepare PAA films, which must eventually be separated from the Al foil for later applications, and the separated PAA film is brittle, so great care needs to be exercised in the preparation and manipulation of the PAA film. This is apparently not suitable for most integrated device applications. Recently, nanoporous PAA has been prepared directly by Al films coated on silicon wafers for integrated microelectronic applications [6,7]. Moreover, in PAA membrane, there is a continuous and dense alumina barrier layer at the pore base, which prevents the direct electrical and chemical contact between the pore and the base conducting substrate. Therefore, it is necessary to develop low-cost and simple methods to remove the barrier layer and get throughout channels to the substrate.

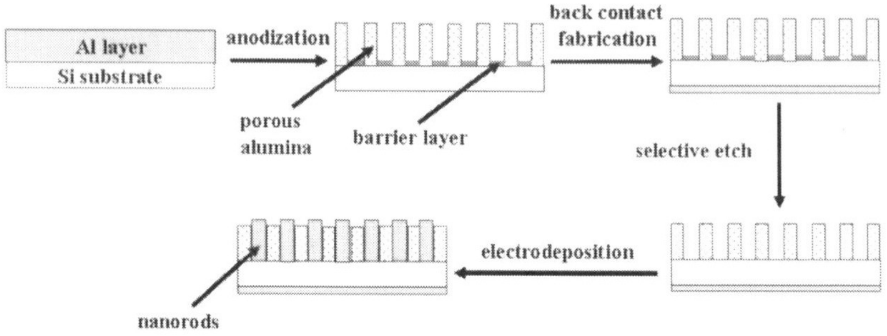

Figure 1. Schematic diagram describing the fabrication of PAA films and nanorod arrays on silicon substrate and other rigid substrates.

In this paper, we present a procedure that combines an anodization process to produce template arrays and a reactive ion plasma etching (RIE) method to get throughout channels to the substrate. Large-area PAA membranes with large cells on silicon substrate were prepared by anodizing evaporated aluminium film in phosphoric acid solution at 160V. RIE was employed to remove the barrier layer in the channel base, and copper-indium nonorods were fabricated by electrodeposition in the PAA template. Figure 1 illustrates the fabrication process in this work.

2. Experiment

An aluminium film with thickness approximately 2 μm was deposited on an N type silicon substrate with resistivity of 1~3 Ω · cm by electron-beam evaporation. A two-step anodization was then carried out in a 5wt% phosphoric acid solution at 1 °C and a DC potential of 160 V. The alumina yielded in the first anodization step was dissolved by immersing the sample in a mixed acid solution with phosphoric acid and chromic oxide solution (6wt%H_3PO_4+1.8wt%H_2CrO_4) at 60 °C. The second anodization was then carried out until the current in the circuit dropped significantly. To ensure complete conversion of the entire aluminium film into its oxide, the voltage was maintained for 60 s after the current had stabilized at this low value. By controlling the time of the second step, we can control the ultimate thickness of the PAA film. Subsequently, the AAO sample was subjected to chemical etching (in 5wt. % H_3PO_4 for 30~50min at 30℃) to enlarge the pores and thin the barrier layer at the bottom of PAA template. Next, the sample with the pore widened alumina on the silicon substrate was subjected to reactive ion etching. The gases used were CHF3 and Ar with flow rates of 5 sccm each. The power was set at 250 W and the chamber pressure was maintained at 25 mTorr. The etching duration was 5 min. Co-depositions of copper and indium through the nanometer-scale windows in the PAA were performed by electrodeposition technique. A three electrode system with platinum counter electrode and silver chloride reference electrode was used. An aqueous solution contains 3mM $CuCl_2$, 10mM $InCl_3$, the solution pH was adjusted to 1.8 with HCL solution. The electrodeposition process was conducted in constant voltage mode at 0.63V at room temperature, and the deposition time was 15 min.

The morphology and composition of the PAA and Cu-In nanorods were characterized, respectively, by a JEOL JSM-6390LA scanning electron microscope (SEM) and a JED-2300 energy dispersive X-ray analyzer (EDS).

3. Results and discussion

To obtain a PAA template with straight channels, Al film of a high purity and smooth surface is needed, so the substrate was rotated and no heating was applied during e-beam evaporation. In the case of phosphoric acid electrolyte, it was not easy to continue electrolysis for a long time without burning, namely. The temperature of the electrolyte need to be maintained at 1℃ during anodization using a cooling system and the solution was stirred vigorously using a stirrer in order to accelerate the diffusion of the heat that evolved from the sample, which is important or preventing localized temperature increases and to maintain the stable growth of the anodic oxide film.

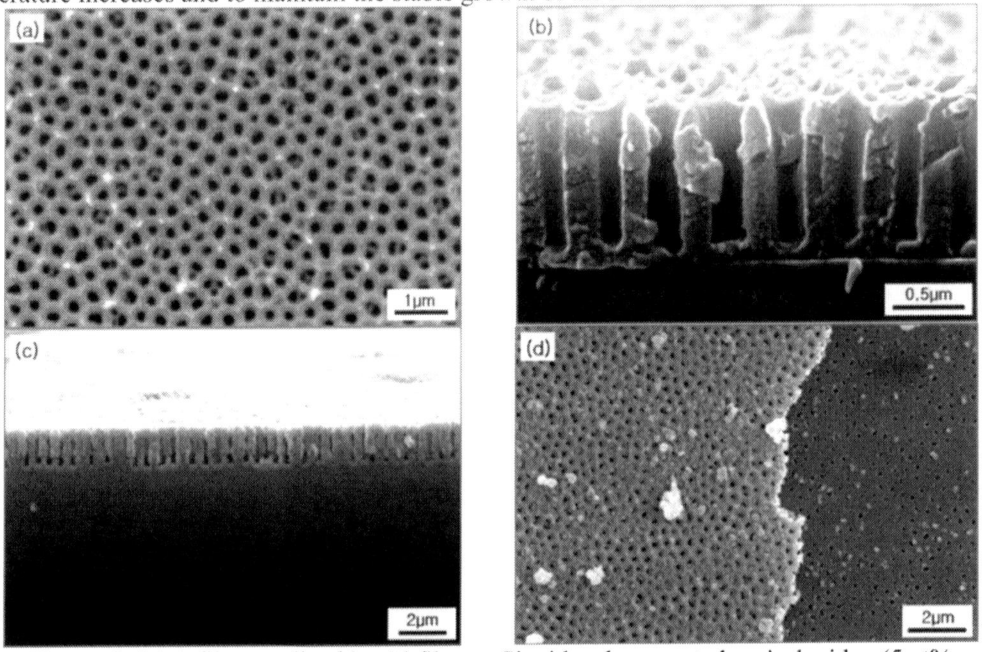

Figure 2. PAA film anodized by Al film on Si with subsequent chemical widen (5wt% H3PO4, at 30℃, 30min.) (a) plane view, (b) cross-sectional view, (c) after RIE etching, and (d) top-view of PAA template filled by Cu-In nanorods.

Figure 2(a) and (b) show the surface morphologies and cross-sectional images of the prepared PAA that were widened in 5wt% H3PO4 solution at 30℃ for 30 min, respectively. The diameters of the nanopores are around 150-250 nm and pore interval distances are 330-520 nm, while the template thickness is approximate 1200 nm. The homogeneity of cell arrangement is rather inferior to those produced in other electrolytes, such as oxalic and sulfuric acid solutions [8,9], because the high voltage tends to induce local events such as electric breakdown, local thickening of the barrier layer and pore branching.

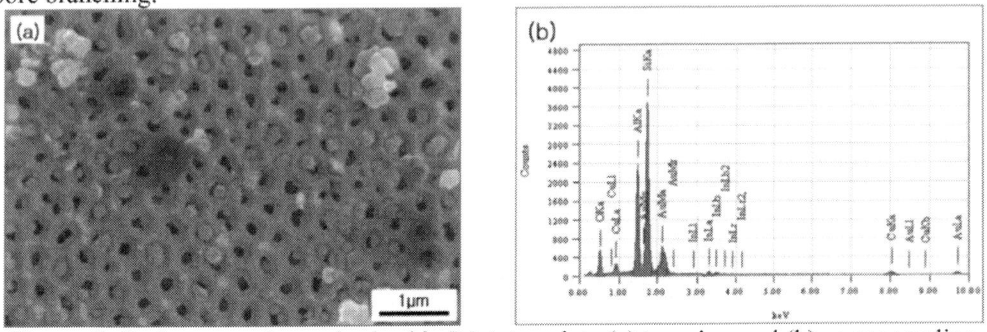

Figure 3. Cu-In nanorods synthesized in PAA template (a) top view and (b) corresponding EDX spectroscopy of the template and nanorods.

Figure 2(c) shows SEM cross-sectional image of the PAA template on silicon after RIE etching. It is obvious that the barrier layer was etched away and the substrate also was etched due to long etching duration. Cu-In alloy nanorods were deposited in the PAA template on Si substrate as shown in figure 2(d). From the experimental results on EDS measurement in figure 3(a) and (b), the peaks of the Cu and In component were observed confirms that the nanorods are Cu-In alloy. The experimental data above clearly show that the incorporation of porous anodic alumina membrane and the template-assist growth of nanorods arrays were achieved, also the PAA template is a good etching mask for fabrication of other nanostructures. Forfuture work a better control of the etching and electrodepsition would be desirable.

4. Conclusion

We report a simple approach for the synthesis copper-indium nanorods on rigid substrate, by combing the direct anodization of aluminum on Si substrate, RIE etching and electrodeposition technique. The pore diameter, interval distance and thickness of the PAA template can be controlled by selecting proper anodization conditions, which makes the alumina nano template particularly well suited to the etching mask and template-assisted growth of nanostructures to be integrated into rigid substrate. Since there is a support of substrate, the resulting structure is robust enough to be grown in large area. The Cu-In nanorods grown on the substrate also have the potential applications in copper indium diselenide based solar cells.

Acknowledgments

This work was supported by Knowledge Innovation Project of The Chinese Academy of Sciences (NO. KJCX2-YW-M11).

References

[1] Mallet J, Zhang K Yu, Matefi-Tempfli S, Matefi-Tempfli M and Piraux L 2005 *J. Phys. D* **38** 909-14.
[2] Yin A J, Li J, Jian W, Bennett A J and Xu J M 2001 *Appl. Phys. Lett.* **79** 1039-41
[3] Tian B Z, Zheng X L, Kempa T J, Fang Y, Yu N F, Yu G H, Huang J L and Lieber C M 2007 *Nature* **449** 885-90
[4] Sullivan J P O and Wood G C 1970 *Proc. Roy. Soc. Lond. A.* **317** 511-43
[5] Masuda H and Fukuda K 1995 *Science* **268** 1466-68
[6] Rabin O, Herz P R, Lin Y M, Akinwande A I, Cronin S B and Dresselhaus M S 2003 *Adv. Funct. Mater.* **13** 631-38
[7] Michael M Crouse, Albert E Miller, David T Crouse and Ataul Aziz Ikram 2005 *J. Electrochem. Soc.* **152** D167-72
[8] Hwang Sun-Kyu, Lee Junghyun, Jeong Soo-Hwan, Lee Pyung-Soo and Lee Kun-Hong 2005 *Nanotechnology* **16** 850-58
[9] Sachiko O, Makiko S and Hidetaka A 2004 *Electrochem. Solid-State Lett.* **7** B21-24

Journal of Physics: Conference Series **188** (2009) 012029 doi:10.1088/1742-6596/188/1/012029

Light distribution analysis of optical fibre probe–based near-field optical tweezers using FDTD

B H Liu[1], **L J Yang**[2] **and Y Wang**[3]
School of Mechanical and Electrical Engineering, Harbin Institute of Technology, Heilongjiang, Harbin, 150001, China

E-mail: richelaw@163.com

Abstract. Optical fibre probe–based near-field optical tweezers overcomes the diffraction limit of conventional optical tweezers, utilizing strong mechanical forces and torque associated with highly enhanced electric fields to trap and manipulate nano-scale particles. Near-field evanescent wave generated at optical fibre probe decays rapidly with the distance that results a significant reduced trapping volume, thus it is necessary to analyze the near-field distribution of optical fibre probe. The finite difference time domain (FDTD) method is applied to characterize the near-field distribution of optical fibre probe. In terms of the distribution patterns, depolarization and polarization, the near-field distributions in longitudinal sections and cross-sections of tapered metal-coated optical fibre probe are calculated. The calculation results reveal that the incident polarized wave becomes depolarized after exiting from the nano-scale aperture of probe. The near-field distribution of the probe is unsymmetrical, and the near-field distribution in the cross-section vertical to the incident polarized wave is different from that in the cross-section parallel to the incident polarized wave. Moreover, the polarization of incident wave has a great impact on the light intensity distribution.

1. Introduction

Due to the difficulty of nanometre scale fabrication and manipulation, the practical application of this concept to the visible regime was not possible until more recently. Tanaka proposed a novel high-resolution trapping scheme and presented a new methodology for calculating the trapping forces acting on a nano-scale particle in the near-field of optical fibre probe[1]. As one of the most powerful tools for nano-manipulation of nano-scale particles, optical fibre probe-based near-field optical tweezers can provide nano-trapping with a significant reduced trapping volume due to near-field evanescent wave, which decays rapidly with the distance to metal probe. Based on the highly enhanced electric field close to the tip of optical fibre probe and the strong mechanical forces and torque associated with these fields, optical fibre probe-based near-field optical tweezers can manipulate particles in non-contact mode without damage, but its manipulation capability is limited to the intensity enhancement of probe tip, and its capture stability will be reduced due to the thermal effect of probe tip. Being one of the key components to near-field optical tweezers, tapered metal-coated optical fibre probe with nano-scale aperture determines the resolution and sensitivity of nano-manipulation; therefore, it is needed to understand its near-field distribution inside and outside.

There are many reports published about near-field optical probes using numerical method. In early research simpler models are used, including infinity thin metal plate with hole and infinity metal plate

[1] To whom any correspondence should be addressed.

with narrow gap[2,3]. These models are too simple to fit the actual configuration of the fibre probe. In recent research finite difference time domain (FDTD), multi-polar method and boundary method have been used for different configurations, focusing on the contribution of the near-field of the probe[4-7]. However, in their works the problems related to the near-field electric intensity inside and outside of fibre probe and the impact of polarization on intensity distribution have not been concerned. FDTD Method is a applicable numerical method for computing electromagnetic fields, which is a discrete straightforward calculation of Maxwell's electromagnetic equations[8]. To use this method, electromagnetic problems of sub-wavelength structures with relatively complicated geometry and medium materials can be solved in various forms. In this paper a commercial FDTD program is chosen to study the electric intensity distribution inside and outside of fibre probe in near-field region and to analyze the impact of polarization on it numerically.

2. Numerical method

The FDTD method was first introduced by Yee and it is now widely used in computational electrodynamics. The method discretizes Maxwell equations, both space and time are meshed and partial derivatives are replaced by finite differences, so it is a flexible numerical means of solving electromagnetic problems by integrating Maxwell's differential equations. Since FDTD solves Maxwell's equations without any simplifying approximations other than the discretized grid, the method is also well suited to simulating near-field configurations. FDTD solves Maxwell's equations within the grid by approximating both the time and space derivatives in the equations with finite-difference forms. Consider the curl E equation:

$$\nabla \times E = -\mu \partial H / \partial t \ . \tag{1}$$

The finite-difference form of this equation for one component of H in any given Yee cell is

$$\frac{E_x^n - E_{x-1}^n}{\delta y} - \frac{E_y^n - E_{y-1}^n}{\delta x} = \mu \frac{H_z^{n+1} - H_z^n}{\delta t} , \tag{2}$$

where the subscript on each field value gives its polarization and cell location, δx and δy are the lengths of the cell in the x and y directions, respectively, δt is the time increment, and the superscript on each field value denotes the time increment index at which the field value was calculated (i.e., $n + 1$ is one time increment after the time n). Note that this equation allows the latest value of the magnetic field, H^{n+1}, to be easily calculated from the previous value, H^n, along with the previous values of neighbouring E fields. A sinusoidal driving electric or magnetic field is added to the field values in certain cells to provide the source of the wave being modelled. The field calculations are done cell-by-cell throughout the grid at each time step, then time is advanced by δt to the next time step and new field values calculated throughout the grid. Thus, the solution is obtained in a "leapfrog" manner by updating the space solutions at each time increment. The time advance continues until steady state is reached (that is, the fields are not changing in peak-to-peak amplitude).

Typically, FDTD calculations have been carried out for dielectric materials. Recent advances have extended the use of FDTD to the treatment of materials with frequency dependent, complex dielectric constants. Real metals characteristically fall in this category. Leubbers et al. developed a FDTD algorithm for treating dispersive materials in a time-dependent fashion[9]:

$$D(t) = \varepsilon_\infty \varepsilon_0 E(t) + \varepsilon_0 \int_0^t E(t-\tau)\hat{\chi}(\tau) d\tau \ . \tag{3}$$

Here, $\chi(\tau)$ is the Fourier transform of $\chi(\omega)$, and ε_0 is the permittivity of free space.

If the metal is assumed to follow the Drude model, one can derive the pertinent optical constants:

$$\begin{aligned} &\tilde{\varepsilon} = \varepsilon' + i\varepsilon'', \\ &\varepsilon' = n^2 - k^2, \\ &\varepsilon'' = 2nk, \\ &\tilde{\varepsilon}(\omega) = 1 + \frac{\omega_p^2}{\omega(i\nu_c - \omega)} = \varepsilon_\infty + \chi(\omega), \end{aligned} \tag{4}$$

where ε is the permittivity, χ is the susceptibility, ω is the angular frequency, ω_p is the radian plasma frequency, and ν_c is the collision frequency.

The Drude model is appealing in that it is derived from the simple physical picture of a free electron gas[10]. The response of near-field probes has been simulated under the Drude model using FDTD with some success[11]. However, the Drude model does not accurately describe the real and imaginary indices of refraction of noble metals over a wide frequency range. Also, the Fourier transform of $\chi(\omega)$ is often not numerically stable for the Drude model. In order to avoid problems with the Drude model, we instead approximate the experimentally determined optical constants of gold with a modified Debye function[12]. Modified Debye materials are described with equations containing a single pole:

$$\tilde{\varepsilon}(\omega) = \varepsilon_\infty + \frac{\varepsilon_S - \varepsilon_\infty}{1 + i\omega\tau} + \frac{\sigma}{i\omega\varepsilon_0} . \tag{5}$$

Here, ε_S represents the static permittivity, ε_∞ is the infinite frequency permittivity, σ is the conductivity, and τ is the relaxation time.

FDTD simulation that runs for tens of thousands of time steps is generally computationally intensive when high spatial precision is required, in this paper a Intel Core 2 Duo 1.8GHz 2GB computer were used to calculate.

3. Numerical results

3.1. Calculation model

The calculated fibre probe consists of a Al-coated waveguide that extends infinitely to the left and has a nano-wavelength aperture at the right cutting-off end, as shown in figure 1. The waveguide, with a radium of Φ_1=700nm, is filled with homogeneous dielectric material with a relative permittivity ε=2.25. The thickness of metal cover on the surface of probe with a length of H=600nm is denoted by T, T=80nm, and the conductivity σ of metal cover is assumed to be infinity. The aperture diameter Φ_2=100nm is used in the computation that is carried out in Cartesian coordinates, setting the aperture centre of fibre probe as coordinates origin.

The excitation light polarized along axis y is injected from the left hand side of the probe, and it can be expressed by $E=E_y\sin(2\pi ft)=\sin(2\pi ft)$. The wavelength λ of the incident plane wave is 632.8nm. For the calculation, the geometry that has been employed is placed in a volume discretized in mesh cells of 100×99×130, with space step of $\Delta x=\Delta y=\Delta z$=10nm, and time step of $\Delta t=\Delta x/2c$, where c is the light speed in vacuum. Interaction between light and particles is usually caused by electric field, so the calculation mainly explores the characteristics of electric field distribution.

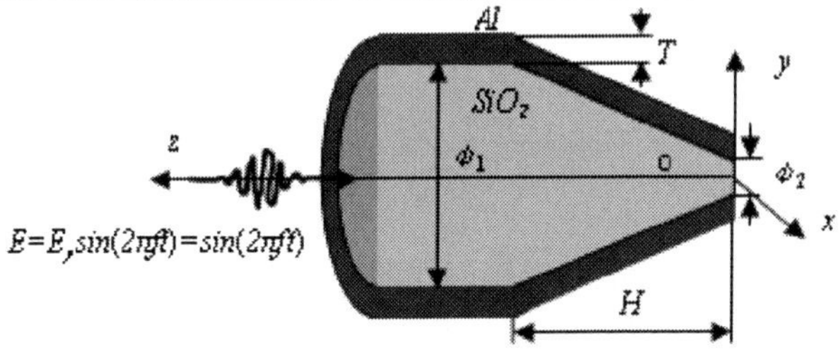

Figure 1. The cross-section of tapered metal-coated optical fibre probe

3.2. Electric field intensity distributions of cross-sections

Figure 2 shows the distributions of total electric field $|E|$ and its components $|E_x|$, $|E_y|$ and $|E_z|$ (from left to right) in cross-sections as z=0nm, -10nm and -70nm. It can be seen that the near-field of metal-coated probe becomes depolarization which obviously affects the electric field intensity distribution. The characteristics of field distributions are described as follows: the distribution of $|E_x|$ is symmetrical along axis x and axis y with four part flower pattern, moreover, the electric intensity at each axis is zero and the electric intensity from the probe tip decreases gradually with the increasing distance to aperture centre. The maximum of $|E_y|$ is smaller than $|E_x|$ and the decrease of $|E_y|$ is slower than $|E_x|$ and $|E_z|$. The distribution of $|E_z|$ is zero along the axis x with two semi-circles pattern outside of the aperture, and the near-field enhancement effect is apparent at the edge of aperture. The width of total electric field is larger than the aperture diameter, and $|E_y|$ is the main electric intensity in total electric field while $|E_x|$ and $|E_z|$ decrease quickly because of depolarization. In z= -70nm plane, the distribution of total electric field appears a ellipse optical spot, meaning the near-field enhancement effect in the edge of aperture can be neglected.

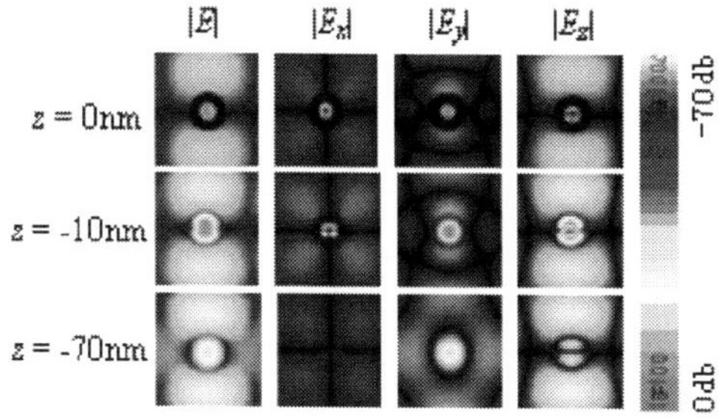

Figure 2. Distributions of total electric field and its components in all cross-sections

3.3. Electric field intensity distributions of longitudinal sections

Figure 3 gives the distributions of total electric field and its components in x=0 and y=0 plane. Analysis shows that the magnitude of $|E_y|$ is larger than $|E_x|$ and $|E_z|$ in each plane with $|E_x|$ being the smallest and being zero at probe axis. $|E_y|$ spreads along all directions after exiting from probe tip, while $|E_x|$ and $|E_z|$ mainly spread along the direction of 45° to probe axis. In addition, in the transmission of light to probe tip most of light is reflected inward by metal cover and then superimposes on the incident light forming a strong standing wave field along probe axis. Based on the waveguide theory, light transmits in single-mode optical fibre waveguide in HE11 mode, when the light injects into the metal-coated optical probe, it will associate to the problem of transmission in metal waveguide where light is coupled to TE11 mode. As can be seen from the calculation, the internal electric field of optical probe is divided into two parts by the cut-off face. One part of light superimposes on the incident light forming a strong standing wave field, while other part of light decays exponential and then exits from the aperture.

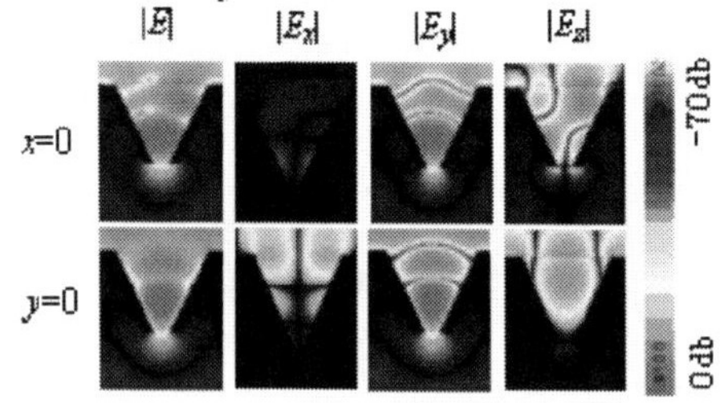

Figure 3. Distributions of total electric field and its components in two longitudinal sections

Figure 4 shows the corresponding vector distribution of electric field. Seen from the vector distribution, the light intensity decays quickly after exiting from the aperture, and the clear diffraction of emergent light occurs in x=0 plane. Because of the polarization of incident wave, the light intensity around probe tip is stronger than that in y=0 plane.

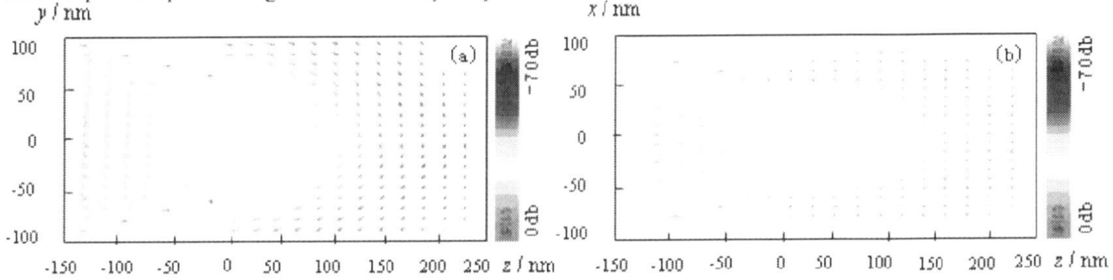

Figure 4. Vector distributions of total electric field in (a) x=0 plane and (b) y=0 plane

3.4. The impact of polarization on electric field intensity distribution

Figure 3 and figure 4 qualitatively show the impact of polarization on electric field intensity distribution. For the different distribution of electric field intensity on x=0 plane and y=0 plane, the former appears depolarization more obviously than the latter. First, the formed optical spots are non-circular symmetrical as a result of polarization, and the total field distributions of two planes are different. The emergent light in y=0 plane can only spread forward along axial direction of probe, while the emergent light from the tip in x=0 plane can not only spread forward, but also spread along all directions, rounding the metal cover and forming wide spatial distribution. Second, due to the discontinuous normal component of electric field, maximum value appears in the interface between waveguide and metal cover in the cross-section (that is, x=0 plane) parallel to the incident polarized light, therefore the interface of metal cover can be seen clearly. In virtue of the continuous tangential component of electric field, the interface between waveguide and metal cover is parallel to the incident electric field in the cross-section (that is, y=0 plane) perpendicular to the incident polarized light, thus the metal cover can not be seen.

Figure 5 quantitatively shows the impact of polarization on electric field intensity distribution. The numerical results for the total electric field distributions along axis x and axis y in all cross-sections of plane nanometre hole and fibre probe are given. It can be seen that the distribution patterns of electric field along axis x and axis y are different without circular symmetry. The solid lines in figure 5(b) show the electric field distributions at the end of fibre probe which present differently as the distance to aperture increasing. As the total electric intensity decreases, the ring formed by metal cover begins to contain electric field and the total electric field distribution becomes smooth from horn to saddle pattern. In z= -70nm plane the electric intensity is below 0.02V/m. The electric field distribution along axis x gradually decreases from centre to boundary, and no near-field enhancement appears in the edge of aperture, in accord with the boundary conditions of electromagnetic field that the tangential electric components in boundary of two mediums are equal to each other. Whereas the electric field distribution along axis y shows significantly edge enhancement in near-field range.

As shown in figure 5(b), the electric field along axis y remains unchanged from the centre to the boundary, then begins to increase quickly and reaches the maximum in the internal boundary of metal cover. According to the boundary conditions of electromagnetic field, electric field will diffract along the outer surface of metal cover resulting in it rapidly drops to zero. After electric field reaching the outside of metal cover, the secondary near-field enhancement will appear in the angularity of metal cover because of point effect. As shown in figure 5(a), the near-field enhancement accordingly becomes weaker as metal cover being thicker. Especially, it won't occur with infinitely thickness of metal cover. The edge enhancement of electric field only occurs in the range of a few nanometres from probe tip. With the increasing distance to probe tip, the total electric field distribution along axis y becomes smooth. In z=-70nm plane, the edge enhancement does not exist, suggesting that the near-field enhancement influenced by near-field characteristics.

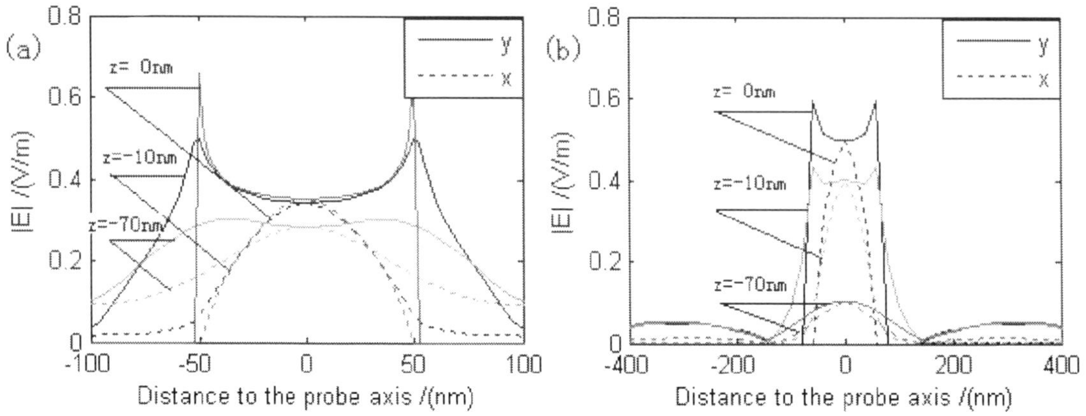

Figure 5. Total electric field distribution along axis x and axis y in all cross-sections, (a) a circular hole on plate, (b) metal-coated optical fibre

4. Conclusion

The electric intensity inside and outside of fibre probe in near-field region and the impact of polarization on intensity distribution are calculated using FDTD. The principle of FDTD is discussed, and through the use of the modified Debye model, the capabilities of the FDTD method to accurately simulate the response of tapered metal-coated optical fibre probe are demonstrated. Known from the near-field distributions in longitudinal sections and cross-sections, the incident polarized wave becomes depolarized after exiting from the nano-scale aperture, and the near-field distributions in two longitudinal sections are different and unsymmetrical. The generated evanescent field is limited to the tip surface in range of sub-wavelength with apparent near-field enhancement at the edge of aperture. Furthermore, the electric intensity distribution is largely influenced by polarization of incident wave through qualitatively and quantitatively analysis. This calculation is very helpful for understanding the radiation mechanism of light in near-field tweezers and the behaviour of light propagation in the optical probe that helps to find optimized tip design.

Acknowledgements

This research is supported by Opening Fund of Mechanical Electronic Engineering, Zhejiang Province, and Open Fund Project of Key Laboratory of Special Equipment Manufacturing and Advanced Process Technology, Ministry of Education, China.

5. References

[1] Tanaka M 1996 *Electronics and Communications in Japan* **79** 101–108
[2] Roberts A 1987 *J. Opt. Soc. Am.* A **4** 1970–1983
[3] Betzig E, Harootunian A, Lewis A and Isaacson M 1986 *Appl. Opt.* **25** 1890–1900
[4] Tanaka M and Tanaka K 1998 *J. Opt. Soc. Am.* A **15** 101–108
[5] Novotny L, Pohl D W and Hecht B 1995 *Opt. Lett.* **20** 970–972
[6] Vaccaro L, Aeschimann L and Staufer U 2003 *Applied physics letters* **83** 584–586
[7] Christensen D A 1995 *Ultramicroscopy* **57** 189–195
[8] Yee K S 1966 *IEEE Trans. Antennas Propag.* **14** 302–307
[9] Luebbers R J, Hunsberger F and Kunz K S 1991 *IEEE Trans. Antennas Propag.* **39** 29
[10] Bian R X, Dunn R C, Xie X S and Leung P T 1995 *Phys. Rev. Lett.* **75** 4772
[11] Ashcroft N W and Mermin N D 1976 *Solid State Physics* (Philadelphia: Saunders W B)
[12] Kunz K S and Luebbers R J 1993 (CRC, Boca Raton, FL)

Third-order nonlinear optical properties and its limiting behavior of polyaniline/SiO_2 composite materials

Y D Zhang[1], L Ma[1], C B Yang and P Yuan[1]

National Key Laboratory of Tunable Laser Technology, Institute of Opto-Electronics Harbin Institute of Technology, Harbin 150080, P. R .China

E-mail: ydzhang@hit.edu.cn

Abstract. Polyaniline are promising molecules which present very attractive properties for nonlinear optical applications. The nonlinear optical properties of polyaniline composite materials(PANI/SiO_2) were studied using the single beam Z-scan technique. The nonlinear absorption coefficient (β) and nonlinear refractive index (n_2) was retrieved by fitting experimental data. These molecules exhibit a strong reverse saturable absorption and an interesting optical limiting performance with a nanosecond Nd: YAG laser pulses at 532 nm.

1. Introduction

Nonlinear optical materials continue attracting attentions because of their potential application in optical communications, optical storage, optical computing, optical switching, optical limiting, etc.[1,2]. Amongst the new classes of materials conjugated polymers are especially interesting because they combine the optical and electronic properties of semiconductors with the processing advantages and mechanical properties of polymers[3]. The polyaniline(PANI) is an unique example since its electrical properties can be reversibly controlled both by changing the oxidation state of the main chain and by protonation of the imine nitrogen atoms. Besides, in addition to a wide range of desirable electrical, electrochemical and optical properties, PANI exhibits excellent environmental and thermal stability [4, 5].

Optical limiters are devices that strongly attenuate optical beams at high intensities while exhibiting higher transmittance at low intensities. Such devices are useful for the protection of the human eye and optical sensors from intense laser beams[6-9]. The development of materials for optical limiting is based on various mechanisms such as thermal refractive beam spreading, two-photon absorption (TPA) and excited state absorption (ESA) et al.

In this study, we investigate the nonlinear optical properties of PANI/SiO_2 using nanosecond doubled frequency Nd: YAG laser pulses at 532 nm. This material showed its large nonlinear absorption coefficient and strong reverse saturated absorption behaviour.

2. Experimental

2.1. Sample preparation and linear characterization

Ethanol was used as a solvent and homogeneous solutions were obtained, after stirring for several hours. In order to avoid molecular aggregation we analyses their linear optical properties by evaluating the solutions absorbance spectra. Fig. 1 shows the absorption spectrum of samples from 300 to

1100 nm. The band at ~400 nm is due to the band-gap transition while the band at ~900 nm is related to the formation of molecular excitons that correspond to a charge transfer into the quinoid ring from each of the neighboring benzenoid rings [10].

Transmission electron microscope (TEM) image of PANI/SiO_2 material is shown in Figure 2. The presence of SiO_2 particles is revealed as black spots and ribbon is the PANI network. Image reveals uniform particles size distribution, SiO_2 particles connected to the network structure of PANI through chemical bonds.

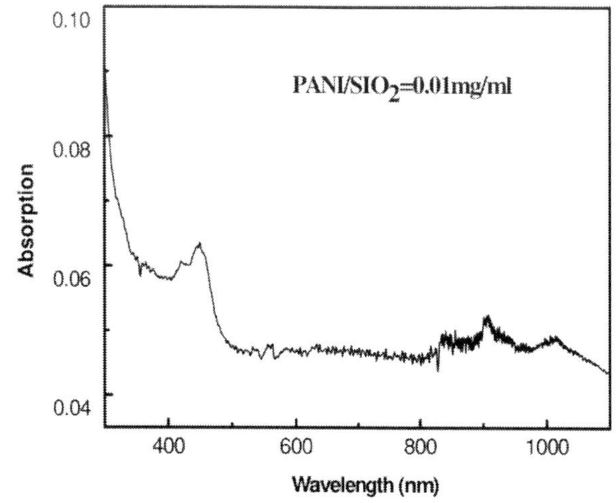

Fig. 1. Linear optical absorbance spectra of PANI/SiO_2 solution

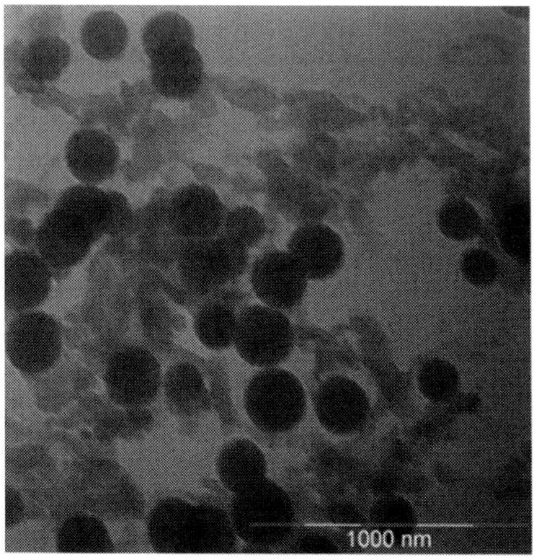

Fig. 2. TEM image of PANI/SiO2

2.2. Z-Scan measurements

The single beam Z-scan technique was used to measure the nonlinear optical properties. Experimental setup is shown as Fig.3. A doubled frequency Nd:YAG laser was used as the light source with a repetition rate of 10 Hz, its pulse width of 6.4 ns and wavelength of 532nm. While the sample was moved through the focus along the z-axis, transmitted open- and closed-aperture signals were detected simultaneously with two detectors. A lens with focal length of 30cm was used to focus the laser pulses into a 1 mm quartz cuvette, which contained the PANI/SiO_2 sample solution. The beam waist radius w_0 was found to be about 48μm.

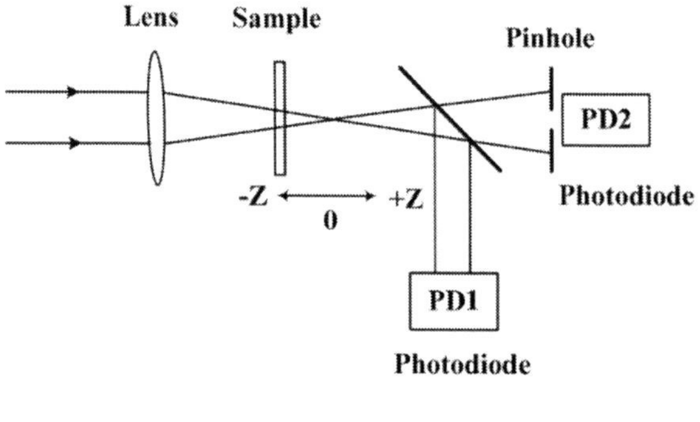

Fig. 3. Simple Z-scan experimental setup

3. Results and discussion

The measured transmittance as a function of sample position along the z-axis is shown in Figs. 4 and 5 for the polyaniline/SiO_2 composite materials.

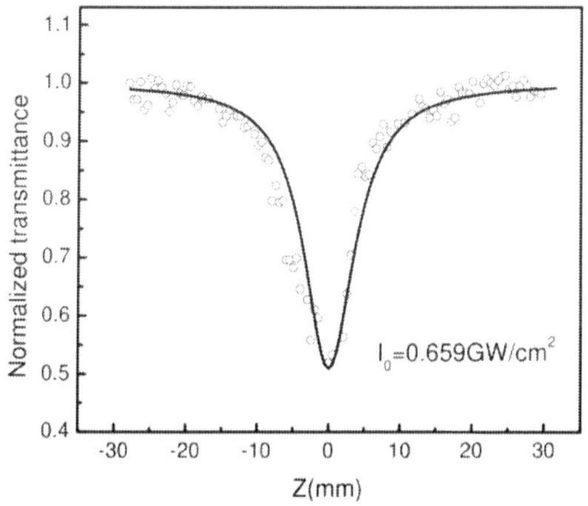

Fig. 4. Open aperture z-scan signal

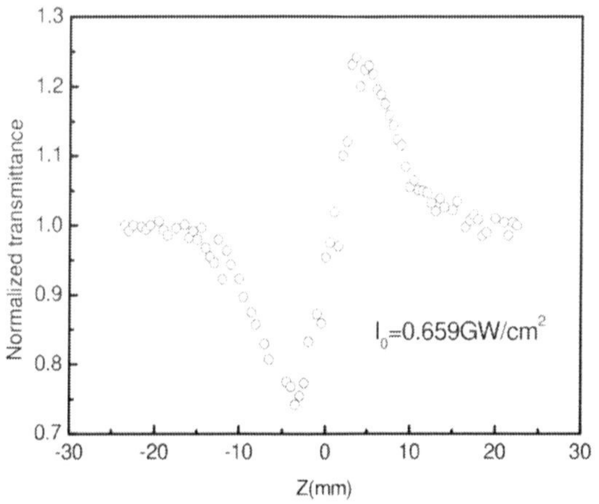

Fig. 5. Close aperture z-scan signal

The normalized transmittance as a function of the z-axis coordinate in the open aperture z-scan measurements can be written as:

$$T(z,S=1)=\frac{\ln[1+q_0(z)]}{q_0(z)} \tag{1}$$

$$q_0(z)=\beta I_0 L_{eff}/(1+z^2/z_0^2) \tag{2}$$

I_0 is the intensity of laser beam at focus ($z = 0$), $L_{eff}=\left(1-e^{-\alpha_0 L}\right)/\alpha_0$ is the effective length with α_0 the linear absorption coefficient and L is the thickness of the sample, z_0 is the diffraction length of the beam. The solid lines in Fig. 3 are the theoretical fits using Eqs. (1) and (2) to obtain the nonlinear absorption coefficient β. So the nonlinear absorption coefficient β can be deduced. The effect of nonlinear absorption in the investigated samples is responsible for transmittance decreasing close to the beam focus. A property of peak–valley indicating a positive type of nonlinearity from closed aperture and reverse saturable absorption type behavior from open aperture scan was observed. The nonlinear refractive index (n_2) was calculated using the standard relations [11]:

$$\Delta T_{p-v}=0.406(1-S)^{0.25}\left|\Delta\phi_0\right| \tag{3}$$

$$n_2=\frac{\Delta\phi_0}{kI_0L_{eff}} \tag{4}$$

where ΔT_{p-v} the peak–valley distance, $\left|\Delta\phi_0\right|$ is the on-axis phase-distortion, S is the linear transmittance of the aperture given by $S=1-\exp(-2r_a^2/w_a^2)$ where r_a is the radius of the aperture and w_a is the radius of the laser spot before the aperture. Using the peak–valley distances measured from the closed aperture z-scans we estimated the values of n_2. Apparently, the nonlinear refractive indices of the samples are positive, and exhibit strong self-focusing effect due to the larger peak-valley difference value. Sample data, together with the obtained results are summarized in Table 1.

Table 1. Summary of the nonlinear optical properties of samples evaluated from Z-scans.

Sample	concentration (mg/ml)	λ (nm)	Linear transmission	β (cm/GW)	n_2 (cm^2/GW)
PANI/SiO_2	0.01	532	66.5%	21.7	1.66×10^{-4}

For optical limiting experiments the sample solution in a 1 mm quartz cuvette was placed just after the focus of lens. Fig.6 depicts the optical limiting curve in ethanol solutions recorded for concentrations of 0.01mg/ml for $PANI/SiO_2$. It is evident from the figure that the threshold was low and was estimated to be ~ 0.12 GW/cm^2. We could not verify the behaviour (damage or breakdown threshold) above 0.7 GW/cm^2 due to limitations with our laser source power. The limiting behaviour observed in the samples is attributed mainly to nonlinear refraction. Since the samples were pumped with laser beam the arising nonlinearities are predominantly thermal in nature.

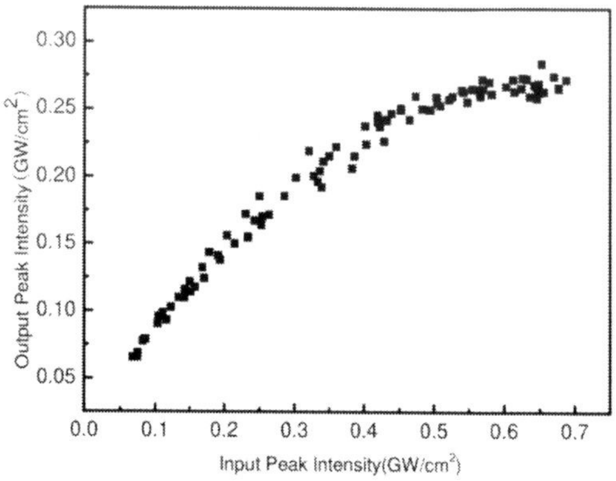

Fig. 6. Optical limiting properties of $PANI/SiO_2$

Conclusions

We have reported the nonlinear optical properties of $PANI/SiO_2$ composite materials using nanosecond doubled frequency Nd: YAG laser pulses at 532 nm. The observed nonlinearities were large and positive in nature. Theoretical fit of the experimental data, i.e. input versus output intensity, was obtained the nonlinear optical coefficient, was calculated to be 21.7 cm/GW and 1.66×10^{-4} cm^2/GW, respectively for β and n_2. The good optical limiting properties of $PANI/SiO_2$ have been established. The optical limiting performance mainly originates from nonlinear refraction. Based on all the observed properties, we deem that $PANI/SiO_2$ both could be promising candidates for optical limiting, optical storage.

Acknowledgement

The research is supported by the National Natural Science Foundation of China under Grant Nos.60272075 and 60478014.

References

[1] Li F, Zheng Q G, Yang G, Dai N L and Lu P X 2008 *Mater. Lett.* **62** 3095
[2] Marder S R 2006 *Chem. Commun.* **2** 131
[3] Maciel G S, de Arau'jo C B and Correia R R B 1998 *Opt. Commun* **157** 187
[4] Dispenza C, Leone M and Presti C L 2006 *J Non-Cryst Solids.* **352** 3835
[5] Bicak N, Senkal B F and Sezer E 2005 *Synthetic. Met.* **155** 105
[6] Tong R, Wu H X and Li B 2005 *Physica B.* **366** 192
[7] Guang S H, Gen C H and Paras N P 1995 *Opt. Lett.* **20** 435
[8] Ji C L, Chuan K W and Faris G 2007 *Phys. Rev. A.* **76** 053804
[9] Sunita S, Devendra M and Ghoshal S K 2008 *Opt. Commun.* **281** 2923
[10] Duke C B, Conwell E M and Paton A 1986 *Chem. Phys. Lett.* **131** 82
[11] Sutherland R L 2003 *Handbook of Nonlinear Optics*(New York) p 627

Preparation of Surface dendrimer-modified gold nanorods by Round-Trip Phase Transfer Ligand Exchange

Peng Huang[1], Zhiming Li[1], Jing Lin[2], Daxiang Cui[1]

[1]Key Laboratory of Nano/Micro Fabrication Technology, Key Laboratory for Thin Film and Microfabrication of Ministry of Education, Institute of Micro-Nano Science and Technology, Shanghai Jiao Tong University, 800 Dongchuan Road, Shanghai 200240, China
[2]College of Chemistry, Chemical Engineering and Biotechnology, Donghua University, Shanghai 201620, China
E-mail: dxcui@sjtu.edu.cn

Abstract. In this work we firstly used partially thiolated polyamidoamine (PAMAM) dendrimer to replace CTAB molecules on the surface of gold nanorods (GNRs) by round-trip phase transfer ligand exchange, thus eliminating their cytotoxicity and improving their biocompatibility. The dendrimer-modified gold nanorods (dGNRs) were characterized by transmission electron microscopy (TEM), high-resolution transmission electron microscopy (HRTEM), UV-vis spectrometer, thermogravimetric analysis (TGA) and Fourier transformation infrared (FTIR) spectroscopy. The results indicate that dGNRs were successfully prepared.

1. Introduction

Gold nanorods (GNRs), because of their small size, strongly light enhanced absorption in near infrared region (NIR), and plasmon resonance enhanced properties, have tremendous applications such as photothermal therapy, cell imaging, and gene delivery [1]. However, cetyltrimethylammonium bromide (CTAB) is one of the widely used surfactant for the synthesis of GNRs. The double layer of CTAB on the surface of GNRs is problematic for bioconjugation, nonspecific adsorption of DNA, cytotoxicity, and stability. Therefore, it has been an extensive interest in modification of the surface of GNRs [2].

Dendrimers are a relatively novel class of polymers with highly ordered structure. The accretion of functional groups, symmetry perfection, nanosize, and internal cavities provide these novel dendritic nanocomposites many potential applications in biochemistry, gene therapy, and nanomedicine. Dendrimer coating of nanoparticle surfaces can alter the charge, functionality, and reactivity, as well as enhance the stability, dispersion and biocompatibility of the nanoparticles [3]. Hence, we develop a novel method to use partially thiolated polyamidoamine (PAMAM) dendrimer to replace CTAB molecules on the surface of GNRs by round-trip phase transfer ligand exchange, thus eliminating their cytotoxicity and improving their biocompatibility.

2. Experiments

Chloroauric acid ($HAuCl_4{\cdot}3H_2O$), Cetyltrimethylammonium bromide (CTAB), sodium borohydride ($NaBH_4$), and ascorbic acid were obtained from the Aldrich Company. Methyl mercaptoacetate, dodecanethiol (DDT), and mercaptoundecanoic acid (MUDA) were obtained from the Alfa Aesar Company. G4.0-PAMAM dendrimers with surface amine groups were stored in our lab.

GNRs were synthesized according to the seed-mediated template-assisted protocol [4]. Functional PAMAM dendrimers, with a small number of thiol groups were prepared by reacting G4.0 PAMAM dendrimers with methyl mercaptoacetate[5].

Ligand exchange was achieved by a two phase extraction, which was performed according to reference [2]. First, CTAB was removed from the GNR-CTAB by extracting the GNRs into an organic phase via the ligand dodecanethiol (DDT). Then, the GNR-DDT was extracted into an aqueous phase by mercaptoundecanoic acid (MUDA). Ligands can be further customized to partially thiolated polyamidoamine (PAMAM) G4.0 dendrimer. Free unreacted dendrimers were removed by filtration with a 100K MWCO microcon device. The unreacted thiol groups on the GNRs were quenched with *N*-ethylmaleimide to minimize the dimer formation.

3. Characterization

HR-TEM (Hitachi H-700H) was used to confirm particle size and observe dendrimer coating. UV–vis spectra were measured at 20°C with a Shimadzu UV-2450 UV-visible spectrophotometer equipped with a 10-mm quartz cell. The 400–1100 nm wavelength region was scanned, since it includes the absorbance of the GNRs. TGA measurement was performed by a TGA 2850 thermogravimetric analyzer (TA Instruments) under N_2 in the temperature range of 30-700°C with an increasing rate of 5°C/min. The fourier transform infrared (FTIR) spectra were recorded on a Perkin Elmer Paragon-1000 FTIR Spectrometer.

4. Results and Discussion

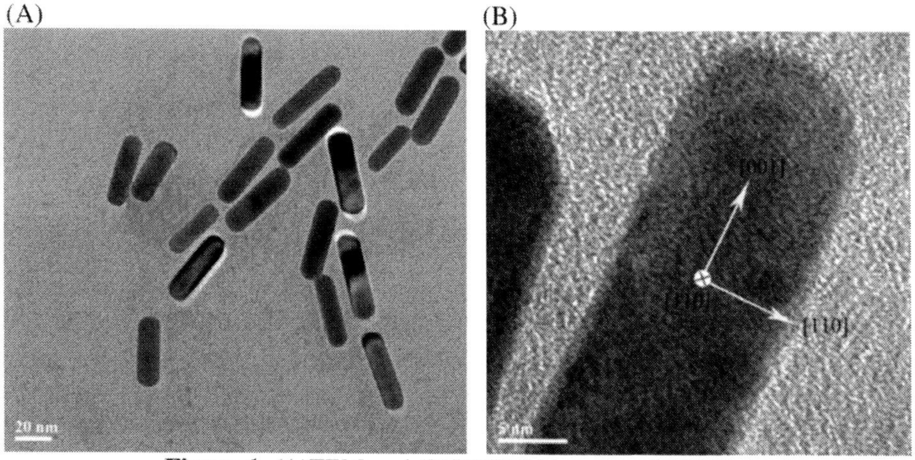

Figure 1. (A)TEM and (B)HRTEM images of dGNRs.

Figure 1 shows the results of G4.0 PAMAM dendrimer functionalized GNRs characterized by TEM and HRTEM. As shown in Figure 1A, synthesized dGNRs are 42 nm in length, and 10 nm in width, indicating no significant size change after dendrimers modified. HRTEM imaging has been carried out to investigate the crystal structure of individual dGNRs in Figure 1B. It showed that the GNR with [110] orientation is single crystalline without twin planes and exhibited good crystalline and clear lattice fringes. The experimental lattice fringe spacing is consistent with the interplanar spacing of the [001] plane of GNRs.

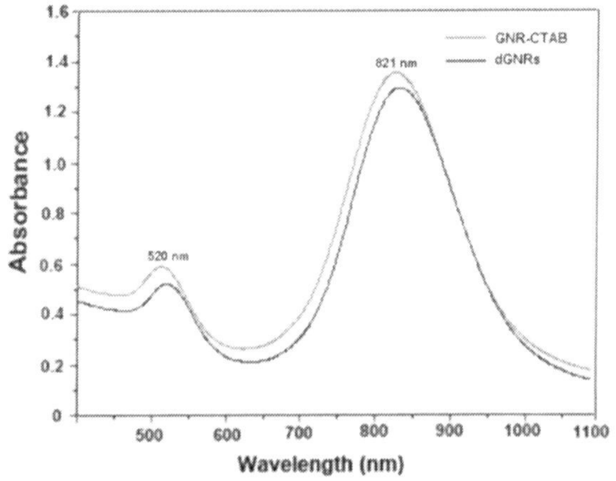

Figure 2. UV–vis absorbance spectra of GNRs before and after dendrimers modification.

Figure 2 shows the UV–vis absorbance spectra of GNR-CTAB and dGNRs in the 400–1100 nm wavelength range, using deionized water as the solvent. The spectrum of GNR-CTAB showed that GNRs-CTAB have two absorption bands, a weak short-wavelength band around 520 nm and a strong long-wavelength band around 821 nm. Moreover, we found that the plasmon peaks of dGNRs exhibited no significant changes in peak width or position.

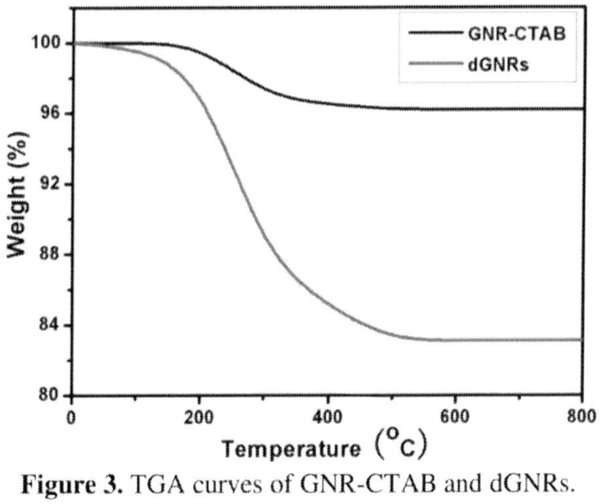

Figure 3. TGA curves of GNR-CTAB and dGNRs.

The G4.0 dendrimer coating layer on the surface of GNRs can be further confirmed by TGA analysis as shown in Figure 3. The grafted dendrimer content (%) of the dGNRs was calculated by TGA from the weight loss between 20°C and 800°C. For comparison, the TGA curve of GNR-CTAB without dendrimer coating is also shown. In the case of the GNR-CTAB, between 200°C and 800°C there was a continuous but not very obvious decrease (~4%) in weight, which is a typical weight loss for the CTAB layer on the surface of GNRs. The quantity of dendrimer attached to the GNRs is also shown in Figure 3. There has been a weight loss of ~17% for the dGNRs. All of these reveal the existence of G4.0 PAMAM dendrimer on the surface of GNRs.

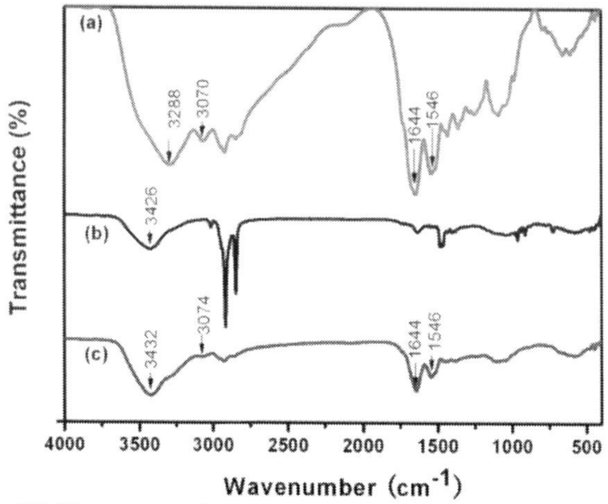

Figure 4. The FT-IR spectra of (a) pure G4.0, (b) GNR-CTAB, and (c) dGNRs.

To further study the replacement of CTAB molecules on the surface of GNRs by G4.0 dendrimer, the FT-IR spectra of (a) pure G4.0, (b) GNR-CTAB, and (c) dGNRs were determined. The IR peaks of pure G4.0 at 3288, 3070, 1644, 1546 cm^{-1} are assigned to the stretching vibration of –OH, amide A (mainly –NH stretching vibration), amide I (mainly C=O stretching vibrations)and amide II (the coupling of bending vibrate of N-H and stretching vibrate of C-N) bands, respectively. The difference between the IR spectrum of pure G4.0 and that of GNRs is obvious. Comparing the IR spectra of dGNRs with those of pure G4.0, the characteristic peak of –OH groups shifts to a high wavenumber of about 144 cm^{-1} and the other characteristic peaks are negligible variations. The results also indicate that the existence of G4.0 PAMAM dendrimer on the surface of GNRs.

5. Summary

The partially thiolated polyamidoamine (PAMAM) G4.0 dendrimer modified GNRs were prepared by round-trip phase transfer ligand exchange. The synthesized dGNRs were 42 nm in length, and 10 nm in width, whose plasmon peaks exhibited no significant changes in peak width or position. TGA curves and FT-IR spectra revealed the existence of G4.0 PAMAM dendrimer on the surface of GNRs. The novel nano-composite has potential versatility in biological applications such as therapy, sensing, and imaging.

Acknowledgements

This work is supported by National Key Basic Research Program (973 Project) (No.2005CB723400-G and 2010CB933900), National 863 Hi-tech Project (2007AA022004), National Natural Scientific Fund (No.20771075), Special project for nano-technology from Shanghai (No.0752nm024), The science and technology commission of shanghai (No.072112006-6), and Important National Science & Technology Specific Projects (2009ZX10004-311). The authors thank the Instrumental Analysis Center of Shanghai Jiao Tong University for the Materials Characterization.

References

[1] Yang DP & Cui DX, 2008 Chemistry-An Asian Journal **3** 2010
[2] Wijaya A & Hamad-Schifferli K, 2008 Langmuir **24** 9966
[3] Shan J & Tenhu H, 2007 Chemical Communications **34** 4580
[4] Murphy CJ & Jana NR, 2002 Advanced Materials **14** 80
[5] Pan B, et al. 2005 Colloids and Surfaces A: Physicochem. Eng. Aspects **259** 89

Aptamer-conjugated dendrimer-modified quantum dots for glioblastoma cells imaging

Zhiming Li[1,2,3], Peng Huang[1], Rong He[1], Xiaomin Zhang[2], Chenchen Bao[1], Qiushi Ren[2*], Daxiang Cui[1*]

[1]National Key Laboratory of Nano/Micro Fabrication Technology, Key Laboratory for Thin Film and Microfabrication of Ministry of Education, Institute of Micro-Nano Science and Technology, Shanghai Jiao Tong University, Shanghai 200240, China.

[2]Institute for Laser Medicine & Biophotonics, School of Life Sciences and Biotechnology, Shanghai Jiao Tong University, Shanghai 200240, China.

[3]Institute of Dermatology & Department of Dermatology at No. 1 Hospital, Wenzhou Medical College, Wenzhou, Zhejiang 325000, China.

E-mail: dxcui@sjtu.edu.cn

Abstract. Targeted quantum dots have shown potential as a platform for development of cancer imaging. Aptamers have recently been demonstrated as ideal candidates for molecular targeting applications. In present work, polyamidoamine dendrimers were used to modify surface of quantum dots and improve their solubility in water solution. Then, dendrimer-modified quantum dots were conjugated with DNA aptamer, GBI-10, can recognize the extracellular matrix protein tenascin-C on the surface of human glioblastoma cells. The dendrimer-modified quantum dots exhibit water-soluble, high quantum yield, and good biocompatibility. Aptamer-conjugated quantum dots can specifically target U251 human glioblastoma cells. High-performance aptamer-conjugated dendrimers modified quantum dot-based nanoprobes have great potential in application such as cancer imaging.

1. Development of Apt-dQDs nanoprobes

In recent years, molecular imaging of tumors has become a hotspot. Most probes for molecular imaging conjugate a targeting molecule to a reporter moiety. This study was aimed at developing a novel type of molecular imaging probes composed of aptamers (Apts), [1]quantum dots (QDs), and polyamidoamine (PAMAM) dendrimers, for targeting to tumor cells.

---

[1] *Correspondence should be addressed to qsren@sjtu.edu.cn or dxcui@sjtu.edu.cn

A previous SELEX experiment targeting U251 glioblastoma cells identified GBI-10 Apts that binds to tenascin-C, an extracellular matrix protein playing a important role in cancer cells proliferation and migration [1-2]. Therefore, we selected the GBI-10 as the targeting molecules for glioblastoma cells targeting and imaging.

QDs have been widely studied due to their unique optical properties, and have become a novel functional platform in bio-analytical science and molecular imaging [3-5]. However, some reports showed that QDs exhibit cellular toxicity [6]. How to decrease their toxicity and enhance their biocompatibility is a great challengeable problem. Dendrimers are a class of polymers with highly ordered structure. Our previous results demonstrate that the dendrimers functionalized nanomaterials such as CNTs, QDs and magnetic nanoparticles, markedly enhance the biocompatibility and cellular uptake of nanoparticles [7-9].

Herein, we used QDs and PAMAM dendrimers as raw materials, and fabricated dendrimer-modified QDs, then we selected tenascin-C over expressed in glioblastoma cells as research target, and conjugated dQDs with Apt, the resultant Apt-dQDs nanoprobes were used to investigate the feasibility of targeting and imaging U251 glioblastoma cells.

To obtain Apt-dQDs nanoprobes, we firstly prepared partially thiolated PAMAM dendrimers (see Figure 1), then we used ligand replacement method to obtain partially thiolated PAMAM dendrimers-modified QDs, then, surface amine groups on the surface of dendrimers were converted to carboxylic acids by reacting with excess glutaric anhydride, and then 5'-amino modified Apt were conjugated with dendrimer-modified QDs, the resultant Apt-dQDs nanoprobes were characterized by HR-TEM (see Figure 2a) and photoluminescence spectra(see Figure 2b). Results demonstrated that as-prepared nanoprobes displayed good dispersibility and water-solubility, and have strong fluorescent signals.

To determine if Apt were successfully conjugated to PAMAM dendrimers, we used agarose (2%) gel electrophoresis to separate the components in the conjugations since agarose gel electrophoresis can separate ssDNA Apt because of its secondary double strands structure(see Figure 3). The shift results demonstrate Apt-dQDs nanoprobes are unable to run on the gel, indicating Apt could be successfully conjugated with PAMAM dendrimers modified nanoparticles. Moreover, we could not observe the other bands in Apt-dQDs probes, suggesting that the byproducts and unconjugated Apt were removed successfully.

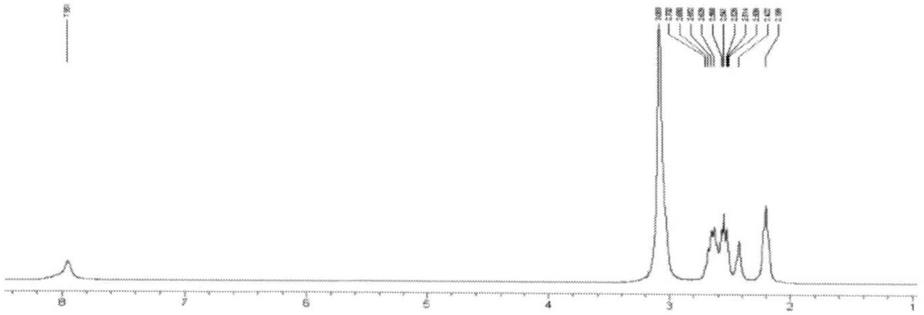

.PAMAM G4.0, ^{1}H NMR (300 MHz, d^6-DMSO)

δ= 7.96 (br s, 60H), 3.08 (br s, 324H), 2.68 -2.42 (br s, 160H), 2.20 (br s, 64H)

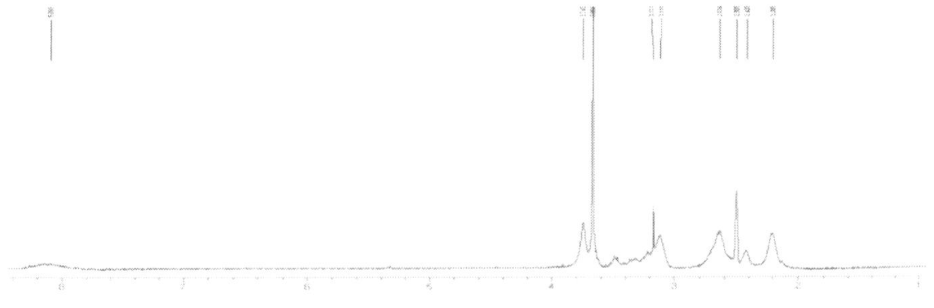

PAMAM-SH, ^{1}H NMR (300 MHz, d^6-DMSO)

δ= 8.15 (br s, 63H), 3.74 (br s, 6H), 3.66 (br s, 12H), 3.16 (br s, 312H), 2.63-2.42 (br s, 160H), 2.20(br s, 58H)

Fig.1 The ^{1}H NMR spectra of the G4.0 PAMAM dendrime and partially thiol-terminated G4.0 dendrimers.

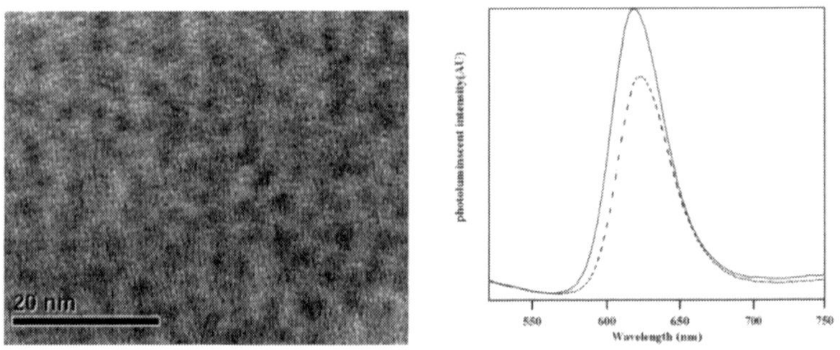

Fig.2 Characterization of Apt-dQDs nanoprobes. (a) HR-TEM images of Apt-dQDs nanoprobes. (b) Photoluminscence spectra of QDs (solid line) and Apt-dQDs (dash line).

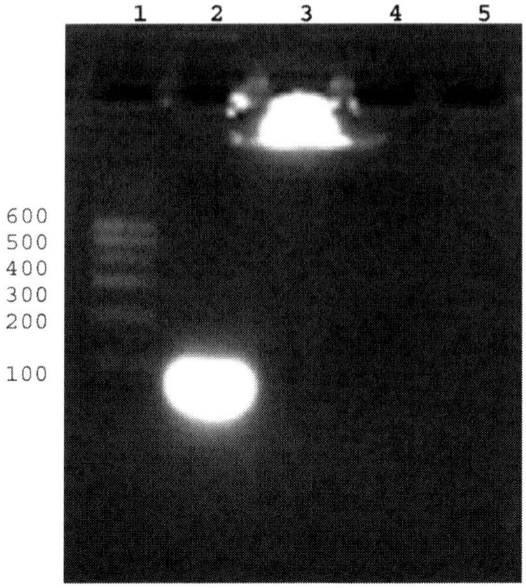

Fig.3 Gel electrophoresis results of Apt-dQDs nanoprobes after staining with ethidium bromide. Lanes 1, 2, 3, 4, and 5 represent the 100 bp DNA ladder, Apt, Apt-dQDs nanoprobes, dendrimer, and dQDs alone, respectively.

2. The cell imaging

To evaluate the targeting of Apt-dQDs nanoprobes, we firstly measured the binding ability of as-prepared nanoprobes to U251 glioblastoma cells. As shown in Fig.4a, U251 glioblastoma cells incubated with Apt-QDs nanoprobes, exhibit strong red color, and could not be washed away, which suggest that the Apt-QDs nanoprobes can bind with U251 glioblastoma cells with high binding affinity. Conversely, control Scb10-dQDs conjugation can not bind with U251 glioblastoma cells (see Figure 4b). Therefore, we considered that Apt-QDs nanoprobes can target U251 glioblastoma cells specifically.

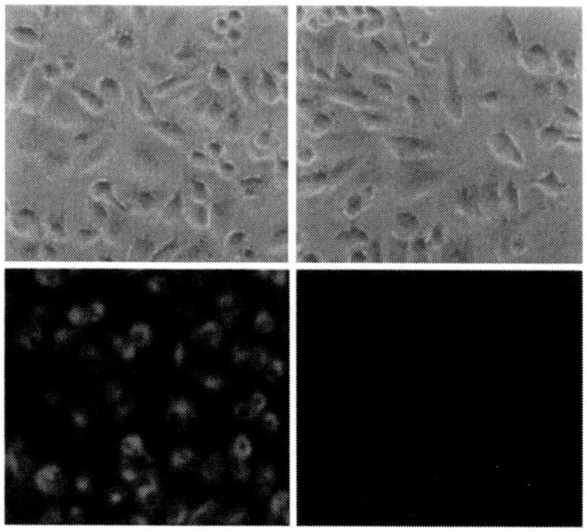

Fig. 4 *In vitro* staining of U251 glioblastoma cells using Apt-dQDs nanoprobes. (a) Cells with Apt-dQDs nanoprobes. (b) Cells with Scb10-dQDs conjugation.

3. Conclusions

We have developed a class of Apt-conjugated nanoprobes, which could specifically bind to cancer cells and exhibits *in vitro* molecular imaging. These nanoprobes have great potential in applications such as *in vitro* and *in vivo* glioblastoma molecular imaging in near future.

Acknowledgements

This work is supported by China National 973 Project (No.2005CB723400-G), China National 863 Project (2005AA001070 No. 2007AA022004), National Natural Scientific Fund (No.20471599 and No.30672147), Shanghai Scientific Fund (0752nm24) and Wenzhou Fund of Science and Technology (No.20070034 and No.07211206).

References

[1] D. Daniels, H. Chen, B. Hicke, K. Swiderek, and L. Gold, Proceedings of the National Academy of Sciences of the United States of America 100 (2003) 15416.

[2] C. Herold-Mende, M. Mueller, M. Bonsanto, H. Schmitt, S. Kunze, and H. Steiner, International journal of cancer. Journal international du cancer 98 (2002) 362.

[3] X. You, R. He, F. Gao, J. Shao, B. Pan, and D. Cui, Nanotechnology 18 (2007) 35701.

[4] X. Chen, Y. Deng, Y. Lin, D. Pang, H. Qing, F. Qu, and H. Xie, Nanotechnology 19 (2008) 235105.

[5] W. Chan and S. Nie, Science (New York, NY) 281 (1998) 2016.

[6] A. Derfus, W. Chan, and S. Bhatia, Nano Letters 4 (2004) 11.

[7] B. Pan, D. Cui, F. Gao, and R. He, Nanotechnology 17 (2006) 2483.

[8] B. Pan, F. Gao, R. He, D. Cui, and Y. Zhang, Journal of colloid and interface science 297 (2006) 151.

[9] B. Pan, D. Cui, Y. Sheng, C. Ozkan, F. Gao, R. He, Q. Li, P. Xu, and T. Huang, Cancer research 67 (2007) 8156.

8th China International Nanoscience and Technology Symposium (CINSTS09) IOP Publishing
Journal of Physics: Conference Series **188** (2009) 012033 doi:10.1088/1742-6596/188/1/012033

The stable and uniform characteristics of nitrogen in nitrogen doped titanium dioxide nano photocatalytic particles

Zhonghua Zhou [1*], Yue Huang [2]

[1]Department of Material Science and Engineering, College of Materials, Xiamen University, Xiamen 361005, P. R. China
[2]Covalent Materials Corporation, Nissei Bldg., 6-3, Osaki 1-chome, Shinagawa-ku, Tokyo 141-0032, Japan

*E-mail: zzh@xmu.edu.cn

Abstract. N doped TiO_2 nano particles were prepared by heating TiO_2 nano particles in NH_3 + C_3H_8 5vol% atmosphere. The stability and uniformity characteristics of the doped N in the prepared N doped TiO_2 nano particles were studied by X-ray photoelectron spectroscopy (XPS) and temperature programmed desorption (TPD). We concluded that the doped N was uniform from the surface to the inside of TiO_2 nano particles and found that the desorption peak of the doped N was more than 900 K in vacuum, showing its high stability and potential of visible light photocatalytic application for N doped TiO_2 nano particles.

1. Introduction

Visible light activated photocatalysts are important due to their ability to efficiently utilize solar irradiation or indoor-light. Recently, studies about doping or modification with non-metals of TiO_2 have attracted much interest. N doped TiO_2 [1-4], C doped TiO_2 [5-7], S doped TiO_2 [8], C and N co-doped TiO_2 [9] show that the photocatalytic activity can be improved through non-metal doping into TiO_2. Our previous studies have also shown that N doped TiO_2 has higher visible light photocatalytic performance than commercial TiO_2 [10, 11]. However, there are few studies focusing on the stable and uniform characteristics of non-metals doped into TiO_2, to the best of our knowledge, though the stable and uniform conditions are very important for their photocatalytic performances not only from the scientific interest but also from the view point of applications. In this study, we prepared N doped TiO_2 nano particles and investigated the stable and uniform characteristics of the doped N by XPS and TPD.

2. Experimental

It was known that N doped TiO_2 nano particles could be prepared by heating TiO_2 primary nano particles in an ammonia atmosphere [1]. Under the consideration of the engineering application of powder heating processes, we built ammonia heating experimental equipment with rotary and pre-desorption processes as shown in Fig. 1. TiO_2 nano particles were set into a furnace tube which was rotated at a rate of 5 rpm during the whole heating process and were heated under air atmosphere to 573K. The particles were then desorbed by vacuum pump to 5 Torr and kept for 0.5h. NH_3 + C_3H_8 5vol% ammonia gas was then introduced and the furnace heated to 873K and kept for 3h, then switched off and cooled down to room temperature. A quartz tube was used for the furnace tube to avoid metal contaminations. The particles of TiO_2 used for the heat-treatment had an average particle diameter of 7 nm and were anatase-type crystalline.

The stable and uniform characteristics of the prepared N doped TiO_2 fine particles were studied by XPS and TPD. The measuring conditions of the XPS analysis were as follows. Apparatus: AXIS-Ultra manufactured by Kratos Co. Excited X-ray source: monochromatic Al Kα line (1486.6 eV). Photoelectron detection angle: 90° (from the surface of the sample). Detection depth: < 10 nm. Charged neutralization gun: used. The measuring conditions of the TPD analysis were as follows. Equipment: temperature programmed desorption gas analyzer (Model: Thermo plus TPD type V). Sample: a powder sample filled in a platinum cell (diameter 6mm, height 2.5mm). Measuring temperature range: from 313 K to 1173 K. Heating rate: 10 K /min. Measuring mode TIC (scanning): m/e = 1 to 100. Vacuum degree at the start of measurement: < or = 2.0×10^{-6} Pa.

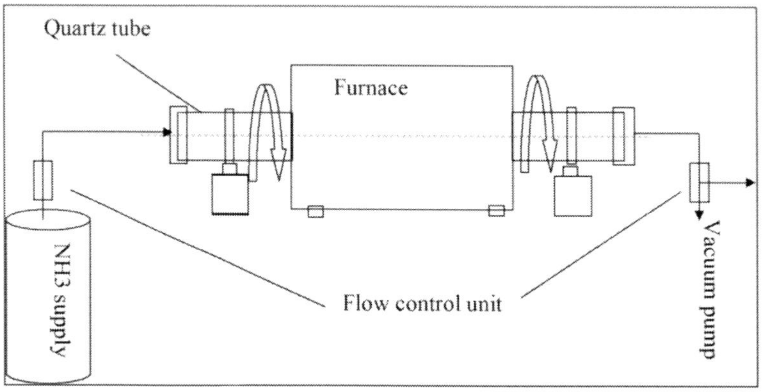

Fig. 1. Sketch of experimental equipment for preparing N doped TiO_2 photocatalytic nano particles.

3. Results

The prepared N doped TiO_2 nano particles have the shape of ellipsoidal sphere with a minor axis length of about 20 nm and major axis length of about 40 nm. The crystal structure of the prepared N doped TiO_2 was confirmed by XRD to be anatase. The photocatalytic activity induced by visible light irradiation was evaluated by IPA (isopropanol) photochemical oxidation and found to be higher than commercial TiO_2 [10, 11].

Fig. 2 shows the N1s XPS analytical results of chemical bonding of the doped nitrogen in the prepared sample. It can be seen from the experimental data points that there are a peak at the binding energy of 400 eV and another peak at 396 eV with a shoulder at about 398 eV, which are in the range of 395 - 401 eV observed by several other researches and are typical of N doped TiO_2 [1-4]. After fitting the experimental dots with Gaussian curve, it was obvious that the shoulder at about 398 eV could be fitted by a peak at the same binding energy. According to the studies about N1s XPS spectra of N doped TiO_2 [1-4, 12], the peak at 400 eV belongs to the NHx adsorbed at the surface of the particles, and the peaks at 398 and 396 eV are attributed to the bond of Ti-N, represented by two types of Ti-N bonds, labeled as TiN (1) and TiN (2) in Fig. 2, respectively. The details of the different bonding of TiN (1) and TiN (2) are discussed in the Discussion section. The N concentration was calculated from the peak areas, and found to be 1050 ppm of N for TiN (1) and 2270 ppm of N for TiN (2), respectively.

The N doped into TiO_2 particles is desorbed as N_2 when heating under vacuum. The N_2 TPD spectrum of the prepared N doped sample is shown in Fig. 3. It can be seen that there is a broad N_2 desorption peak from 700 to 1200 K, with a peak at about 1000 K and a shoulder at about 900 K. Considering the XPS results of two types of Ti-N bonds being present, the experimental data was fitted with two Gaussian peaks. The results were peak-1 at 908 K and peak-2 at 1020 K, labeled as TiN (1) and TiN (2) in Fig. 3, respectively. The N concentration was calculated from the peak areas, and was found to be 759 ppm of N for TiN (1) and 2041 ppm of N for TiN (2). Also, the NH_3 adsorbed on TiO_2 particles is desorbed as NH_3 gas when heating under vacuum. NH_3 desorption of the prepared N doped TiO_2 particles was confirmed by NH_3 TPD spectrum (not shown) below 773 K, and no NH_3 desorption was found around high temperature above 773 K.

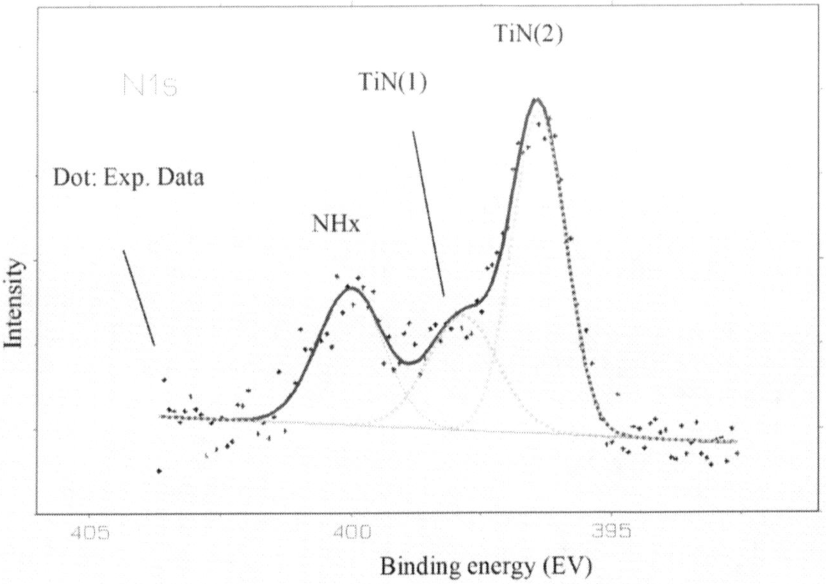

Fig. 2. N1s XPS spectra of the prepared N doped TiO_2 nano particles.

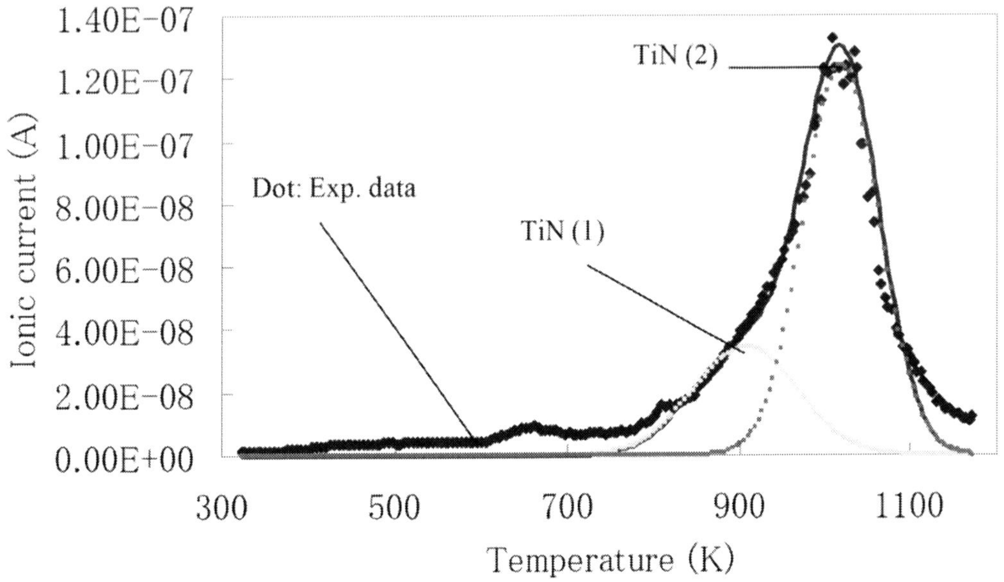

Fig. 3. N_2 TPD spectra of the prepared N doped TiO_2 nano particles.

4. Discussion

For the two types of bonding of the N doped into TiO_2 observed in the N1s XPS spectra as shown in Fig. 2, the main peak observed at binding energy 396 eV was consistent with the results of Asahi et al. [1], who found that the visible light photocatalytic activity increased with the increase of the 396 eV peak in N1s XPS spectra, and concluded that the 396 eV peak was attributed to the substitutional site of N. Furthermore, Saha at el. [12] investigated the N1s XPS spectra during the oxidation process of TiN nitride, found that the main peak of N1s XPS was at 397 eV for TiN nitride and shifted to lower binding energy with oxidation time, and assigned the peak at binding energy 396 eV as chemisorbed atomic N. Therefore, we could derive that the peak of 396 eV labelled as TiN (2) in Fig. 2 was attributed to the bonding of Ti-N-Ti. The Ti-N-Ti bond means that the doped N substitutes the oxygen site in the TiO_2 crystal. For the small peak at 398 eV labelled as TiN (1) in Fig. 2, it was hard to assign it to molecularly chemisorbed N_2 which was observed at above 400 eV by Asahi et al.[1] and Saha et al.[12]. The peak could be due to the N invading in crystal lattice interstices of TiO_2, expressed as the Ti-N-O bond, which was supported by our TPD results discussed following.

For TPD detection of N doped TiO_2, the degree of the peak temperature of N desorption represents the degree of strength of N bonded to Ti. The higher the peak temperature of N_2 desorption is, the stronger the N is bonded to Ti. In general, the doped N is more stable in substitution site than in lattice interstice site. Therefore, it was reasonably derived that the N_2 TPD peak at 908 K in Fig. 3 was due to the Ti-N-O bond and the peak at 1020 K was due to the Ti-N-Ti bond. The labels of TiN (1) and TiN (2) in Fig. 3 were consistent with those in Fig. 2. The desorption peak of the doped N was more than 900 K in vacuum, showing its stability and high potential of visible light photocatalytic application for N doped TiO_2 nano particles.

Whether the concentration of the doped N is uniform from the surface to the inside of the particles or not is an important factor to discuss. XPS detection provides structure information about the surface, while TPD detection gives total information including the surface and bulk. Therefore, by comparing the results of XPS with those of TPD, we can get information about the uniform characteristics of the doped N. First, we compared the total concentrations of the doped N calculated from the sum of TiN (1) and TiN (2), 3320 ppm for XPS detection and 2800 ppm for TPD detection. Considering the error of about 10% ~ 20% for the XPS experiment, the results of XPS were derived to be consistent with those of TPD, so we came to the conclusion that the doped N in TiO_2 nano particles was uniform from the surface to the inside of the particles. Next, we compared the Ti-N-O bond and the Ti-N-Ti bond from XPS results with those from TPD results. The concentration of the doped N in Ti-N-O bonds was 1050 ppm from XPS, which was consistent with 759 ppm from TPD, considering the error stated above. Meanwhile, the doped N in Ti-N-Ti bonds was 2270 ppm from XPS, which was also consistent with 2041 ppm from TPD. Therefore, the doped N in the Ti-N-O bonds and in the Ti-N-Ti bonds was uniform from the surface to the inside of the prepared N doped TiO_2 nano particles.

5. Conclusion

N doped TiO_2 nano particles were prepared by heating TiO_2 nano particles in NH_3 + C_3H_8 5vol% atmosphere. We concluded that the doped N was uniform from the surface to the inside of TiO_2 nano particles and found that the desorption peak of the doped N was more than 900 K in vacuum by XPS and TPD detection.

Acknowledgement: The corresponding author thanks the help by R & D of Covalent Materials Corporation.

References

[1]Asahi R, Morikawa T, Ohwaki T, Aoki K and Taga Y. 2001 *Science* **293** 269
[2]Chen XB and Burda C. 2004 *J. Phys. Chem. B* **108** 15446
[3]Sathish M, Viswanathan B, Viswanath RP and Gopinath CS. 2005 *Chem. Mater.* **17** 6349

[4]Li HX, Li JX and Huo YN. 2006 *J. Phys. Chem. B* **110**: 1559
[5]Ohno T, Tsubota T, Nishijima K and Miyamoto Z. 2004 *Chem. Lett.* **33** 750
[6]Li Y, Hwang DS, Lee NH and Kim SJ. 2005 *Chem. Phys. Lett.* **404** 25
[7]Sakthivel S and Kisch H. 2003 *Angew. Chem. Int. Ed.* **42** 4908
[8]Ohno T, Mitsui T and Matsumura M. 2003 *Chem. Lett.* **32** 364
[9]Cong Y, Chen F, Zhang JL and Anpo M. 2006 *Chem. Lett.* **35** 800
[10]Zhou ZH, Tokuoka F, Ito Y, Ishii T, Shimai S, Yamaguchi H, et al. 2007 *US Patent* 7175911.
[11]Zhou ZH, Tokuoka F, Ito Y, Ishii T, Yao F 2004 Proc. of 9th Int. Conf. on TiO_2 Photocatalyst, San Diego, CD-ROM ed. by H. Al-Ekabi, Redox Technologies, Inc.
[12]Saha NC and Tompkins HG. 1992 *J. Appl. Phys.* **72** 3072

Shape tuning of ZnO with ammonium molybdate and their morphology-dependent photoluminescence properties

S P Wang[1], S L Zhong[1,3], H L Xu[2]

[1] College of Chemistry and Chemical Engineering, Jiangxi Normal University, Nanchang, 330022, PR China

[2] Institute of Physics and Communication Engineering, Jiangxi Normal University, Nanchang, Jiangxi 330022 ,China

E-mail: zslxhx@yahoo.com.cn

Abstract. In this work, several nano/microscale structures of ZnO including nanospindles, microbundles, microrods, and nanoflowers have been prepared via a facile template and surfactant-free hydrothermal method mediated with ammonium molybdate. It is found that the morphology of ZnO could be readily tailored by simple variation of the concentration of ammonium molybdate. The effects of the concentration of ammonium molybdate and reaction temperature on the synthesis were investigated. X-ray diffraction (XRD), scanning electron microscopy (SEM), energy disperse spectroscopy (EDS), and spectrofluorometer (FL) were used to characterize the products. Results reveal that the photoluminescence (PL) properties of the as-prepared ZnO are strongly dependent on their morphologies.

1. Introduction

As a typical II-VI wide bandgap semiconductor, zinc oxide (ZnO) have been used as transparent electronics, ultraviolet (UV) light emitters, piezoelectric devices, chemical sensors and spin electronics due to its larger excitonic binding energy (60 meV) and high mechanic and thermal stabilities[1-3]. It is well known, the properties of nano/microscale materials depend greatly on their shape, size, and crystallinity. In recent years, morphology control of ZnO received intensive attention. Various morphologies of ZnO such as nanorods[4-6], nanotubes[7,8], nanowires[9], porous nanoblets[10], nanosheets[11,12], flowerlike structures[13], and bundles[14-16] have been synthesized successfully. It turns out that the use of organic molecules such as surfactants, polymers, and capping ligands is an effective method to control the size, morphology, and crystal structure of ZnO [15, 17-19].

In recent years, the controlled synthesis of nanomaterials by the addition of inorganic species received increasing research and many interesting results were obtained. Various morphologies of Ag nanoparticles have been prepared by Xia and co-workers [20] using inorganic ions as capping agent. Silver nanowires have been synthesized by controlling the amount of NaOH in the reaction without the use of a polymer or a surfactant [21]. Controlled synthesis of copper nanostructures has been prepared in a micelle system in the present of inorganic salts [22]. Crystalline 3C−SiC nanobelts have been synthesized by a lithium-assisted synthesis method [23]. EuF_3 with different crystalline phases and morphologies have been prepared by reactions between $Eu(NO_3)_3\cdot 6H_2O$ and fluoride sources XF (X = K^+, H^+, NH_4^+, Na^+, Rb^+, and Cs^+) under ambient conditions[24]. To the best of our knowledge, the morphology control of ZnO mediated by the addition of inorganic species has been rarely concerned.

In this work, ZnO with different morphologies was obtained by a facile hydrothermal process with the assistance of ammonium molybdate. It turns out that several morphologies of ZnO can be readily obtained by only tuning the ammonium molybdate concentration in the synthesis system. Furthermore, their morphology-dependent photoluminescence properties were investigated.

[3] To whom any correspondence should be addressed.

2. Experimental section

2. 1. Synthesis. All the reagents were of analytical grade and used without further purification. In a typical synthesis, 1 mmol $Zn(NO_3)_2{\cdot}6H_2O$ and 0.00166 mmol $(NH_4)_6Mo_7O_{24}{\cdot}4H_2O$ were dissolved in 32 mL deionized water, then 1 mL diethanolamine (DEA) was introduced into above mixture under stirring and a white colloidal precipitate was obtained. After additional 30 min of stirring, the white colloid was transferred into a 45 mL Teflon-lined autoclave, sealed and heated at 160 ℃ for 24 h. After being cooled to room temperature naturally, the white precipitates were centrifuged and washed with alcohol and deionized water in sequence, and then dried at 80 ℃ for 6 h.

2. 2. Characterization. The crystalline phase of the as-prepared products were measured on a Philips X' Pert Pro Super powder X-ray diffractometer equipped with graphite monochromatized Cu Kα radiation (λ = 1.541 78 Å). The morphologies and sizes of the samples were obtained using an environmental scanning electron microscope (QUANTA 2000). EDS analysis was obtained with an Oxford EDS detector installed on an environmental scanning electron microanalyzer (QUANTA 200). The PL spectra were recorded with a Fluorolog3-TAU-P spectrofluorometer at room temperature.

3. Results and discussion

Figure 1: Typical SEM photos of the products prepared at 160 ℃ for 24 h with the concentration of ammonium molybdate is 0.005 mol/L.

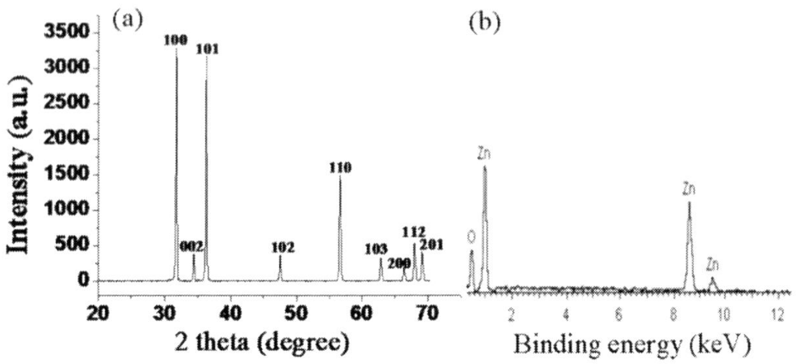

Figure 2: (a) Typical XRD pattern of the as-synthesized bundle-like ZnO; (b) typical EDS spectrum of the as-synthesized bundle-like ZnO.

Figure 1 shows the typical SEM images of as-obtained product prepared at 160 ℃ for 24 h when the concentration of ammonium molybdate is 0.005 mol/L. Low-magnification SEM image of as-prepared product is shown in the figure 1a, which clearly reveals that the as-synthesized ZnO with average length of ca. 60 μm is obtained in large scale. From the enlarged image shown in figure 1b, it is interesting that an individual bundle-like structure is composed of many hexagonal microwires with a diameter of about 800-1000 nm and length of around 30 μm, and the microwires align in a radial way from a center. Furthermore, every microwire is with a sharp top and has a hexagonal cross section, which is different from the previously reported ZnO bundle-like structures [14-16].

The crystal phase and chemical composition of the as-prepared product were determined by XRD and EDS. A typical XRD pattern of the as-obtained bundle-like ZnO is shown in figure 2a. The position of all diffraction peaks for the as-prepared products can be indexed to pure hexagonal wurzite ZnO structure, which agrees well with the standards JCPDS file (No.36-1451). No additional peaks have been found in the XRD pattern, indicating high purity of the sample. It is worth noting that compared to the standard file of hexagonal wurzite ZnO, the relative intensity of (100) is much stronger that of (002), which may be attributed to the preferential growth of the microwires. The composition of the ZnO bundles was further confirmed by EDS. Figure 2b shows the corresponding EDS spectrum of the as-synthesized bundle-like ZnO. From it can be seen, no other elements other than Zn and O are found, and the Zn: O molar ratio is around 1.09: 1, which agrees well with the theoretical composition of ZnO. It also reveals that the molybdate anions can be completely eliminated away from the product after washing with deionized water and ethanol.

Figure 3: SEM images of the as-synthesized ZnO prepared at different concentration of ammonium molybdate. (a) 0 mol/L; (b) 0.0025 mol/L; (c) 0.01mol/L. All the sample were hydrothermally treated at 160 ℃ for 24 h.

Figure 4: SEM images of ZnO prepared at different temperature for 24 h when the concentration of ammonium molybdate is 0.005 mol/L. (a) 100 ℃; (b) 120 ℃; (c) 180 ℃.

To investigate the effect of the addition of ammonium molybdate on the morphology of the products, experiments were carried out at different concentration of ammonium molybdate with other experimental conditions unchanged. In the absence of ammonium molybdate, ZnO nanospindles with average diameter about 150 nm and length around 400 nm were obtained in large scale (figure 3a). Figure 3b show the SEM image of as-prepared ZnO when the concentration of ammonium molybdate is 0.0025 mol/L, which clearly exhibits that the as-prepared ZnO is microwires with lengths of ca. 15

μm and diameters of around 600-700 nm. Increasing the concentration of ammonium molybdate to 0.01 mol/L, flower-like ZnO is formed (figure 3c). The ZnO flowers are in the diameter of ca.20 μm. And the flowers are constructed by a large quantity of nanopetals with a thickness about 70-100 nm. Based on above results, it is clear that the concentration of ammonium molybdate has great effect on the morphology of the final products. With the increasing of concentration ammonium molybdate, the final morphologies of ZnO transform from nanospindle to nanoflower. However, the exact role of ammonium molybdate plays on the synthesis is still not clear, further work is needed.

It was found that the morphology of the as-prepared ZnO was dependent on the reaction temperature. Temperature-dependent experiments were carried out when the concentration of ammonium molybdate was 0.005 mol/L, as shown in figure 4. ZnO hexagonal plates are obtained when the reaction temperature is 100 ℃ (figure 4a). The individual hexagonal plate is in the size varying from 1 μm to 2.5 μm. Bundle-like ZnO architectures also can been obtained at 120 ℃ and 180 ℃. A detail morphological observation of bundle-like ZnO architectures obtained at 120 ℃ is shown in figure 4b, which clearly reveals the ZnO bundles is majorly composed of microbelts with a width of ca. 1 μm. Increasing the reaction temperature to 180 ℃, bundles still can be obtained, as shown in figure 4c. However, the microwires are not as the microwires prepared at 120 ℃, and some nanobelts are found in the product.

On the basis of above results, a possible for the formation mechanism for the crystal growth of ZnO was proposed. The possible involved reaction during the synthesis is as below:

$HN(CH_2CH_2OH)_2 + H_2O \rightarrow H_2N(CH_2CH_2OH)_2^+ + OH^-$ (1)

$Zn^{2+} + 2OH^- \rightarrow Zn(OH)_2$ (2)

$Zn(OH)_2 + 2OH^- \rightarrow Zn(OH)_4^{2-}$ (3)

$Zn(OH)_4^{2-} \rightarrow ZnO + 2OH^- + H_2O$ (4)

Firstly, with the addition of DEA into the reaction system, it hydrolyze to produce OH^-, and the Zn^{2+} reacts with OH^- to form white Zn $(OH)_2$ colloid precipitate. With the prolonging of reaction time, more DEA hydrolyze and provide more OH^-. Secondly, the newly formed Zn $(OH)_2$ colloids gradually dissolves to form Zn $(OH)_4^{2-}$. Thirdly, the ZnO seeds are formed through reaction (4) [25]. Pileni et al. find anions Cl- and Br- in the added salts can control the morphology of the products by selectively adsorbing on the different crystal faces of copper particles [22]. Wang et al have suggested the action of different cations absorbed on the EuF_3 crystal faces of the initial particles is probably different on the morphology of EuF_3 [24]. So the adsorption of molybdate on the surface of ZnO seeds also affects the

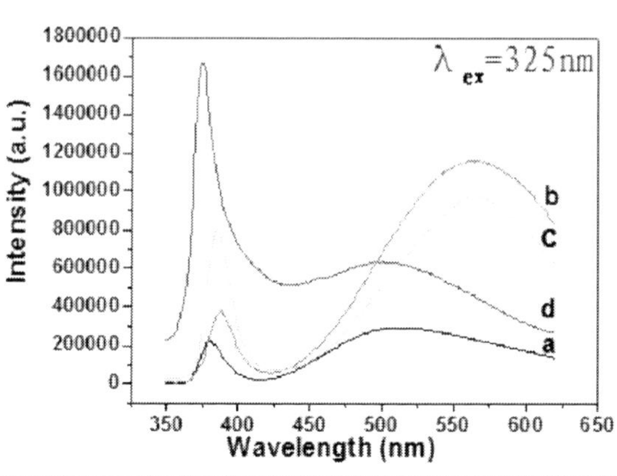

Figure 5: Room temperature PL spectra of the as-prepared samples prepared at different concentration of ammonium molybdate. (a) 0 mol/L (nanospindles); (b) 0.005mol/L (microbundles); (c) 0.0025mol/L (microwires); (d) 0.01mol/L (nanoflowers). All the sample were hydrothermally treated at 160 for 24 h.

crystal growth of ZnO. With different concentration of the ammonium molybdate, the selective adsorption of molybdate on certain crystallographic planes results in the morphology variations.

The photoluminescence properties of the four samples with different morphologies prepared at different concentration of ammonium molybdate were investigated, as shown in figure 5. From it can be seen, the ZnO nanospindles, microbundles, microwires, and nanoflowers exhibit the typical characteristic peaks of ZnO under 325 nm excitation [17, 28]. There are two emission peaks in the spectrum: one is in the ultraviolet region; the other is in the visible region. The emission in the ultraviolet comes from the recombination of free excitons of ZnO [26, 27], while the visible emission is attributed to the presence of singly ionized oxygen vacancies [28]. The emission in the ultraviolet region for the ZnO nanospindles, microbundles, microwires and nanoflowers is centered at 375 nm, 385 nm, 387 nm and 379 nm, respectively. And the visible emission for the ZnO nanospindles, microbundles, microwires and nanoflowers is centered at 507 nm, 563 nm, 562 nm and 507 nm, respectively. From it can be seen, obvious shift was realized in these products, especially at in the visible emission region. And the emission intensity of the four samples increases in the order of nanospindles, microbundles, microwires, and nanoflowers. It is also found that the emission intensity for the ZnO nanospindles, microbundles, and microrods in the visible region is stronger than that of the ultraviolet region. However, for the ZnO nanoflowers, the emission intensity in the ultraviolet region is much stronger than that of the visible region. From above it can be drawn a conclusion that the photoluminescence properties of the as-prepared ZnO depend greatly on their morphologies.

4. Conclusion

In summary, a variety of morphologies of ZnO have been prepared though a facile hydrothermal method in the presence of ammonium molybdate. It turns out that the addition of ammonium molybdate has great effects on the final morphologies of the products. The photoluminescence properties of the as-prepared ZnO are strongly dependent on their morphologies.

References

[1] Wang Z L and Song J H 2006 *Science* **312** 242
[2] Huang M H, Mao S, Feick H, Yan H, Wu Y, Kind H, Weber E, Russo R and Yang P 2001 *Science* **292** 1897
[3] Wang X D, Song J H, Liu J and Wang Z L 2007 *Science* **316** 102
[4] Alammar T and Mudring A V 2009 *Mater. Lett.* **63** 732
[5] Liu B and Zeng C H 2003 *J. Am. Chem. Soc.* **125** 4430
[6] Wang X D, Summers C J. and Wang Z L 2004 *Nano Lett.* **4** 423
[7] Wang H Q, Li G H, Jia L C, Wang G Z and Tang C J 2008 *J. Phys. Chem. C* **112** 11738
[8] Elias J, Zaera R T, Wang Guillaume Y S and Lévy-Clément C 2008 *Chem. Mater.* **20** 6633
[9] Cheng H M, Chiu W H, Lee C H, Tsai S Y and Hsieh W F 2008 *J. Phys. Chem. C* **112** 16359
[10] Shen J B, Zhuang H Z, Wang D X, Xue C S and Liu H 2009 Crys. Growth Des. **9** 2187
[11] Zhou X F, Hu Z L, Fan Y Q, Chen S, Ding W P and Xu N P 2008 *J. Phys. Chem. C* **112** 11722
[12] Chen S J, Liu Y C, Mu R, Lu Y M, Zhang J Y, Shen D Z and Fan X W 2005 *Adv. Mater.* **17** 586
[13] Wang Z, Qian X F, Yin J and Zhu Z K 2004 *Langmuir* **20** 3441
[14] Jiang C L, Zhang W Q, Zou G F, Yu W C and Qian Y T 2005 *J. Phys. Chem. B* **109** 1361
[15] Zeng Y, Zhang T, Wang L J, Wang R, Fu W Y and Yang H B 2009 *J. Phys. Chem. C* **113** 3442
[16] Yu Q J, Fu W Y, Yu C L, Yang H B, Wei R H, Li M H, Liu S K, Sui Y M, Liu Z L, Yuan M X and Zou G T 2007 *J. Phys. Chem. C* **111** 17521
[17] Liang J B, Bai S, Zhang Y S, Ming Li, Yu W C and Qian Y T 2007 *J. Phys. Chem. C* **111** 1113
[18] Tang H, Chang J C, Shan Y Y and Lee S T 2008 *J. Phys. Chem. B* **112** 4016
[19] Shen L M, Bao N Z, Yanagisawa K, Domen K, Grimes C A. and Gupta A 2007 *J. Phys. Chem. C* **111** 7280
[20] Im S H, Lee Y T, Wiley B and Xia Y N 2005 *Angew. Chem. Int. Ed.* **44** 2154
[21] Caswell K, K Bender C and Murphy B C J 2003 *Nano Lett.* **3** 667
[22] Fillankembo A, Giorgio S, Lisiecki I and Pileni M P 2003 *J. Phys. Chem. B* **107** 7492
[23] Xi G C, Peng Y Y, Wan S M, Li T W, Yu W C and Qian Y T 2004 *J. Phys. Chem. B* **108** 20102
[24] Wang M, Huang Q L, Hong J M, Chen X T and Xue Z L 2006 *Crys. Ggrowth Des.* **6** 2169

[25] Li W J, Shi E W and Zheng Y Q 2001 *J. Mater. Sci. Lett.* **20** 1381
[26] Huang M H, Wu Y, Feick H, Tran N, Weber E and Yang P D 2001 *Adv. Mater.* **13** 113
[27] Chen Y, Bagnall D M, Koh H J, Park K T, Hiraga K, Zhu Z and Yao T J 1998 *J. Appl. Phys.* **84** 3912
[28] Vanheusden K, Warren W L, Seager C H, Tallant D R, Voigt J A and Gnade B E 1996 *J. Appl. Phys.* **79** 7983

An effective method of increasing production rate of onion-like fullerenes

Wen Liu[1,2,3], Qing Sen Meng[1]
1. College of Materials Science and Engineering, Taiyuan University of Technology,Taiyuan ,030024 ,China
2. Institute of Chemical and Biological Technology, Taiyuan University of Science and Technology,Taiyuan, 030021, China
3. Key Laboratory of Interface Science and Engineering in Advanced Materials of Technilogy, Ministry of Education, Taiyuan, 030024, China

E-mail: lw915136@sina.com.cn

Abstract. Onion-like fullerenes (OLFs) were synthesized by arc discharge in benzene using graphite as electrode and ferrocene as catalyst. The effect of ferrocene on the morphologies and structures of the OLFs was investigated by HRTEM and XRD. Results show that ferrocene directly influenced the morphology and yield of OLFs: The degree of graphitization is better. The diameters of the OLFs can be controlled in the range between 10 and 30 nm. The method described here suggests a novel and promising route to synthesize OLFs in large scales.

1. Introduction

Onion-Like Fullerenes (OLFs) have unique hollow cage and concentric shells structure,They can be promisingly applied as source materials, high-powered and high temperature wearable materials, superconduct materials and bio-medical materials[1-2].The main methods of synthesis of Onion-like Fullerenes (OLFs) are by arc discharge[3-6], electron irradiation [7], radio frequency plasma[8],chemical vapor deposition[9-10], arc in water or liquid nitrogen. [11-15] etc.'Arc in liquid' method is a cost-effective technique lately proposed to fabricate carbon nanomaterials, which does not require vacuum facilities.However, the production rates of the resulted materials are limited by the evaporation rate of the graphite electrodes. This restriction may cause a limit to apply the'arc in liquid' method to large-scale production[16].

In order to increase the production rate of carbon nanomaterials, Sano[16] first used liquid benzene (C_6H_6) as the quenching media because organic liquid is expected to play a role of carbon source[16]. As a result, CNTs and OLFs are not formed in this condition. In this paper, the author also used liquid benzene (C_6H_6) as the quenching media,the graphite as the electrode and ferrocene as catalyst. The result indicates that OLFs could be efficiently synthesized in high yield.

2. Experimental

Direct current (dc) arc discharge was generated between a graphite cathode (diameter=20mm) and an anode (diameter = 6mm) submerged in liquid benzene (1000cm3) of ferrocene (10g) in a glass container. Using the self-sustained arc-discharge mode, the discharge voltage and current were 22–25V and 50A, During arc discharge, an anode was moving with certain speed,The current 50A was sustained for 5min by adjusting the cathode–anode gap to be approximately 1 mm. Deionized water was continuously supplied into the container to cut the air off from the reaction field to avoid explosion of evaporated C_6H_6. After completion of the arc discharge, the fine powders suspended in the residual liquid C_6H_6 was collected using cyclic filter, the fine powders were put in resistance furnace to heat up to 500℃ under the argon atmosphere ,maintain 30min , the elimination condenses organic matter, cooling spare. The products were characterized by high resolution transmission electron microscopy (HRTEM, JEM-2010 at 200 kV) and X-ray diffraction (XRD).

3. Results and discussion

3.1 morphology and structure

Fig.1 shows TEM images of product prepared using benzene as carbon source and ferrocene as catalyst. Fig. 1(a) focused many carbon nanoparticles by HRTEM, mainly is hollow spherical and polyhedron OLFs, The diameters of OLFs are in the range from 10 to 30nm. The degree of graphitization is higher. CNTs clearly could be observed.Fig. 1(b) shows the magnification image of marked part by a white square in (a) and represent hollow spherical OLFs with diameters of about 20nm ,hollow diameter is bigger and about 10nm, Shelled layer constituted by the 12 concentric graphite layer. The degree of graphitization is better. The particle shape was considered to be nearly spherical, though most were not perfect spheres.

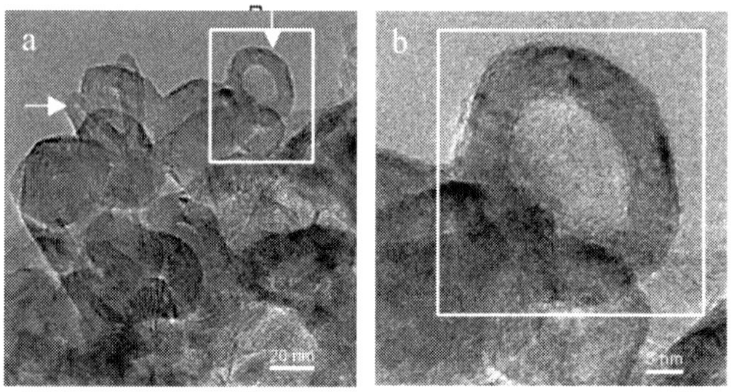

Figure 1. HRTEM images of produces using ferrocene as catalyst

Fig.2 shows TEM images of product prepared by arc discharge in benzene.Obviously, the purity of OLFs in Fig.1was higher than in Fig.2. Here, it must be reminded that the fine powders suspended in the residual C_6H_6 liquid contains OLFs and MW-CNTs. the cathode deposit did not formed by arc in mixture of benzene and ferrocene. Therefore, the reaction field at the arc zone in C_6H_6 is considered to be essentially different from that in mixture of benzene and ferrocene environment. Importantly, the production rate of OLFs and CNTs can be increased by use of ferrocene.

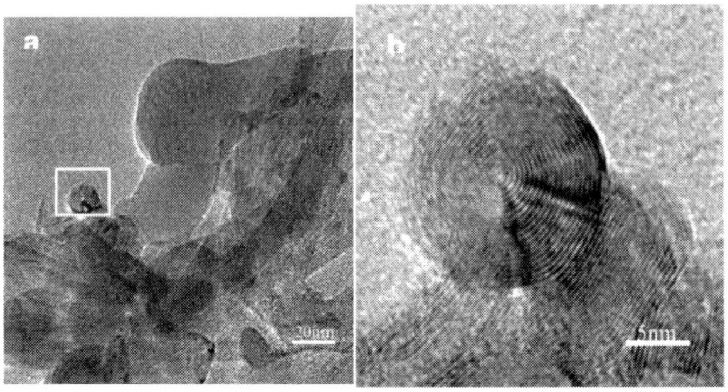

Figure 2. Images of products not using ferrocene as catalyst

3.2XRD analysis

The XRD spectra of the samples are shown in Fig. 3. There was a strong peak at 2θ 26.1° ,close to 2θ 26.41 for graphite, with the dvalues being about 0.351nm and 0.348 nm, a little larger than 0.336nm for graphite. The peak in the curve is for plane (002) of OLFs. Meanwhile appeared a strong peak in this peak behind, It may be due to the presence of structure curvature or some structure

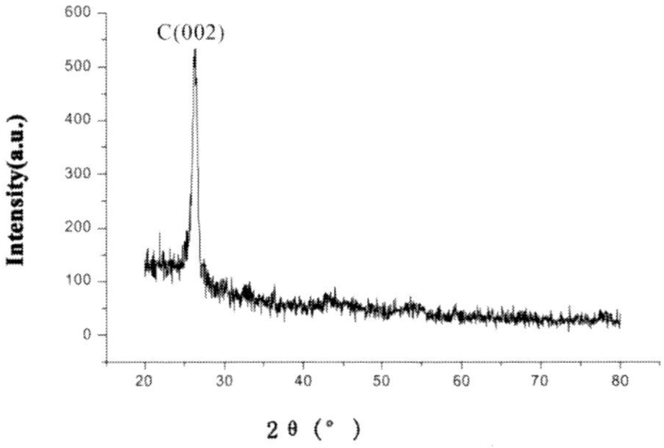

Figure 3. XRD pattern of OLFs prepared from mixture of benzene and ferrocene

3.3 Influence on yield and quality of the onion-like fullerenes by ferrocene.

The catalyst is an essential condition of prepareing OLFs. Many kinds of catalyst are used to synthesizing Onion-like Fullerenes (OLFs) by arc discharge, like Fe,Co,Ni,Yi and so on. Used the benzene as the discharge medium to play a role of carbon source, the catalyst activity is considered when choosing catalyst, meantime, it is considered that whether catalyst can dissolve in the liquid benzene, metal organic compounds eg ferrocene is quite appropriate for catalyst. Using ferrocene as catalyst, the reaction field at the arc zone in C_6H_6 is as follows:

$$C_6H_6 + Fe(C_2H_5)_2 \xrightarrow[5min \pm 0.08min]{3727℃} C_n + C_aH_b + H_2 + Fe$$

(in the formula: C_6H_6- benzene $Fe(C_2H_5)_2$- ferrocene C_n- fullerence C_aH_b- the organic matter H_2- hydrogen).

Direct current arc discharge was generated between a graphite cathode and an anode ,temperature can reach 3727 ℃,C_6H_6 will decompose to play a role of carbon source. ferrocene will decompose into Fe and C_2H_5, Fe maybe act as the catalyst of OLFs growth. There are numerous studies on the catalytic growth of fullerenes using metallic nanoparticles[17,18].Among these studies, a plausible mechanism for the formation of nanotubes and onion-like fullerenes on metallic catalysts could be summarized as carbon-solvation and diffusion in the catalyst particle or surface diffusion, and precipitation growth of graphitic structures. Here, it should be noted that if carbonaceous precursors were not continuously supplied to the Fe nanoparticles, fullerenes could not grow along length. Thus spherical onion-like fullerenes were preferably formed. Therefore, adding ferrocene increased the production rate of OLFs in the same experimental parameters.The result was shown in Table 1.

Table 1. Products synthesized by arc discharge in benzene and in mixture of benzene and ferrocene under the condition of same current.

Current (I/A)	formation rate of product collected in benzene (mg/s)	Anode consumption rate (mg/s)	A formation rate of cathode deposit (mg/s)
50	20.5	1.5	No
50	9.2	1.5	9.5

Arc discharge was generated between a graphite cathode and an anode, which can provide enough energy to decompose C_6H_6 intosome aromatic fragments, some aromatic fragments will be further broken up into C_1 or C_2 carbon units, surfaces of the Fe play a role of active sites, the aromatic fragments were nucleated at the surfaces of Fe. Before the chemical bonds are broken up, the aromatic fragments have obtained enough energy to reorganize on the surface of the catalysts and curve with the help of the pentagonal rings. If the smaller aromatic fragments or C_1 and C_2 carbon units act as building blocks in the innermost shell, the central hollow will be smaller,or the central hollow will be larger. After the nucleation is formed and excited to vapor, the aromatic fragments in the vapor continue to bond with the nucleation. Concerning the possible arrangement of concentric defect-free graphite cages, polyhedral OLFs have all the shells in the same orientation, so that all the corners (pentagons) are perfectly superimposed. If the pentagons of the concentric shells are not aligned, the final shapes of OLFs should be much closer to quasi-spherical morphology [19].

4. Conclusions

In summary, The OLFs with high yields have been first synthesized by arc discharge in benzene using graphite as electrode and ferrocene as catalyst. The OLFs display polyhedral or spherical morphology with hollow center. The diameters of the OLFs can be controlled in the range between 10 and 30 nm. The method described here suggests a novel and promising route to synthesize OLFs in large scales.

5. Acknowledgments
This study works were supported by the projects of the NSFC (No.10872197).

Referenc

[1] Ugarte D, 1992 Curling and closure of graphitic networks under electron beam irradiation *Nature* **359** 707

[2] Xu B S and Tanaka S I 1998 Formation of giant onion-like fullerenes under Al nanoparticles by electron irradiation. *Acta Mater.* **46** 5249

[3] Xu B S,Yan X Q and Wang X M 2001 Studies on the growth mechanism of nano-onion-like fullerenes in arc discharge *Transactions of Materials And Heat Treatment* **22(4)** 9

[4] Wang H Y, Wang X M and Zhang H X 2003 Studies on the preparation of metal encapsulated nanosized onion-like fullerenes by arc discharge *Transactions of Materials And Heat Treatment* **24(4)** 41

[5] Zhang H X, Wang X M and Xu B S 2004 The preparation of nano-onion-like fullerence by arc discharge *New Carbon Material* **19(1)** 612

[6] Wang X M, Xu B S, Liu X G and Hideki Ichinose 2004 Structural characteristics of Onion-Like Fullerenes synthesized by metal nanoparticle catalysis *New Carbon Material* **19(3)** 209

[7] Xu B S and Tanaka S I 1997 Multiple-nuclei onion-like fullerenes cultived by electron beam irradiation. *Proc Int Conf International Contennial Symposium on Electron* 355

[8] Du A B, Liu X G, Xu B S 2005 Synthesis and structure characterization of coal-based nano-structured Onion-Like Fullerenes **20(4)** 779

[9] Li T B, Xu B S and Han P D 2005 Morphology analysis of onion-like fullerenes prepared by CCVD. *New Carbon Materials* **20(1)** 23-26

[10] Yang Y Z, Liu X G and Xu B S 2006 Preparation of vapor grown carbon fibers from deoiled asphalt. *Carbon* **44** 1661

[11] Sano N, Chhowalla M and Wang H 2003 Synthesis of fullurene-like particles and nanotubes by arc discharge in cold liquid. In: 16th *International Symposium on Plasma Chemistry, Taormina*, Italy **680th** file in CD-Rom.

[12] Sano N, Wang H and Alexandrou I 2001 Synthesis of carbon 'onions' in water *J Nature* **414** 506

[13] Sano N, Wang H and Alexandrou I 2002 Properties of nano carbon particles produced by an arc discharge in water *J Appl Phys* **92** 2783

[14] Sano N, Wang H and Chhowalla M 2003 Fabrication of Inorganic molybdenum disulphide fullerenes by arc in water. *Chem Phys Lett* **368** 331

[15] Guo J J, Wang X M and LI T B 2006 Synthesis of onion like fullerenes by arc discharge in water *J New Carbon Materrials* **21(2)** 171

[16] Sano N 2004 Formation of multi-shelled carbon nanoparticles by arc discharge in liquid benzene.Materials *Chemistry and Physics* **88** 235

[17] M. Endo, T. Takeuchi, S. Igarashi, K. Kobori, M. Shirashi and H.W. Kroto 1993 *J Phys Chem Solids* **54** 1841

[18] R. Marangoni, P. Serp, R. Feure, Y. Kihn, P. Kalck and C. Vahlas 2001 *Carbon* **39** 443

[19] D.Ugarte 1995 Onion-like graphite particles *J Carbon* **33** 989

Water Jet Shooting Each Other and Nano-grade Microbe Cells Crusher

Hongwen Yao[1] Shuzhong Liu[1] junying Li[1] suge Guo[2] xuewu Chen[1] wenge Zhou[3]

(Hebei University of Science and Technology. Shijiazhoang, China yaohw163@sina.com)[1]
(Weisheng Pharmaceutical (Shijiazhuang) Co. Ltd. Shijiazhoang, China. gsg1027@sohu.com)[2]
(301 Hospital of the PLA 102208)[3]

Abstract: Through researching the special nozzles, we shoot fan-shape water jet and shoot each other to crush microbe cells and achieve control of particle size. In the test of crushing microbe cells, we find that density of water jet and the size of the nano-particle is in proportion, for the expression of its function: y = f (x). We create the expansion distance between water molecules formulation: $r = r_0 \times \eta$ and other mathematical models. When the expansion of multiples $\eta_n = 1$, the distance between water molecules $r_0 = 0.4$ nm, the size of particle is less than nano-grade, when $\eta_n = 2.6 \sim 250$, $r_n = 1 \sim 100$ nm, get the broken nano-particles; when $\eta_n > 250$, $r_n > 100$ nm, the size of broken particle is greater than nano-grade. Accordingly we developed a water jet microbe cells crusher. We get the Chinese invention patents with No.ZL03143321.9.

Key words: water jet, jet generator, microbe cells

Ⅰ. Introduction

Water jet technology has 30 years of history in exploitation of coal, ores, and in factories, mines cleansing[1,2]. In the late 1960s, the University of Missouri in the United States developed of water jet cutting (water saws, water knife) [1,2] and agreed invention patents. But there is no report about water jet crushing microbe cells (microsubstance) at home and abroad [1, 2, 3]. Water Jet technology and nano-microbe cells crusher are urgent need of special equipment and core technology in microorganism fermentation, extraction of cells and ultra-fine processing industries. The invention of the water jet technology has improve to a new level.

Ⅱ.Research Methods

A. Research and design the nozzle of water jet Crusher [4,5,6,7,8,9,10,11,15]

The nozzle is the core engine of the water jet crusher and can launch a fan-shaped water jet and control the density of water jet (that is, the distance between water molecules). If the needs of smaller particles, the distance between water molecules should be smaller. If the biological needs of larger particles, the distance between water molecules should be bigger, with direct proportion of its function: y = f (x). The fan-shape water jet is fired through nozzles. Please see Figure 1.

In Figure 1, we can see from d4, d3, d2, d1 the water density is different. We summarised the distance between water molecules expansion multiplier formulation[6,7,8] in accordance with the relevant principles of mathematics.

the expansion of multiples η_n wher dn,

$$\eta_n=\frac{Sn}{S_0}=\frac{\pi\left(\frac{d_n}{2}\right)^2}{\pi\left(\frac{d}{2}\right)^2}=\frac{d_n^{\,2}}{d^{\,2}}\quad n=1,2,3 \tag{1}$$

The expansion formula of the distance between water molecules [6,7,8]

$$r_n=r_0\times\eta n \tag{2}$$

FIG.1 A schematic diagram of water intermolecular distance enlargement

FIG.2 A sketch map of parts assembly and shoot water jet

r_n denotes the distance between water molecules after expansion, r_0 denotes the original distance between water molecules before expansion, r_0 = 0.4 [7] nm; η_n for the expansion of multiples. According to Figure 1 and calculation we can know $r_1 < r_2 < r_3 < r_4$,and r controls particle size for r denotes the size of the place where water jet fail to impact of microbe cells, as for the control of the production of biological particles broken the basis of data.

Now, living examples of diameters of biologic particles obtained by the machine were illustrated. Structure of the spray nozzle was presented in figure1, Water jet in left and right spray pipes where the interval between the water molecules r_0, r_0=0.4 nm, was interfused 10%-12% of microbe cells and given strong pressure P which was strong enough to crush microbe cell. The size of obtained biologic particles was calculated as follows:

With the practical measuring values:

d=1.2mm, d_1=2.4mm, d_2=4.8mm,d_3=9.6mm, d_4=19.2mm.

The interval between the water molecules at the position d_1 may be calculated: r_1,

$$r_1=\eta_1\times r_0=\frac{d_1^{\,2}}{d^2}\times r_0=\frac{2.4^2}{1.2^2}\times 0.4=1.6\text{nm}$$

By the same way we can know, r_2=6.4nm, r_3=25.6nm, r_4=102.4nm.

When η_n =1， r_n =0.4 nm, the size of particle is under nano-grade. When η_n =2.6～250, r_n =1～100 nm, the particles are just nanograde. When η_n > 250， r_n > 100 nm, the particles are upper nano-grade.

B. Determine the structure of water jet microbe cells crusher [4,5,6,7,8,9,10,16]

Injector nozzle assembly in 3 suites, the waste microbe cell evenly mixed into water. When fired through the nozzles the two fan-shape water jets carrying secretly the microbe cell shoot each other and toward the waste microbe cell at the same time. The microbe cell can be crushed into nano-grade particle. Nozzle assembly diagram 2.

In figure 2, 1and 2 denote left and right spray nozzles, 3and 4 denote left and right spray pipes, 5 and 6 denote left and right water jet flows of carrying secretly microbe cells, and 7 denotes the crushed particles from water jet flow shooting each other. The suite (1) means that the machine crushes the microbe cells into particles under nano-grade, the suite (2) means that the machine crushes the microbe cells into particles just nano-grade and the suite (3) means that the machine crushes the microbe cells into particles upper nano-grade. The three suites of spray nozzle and spray pipe are assembled one machine and we call it nano-grade microbe cell crusher.

C. **The design for water jet microbe cells crusher**[4,5,6,7,8,17]

We do some experiments with the developed model machine and establishe of mathematical models and then zoom in to develop production water jet machines.

The calculation formula of the diameter d of the nozzle can be determined. It was described in following sections.

$$\Phi = 1.233 \times \sqrt{\frac{Q}{\sqrt{P \times n}} \times \omega \times \Psi} \qquad (3)\ [4]$$

d: diameter of spray pipe, unit: mm. Q: spray flux, unit: L/min. P: jet flow pressure, unit: Mpa. n: number of spray nozzle, unit: quantity. ω: quotient of water flux quotien, ω=0.60~0.85. Ψ: quotiety of spray nozzle efficiency, Ψ=0.75 ~0.95.

The calculating formula of velocity of jet flow (v) of the machine was described as the following:

$$V = \frac{Q}{\pi \gamma^2} \qquad (4)\ [2]$$

V: velocity of jet flow , unit: m/s. Q: flux of jet flow, unit: m^3/s. r: semidiameter of spray nozzle, unit: m.

Calculating formula of impulsive force (F) of jet flow effect of the machine was described as following:

$$F = \frac{m(v - v_0)}{t} \qquad (5)$$

F: an impulsive force of receiving 1g of particles. unit: kgm/s^2. m: 1g of particles, unit: g. v: velocity of jet flow as striking 1g of particles, unit: m/s. v_0: muzzle velocity of jet flow, unit: m/s, v_0=0. t: expending time of jet flow as striking 1g of particles, Unit: s. t=6.0x10^{-3}/s.

The calculating formula of sectional area (S) of spray pipe in the machine was described as the following:

$$S=\pi r^2 \qquad (6)\ [3]$$

S: sectional area of spray pipe, unit: m^2. π: 3.14. r: semidiameter of spray pipe, unit: m.

The calculating formula of jet flow (Q) of the machine was described as the following:

$$Q=S\times V \qquad (7)\ [2]$$

Q: jet flux, unit: m^3/s. S: sectional area of spray pipe, unit: m^2. V: velocity of jet flow, unit: m/s.

The calculating formula of Electromotor power (N) of nanograde microbe cell crusher is as the following:

$$N = \frac{\psi P_N Q_N}{600\eta_p} \qquad (8)\ [12]$$

N: power of engine, unit: kw. ψ: Coefficient of transformation $\frac{p_{max}}{p_N}$, $\frac{p_{max}}{p_N}$=0.52-0.54. P_N: rated working pressure, unit: Mpa. Q_N: rated working flux, unit: L/min. η_p: efficiency of engine, unit: 80%-90%.

According to the above calculation, The machine consists of 3 parts: the motivity source, switch cabinet, jet generator and the system of jet shooting each other. The working situation of the machine was recorded in Table 1,and modeling can be seen figure 3.

Water Jet microbe cells crusher its specifications are: CHKD4000L, CHKD3600L, CHKD3000L, CHKD2400L, productivity is per hour for 4000 kg, 3600kg, 3000kg, 2400kg. Production is applicable for industrial production.

Table 1.The general working situations of the machine

PS(L/h)	DCP(Mpa)	Φ(mm)	V(m/s)	1g（Mpa）	S（m^2）	N（kw）
4000	70-140	1.57	578.9	96.5	$1.9x10^{-6}$	50-100
3600	70-140	1.49	588.2	98.0	$1.7x10^{-6}$	45-90
3000	70-140	1.36	592.8	98.8	$1.4x10^{-6}$	38-75
2400	70-140	1.21	609.1	101.5	$1.1x10^{-6}$	30-60

1. PS—productive scales. 2.DCP—designment crushing pressure. 3.φ—diaeter of spray pipe of productive machine. 4. v—velocity of jet flow of productive machine. 5. 1g—receiving impulsive force of 1g particles. 6. S—sectional area of spray pipe of productive machine. 7. N—power of engine.

D. **Water jet microbe cells crusher visual effects** [7,8,9,10,11,12,13,14,18,19]:

In Figure 2 (1), when η_n = 1, r_0= 0.4 nm, under 100MPa pressure, the size of particle of beer yeast cell after broken is about 0.3～0.9nm (the original size 3 ~ 8 μ m). It can be seen with Hitachi H-700 electronic lens zooming 100,000 times, its visual effect can be seen in Figure 4.

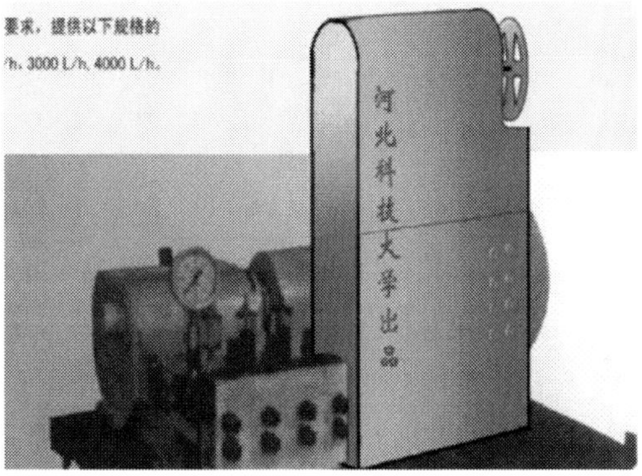

FIG.3 . Water jet microbe cells crusher

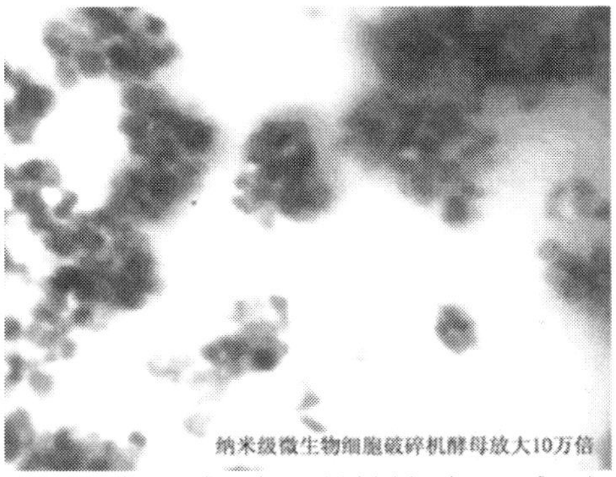

FIG.4. The enlarging 100000 times facsimile picture of yeast after crushed by the machine.

When η_n = 2.6 ~ 250, r_n= 1 ~ 100 nm, under 100MPa pressure, the size of particle of pseudomonas and acremonium after broken is about 20～40nm(the original size of pseudomonas 0.5~1.5μm and acremonium 2 ~ 4μm), its visual effect can be seen in Figure 5.

When η_n>250, r_n= >100nm, under 100MPa pressure, the size of particle of beer yeast cell after broken is range from 1～6nm(the original size 3 ~ 8 μ m). Its visual effect can be seen in Figure 6.

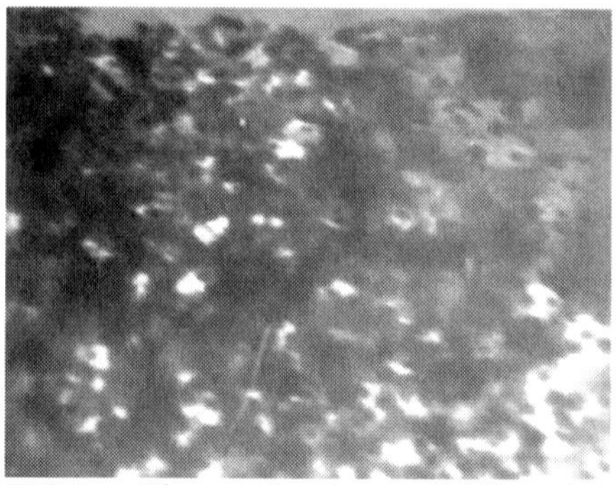

FIG.5. The enlarging 48000 times facsimile picture of pseudomonas and acremonium after crushed by the machine.

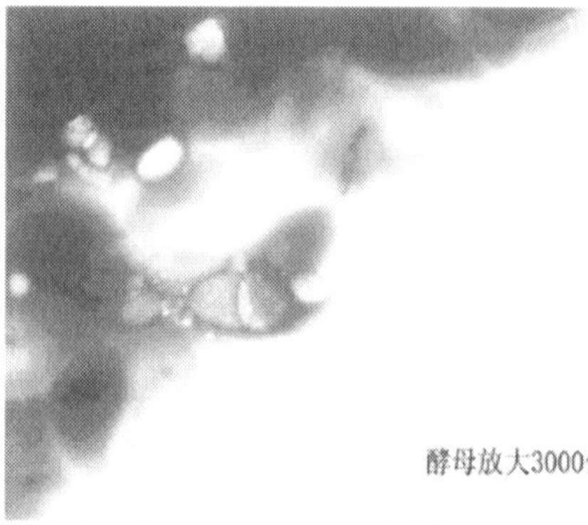

FIG.6. The enlarging 3000 times facsimile picture of yeast after crushed by the machine.

Ⅲ. Results and discussion

1. We developed water jet microbe cell crushing machine in accordance with the theory of enlarging the distance between water molecules. Its specifications are: CHKD4000L, CHKD3600L, CHKD3000L, CHKD2400L. It is possible to realize industrial production that hourly production capacity 4000kg,3600kg, 3000kg, 2400kg, The development of the machine ended the history that the microcosmic materials cannot be crushed, and the machine has become important member of crushing family.

2. The role of time when water jet crusher crushed microbe cells into nano-grade particles was 3 / 1000 second for an instant broken. The material doesn't heat easily and is sure of protecting its living creature.

3. The biological particles of nano-grade produced by the machine contain the protein content of 10-50 percent. They can be used in the fermentation industry instead of the nitrogen materials. The machine can realize circle economy of waste microbe cell to ferment industry and cleansing production and can be the professional tools for the extraction of cells within the industry to provide more, faster, better.

4. The formula of enlarging the interval between the water molecules: $r_n = r_0 \times \eta$ was set up, when $\eta_n = 1$, $r_0 = 0.4$ nm, the size of particle is smaller than nano-grade. When η_n=2.6-250, r_n=1-100 nm, the particle is nano-grade. When $\eta_n > 250$, $r_n > 100$ nm, the particle is bigger than nano-grade. According to this discovery, we developed nano-grade microbe cell crusher and ended a history that the microcosmic substance cannot be crushed.

Ⅳ. References

(1) W.Zhao, X.B.Liu, P.L.Zhu. Processing high-pressure water jet technology, The world papers. 11, 24 (2007).
(2) S.X.Xue.The U.S. high-pressure water jet cutting, fluid machinery. 5 (1996).
(3) L.Hua, et al.. Laser cutting and water jet cutting, fluid machinery. 3 (1996).
(4) Z.Y.Dong. Jet Flow Mechanics. (2005).

Journal of Physics: Conference Series **188** (2009) 012036 doi:10.1088/1742-6596/188/1/012036

(5) H.M.Wang. Engineering Fluid Mechanics. (2005).
(6) H.J.Wang. Technical Mat hematics. (2003).
(7) W.M.Huang. Technical Physics Foundation.1, (2002).
(8) H.Y.Yao, et al.. Nanometer grade of microorganism cell crushing machine and applies in the beer industry. 7, 48-50 (2004).
(9) G.Q.Liu. Bio-Engineering Downriver Technology. **9**, 65–69(2003).
(10) D.C.Fan. Handbook of Hydraulic Pressure Technology. (2004).
(11) W.J.Zhao, et al.. Management and Rule of Law of Solid Rubbish. (2004).
(12) M.F.Hua, et al.. Machine Manufacture Technology. (2004).
(13) M.Y.Wang, et al.. (Translate). Handbook of Nanotechnolog. (2005).
(14) D.Y.Guan. Handbook of Beer Industry. (1982).
(15) Sreeganga chandra, Gilbert Gallardo---. A-Synuclein Cooperates with CSP α in Preventing Neurodegeneration. Cell vol 123november. **4**, 383-396 (2005).
(16) Chaitan Khosla and Yi Tang Anew Route to Designer Antibiotics. Science. **308**, 367-368 (2005).
(17) S.O.Valenzuela, M.Tinkham. Direct electronic measurment of the spin Hall effect. Nature **442**, 176-181 (2006).
(18) Masahiko Tamura, Osamu Matsumoto. Production of Cerebroside from Beet molasses by the Yest Saccharomyces Kluyvri. Food Biote chnlogy. **19**, 95-105 (2005).
(19) Xuewen pan, Ping Ye, Daniel Yuan. A Dna Integrity Networ in the Yeast Saccharomyces cerevisiae. Cell **124**, 1069-1081 (2006).

Journal of Physics: Conference Series 188 (2009) 012037 doi:10.1088/1742-6596/188/1/012037

Preparation and Characterization of CdSe QDs

XIA Jiao-yun *, XU Wan-bang, CAO Zhong

(College of Chemistry and Biological Engineering, Changsha University of Science and Technology, No.960, Wanjiali Road, Yuhuaqu, Changsha, Hunan 410114, China)

Abstract: An optimized synthesis route of CdSe quantum dots (QDs) in an aqueous solution was obtained. The shape, crystal structure and optical property of CdSe QDs were also characterized by XRD, UV-Vis and TEM. The result showed that the high-quality cubic CdSe QDs with 3 nm was obtained. The mechanisms about the influence factors were also presented.

Keywords: CdSe QDs; Optimized Synthesis; Characterization

1. Introduction

Quantum dots (also known as colloidal semiconductor nanocrytals, QDs) are generally composed of II-VI and III-V elements, which exhibit strongly size dependent optical and electrical properties[1-4]. Highly quality semiconductor nanocrystals have shown potential applications in thin film light-emitting devices, non-linear optical devices, solar cells and life science[5-8]. Quantum dots synthesized by organometallic approach have high quantum yield, sharp emission spectra, monodispersity and good crystallinity. However, key chemicals used in this route are toxic, expensive, pyrophoric, and even explosive. In this thesis, we synthesized a series quantum dots in aqueous solution by using thiols as the stabilizers. This method is simple and controllable; as-prepared quantum dots show good optical properties.

2. Experimental section

2.1 Reagents

Cadmium chloride ($CdCl_2\cdot5/2H_2O$), Sodium hexametaphosphate ($(NaPO_3)_6$, HMP), Mercaptoacetic acid (MAA), Thioglycolic acid (TGA) and Se powders were purchased from

* Correspondence author: College of Chemistry and Biological Engineering, Changsha University of Science and Technology, No.960 Wajiali Road,Yuhuaqu, Changsha, Hunan 410114, P. R. China; Tel: (86)731-82618702, 86-13975179430; E-mail: xiajy625@yahoo.com.cn

Sinopharm Chemical Reagent Co. Ltd.. Octyl-mercaptan (OMT) was purchased from J&K China Chemical Ltd. All the chemicals were of analytical grade and were used without further purification. Deionized water was used as a solvent.

2.2 Characterization

UV-Visible absorption spectra were measured with an UV-240 UV-visible spectrophotometer (Shimadzu, Japan). The morphology and size of the CdSe QDs were observed by a Hitachi Model-800 transmission electron microscope (TEM, Japan). The crystal structure of the as-prepared products were examined by Rigaku D/max-2200 X-ray diffraction (XRD) with Cu K_a radiation (λ = 0.15418 nm).

2.3 Preparation of CdSe QDs

The preparation of CdSe QDs by the colloidal method had been described in detail elsewhere [9], and we made a little modification. The black Se powders (0.001mol) were added into 100ml sodium sulfite (0.1M) solution and refluxed the solution under stirring with N_2 protecting for several hours to formation the 10^{-2} M sodium selenosulfate solution (Na_2SeSO_3). Then, several Cd precursors prepared by mixing 10 ml $CdCl_2$ solution (10^{-3}M) with different stabilizer solution are adjusted to different pH values with 1M NaOH or HCl, respectively. And finally, designed volume as-prepared non-oxygen Na_2SeSO_3 solution was added into the Cd precursor solution with stirring and refluxing 30 min under N_2 protection.

3. Results and discussion

3.1 Morphology analysis of CdSe QDs

The shape and size of CdSe QDs was observed by TEM. The TEM confirmed the spherical CdSe QDs were around 3 nm in size and dispersed well.

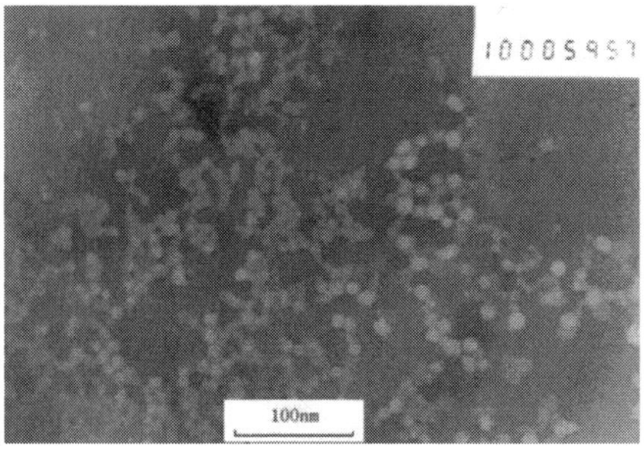

Figure1 TEM image of CdSe QDs

3.2 UV-Vis analysis

Generally, the electronic state is an important property and can be described in terms of valence and conductivity bands and a gap between these bands. However, as the particles become smaller, the wavelength of the electrons is closer to the range of the particle sizes and the laws of classical physics have to be substituted by quantum confinement or quantum size effect (QSE). Besides, the literature [10] reported that the QSE in direct-gap semiconductors is well understood, such as a shift of the optical absorption edge to higher energies with decreasing size. And it could explain the PL and UV-Vis blue-shift effect.

The UV-Vis absorption spectrum showed that the absorption peak of as-obtained CdSe QDs in aqueous solution is 543 nm(2.28 eV), relative blue-shift to the band gap of bulk cubic CdSe (1.78 eV, 698 nm). The results clearly showed the effect of QSE. The estimated particle size was about 3 nm according to Brus-Equation.

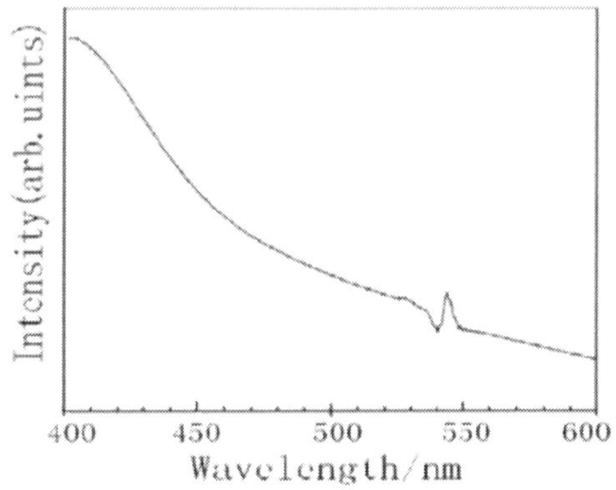

Figure 2 UV-Vis absorption spectra of CdSe QDs

3.3. X-ray diffraction analysis

The XRD patterns in Figure 3 indicated that the as-prepared CdSe QDs are of cubic crystal. According to the standard JCPDS card No. 19-0191, the diffraction peaks correspond to the (111), (220) and (311) crystal plane. The wide diffraction peaks suggested that the ultrafine particles were in poor crystallization. Average crystalline size of the CdSe QDs was estimated at 3 nm by Debye–Scherrer equation after modified instrument's factors.

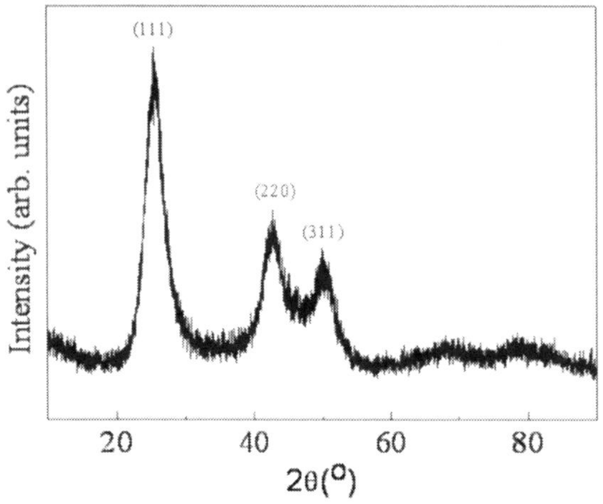

Figure 3 XRD patterns of the CdSe QDs

4. Measurement of quantum yield (QY) of CdSe QDs

The QY of CdSe QDs was measured according to the method described in Ref. [11,12]. Briefly, Rhodamine B was chosen as a reference standard, the absorbance for the standard and the as-prepared CdSe aqueous samples at the excitation wavelengths and the fluorescence spectra of the same solutions were measured, respectively. The integrated fluorescence intensity (IFI, which is the area of the fluorescence spectrum) from the fully corrected fluorescence spectrum was calculated. Six respective different concentration solutions of Rhodamine B (in ethylene glycol) and CdSe QDs aqueous solutions were used in the measurements. Optical densities of all solutions were adjusted to values between 0.1 and 0.25 at the excitation wavelength to avoid re-absorption effects. The areas of IFI VS absorbance were plotted. The plot obtained should be a straight line with a gradient, which was used to calculate the quantum yield according to the following equation:

$$QY_u = QY_s \times \frac{F_u}{F_s} \times \frac{A_s}{A_u} \qquad (1)$$

Where the subscripts s and u denote standard (Rhodamine B) and test samples, respectively, and F is the IFI of the solvent. It should be noted that the excitation wavelength for measurements of the QY was set at the excitonic absorption peak of the CdSe QDs samples in our experiments. With the optimized routes, the maximum quantum yields of CdSe QDs up to 32.3%.

5. Conclusions

In this work, spherical cubic CdSe QDs with better FL properties and dispersedly had been obtained successfully by an optimized and modified-synthesis process. The result showed that the high-quality cubic CdSe QDs with 3 nm was obtained. With the optimized routes, the maximum quantum yields of CdSe QDs up to 32.3%. The CdSe QDs has more possibilities for important implications in environmental, biological and other applications.

6. Acknowledgements

This work was financially supported by the National Natural Science Foundation of China (20775010).

References

[1] Chan WCW, Maxwell DJ, Gao XH, 2002 Luminescent quantum dots for multiplexed biological detection and imaging *Curr. Opin. Biotech.*1 40-46

[2] Melton L 2005 Endothelial cells and VEGF in vascular development *Nature* **37** 775-779

[3] Chan W C W, Nie S M 1998 Quantum Dot bioconjugates for ultrasensitive nonisotopic detection *Science* **281** 2016-2018

[4] Ballou B, Lagerholm B C, Ernst A L 2004 Noninvasive imaging of quantum dots in mice *Bioconj. Chem.* **15** 79-86

[5] Tanke H J, Dirks R W, Raap T. 2005 FISH and immunocytochemistry: towards visualising single target molecules in living cells *Curr. Opin. Biotech.* **16** 1-6

[6] Dhruba J. Bharali, Derrick W. Lucey, Harishankar Jayakumar 2005 Folate-Receptor-Mediated Delivery of InP Quantum Dots for Bioimaging Using Confocal and Two-Photon Microscopy *J. Am. Chem. Soc.* **127** 11364-11371

[7] Santra S,Dutta D,Walter G A 2005 Fluorescent nanoparticle probes for cancer imaging *Technol. Cancer Res. Treat.* **4** 593-602

[8] Alex G. Cuenca, Huabei Jiang, Steven N. Hochwald 2006 Emerging implications of nanotechnology on cancer diagnostics and therapeutics *Cancer*, **107** 459 – 466

[9]Yan Yu-lin, Li Yi, Qian Xue-feng 2003. Preparation and characterization of CdSe nanocrystals via Na_2SO_3-assisted photochemical route *Mater Sci Eng B-Solid* **103** 202-206

[10] Zhao X. W., Komuro S., Fujita S. 1999 Fabrication and stimulated emission of Er-doped nanocrystalline Si waveguides formed on Si substrates by laser ablation *Appl Phys Lett*;

74 120

[11] Wang Y., Tang Z., Correa-Duarte M. A. 2004 Mechanism of Strong Luminescence Photoactivation of Citrate-Stabilized Water-Soluble Nanoparticles with CdSe Cores *J. Phys. Chem. B* **108** 15461-15469

[12] Chen GZ, Huang XZ, Zhang ZZ 1990 Fluorescence Analysis, *Science Publishing House, Beijing, China*, 15 – 18.

Preparation and magneto-nanocrystallization of amorphous alloy $Fe_{60}Co_{26}Hf_7B_6Cu_1$

Yue-Sheng Chao[1, 2], Han-Xian Zhu, Zheng-Ming Li, Yan-Hui Zhang and Zhuang-Hui Zou

[1] Institute of Physics and Chemistry of Materials, Northeastern University, Shenyang 110004, China

[2] Project supported by the National Natural Science Foundation of China (Grant No.50771025)

E-mail: yschao@mail.neu.edu.cn

Abstract. Amorphous alloy ribbons $Fe_{60}Co_{26}Hf_7B_6Cu_1$ (3mm in width and 30μm in thickness) were prepared using the single-roller melt-spinning technique. Amorphous specimens were treated by pulsed magnetic field with lower frequency. The microstructure and crystallization of amorphous specimens were investigated by Mössbauer spectroscopy, XRD and TEM etc. The results show that the magnetic pulsing can induce the crystallization of amorphous alloy at low temperature. With the increasing of pulsing frequency, the precipitated fraction and corresponding hyperfine magnetic field of the base metal phase α-Fe(Co) is increased.

1. Introduction

Nanocrystalline alloys obtained by primary crystallization of Fe-based amorphous alloys have been extensively investigated in the last 15 years. FINEMET- Fe-Co-M-B-Cu (M =Zr, Nb, Hf) alloys are examples of such materials, which have been developed from Fe–Zr–B glass-forming alloys. Cu is added as a nucleating agent and Nb/Zr/Hf restrict the size of the nanocrystalline bcc phase. Partial substitution of Co by Fe in Fe-Co-M-B-Cu (M =Zr, Nb, Hf) was proposed to extend the soft magnetic behavior of nanocrystalline alloys up to higher temperatures. This new nanocrystalline alloy is one of the excellent candidates for soft magnetic materials used at high temperature applications with the range of 773~873K. In the present work, nanocrystallization behavior and evolution of microstructure on low-frequency magnetic pulsing treatment of amorphous $Fe_{60}Co_{26}Hf_7B_6Cu_1$ have been investigated by using transmission electron microscopy (TEM), X-ray diffraction (XRD) and Mössbauer spectroscopy (MS).

2. Experimental

Mixed Fe、Co、Hf、B and Cu ingredients (purity above 99.99wt%), and then melted in a quartz tube with nozzle. The alloys were prepared in a vacuum induction melting furnace (Hechigen). Rapidly solidified ribbons were produced using a single roll melt spinner with the surface linear velocity of a copper wheel at a speed of 39m/s (the equipment named Edmund Bühler in Institute of Metal Reserch Chinese Academy of Sciences), in an argon atmosphere at the temperature of 1300℃. Rapidly solidified ribbons of 3mm width and 30μm thickness were thus obtained. The samples were characterized by X-ray diffraction (XRD), Differential thermal analysis (DTA) and Mössbauer spectroscopy (MS). The samples were treated on the low-frequency magnetic/current pulsing equipment, the treatment conditions were given in table 1.

Table 1. Magneto-pulsing conditions of $Fe_{60}Co_{26}Hf_7B_6Cu_1$ amorphous alloys.

sample	Field intensity/Oe	frequency/Hz	Acting time/s
A0	0	0	0
A1	222	10	240
A2	222	20	240
A3	222	30	240
A4	222	40	240

3. Results and discussion

Figure 1 shows the XRD patterns of $Fe_{60}Co_{26}Hf_7B_6Cu_1$ alloy ribbons, and from the free and wheel-contacted surfaces of the alloy ribbon presented a broad peak. The as-quenched (A0) alloy was confirmed integral amorphous. Figure2 (a) shows TEM images and corresponding selected area diffraction (SAD) patterns of as-quenched (A0) alloy sample. The featureless contrast in microstructure and diffused rings in SAD pattern confirms the presence of amorphous phase in the as-quenched sample. Combined the two results, the as-quenched (A0) alloy was further confirmed integral amorphous.

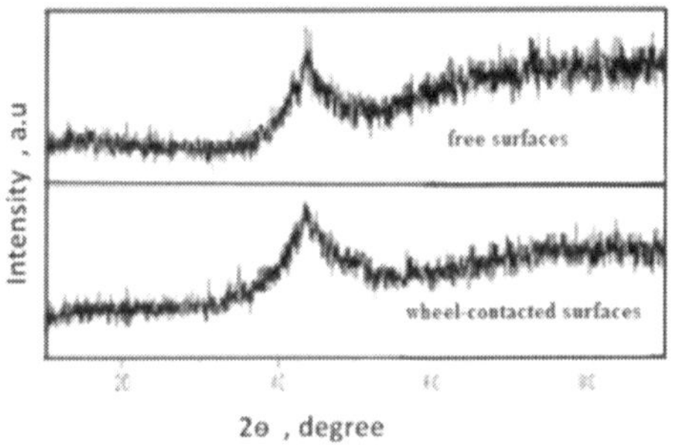

Figure 1. The XRD patterns of $Fe_{60}Co_{26}Hf_7B_6Cu_1$ alloy ribbons.

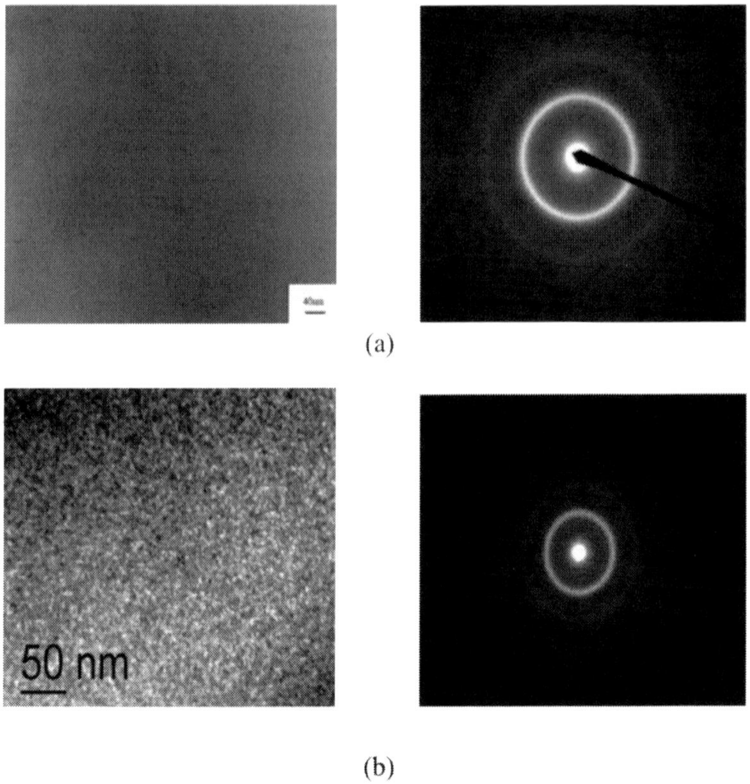

Figure 2. TEM morphology and SAD pattern of specimen $Fe_{60}Co_{26}Hf_7B_6Cu_1$.

Figure 3 shows the Mössbauer spectrum of as-quenched (A0) alloy and the samples after the low-frequency magnetic pulsing treatment (because the trait of Mössbauer spectrum of amorphous have no big difference, just the spectrum of A0, A2 and A4 are given), the fitting results are shown in Table 2.

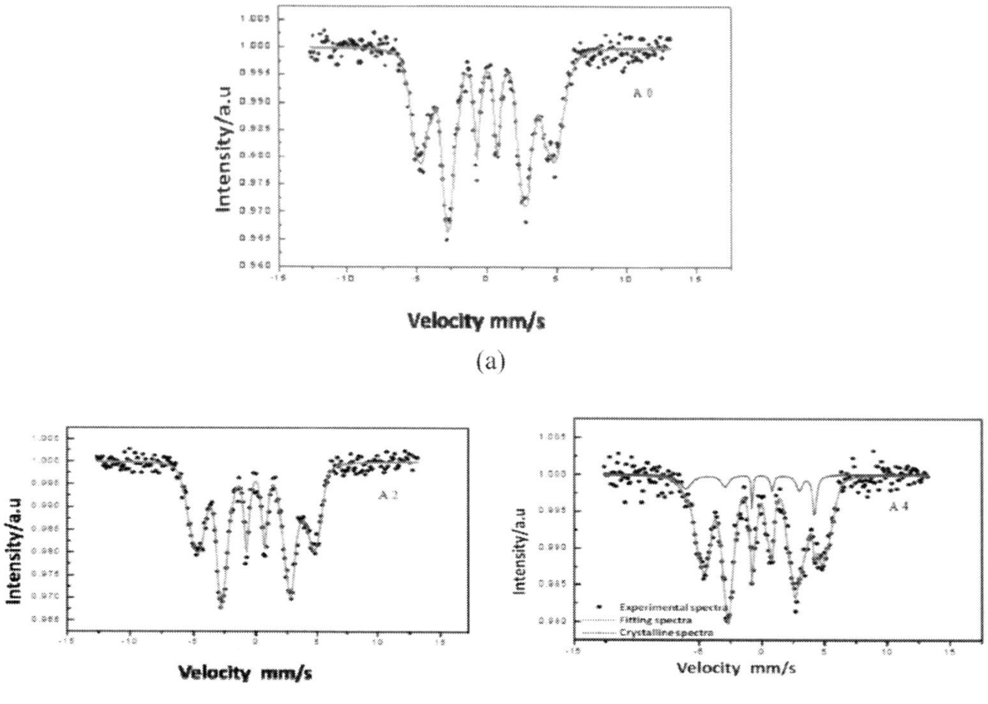

(b) (c)

Figure 3. Mössbauer spectra of amorphous alloy specimens before and after treatment.
(a) as-quenched A0; (b) after treatment A2; (c) after treatment A4

The Mössbauer spectra of samples (A0) are shown in Figure 3. It is immediately obvious from this figure that the spectrum is a non-symmetrically broadened sextet, typical of an amorphous material. After the treatment, the spectra have no big change. However, when the frequency change from 10Hz to 40Hz, the relative intensity ratio (A2,5/A3,4 of peak 2 (or 5) and peak 3 (or 4)) is from 2.73 to 1.74 respectively which given in Table 2. By the formula:

$$b = A_{2,5} \Big/ A_{3,4} = \frac{4\sin^2\theta}{1+\cos^2\theta} \quad (1)$$

where θ is the angle between the γ-ray propagation direction and the direction of the magnetic hyperfine field. On these grounds, the decrease in b indicates that a rotation of the average magnetic moment directions from the in-plane (θ = 90 °) to the out-of plane orientation has taken place. The transmission Mössbauer γ-ray experiment is the vertical incidence of the sample surface, so θ from 0 ° → 90 °, corresponding to the b from 0 → 4, the corresponding hyperfine magnetic field direction of the transition from the vertical surface parallel to the surface. For ferromagnetic polycrystal of disorder magnetic moments, b=A2,5/A3,4=2.0. While for the alloy ribbon prepared using a single roll melt spinner that the magnetic moment is parallel to the swing direction, the value of b is bigger than 2.0 (Table 2). The increase of the frequency, consisting in a decrease of the relative intensity ratio of the second (fifth) to the third (fourth) lines. In other words, the microstructure manipulation had taken place. In Table 2, with field strength of 222Oe, the acting time of 240s, the frequency of 20~30Hz, initial crystallization appeared-the volume fraction of precipitated phase (α-Fe(Co)) is 0.5%. When the frequency increased by 40Hz, the volume fraction of precipitated phase (α-Fe(Co)) got to 3.5%. So the volume fraction of crystallization phase increased with the increase of the frequency of magnetic field.

Table 2. The fitting results of Mössbauer spectra of specimens before and after applied by pulse magnetic fields.

samples	Phase structure	A2,5/A3,4	Hhf(KOe)	H*hf(KOe)
A0 (as-quenched)	amorphous	2.73	269.47 308.38 267.48 229.50	—
A1 (222Oe、240s、10Hz)	amorphous	2.66	232.04 292.43 250.44 300.85	—
A2 (222Oe、240s、20Hz)	amorphous	2.69	220.91 294.38 249.02 303.27	—
A3 (222Oe、240s、30Hz)	Residual amorphous+0.5% α-Fe(Co)	2.35	170.08 150.38 210.94 —	262.03
A4 (222Oe、240s、40Hz)	Residual amorphous+3.5% α-Fe(Co)	1.74	261.74 263.48 284.35 —	319.44

The results show that the crystallization of the $Fe_{60}Co_{26}Hf_7B_6Cu_1$ amorphous alloy can be promoted by the low-frequency pulse magnetic field at the room temperature, a temperature rise of less than 20℃. The DTA curve shows that crystallization only can occur at the temperature of Tp1=515℃. From the TEM results in Figure 2 (b), grain size of the crystallization phase is about 10 nm. In short, the amorphous alloy treated by pulsating magnetic field could be nanocrystallized at lower temperature.

Magnetic moments of Fe atoms tend to orient towards the direction of external magnetic field ($\vec{H}_p$) due to the magnetic interaction ($E_H = -\mu_0 M_s H_p \cos\phi$, where μ_0 is vacuum permeability, M_s is spontaneous magnetization, ϕ is the angle between the atomic magnetic moment direction and the direction of pulsing magnetic field $\vec{H}_p$) between the atomic magnetic field and hyperfine magnetic field. The magnetic moments constantly changed in orientation and vibrated in the process of magnetic plusing. It subsequently led to the change of the distance between each atom. Namely the phenomenon was associated with the macroscopic magnetostriction ($E_\lambda = \frac{3}{2}\lambda_s\sigma\sin^2\theta$, where λ_s is saturation magnetostriction, is the angle between magnetization direction and the direction of stress). The atom would across potential barrier of energy and change from metastable amorphous state to stable crystalline state in proper condition. The periodical vibrations of atoms due to the pulsating magnetic-field action increase the initial energy of atoms,

$$E = E_H + E_\sigma = -\mu_0 M_s H_p \cos\phi + \tfrac{3}{2}\lambda_s\sigma\sin^2\theta \quad (2)$$

as a result, the potential barrier u of nucleation of amorphous alloy was relatively reduced hence the nucleation rate of matrix metal phase was increased. That is the low-frequency magnetic-field, which promote the crystallization. The increase of frequency of pulsing led to the increase of the vibratory frequency, so that the diffusion factor D($D = D_0\exp[-u/KT]$, where D_0 is a constant, $D_0 = a^2v/6$) increased. The nucleation rate determined by the following equation:

$$I = I_0\exp[-\Delta G/KT]\exp[-u/KT] \quad (3)$$

This suggests that increase in pulse frequency will increase the diffusion coefficient of atoms to fill the gap of amorphous structure of non-ferrous atom, so the degree of precipitation is increased. The metal matrix phase α-Fe(Co) crystallization precipitation increased, at the same time, the crystalline precipitate of the corresponding value of hyperfine magnetic field increased (from 262.0 kOe to 319.4 kOe) due to weakened metal iron and other non-iron hybrid electronic orbital effect.

Amorphous microstructure, crystallization precipitation situation to deal with the magnetic pulse parameters (frequency, field strength and the role of time) for the change remains to be further study.

4. Conclusions

(1) $Fe_{60}Co_{26}Hf_7B_6Cu_1$ alloy ribbons prepared by single-roller melt-spinning were confirmed integral amorphous.

(2) Low-frequency magnetic pulsing can lead to the structural transformation of Fe60Co26Hf7B6Cu1 alloy ribbons and the precipitation of crystallization phase-α-Fe(Co) .

(3) With the increase of pulsing frequency, the volume fraction of precipitated phase α-Fe(Co) is increased from 0.5% to 3.5% and the hyperfine magnetic field is increased from 262.0 kOe to 319.4 kOe.

References

[1] Lu K, Zhou F 1997 *Acta Metall Sin.* **33** 99 (in Chinese)
[2] H. Iwanaabe, Bin Lu, M. E. Mchenry and D. E. Laughlin 1999 *J. Appl. Phys.* **85** 4424
[3] X. B. Liang, T. Kulik, J. Ferenc, et al. 2007 *Mater. Charact.* **58** 143

[4] B. Majumdar, S. Bysak, B. Akhtar 2007 *J. Magn. Magn. Mater.* **309** 300

[5] T. C. Gibb, 1976 *Principles of Mössbauer Spectroscopy* (London: Chapman and Hall) p 41

[6] Yuesheng Chao, Zuhan Lai, Kaiyuan He and Hongru Er 1994 *Mater. Sci. Eng.* **A181/A182** 982. (in Chinese)

[7] Chao Yuesheng, Zhang Yanhui, Guo Hong, et al. 2007 *Acta Metall Sin* **43** 23 (in Chinese)

[8] M. A. Willard, D. E. Laughlin, and M. E. Mchennry, et al. 1998 *J. Appl. Phys.* **84** 6773

[9] J. Ferenc, J. Latuch, T. Kulik 2004 *J. Magn. Magn. Mater.***1469** 272-276

[10] M. A. Willard, J.C. Claassen, R. M. Stroud, T. L. Francavilla, V. G. Harris 2002 *IEEE Trans. Magn.* **38** 3045

[11] Y. H. Zhang, Y. S. Chao 2007 *Mater. Sci. Eng.* **251** 460-461

[12] S. Stenger, M. Sorescu and U. Gonser. *J.* 1992 *Non-Crys. Sol.* **66** 151

[13] J. S. Blazquez, V. Franco, C. F. Conde, A. Conde, S. Roth. *J.* 2005 *Appl. Phys.* **97** 044308

[14] Wang F X. 1981 *Metal Physics* (Beijing: China Machine Press) p 183

[15] C. F. Conde, J. S. Bla′ zquez, A. Conde, B. Idzikowski, et al. (Eds.) 2005 *Properties and Applications of Nanocrystalline Alloys from Amorphous Precursors* (Kluwer Academic Publishers: Dordrecht) p 111.

Journal of Physics: Conference Series **188** (2009) 012039 doi:10.1088/1742-6596/188/1/012039

Development of near infrared responsive material based on silica encapsulated gold nanoparticles

Huang Chih Lu,[1] I Shou Tsai,[1,2] Yu Hsiang Lin[2]

1. Nanotechnology Research Center, Feng Chia University, Taiwan
2. Department of Fiber and Composite Material, Feng Chia University, Taiwan

Abstract

In this research, the silica encapsulated gold nanoparticles with optical resonance located in 700-850 nm spectral region were synthesized by combining gold core and the silica shell. The synthesized composite particles can be potentially used in biological fields, due to the biocompatibility of gold, silica and biopolymers. Nano-gold, which can act a raw material of composite particles, was fabricated by stand citrate reduction method. Then, the polyvinyl pyrrolidone (PVP) and (3-mercaptopropyl)trimethoxysilane (MPTMS) are added to deposit silica shell and control particle size. Finally, a complete silica nanoshell was formed on the gold surface by the one-step method, without a repeated coating process. The controllable absorption wavelength of nano-composite particles can be easily controlled by the concentration ratio of PVP and MPTMS. Transmission electron microscopy (TEM) images and optical absorption spectra clearly showed that silica was successfully deposited onto the gold surface by this novel method. As the decreasing aggregation of gold core by adding PVP, the optical plasmon peaks became blue-shifted, but the optical plasmon resonance became red-shifted and the absorption spectra were functions of addition ratios by MPTMS. We expect to extend these functional nanoparticles for real sample applications in the near future.

Keywords : nanoparticle; optical plasmon resonance, PVP, MPTMS, gold , silica

Journal of Physics: Conference Series **188** (2009) 012039 doi:10.1088/1742-6596/188/1/012039

1. Introduction

Nanostructured materials are of intense interest for the development of the nanofabrication of functional nanodevices, high surface area materials, consolidated materials and biotechnological applications [1-3]. Nanoscale metal particles are well known for their ubiquitous optical plasmon resonance, which allows these materials to be used in the optoelectronic and biological fields such as ultra-fast optical switching, optical stability, chemical and biological sensing, optical filters, surface-enhanced raman spectroscopy, and other surface-enhanced spectroscopies [4-5]. Among them, Au nanoparticles display good biocompatibility as well as fascinating optical and electronic properties, which make it possible to be used for wide biological applications [6-8]. However, the utility of Au nanoparticles is limited because their plasmon resonance is confined to relatively narrow wavelength ranges and cannot be readily shifted.

Until now, there have been several approaches to manipulate the optical resonance of metal nanoparticles by tailoring the particle geometry to metal nanorod or metal nanoshell. Metal nanoshells are composite nanoparticles consisting of a dielectric core coated with a thin metal shell. The plasmon resonance of a metal nanoshell may be shifted throughout the visible and near infrared by varying the ratio of the core radius to the shell thickness. By controlling the core/shell ratios, the optical resonance of gold nanoshell can be extended from the visible region to the infrared region. This spectral range includes the 800–1300 nm and 1600–1850 nm, a region of high physiological transmission which has been demonstrated as the spectral region best suited for optical bio-imaging and biosensing applications [9-11].

In this article, we fabricated completely silica-encapsulated gold particles by combining the standard reduction seeding with one step shell-growing method. Owing to the addition of PVP and MPTMS for nanoshell fabrication, continuous silica nanolayers were easily deposited on gold particles. In this case, the different concentrations of PVP formed around the gold particles acted as a protecting agent to avoid aggregation, and the MPTMS as well as an efficient catalytic surface agent for the formation of silica onto the gold surface. The complete nanoshell was easily formed by the one-step shell growing method, without a repeated coating process for controlling the thickness of the layer. And the optical properties of the composite particles could be easily controlled so that the optical plasmon resonance is located in the near infrared region [12-13]. The spectrum range is ideally suited for in vivo imaging because of minimal light absorption by hemoglobin and water. We expect that the near infrared responsive silica-encapsulated gold particles can be potentially used in targeted bio-imaging and therapeutics applications, due to their good biocompatibility and easiness of bio-conjugation.

2. Experimental

2.1. Materials and preparation of composite particles

Tetraethoxysilane (TEOS, 99.999+%), ammonium hydroxide (27%), ethyl alcohol, hydrogen tetrachloroaurate(III) ($HAuCl_4$, 99.9%), sodium citrate (97%), polyvinyl pyrrolidone (PVP, 99.5%), and (3-mercaptopropyl)trimethoxysilane (MPTMS, 95%) were purchased from Aldrich Chemical Co. The deionized water was obtained by Millipore System. All glassware were cleaned with chromic acid and rinsed with deionized water. Gold colloids were prepared by sodium citrate reduction of $HAuCl_4$ as reported earlier. A 250 mL sample of 1 mM $HAuCl_4$ was brought to a vigorous boil with stirring in a round-bottom flask fitted with a reflux condenser, and 25 mL of 38.8 mM sodium citrate was rapidly added to the solution. The solution was boiled for another 15 min, during which time the solution changed color from pale yellow to deep red. The solution was allowed to cool to room temperature with continued stirring. The suspension was filtered using a 0.22 μm filter (Corning, NY) and stored at 4 °C until further use.

To synthesize silica-encapsulated gold particles, 12 ml of above gold colloid, 1.6 g of PVP, 6.4 ml of ethanol, and 50 μl of MPTMS were mixed and stirred 12 hours under dark situation. Then the 200 μl of TEOS and 200 μl of ammonium hydroxide were added to the above mixture for further reaction. After 24 hours continuing stir at 4 °C, this mixture was allowed to age in the dark for one day. The different concentration ratios of PVP and MPTMS were prepared to control the size of gold core and the thickness of silica shell. Finally, these composite particles were centrifuged and washed with ethanol for three times to remove excess residuals.

2.2. Characterization of composite particles

The samples for transmission electron microscopy (TEM) were prepared by placing a drop of dilute solution of the sample on a copper grid. The bright field images and energy dispersive analytical X-ray (EDAX) were obtained using JEOL JEM-2010 TEM operating at 200 kV. UV-visible absorption spectra were obtained with a Hitachi U2000 UV-visible scanning spectrometer in the wavelength range from 300 to 1100 nm. Solution spectra were obtained by measuring the absorption of a dilute solution in a methacrylate cell with a path length of 1 cm.

3. Results and discussion

Figure 1a shows the schematic representation of the fabrication process of silica encapsulated gold particles by stand citrate reduction, gold-thiol assembly, and sol-gel method.

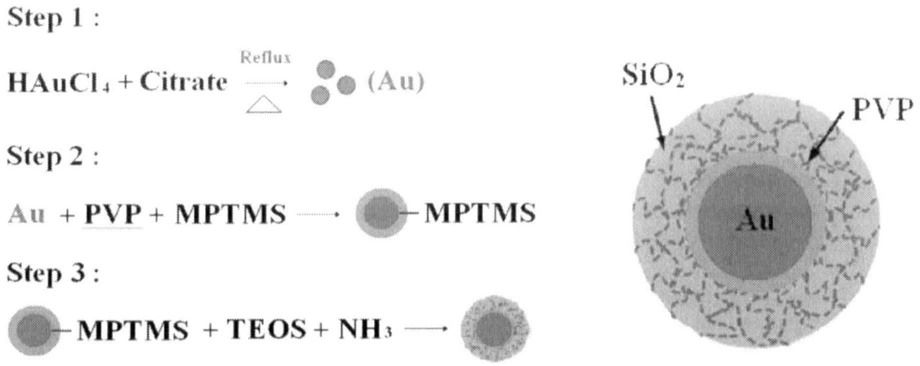

(a) (b)

Figure 1. (a)Schematic representation for the synthesis of silica encapsulated gold particles by stand citrate reduction (step 1), gold-thiol assembly (step 2), and sol-gel method (step 3). (b)The detailed construction of composite particle in this research.

The gold nanoparticle was synthesized in the size of 12 nm by reacting sodium citrate and tetrachloroauric acid; details on the synthesis are in the experimental section and supporting information. Both the shape and size of gold nanoparticle were carefully quantified by TEM and DLS (data not shown), because both variables play an important role for further nanoparticle fabrication. In addition, UV-visible absorbance spectroscopy was used for optical characterization of the gold nanoparticle in solution. The spectra of nanoparticles were measured in aqueous media between 350 and 850 nm, and the maximum absorption wavelength was observed at 520 nm with a narrow bandwidth. This result shows that our gold nanoparticles are well dispersion in water solution, and this material is suitable for further use.

A primary purpose of the introducing PVP was to protect the gold nanoparticles from growing and agglomerating. From our previous study, the diameter of the gold particles should be controlled shorter than 50 nm for various applications, and with the increase in PVP, the particles were dispersed better. This is because that gold particle would coordinate with nitrogen or oxygen atom in PVP, and a covered layer would generate on the surface of the particles. The layer inhibited the growth and agglomeration of the particles. For the gold nanoparticle prepared in this research, both of the two schemes are possible for PVP protection with diameters of about 50 nm. Though the pyrrolidyl has steric effect on the coordination between nitrogen atom and gold, but the electronegativity of nitrogen atom was lower than that of oxygen, and the capability of electron donating of nitogen was higher than that of oxygen. For the particles with the diameter shorter than 50 nm, the steric effect was not the main factor that influenced the reaction between PVP and gold. As this result, the coordination between PVP and gold was not on oxygen atom, but on nitrogen atom.

The immobilization of silica onto gold nanoparticles from an alcohol solution is easily achieved by exploiting the affinity of gold for thiol groups. In this research, the MPTMS was used to present thiol groups for the surface of gold nanoparticles. Overnight incubation of the thiol-functionalized silane with gold nanoparticles then results in the self-assembly of the MPTMS onto the surface of gold nanoparticles, driven by chemisorption between thiol group and gold nanoparticles. Moreover, MPTMS can perform in the role of linking reagent through its silane terminal. Then, the following addition of TEOS and ammonia hydroxide can help to encapsulate the gold nanoparticles into silica shell by sol-gel reaction.

In order to monitor the surface characteristic of synthesized composite particles, we obtained TEM image and the corresponding EDAX result for the silica encapsulated gold particles (shown in figure 2). The average size of composite particles is **R** = 120 ± 5 nm, which was taken from many different TEM images. After the hydroxylated silica surface was formed by alkaline solution with TEOS, the only visible situation is the presence of small darker dots (gold nanoparticles) within the silica surface. The corresponding EDAX result for composite particle confirms that the silica layer is successfully deposited onto gold particles by our fabrication scheme. Both the Si atom and O atom peaks are clearly shown, which result from silica outer shell without the inner gold particles.

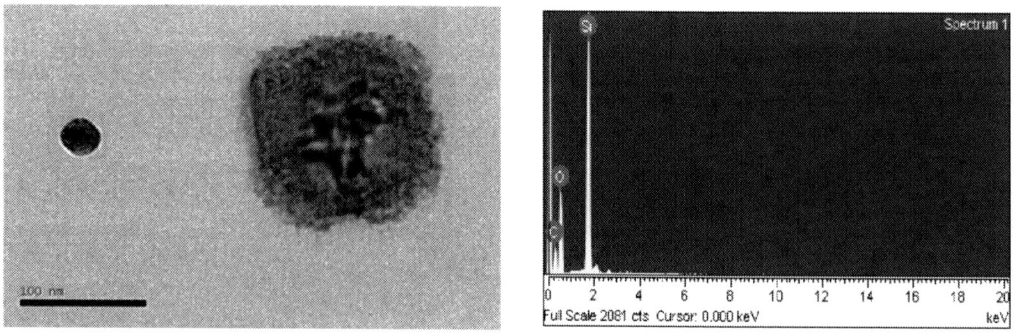

Figure 2. TEM image (left) and their corresponding EDAX data (right) for silica encapsulated gold particles.

As encapsulated silica grew and emerged, the particles had a distinct optical spectrum that could be monitored with an UV-visible spectrometer. Figure 3a and 3b shows the optical spectra of silica-coated gold particles fabricated by a different amount of PVP and MPTMS. From the previous discussion, the concentration of PVP may protect gold nanoparticles from aggregating. Figure 3a showed that the increasing addition of PVP could efficiently avoid the aggregation of the gold colloid. And then, the optical plasmon peaks became blue-shifted, which was consistent with theoretical predictions on optical properties of metal coated particles. In addition, MPTMS is another control factor. When the silane based MPTMS attached on gold surface grew, their optical plasmon peak became slightly red-shifted (shown in figure 3b). The peak became distorted into a broad shoulder which was similar to that previously observed in spectroscopic studies of gold platelets in solution, as the growing silica layer began to coalesce and encapsulated the different amounts of gold nanoparticles. This fact also suggests that the silica shell is completed cover and its absorption spectra are functions of the addition ratios by MPTMS. Furthermore, the completed nanoshells, whose optical plasmon resonance peak ranges in the 750–900 nm regions, can be used as a powerful tool in bio-imaging and biosensing applications. This spectrum range is ideally suited for in vivo imaging because of minimal light absorption by hemoglobin and water. The optical absorption spectrum range is relatively broad compared with that of pure gold colloid. Mie scattering theory for the nanoshell geometry quantitatively accounts for the observed plasmon resonance shifts and line-widths. In addition, the plasmon line-width is dominated by electron surface scattering.

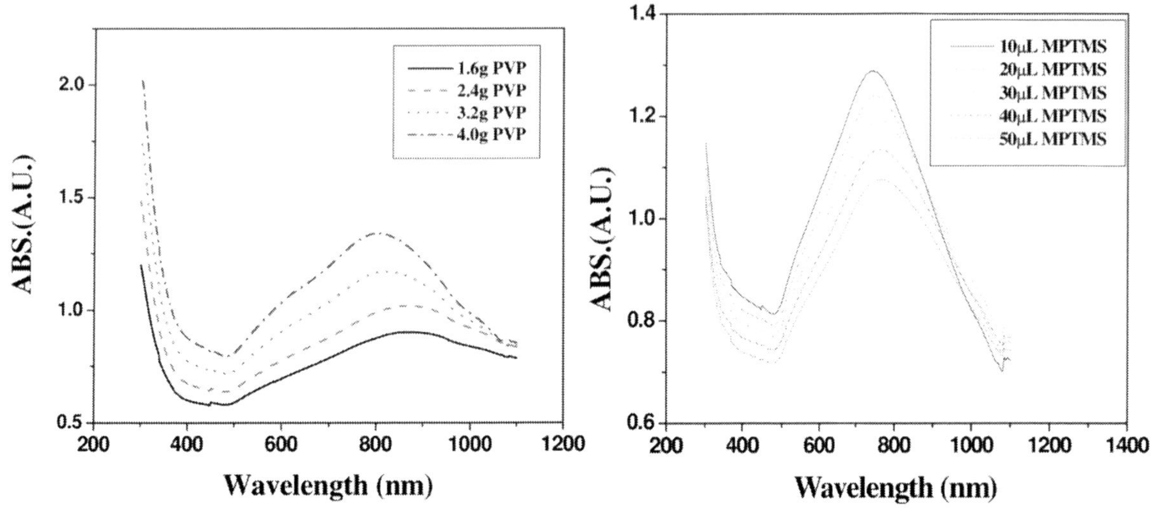

Figure 3. UV-visible absorption spectra of silica-coated gold nanoparticles with a different amount of addition for (a)PVP and (b)MPTMS.

4. Conclusions

In conclusion, we have prepared silica nanolayer-encapsulated gold particles by combining stand citrate reduction, gold-thiol assembly, and sol-gel methods. Because the PVP and MPTMS attached onto a gold surface could act as protection agent and linker molecule for the coating of silica layers, only two pretreatment steps were need to fabricate a continuous outer silica shell. It was also demonstrated that we could control the amount of silica deposition by changing the functional agents to the metal colloid solutions, without repeating the coating step. Because this method is easy, novel and effective in producing silica-encapsulated particles, the fabrication technique can be also applied to produce other silica-encapsulated metal particles for various potential applications such as photonic devices, catalysis, and biosensing materials.

Acknowledgments

The authors are thankful for the financial supports by the Feng Chia University.

References

[1] Tasis, D.; Tagmatarchis, N.; Bianco, A.; Prato, M. *Chem. Rev.* **2006**, *106*, 1105–1136.
[2] Bianco, A.; Kostarelos, K.; Prato, M. *Curr. Opin. Chem. Biol.* **2005**, *9*, 674–679.
[3] Duncan, R. *Nat. Rev. Drug Discov.* **2003**, *2*, 347–360.
[4] Rosi, N.L.; Mirkin, *C.A. Chem. Rev.*, **2005**, 105, 1547-1572.
[5] Smith, P.W., *J. Cell Biochem.(Suppl)*, **2002**, 39, 54-59.
[6] Doering, W.E.; Nie, S. *Anal. Chem.* **2003**, *75*, 6171–6176.
[7] Moore, B.D.; Stevenson, L.; Watt, A. *Nat. Biotechnol.* **2004**, *22*, 1133–1138.
[8] Faulds, K.; Smith, W.E.; Graham, D.; Lacey, R.J. *Analyst* **2002**, *127*, 282–286.
[9] Connor, E. E.; Mwamuka, J.; Gole, A.; Murphy, C. J.; Wyatt, M. D. *Small* **2005**, *1*, 325-327.
[10] Houten, J. V.; Benaron, D.; Spilman, S.; Stevenson, D. K. *Ped. Res.* **1996**, *39*, 470–476.
[11] Holstrum, N.; Nilsson, P.; Carlsten, J.; Bowland, S. *Biosensors and Bioelectronics* **1998**, *13*, 1287–1295.
[12] Ichihara T.; Sakamoto K.; Mori K.; Akagi M. *Cancer Res.* **1989**, 49, 4357–4362.
[13] Schmidt, H. T.; Gray, B. L.; Wingert, P. A.; Ostafin, A. E. *Chem. Mater.* **2004**, *16*, 4942-4947.

Journal of Physics: Conference Series **188** (2009) 012040 doi:10.1088/1742-6596/188/1/012040

Low-temperature expanded graphite for preparation of graphene sheets by liquid-phase method

Sha Jin, Lin-sheng Xie, Yu-lu Ma, Jing-jie Han, Zhang Xia, Guo-xun Zhang, Shu-mei Dong, Yan-yun Wang

Laboratory of equipment material processing, East China University of science and technology, Shanghai 200237, China

clxw@ecust.edu.cn

Abstract: A facile and scalable preparation of aqueous solutions of isolated, sparingly graphene sheets from low-temperature expanded graphite by liquid-phase method is reported. And the influence of expansion temperature on the property of graphene sheets in expanded graphite is discussed by several characterization methods. The TEM scans also indicate that graphene sheets exfoliated from expanded graphite has been restored to an extended conjugated sp2 network.

1. Introduction

Known as "the thinnest material in our universe" [1], graphene, a single-atom-thick sheet of hexagonally arrayed sp2-bonded carbon atoms, promises a diverse range of applications from composite materials to quantum dots [2-7]. However, just as with other newly discovered allotropes of carbon (fullerenes and single-wall nanotubes), material availability and processability will be the rate-limiting steps in the evaluation of putative applications of graphene. For graphene, the availability is encumbered by having to surmount the high cohesive van der Waals energy (5.9 kJ mol^{-1}) [8]adhering graphene sheets to one another. Herein, the developments of various methods for producing graphene sheets have stimulated a vast amount of research in recent years [9]. As described in the reference [10], certain organic solvents would be suitable for preparation of graphene sheets directly from flake graphite, which has been named as the liquid-phase method. Here, we extend the raw materials of method from flake graphite to low-temperature expanded graphite (EG), which has been produced by thermal shocked in 300℃. The influence of expansion temperature on the property of expanded graphite was also discussed.

2. Experimental Section

The flake graphite powder used in the majority of experiments was purchased from Sigma-Aldrich (Product Number 332461) and sieved through a 0.5 mm mesh sieve to remove the larger particles. All solvents and reagents used in experiment were obtained from Sinopharm Chemical Reagent Co., Ltd and used as received. The raw graphite oxide material was prepared by modified Hummers-offeman method [11]. And expanded graphite was produced by thermal shock of graphite oxide under forming gas. Samples of expanded graphite were produced in three different expansion temperatures: 300℃, 600℃, 900℃ with same exposure time (30 seconds). TGA/DSC, XRD, FT-IR, XPS and Rama spectra have been used to characterize the expanded graphite with flake graphite as control.

All materials, including flake graphite and prepared expanded graphite, were stored in drying basin with P_2O_5 for five days before characterization and preparation of graphene sheets by liquid-phase method. For preparing graphene sheets, flake graphite and expanded graphite were dispersed in the relevant solvent (N-methylpyrrolidone (NMP) and g-butyrolactone (GBL)) at a concentration of 0.1 mg/ml by sonicating with a low power tip sonication (KS900c, Hanshu kesheng Ultrasonic) for 1 hour. Care must be taken as excessive sonication can lead to destruction of the graphene. The resultant dispersion was then centrifuged using a TF-2-B centrifuge for 90 minutes at 500 rpm. After centrifuge, decantation was carried out by pipetting off the top half of the dispersion. TEM measurements were made by drop casting from the dispersion onto holey carbon grids (400 mesh size). Initial sample drying was carried out in a vacuum oven at room temperature at a pressure of ~10^{-3} mbar. Subsequently, annealing was carried out to remove residual solvent. The samples were heated to 210℃ for 4 hours in vacuum.

Thermal gravimetric analysis (TGA) and differential scanning calorimeters (DSC) was carried out in an NETZSCH STA 409 PC/PG under Ar gas flow (100mLmin^{-1}) and at a heating rate of 10℃/min. FT-IR spectra were recorded with a Nicolet Magna-IR 550 spectrometer using pellets in KBr. The recorded spectra were the result of coadding 64 interferograms obtained at a resolution of 4 cm^{-1}. XRD patterns were obtained with Rigaku Miniflex diffractometer (D/MAX 2550 VB/PC, GuKα radiation, λ= 1.5406Å). SEM images were taken with JSM-6360LV (JEOL). XPS experiments were carried out on a RBD upgraded PHI-5000C ESCA system (Perkin Elmer) with Mg Kα radiation (h_v=1253.6 eV) for powder. Bright Field TEM images were taken with a Jeol 2100F. Some bright field images (figure 6(d)) and associated ED patterns were taken using a Jeol 2010.

3. Results and Discussion

As reference reported [10], graphene sheet could be dispersed in a small number of solvents stably. Such exfoliation occurs because the strong interaction between solvent and graphene sheets means that the energetic penalty for exfoliation and subsequent solvation becomes small. Thus, researching the influence of expansion temperature on property of expanded graphite and comparing the prepared expanded graphite with pristine flake graphite would be meaningful.

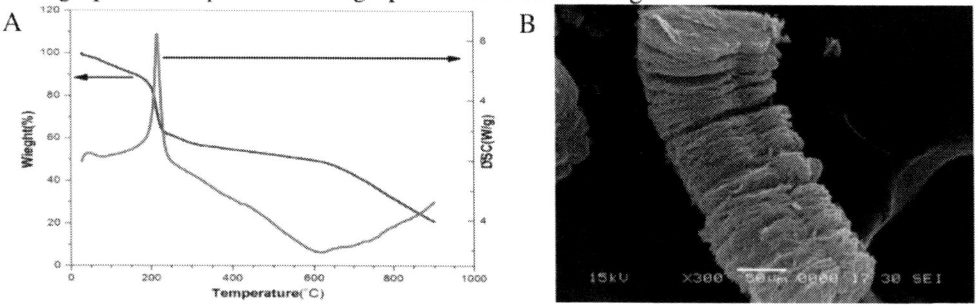

Figure 1. (A) TGA/DSC scans of graphite oxide. There is a sharp mass loss at an onset temperature of approximately 200 ℃ accompanied by a strongly exothermic DSC peak. (B) SEM image of expanded graphite expanded in 215℃, which appeared the classic "worm-like" structure.

In the beginning, TGA/DSC has been used for characterizing the expansion process of graphite oxide. Figure 1 (A) shows the thermogravimetric (TGA) plot of prepared graphite oxide, together with its DSC plot. In agreement with previous reports in the literature for graphite oxide[6], the main mass loss (~30%) takes place around 210°C and is ascribed to the decomposition of labile oxygen-containing groups present in the material. There is also a mass loss (~5%) below 100℃attributed to the removal of adsorbed water and a slower, steady mass loss (~20%) over the whole temperature range above 300℃, which can be assigned to the removal of more stable oxygen functionalities. Further experiment also supports the TGA result: as shown in Figure 1(B), we have got expanded graphite with classic "worm-like" structure at a much low expansion temperature 215℃, which is always believed to be higher than 900℃. As for the endothermic peak of DSC curve at about 600℃ in Figure 1(A), it might be ascribe to the decomposition of more stable oxygen-containing groups, which we will discuss in FT-IR spectra.

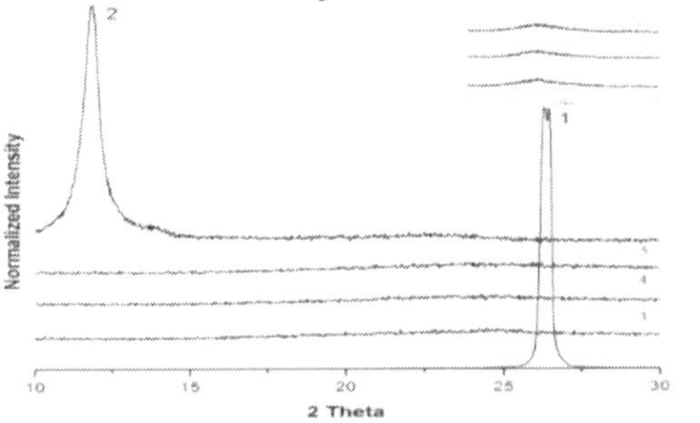

Figure 2. Normalized intensity XRD pattern of flake graphite (1), graphite oxide (2), expanded graphite prepared in 300℃ (3)600℃ (4)900℃ (5). The inset pattern show the XRD curves of expanded graphite prepared in 300℃ (3)600℃ (4)900℃ (5) separately, corresponding a diffraction peak at about 24° .

As TGA/DSC analysis, the main expansion process of graphite oxide happens in the temperature range 200~300℃. For further investigating the expansion result of graphite oxide in different temperature, XRD data was also used to characterize the exfoliation degree of expanded graphite. As Figure 2 shows, pristine flake graphite exhibits a diffraction peak at 2θ=27°. For a successful exfoliation process, we have found it necessary to first increase the *c*-axis spacing of graphite oxide by oxidation to 0.7-1.2 nm and completely eliminate the 0.34 nm graphite interlayer spacing, as shown as line 2. For expanded graphite, there was no significant diffraction peak, which indicates a full exfoliation of graphite flakes. On the other hand, all expanded graphite samples, even prepared at 300℃, exhibits the same diffraction curve as the other samples. The inset XRD curves of expanded graphite indicate the farthing diffraction peak at 2θ=24°, which might be ascribed to the collapse of exfoliated graphite flakes because of the weak interaction. In all, the high expansion temperature shows no significant influence on the exfoliated structure of graphite in XRD pattern.

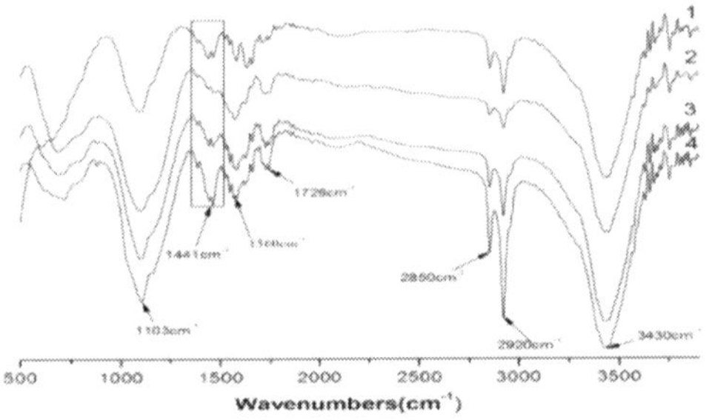

Figure 3. (A)FT-IR spectrum of the flake graphite (1), and expanded graphite prepared in (2) 900℃, (3) 600℃, (4) 300℃.

Considering the oxygen-containing groups remained in expanded graphite, FT-IR was used. Figure 3 shows the FT-IR spectrum obtained from flake graphite and expanded graphite. The most characteristic features are the broad, intense band at 3430 cm^{-1} (O-H stretching vibrations), the bands at 2650cm^{-1} and 2920cm^{-1}(vibrations of O-H in H_2O and dimeric COOH, but no clear distinction seems possible between C-OH and H_2O peaks.), and1103cm^{-1} (C-O stretching vibrations). In addition, there are other band at 1414 cm^{-1}(carboxyl C-O stretching vibrations), 1560 cm^{-1} (skeletal vibrations from unoxidized graphitic domains), and 1726cm^{-1} (C=O stretching vibrations from carbonyl and carboxylic groups). It is well known that the flake graphite would contain some physical absorbed water to maintain the lubrication property. Herein, water molecular can intercalate into the interlayer of graphene sheets in graphite. While for the expanded graphite exfoliated by thermal shock, the most oxygen-containing groups should ascribe to the residual of groups on graphite oxide flakes. Here, a band at 1414 cm^{-1} disappears in curve 2, corresponding to the expansion temperature 900℃. The phenomenon can be explained based on the expansion theory of graphite oxide[12]: the exfoliation takes place when the decomposition rate of the epoxy and hydroxyl sites of graphite oxide exceeds the

Journal of Physics: Conference Series 188 (2009) 012040 doi:10.1088/1742-6596/188/1/012040

diffusion rate of the evolved gases, thus yielding pressures that exceed the van der Waals forces holding the graphene sheets together. Because of the relative short exposure time of graphite oxide, it is hard for all oxygen-containing groups to decompose completely. Therefore, the expansion temperature is hard to determine the amount of residual oxygen-containing groups on expanded graphite. The main difference of the three curves of expanded graphite is the disappearance of absorb band at 1414 cm^{-1}, which might be ascribe to the complete thermal decomposition of carboxyl C-O band. However, it is hard for FT-IR measurement to exactly determine the amount of carbon and oxygen atoms on the graphene sheets of expanded graphite. In this view, XPS experiment was taken for investigate the atomic ratio of carbon and oxygen of graphene sheets. Three samples were taken, including graphite oxide and expanded graphite with expansion temperature 300℃ and 900℃.

Table 1. Area atomic ratio (C/O) from XPS experiment

Material	C1s	O1s	Area atomic ratio (C/O)
Graphene oxide	69.16%	30.84%	2.243
Expanded graphite A(300℃)	84.94%	15.06%	5.644
Expanded graphite B(900℃)	89.96%	10.04%	8.960

As Table 1 shows, the area atomic ratio of carbon and oxygen increased significantly from graphite oxide (2.243) to expanded graphite (900℃) (8.960). The raise of atomic ratio (C/O) indicates that the decomposition degree of graphite oxide increases with the expansion temperature raise. On the other hand, the percentage of carbon atoms does not have great variation, which might be ascribed to the exfoliation process. We suppose that the high expansion temperature would increase the decomposition rate of oxygen-containing groups, and in a certain degree influence the decomposition degree of graphite oxide. Thus, it is interesting to investigate the microstructure changes of expanded graphite in different expansion temperature.

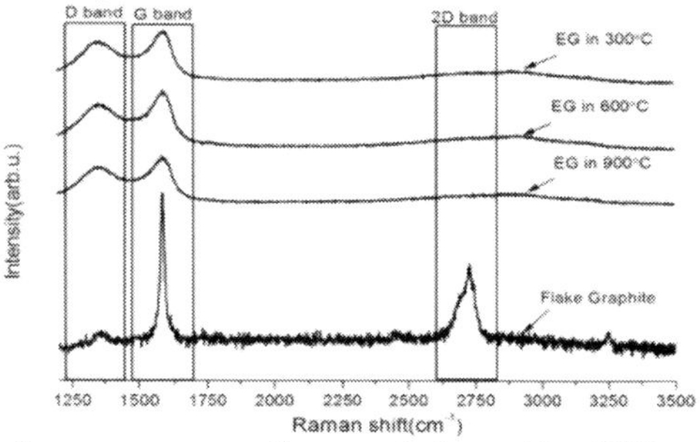

Figure 4. Raman spectra of expanded graphite (EG) prepared in 300℃, 600℃, 900℃.

Raman spectroscopy is also used to investigate the microstructure of expanded graphite. Figure 4 shows the Rama spectrum curves of flake graphite and prepared expanded graphite. The bottom curve, corresponding to flake graphite exhibits prominent G and 2D peaks at 1600 cm^{-1} and 2700 cm^{-1}, respectively. The G peak relates to the E2g vibrational mode within aromatic carbon rings and gives some insight into the degree of graphitization for the area of excitation. The 2D peak represents an out of plane vibrational mode. Near the edge of both single sheets and bulk graphite, an additional peak is present around 1350 cm^{-1}. This D peak is forbidden in pristine systems of π conjugation, but allowed in small and isolated domains of aromaticity, ultimately giving a relative measure of the amount of sp3 carbons in the surroundings[13]. Representative Raman spectra for prepared expanded graphite are presented by the curves on the top of the pattern. We can observe an increase in the D/G ratio and disappearance of 2D band. The disappearance of 2D band can be explained by the exfoliation of graphite oxide, which induces the interlayer spacing of expanded graphite flakes. The increase of D/G ratio can be ascribe either to oxidization of graphene sheets during the preparation of graphite oxide, or to the violent decomposition of oxygen-containing groups in graphite oxide. As the observation of spectra, three expanded graphite samples have exactly the same Rama curves, which might be supposed an indication of the same microstructure of graphene sheets in expanded graphite.

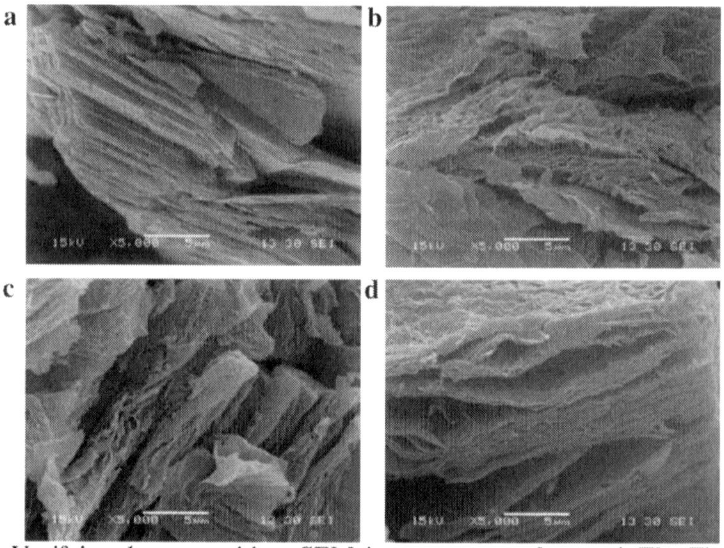

Figure 5. SEM images of flake graphite (a), and expanded graphite prepared in 300℃ (b), 600℃(c), 900℃ (d) with same magnification (× 5000), and all expanded graphite samples show the same "worm-like" morphology regardless the difference of expansion temperature.

Verifying the supposition, SEM images were referenced. The Figure 5 shows the SEM images of flake graphite and prepared expanded graphite in three different expansion temperatures. Image (a) shows the compact layers of graphene sheets, while images (b) (c) (d) show the "worm-like" structure of expanded graphite. It can be obviously observed that the interlayer spacing of graphene sheets has been greatly extended, which can be valued cursorily by the determining the expansion ratio of graphite oxide. But for the thickness of graphene sheets separated in expanded graphite, there is no significant difference for the three samples.

Based on the researches above, it can be inferred that the expansion temperature of graphite oxide has influence on the decomposition of oxygen-containing groups and evaporation of absorbed and intercalated water molecular. But for the exfoliated graphene sheets, the temperature exhibits no significant influence on either the oxygen-containing groups' residues or the structure of graphene sheets in expanded graphite, except the influence of contain of carboxyl C-O band on graphene sheets. Thus, we take the expanded graphite prepared in 300℃ as the raw material for preparation of graphene sheets by liquid-method. There are two reasons, (1) the thermal expansion of graphite oxide would induce a graphitic material crystal structure as pristine flake graphite, or a pre-exfoliation of graphene sheets, which might be useful in mass production of graphene sheets.(2) the optimization of expansion temperature would largely reduce the cost of preparation. Even if the expansion of graphite oxide at 215℃ can reach classic "worm-like" structure, we would like to prefer to take 300℃ as expansion temperature based on the TGA/DSC plots. At the low temperature 300℃, most of oxygen-containing groups would decompose and induce a relatively stable mass residence.

Figure 6. TEM images of graphene sheets (darker gray areas) supported by a holey carbon grid and corresponding Electron Diffraction patterns. (a) Monolayer of graphene sheet prepared in NMP from flake graphite, and corresponding Electron Diffraction pattern (b). (c) Graphene sheet prepared in GBL from flake graphite. (d) Bilayer graphene sheet prepared in NMP from expanded graphite prepared in 300℃ and corresponding Electron Diffraction pattern (e). (f) Graphene sheet prepared in GBL from expanded graphite prepared in 300℃.

By liquid-method, graphene sheets were prepared in NMP and GBL from pristine flake graphite and expanded graphite prepared in 300℃, respectively. As shown in Figure 6, image (a) and (c) corresponds to graphene sheets prepared from pristine flake graphite, image (d) and (f) correspond to graphene sheets prepared from expanded graphite. There are wrinkled sheet structure resulting from reaction sites involved in oxidation and reduction processes [14] for graphene sheets in image (d) and (f), significantly comparing to the smooth surface of graphene sheets in (a) and (c). On the other hand, the ED patterns in image (b) and (e) indicates the crystal structure of prepared graphene sheets. However, (b) indicates a structured single crystal structure, while (e) indicates a crystal structure with defects. The result of ED pattern investigation supports the former Rama Spectrum results, which explains the increased D/G band ratio of expanded graphite. Although it has been observed that there are lots of "wrinkle" structures on the graphene sheets of graphite oxide[12], but in microns scale it is hard to observe "ripples" on the surface of graphene sheets. Thus it is reasonable to suppose that the thermal shock of graphite oxide that induces violent decomposition of oxygen-containing groups should be responsible for the micron scale wrinkle structure of graphene sheets in graphite oxide. Thus, it is interesting to investigate the expansion process of graphite oxide in different expansion temperature, which would be performed in our next work.

4. Conclusions

We discussed the influence of expansion temperature on expanded graphite by thermal shock of graphite oxide. It indicates that the rise expansion temperature improves the expansion ratio of graphite oxide, but has little to do with the amount and species of residual oxygen-containing groups and microstructure of graphene sheets in expanded graphite. Meanwhile, by strong interaction between solvent and expanded graphite flakes, we have demonstrated a scalable method to produce aqueous solutions of isolated, sparingly graphene sheets from low-temperature expanded graphite. Using certain solvents, graphene sheets can be dispersed at concentrations of up to 0.1 mg/ml stably. These dispersions can then be used to deposit flakes by spray coating, vacuum filtration or drop casting. Otherwise, by thermal shock, graphite oxide will be able to transform to defected single crystal structure. The defects might ascribe to the violent expansion of graphite oxide in high expansion temperature, which might induce a wrinkle structure on graphene sheets as observed.

References

[1] Geim A K. and MacDonald, A. H 2007 *Phys.Today*, **60** 35
[2] Ponomarenko L A, Schedin F, Katsnelson M I, Yang R, Hill E W, Novoselov K S and Geim A K 2008 *Science*, **320** 356
[3] Novoselov K S, Geim A K, Morozov S V, Jiang D, Zhang Y, Dubonos S V, Grigorieva I V and Firsov A 2004 *Science*, **306** 666

[4] Novoselov K S, Jiang Z, Zhang Y, Morozov S V, Stormer H L, Zeitler U, Maan J C, Boebinger G S, Kim P and Geim A K. 2007 *Science*, **315** 1379
[5] Geim A K and Novoselov K S 2007 *Nat. Mater*, **6** 183
[6] Stankovich S, Dikin D A, Dommmett G H, Kohlhaas K M, Zimney E J, Stach E A, Piner R D, Nguyen S T and Ruoff R S 2006 *Nature*, **442** 282
[7] Watcharotone S, Dikin D A, Stankovich S, Piner R, Jung I, Dommett H B, Evmenenko G, Wu S-E, Chen S-F, Liu C-P, Nguyen S T and Ruoff R S 2007 *Nano Lett*, **7** 1888
[8] Zacharia R, Ulbricht H and Hertel T 2004 *Phys. ReV. **B*** **69** 155406
[9] Sungjin Park and Rodney S. Ruoff 2009 *Nature Nanotechnology*, **4** 217
[10] Yenny Hernandez, Valeria Nicolosi, Mustafa Lotya, Fiona M. Blighe, Zhenyu Sun, Sukanta De, I T McGovern, Brendan Holland, Michele Byrne, Yurii K. Gun'Ko and et al 2008 *Nature Nanotechnology*, **3** 563
[11] Hummers W S and Offeman R E 1958 *J. Am. Chem. Soc*, **80** 1339
[12] Michael J. McAllister, Je-Luen Li, Douglas H. Adamson, Hannes C. Schniepp, Ahmed A. Abdala, Jun Liu, Margarita Herrera-Alonso and etc 2007 *Chemical material*, **19 (18)** 4396.
[13] Ferrari A C and et al 2006 *Phys. Rev. Lett*, **97** 187401
[14] Hannes C. Schniepp, Je-Luen Li, Michael J. McAllister, Hiroaki Sai, Margarita Herrera-Alonso, Douglas H. Adamson and etc 2006 *J. Phys. Chem. **B*** **110 (17)** 8535

The Mossbauer spectra of carbon nanotubes synthesize using ferrite catalyst

Haiyan Zhang*, Jiapeng Lin, Zuxiong Peng, Guoxun Zeng, Jinshan Pang and Yiming Chen
Faculty of Material and Energy, Guangdong University of Technology, Guangzhou 510006, China

E-mail: hyzhang@gdut.edu.cn

Abstract: The ferrite powder with honeycombed structure obtained by chemical combustion was used as catalyst to synthesize multi-walled carbon nanotubes by chemical vapor deposition. The magnetic components and characters of the the carbon nanotubes synthesized were investigated by X-ray diffraction (XRD), Mossbauer spectra and vibrating-sample magnetometer (VSM). The ferric components of the carbon nanotubes samples can be identified by Mossbauer spectra. The Mossbauer spectra of carbon nanotubes sample after purification contains two ferromagnetic sextet components corresponding to α-Fe species and Fe_3C (cementite) species. While the Mossbauer spectra of the carbon nanotubes sample before purification contains three ferromagnetic sextet components corresponding to α-Fe species, Fe_3C species and γ-Fe_2O_3. The saturation magnetization intensity Ms of carbon nanotubes sample after purification is decreased from 46.61 to 2.94 emu/g, but the coercive force increasd and reached 328Oe.

Keywords: multi-walled carbon nanotubes, Mossbauer spectra, ferrite powder

1. Introduction

Due to their fascinating physical and chemical properties, carbon nanotubes may be used for nanoscale devices [1-6]. At present, chemical vapor deposition (CVD) process [7-10], in which carbon nanotube was synthesized in large-scale and high quality, make this application practicable. However, for all these device applications of CVD carbon nanotubes, it is important to further understand the magnetic properties of the carbon nanotubes sample synthesized in order that controllable growth for carbon nanotubes given magnetism. Previous studies[11-13] about magnetic properties of the carbon nanotubes focused on electron spin resonance (ESR) spectra of purified and unpurified carbon

* Corresponding author: hyzhang@gdut.edu.cn

nanotubes. Other studies about the magnetic properties of the carbon nanotubes of purified and unpurified are few.

In this paper, we have prepared a pure ferrite powder with honeycombed structure by chemical combustion using ferric nitrate and citric acid as raw material. By using the ferrite powder with honeycombed structure as a catalyst, we also have prepared multi-walled carbon nanotubes by chemical vapor deposition. The multi-walled carbon nanotubes prepared may be used as a kind of good electromagnetical wave absorbent. We tried to study the magnetic component and especial magnetic properties of these carbon nanotubes sample synthesized using magnetic ferrite powder as a catalyst by Mossbauer spectra and vibrating-sample magnetometer (VSM).

2. Experimental

2.1 Preparation of ferrite catalyst

$FeNO_3.6H_2O$ and citric acid were dissolved in the adding de-ionized water at 1:1 mol ratio, and then PH value of the solution was adjusted to seven by ammonia. The prepared solution was baked in 120℃, and a red-brown xerogel was gained. A russety and fleecy ferrite is obtained by burning the gel in the air at about 200℃. The ignition point was examined by the thermogravimetric analyses (TGA). TGA experiments were performed at a heating rate 20 ℃/min from room temperature to 350℃ under dry-air atmosphere. The surface morphology of the ferrite was observed by scanning electron microscope (SEM).

2.2 Preparation of carbon nanotubes

The ferrite with nanometer pores was used as catalyst and was loaded on a quartz reactor. First, nitrogen gas with a flow rate of 50ml/min was introduced until the furnace temperature up to 550℃ and then hydrogen gas with a flow of 50ml/min was introduced When the temperature was up to 850℃, acetylene was introduced at a flow rate of 100 ml/min. After 30 minutes reaction, hydrogen and acetylene gas were closed and nitrogen gas was introduced until the equipment cooled down to room temperature. Then multi-walled carbon nanotubes were obtained. The prepared carbon nanotubes are purified by heating reflux with 98% strong nitric acid for an hour. Transition electronic microscope (TEM) and SEM were employed to observe the carbon nanotubes samples. XRD was employed to test the composition and structure of the samples. The Mossbauer spectra of samples are measured by Mossbauer spectrometer using ^{57}Co (Pd) source at room temperature. All isomer shifts are given relative to that of α-Fe. The vibrating-sample magnetometer (VSM) experiments of the samples were also performed at room temperature.

3. Result and discussion

Ferric nitrate and citric acid were reacted according to the chemical equation as follows:

$$2Fe(NO_3)_3+C_6H_8O_7\cdot H_2O+2O_2=Fe_2O_3+6NO_2+6CO_2+H_2O$$

Fig.1 is the SEM image of the ferrite after combustion reaction, from which we can see that ferrite presents a honeycombed structure, and there are many pores with the diameter from tens to hundreds nanometer. When using as a catalyst, this kind of structure with very large specific surface area can contact reacting gas fully in chemical vapor deposition processes of preparing carbon nanotubes.

Fig.2 shows the thermogravimetric (TG) curve of xerogel prepared by citric acid and ferric nitric, from which we can see the sample's weight decrease slowly when the temperature varies from 30℃ to 210℃. This is because the crystalline water in the ferric nitrate and citric acid are removed. The sample's weight sharply decreases when the temperature rises up to 210℃, which is the ignition point of xerogel and the sample starts to be decomposed. When the temperature rises to 250 ℃, the sample is decomposed completely and hybrid metals oxide is formed, so the sample's weight no longer changes when the temperature continues to go up. Due to the strong oxidizing property of nitrate ion (NO_3^-), the xerogel occur oxidation-deoxidization reaction above 100℃ until it is burned entirely. The ferrite having nanometer pores is formed, with high activity and uniform pore size distribution.

Fig.1 The SEM image of ferric oxide sample

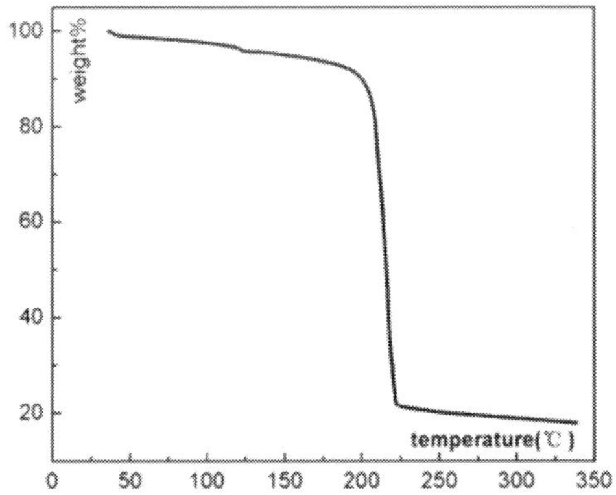

Fig.2 TG curves of the gel prepared by ferric nitrate and citric acid

Fig.3 and Fig.4 show that the SEM and TEM image of the multi-walled carbon nanotubes prepared by using ferrite with nanometer pores as catalyst, and we can see the diameter of carbon nanotubes is between 20-30 nm, and the length of carbon nanotubes is above 20 μm. The diameter of carbon nanotubes prepared is uniform. The growth process of carbon nanotube is probably as follows: In the hydrogen atmosphere, the ferrite is deoxidized, then carbon nanotubes are formed when acetylene is decompounded on the ferrite surface.

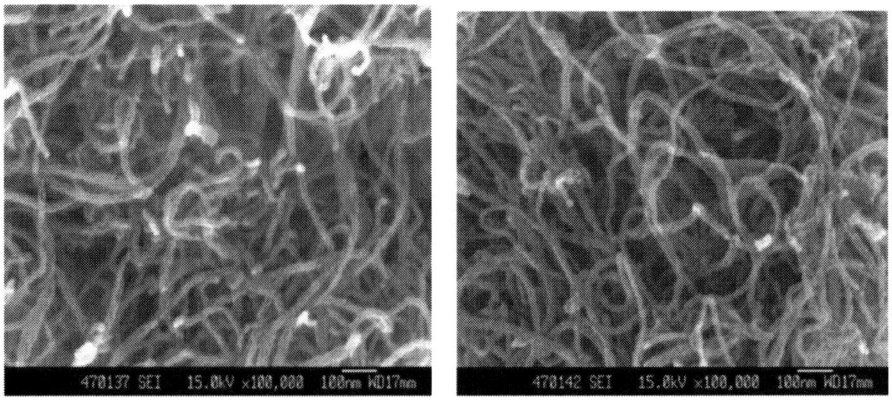

Fig.3 The SEM image of carbon nanotubes

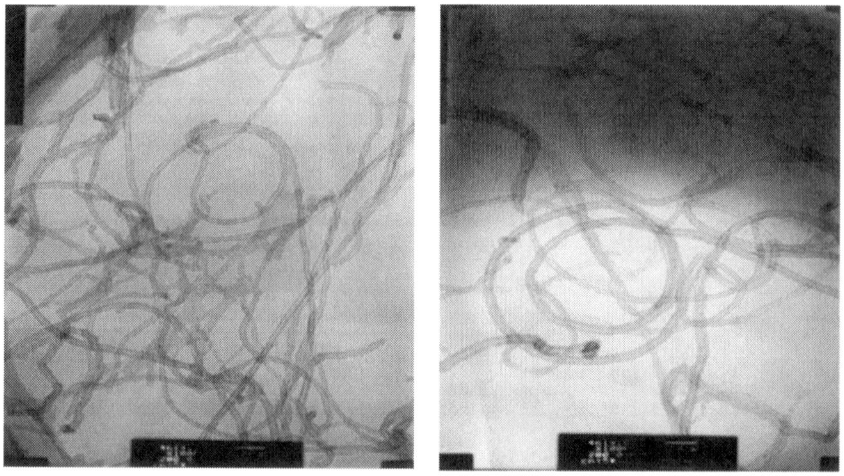

Fig.4 The TEM image of carbon nanotubes

XRD spectra of the carbon nanotubes before and after purification are shown as Fig.5, and we can see that the main components of the sample are carbon, α-Fe and a small quantity Fe_3C. The catalyst is removed from the sample by nitric acid heating reflux. XRD spectra of the carbon nanotubes after purification are shown in fig.6, the main components in the purified sample are carbon, a small quantity of α-Fe and Fe_3C. The relative content of carbon and α-Fe is much larger than before purification from the X-ray diffraction, but the peak of α-Fe is still higher because α-Fe is crystal even of a little content, so its peak of X-ray diffraction is still stronger than that of carbon.

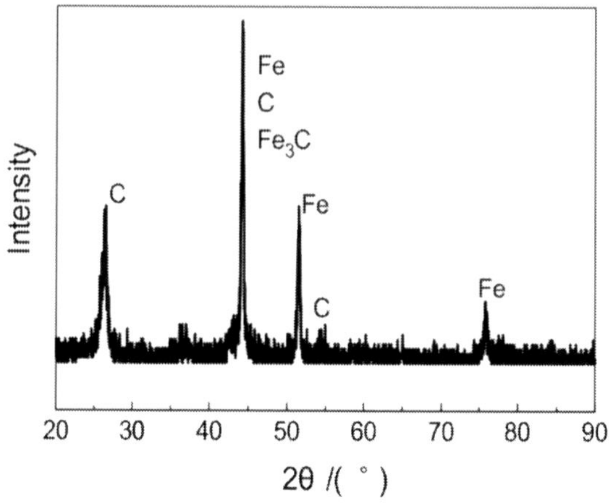

Fig.5 The X-Ray diffraction pattern of carbon nanotubes before purification

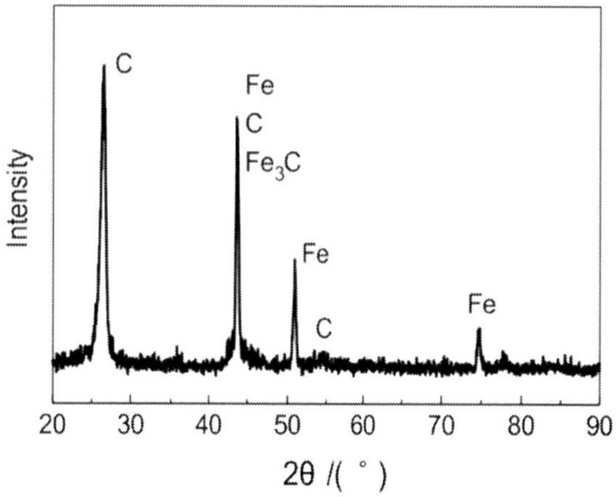

Fig.6 The X-Ray diffraction pattern of carbon nanotubes after purification

Because of its high energy distinguish ability, Mossbauer spectra can be used to test sensitively most tiny energy variety occurred around the atomic nucleus, and it is a effective way to investigate the composition of nanomaterials. There are two samples tested in our experiments: one is carbon nanotubes before purification (named sample A), the other is carbon nanotubes after purification (named sample B).The Mossbauer spectra of carbon nanotubes (sample A and sample B) are shown in

Fig.7. Mossbauer parameters obtained by fitting the curves in Fig.7 are shown in Table 1 and Table 2. Because Mossbauer spectra are only used to check the content of magnetic composition of elements and compound in samples, the magnetic elements and compound containing iron in carbon nanotubes samples can be checked by using Mossbauer spectra, except large numbers of carbon.

Table 1 Mossbauer parameters of sample A

Component	IS (δ) (mm/s)	QS (△) (mm/s)	Hyperfine field (A/m)	WHM (half width of half height)	Relative area(%)
α-Fe	0.01±0	-0.01±0	330.0±0.1	0.16±0	29.7%
Fe_3C	0.20±0	0	209.7±0.1	0.20±0	65.5%
γ- Fe_2O_3	0.28±0.03	-0.06±0.06	485.9±2	0.27±0.02	4.8%

Table 2. Mossbauer parameters of sample B

Component	IS (δ) (mm/s)	QS (△) (mm/s)	Hyperfine field (A/m)	WHM (half width of half height)	Relative area(%)
α-Fe	-0.02±0.01	0.05±0.03	332.6±1	0.13±0.01	9.1%
Fe_3C	0.21±0	0.03±0.01	207.4±0.4	0.25±0	90.9%

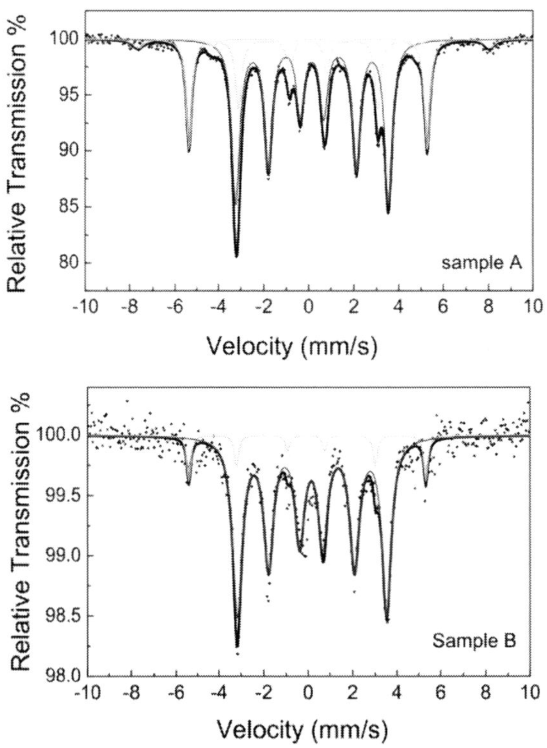

Fig.7 Mossbauer of CNTs before purification (sample A) and CNTs after purification (sample B) at room temperature

According to the Mossbauer spectra data, the composition of carbon nanotubes samples besides carbon can be identified. Mossbauer spectra of sample B were fitted to two ferromagnetic sextet components corresponding to α-Fe species and Fe_3C (cementite) species. The Mossbauer spectra of sample A are fitted to three ferromagnetic sextet components corresponding to α-Fe species, Fe_3C species and γ-Fe_2O_3. From Table 1 and Table 2, the components containing iron in the sample before purification include α-Fe, Fe_3C and Fe_2O_3, however, only α-Fe, Fe_3C after purification. So there is no ferric oxide in the carbon nanotubes sample after purification. On the other hand, the content of α-Fe is less after purification than that before purification. In the high temperature of preparing carbon nanotubes, Fe_3C may forme and not be eliminated in such purification way. XRD spectrum shows that the content of Fe_3C is much less than that of carbon. The α-Fe content in the carbon nanotubes sample after purification is only 9% of Fe_3C by Mossbauer spectra data, soα-Fe content can be neglected in the carbon nanotubes sample after purification.

Fig.8 shows that the hysteresis loop of carbon nanotubes before and after purification measured by using the vibrating-sample magnetometer (VSM).The carbon nanotubes represent soft magnetic property. Comparing the magnetism parameters of carbon nanotubes before and after purification (Table.3), and also compared with that of α-Fe, the magnetism parameters of carbon nanotubes after purification is lower than that before purification. The residual magnetization intensity Mr is about 0.53 emu/g and saturation magnetization intensity Ms is about 2.94 emu/g. This is because a lot of α-Fe and Fe_2O_3 are eliminated after purification. While the coercive force of the carbon nanotubes after purification is 328 Oe , close to that of iron. This shows that α-Fe and Fe_2O_3 magnetic components in carbon nanotubes may decrease the coercive force of carbon nanotubes.

Table.3 the magnetic parameters of carbon nanotubes before and after purification measured by the vibrating-sample magnetometer (VSM)

	Ms (emu/g)	Mr(emu/g)	Hc(Oe)	weight(g)
CNTs before purification	46.61	3.84	113	0.0257
CNTs after purification	2.94	0.53	328	0.0204
α-Fe	113.69	16.71	344	0.0489

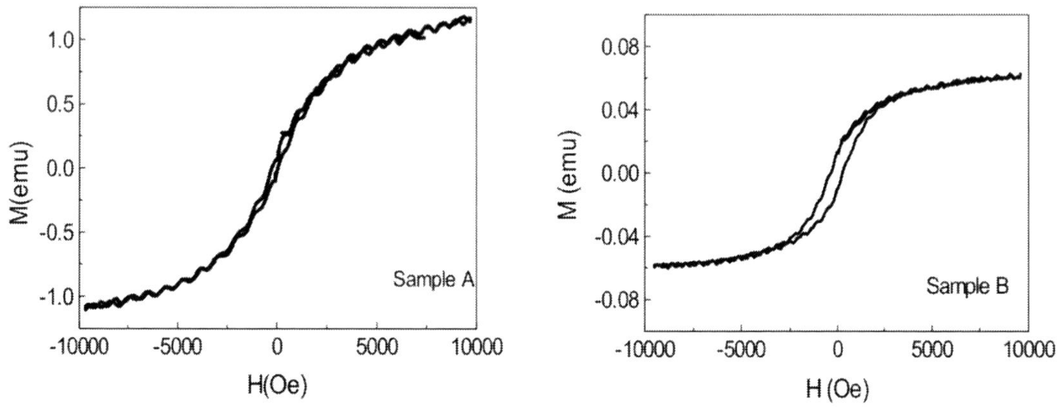

Fig.8 The M (H) loop curves of carbon nanotubes before purification (sample A) and after purification (sample B)

4. Conclusions

The ferrite powder with honeycombed structure can be synthesized by sol-gel process and chemical combustion method using ferric nitrate and citric acid as raw materials. This ferrite powder was used as a catalyst to synthesize successfully magnetic multi-walled carbon nanotubes by chemical vapor deposition. We have studied the magnetic components and characters of the carbon nanotubes synthesized by Mossbauer spectra, XRD spectra and vibrating-sample magnetometer (VSM). The main components of the samples with soft magnetic property are carbon nanotubes, α-Fe, Fe_3C and Fe_2O_3 before purification. After purification, Fe_2O_3 and the mass of α-Fe in the carbon nantubes samples were eliminated, and the main components of the samples is carbon nanotubes and Fe_3C. The magnetism of the carbon nanotubes after purification is decreased and the saturation magnetization intensity Ms of carbon nanotubes sample is from 46.61 to 2.94 emu/g, but the coercive force of carbon nanotubes increases to 328 Oe , close to that of iron.

References

[1] Tombler T W, Zhou C, Alexseyev L, Kong J, Dai H., Liu L, Jayanthi C S, Tang M., Wu S Y, 2000 *Nature*, **405** 769
[2] Bachtold A., Hadley P, Nakanishi T, Dekker C, 2001 *Science* **294** 1317
[3] Zhou G., Duan W, Gu B, 2001 *Phys.Rev.Lett.* **87** 095504
[4] Leonard F, Tersoff J, 2002 *Phys.Rev.lett.*, **88** 258
[5] Parthangal P M， Cavicchi R E and Zachariah M R, 2007 *Nanotechnoloy*, **18** 185605
[6] Lin C L, Chen C F, Shi S C, 2004 *Diamonda and Related Materials*,**13** 1026
[7] Ge M, Sattler K, 1993 *Science* **260** 515
[8] Fujiwara Y, Takegawa H, Sato H, et al., 2004 *Journal of Appl.Phys.*, **95** 7118
[9] Xu F S, Liu X.F, Tse S.D. 2006 *Carbon*,**44** 570
[10] TalaPatra S, Kar S, Pal S K.et al. 2006 *Nature nanotechnology*, **1** 112
[11] Bandow S, Asaka S, Zhao X, Ando Y, 1998 *Appl.Phys.A*,**67** 23
[12] Zhang H, He Y, Liang L, et al., 2000 *Journal of Phys.and Chem.of Solids*, **61** 1123
[13] Macedo CA, Souza AMC， 2004 *Physica B: Condensed Matter* **354** 290

Acknowledgements

This work is supported by the National Natural Science Foundation of China (Grant No.50372013), by the Specialized Research Fund for the Doctoral Program of Higher Education of China (Grant No. 20050562002), by the Guangdong Provincial Natural Science Foundation of China (Grant No. 07001769)

Journal of Physics: Conference Series **188** (2009) 012042 doi:10.1088/1742-6596/188/1/012042

A silicon micromachined resonant pressure sensor

Zhangyang Tang[1,2], Shangchun Fan[1,2], Chenguang Cai[1]

[1]School of Instrumentation Science & Opto-electrionics Engineering, Beihang University, Beijing, 100191, China

[2]State Key laboratories of transducer technology, Shanghai, 200050, China

Email: tangzhangyang1@163.com

Abstract. This paper describes the design, fabrication and test of a silicon micromachined resonant pressure sensor. A square membrane and a doubly clamped resonant beam constitute a compound structure. The former senses the pressure directly, while the latter changes its resonant frequency according to deformation of the membrane. The final output relation between the resonant frequency and the applied pressure is deducted according to the structure mechanical properties. Sensors are fabricated by micromachining technology, and then sealed in vaccum. These sensors are tested by open-loop and close-loop system designed on purpose. The experiment results demonstrate that the sensor has a sensitivity of 49.8Hz/kPa and repeatability of 0.08%.

1. Introduction
Pressure is one of the most important parameters in the aircraft measurement and control. Among various kinds of pressure sensors, the resonant pressure sensor is attractive because of its digital and high performance. The vibrating cylinder pressure sensor is a typical one which has been successfully used on the aircrafts. However, as the development of modern aircraft manufacture technology, the demand for smaller, lighter and better pressure sensors is emerged [1]. The maturation of silicon micromachining technique guarantees the fabrication of this minized sensor. As far, there are a few companies selling their products on the market as the Druck in U.K. and the Yokogawa in Japan [2,3]. For China, the research on this field started since last ten years and is still in the stage of laboratory.

This paper presents the design and realization of a compound structure silicon micromachined resonant pressure sensor, and the test results by the open-loop and closed-loop system designed on purpose.

2. Design and theory

2.1. Working principle
The compound structure of this silicon resonant pressure sensor is illustrated by figure.1. The square membrane is the direct pressure sensing component, and is deformed under the measured pressure. A doubly clamped beam is suspended on the surface at the center of the membrane. Its axial stress is changed as the membrane deformed, and accordingly its resonant frequency is changed. Consequently, the measured pressure can be obtained through the resonant frequency. Figure.2 shows a fabricated sensor.

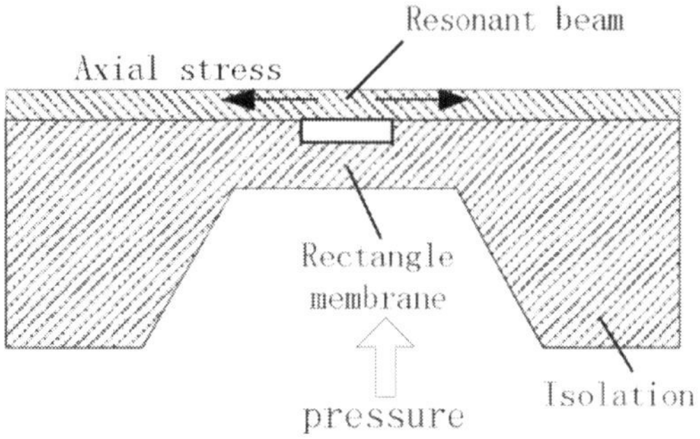

Figure.1 compound structure of the pressure sensor

Figure.2 the fabricated sensor photo

2.2. Theory calculation

Assuming that the membrane side length is A, thickness is H, and the direction along the beam axis is x direction. Then, the displacement along x direction on the surface of the membrane $u(x)$ is as in equation (1).

$$u(x) = \frac{-49P(1-\mu^2)}{96E}\left(\frac{A}{H}\right)^2\left(\frac{x^2}{A^2}-1\right)x \quad (1)$$

Where E is the Young modulus of the material, and μ is the Poisson ratio of the material [4].

Assuming that the beam length, width and thickness are l, b and h. As b and h is relatively small than A, the axial strain of beam ε_x is merely determined by the displacement of the two suspended points as in equation (2).

$$\varepsilon_x = [u(\frac{l}{2}) - u(-\frac{l}{2})]/l \quad (2)$$

From the solution of the vibrating partial differential equation for the resonant beam, the vibrating shape and resonant frequency under the axial force N of each modes can be obtained. To working at the lowest energy point, the first resonant frequency is chose as the working frequency, expressed by equation (3).

$$f_1(N) = f_{n1}(0)\left(1+\gamma_1\frac{Nl^2}{Ebh^3}\right)^{\frac{1}{2}} \quad (3)$$

Where, $f_{n1}(0) = \frac{\alpha_1^2 h}{2\pi l^2}\left(\frac{E}{12\rho}\right)^{\frac{1}{2}}$ is the nature frequency of the beam when $N = 0$, and

$\gamma_1 = 0.295$, $\alpha_1 = 4.730$, which are constants related with the order of mode [5].

Substituting $N = bhE\varepsilon_x$ into equation (3), the relation between the measured pressure and the first order resonant frequency can be obtained by equation (4).

$$f_1(P) = f_{n1}(0)\left[1+\frac{49\gamma_1 l^4}{96Eh^2}(1-\mu^2)\left(\frac{A}{H}\right)^2\left(1-\frac{l^2}{4A^2}\right)\cdot P\right]^{\frac{1}{2}} \quad (4)$$

As for the relatively small pressure range, equation (4) is approximately linear, the sensitivity of resonant frequency to pressure is expressed by equation (5).

$$S_{pf} = \frac{1}{P}\frac{f_1(P)-f_{n1}(0)}{f_{n1}(0)} \approx 0.075\frac{(1-\mu^2)}{E}\left(\frac{Al}{Hh}\right)^2\left[1-\left(\frac{l}{2A}\right)^2\right] (5)$$

3. Experiment analysis

3.1. Test system

As the dimension of the resonant beam and the vibration amplitude is very small, the pickup signal is very weak and is submerged in the strong background noise [6]. To extract the weak vibrating signal, an detection circuit system is designed based on the lock-in amplify principle. Tests for the pressure sensor characteristics are realized by it. Figure.3 is the block diagram of the circuit system.

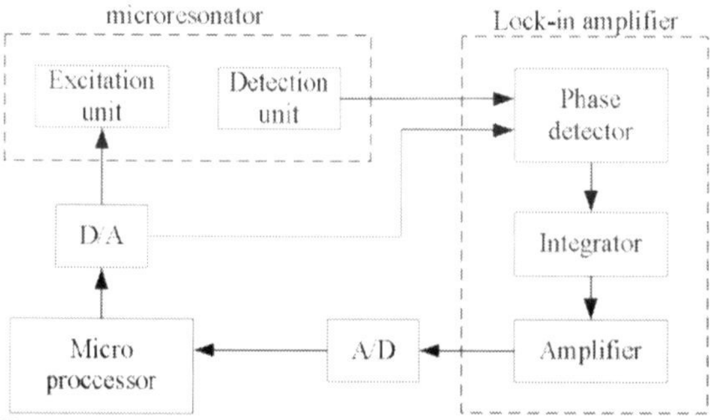

Figure.3 the block diagram of the detection circuit system

Figure.4 the detection circuit board

The vibration is transformed to voltage. This electrical signal is input to the phase detector and further to the low pass filter, which fulfill the lock-in function together. After appropriate amplifying, the prepoccessed signal is transformed to digital values for the microprocessor. Open-loop and closed-loop algorithms are carried out to complete the characteristic test. The open-loop test flow consists of the large scale frequency scanning and the samll scale frequency scanning. The former is to determine the approximate range of the resonant sensor, and the latter adjust the frequency step to locate the resonant frequency precisely. Finally, the resonant peak value, phase shift and Q value are calculated and displayed. Based on these characteristics, the closed-loop judges the present frequency by its responding vibration amplitude. Through the feedback, the control logic tracks and mantains the resonant state. The control model is based on the relation between the frequency and the amplitude differential. Figure.4 is the photo of the circuit board.

3.2. Experiment result

The standard pressure is generated by DHI PPC3 pressure calibrator and the constent temperature is supplied by GDS-50L temperature controllor. The open-loop test undergoes in the pressure of 100kPa and temperature of 25℃. The pressure values are 10kPa, 30kPa, 50kPa, 70kPa, 90kPa, 110kPa, 130kPa for closed-loop test and repeats 3 times in the constant temperature of 25℃. The results are presented by figure.5 and figure.6.

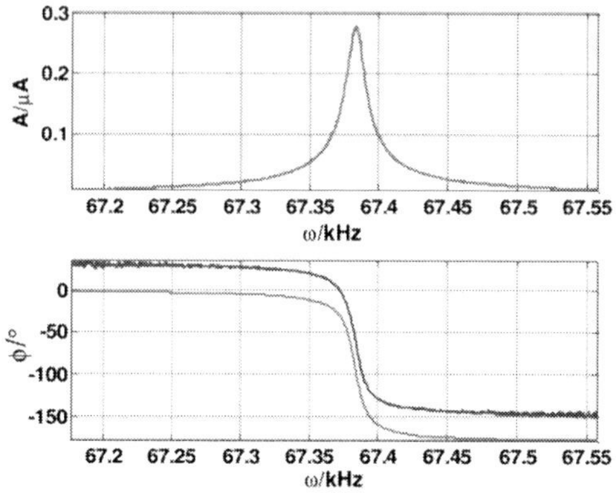

Figure.5 open-loop test result, the blue and the red refer to the actual value and the theory value, respectively.

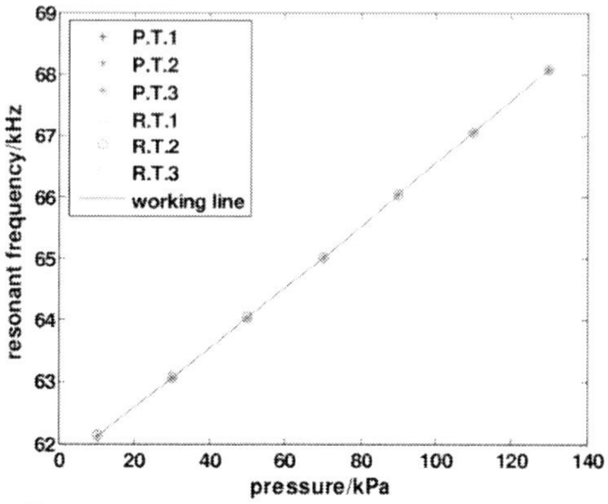

Figure.6 close-loop test result, the note'P.T.' and 'R.T.' refers to positive travel and reverse travel, respectively.

The open-loop test result shows that, the resonant frequency of the silicon micromachined resonant pressure sensor is 67.3829kHz, the Q value 5195, the resonant peak value 0.278uA and phase shift 35.5° . The blue curve is the actual value, while the red one is the calculation value according to the theory formula. The amplitude characteristics accord with the theory very well. The phase characteristics has a constant deviation from the theory, which is result from the error of the analog device.

The close-loop test result shows that, in the pressure range of 10kPa to 130kPa, the full output is 5.972kHz, and the sensitivity is 49.8Hz/kPa, and the repeatablity is 0.08%, the hysteresis is 0.15%, and the total error is 0.17%.

4. Conclusion

The resonant pressure sensor presented in this paper is fabricated by silicon micromachining technique. It has a compound structure, which consists of a square membrane and a doubly clamped resonant beam. A detection circuit system based on the phase-lock amplifier principle is designed. With the pressure and temperature controller, the experiment platform is setup to fulfill the open-loop and the close-loop test. The experiment results demonstrate that the sensitivity of the sensor is 49.8Hz/kPa, the repeatablity is up to 0.08%, and the total accuracy is 0.17%.

Reference

[1] Janson S W and Helvajian H 1999 MEMS, microengineering and aerospace systems *AIAA.J.* **99** 3802

[2] Welham C J, Greenwood J and Bertioli M M 1999 A high accuracy resonant pressure sensor by fusion bonding and trench etching *Sensor. Actuat.* A **76** 298-304

[3] Harada K, Ikeda K, Kuwayama H and Murayama H 1999 Various applications of resonant pressure sensor chip based on 3-D micromachining *Sensor. Actuat.* **73** 261-266

[4] Fan S C 2004 *Sensor technology and applications* (Beijing: Buaapress) p 100

[5] Liu G Y 2008 *Design, fabrication and application of microsensors* (Beijing: Buaapress) p 120

[6] Rödjegård H and Andersson G I 2005 A robust semi-digital readout method for crosstalk elimination in resonant sensors *Sensor. Actuat.* A **121** 72-77

Journal of Physics: Conference Series **188** (2009) 012043 doi:10.1088/1742-6596/188/1/012043

Online Self-compensation for Enhanced the Scale Factor Stability of a Micromachined Gyroscope

Zhou Bin [1], **Zhang Rong, Chen Zhiyong**

Department of Precision Instrument, Tsinghua University, Beijing, 100084, China

E-mail: zhoubin98@tsinghua.org.cn

Abstract. In this paper, an online self-compensation control scheme for micromachined gyroscope has been presented to eliminate the scale factor drift due to temperature influence. Firstly, the error sources of scale factor have been analyzed. According the analysis results, a novel control scheme which contains three loops has been proposed: a phase-locked loop of driving mode is to drive the proof mass oscillation in its' resonant frequency, an AGC loop of driving mode is to keep a constant value of the drive amplitude, an additional scale factor error online detection and cancellation loop is to keep the scale factor stable. A digital hardware prototype has been implemented to perform the precision loop control and self-compensation loop. Scale factor of the gyroscope has been measured in a temperature-controlled turntable. Experiment results show that the scale factor drift is -3.5% to 5.2% over the temperature range of -45℃ to +80℃ without the self-compensation loop, while the scale factor drift decrease to -0.009% to 0.15% after the self-compensation loop is applied.

1. Introduction

The performance improvement of micromachined gyroscope is greatly depend on its' environmental adaptability, especially the temperature dependency of the bias and scale factor. The temperature drift error of a micromachined gyroscope may come from the defects of structure, difference of material properties, inner air pressure change, error of detection and control circuit etc. An effective way to reduce the influence of those errors is to apply precision close loop control on the movement of gyroscope sensing elements. Although the errors from forward channel can be eliminated, the changes out of the loop still influence the stability of the bias and scale factor of a gyroscope. Considering the test cost, temperature compensation for scale factor is more expensive than the same for bias. According the performance of the available MEMS gyroscopes from commercial market, the scale factor drift over temperature range is normally large than 1% (typical value). The application range of those devices will be limited without temperature compensation of scale factor. In this sense, it is quite important to develop a low cost way to greatly improve the scale factor stability [2, 3].

2. System model and scale factor error analysis

---

[1] To whom any correspondence should be addressed.

The dynamic model of a micromachined Coriolis vibratory gyroscope can be described as a two-freedom oscillation mass-spring-damp system [1]. As show in Fig. 1, $Ox_a y_a$ is inertial reference coordinate system and $Ox_b y_b$ is vector coordinate system.

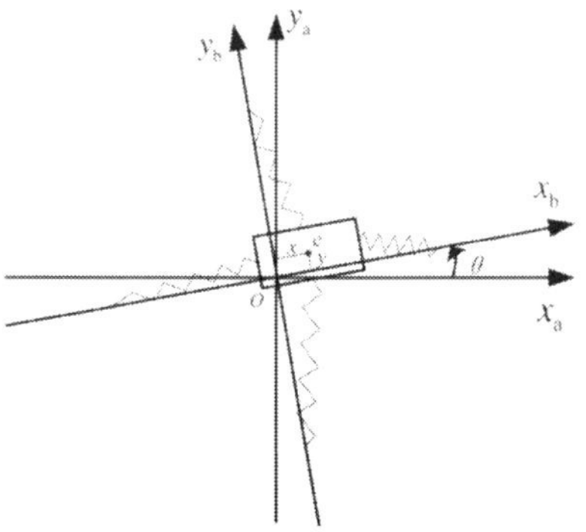

Figure 1. Dynamic model of a micromachined gyroscope.

The ideal time-domain model of gyroscope can be expressed by

$$\begin{cases} \ddot{x}(t)+ \frac{1}{Q_1}\omega_{n1}\dot{x}(t)-2\dot{\theta}\dot{y}(t)-\ddot{\theta}y(t)-\dot{\theta}^2x(t)+\omega_{n1}^2x(t)=f_{ex}(t) \\ \ddot{y}(t)+ \frac{1}{Q_2}\omega_{n2}\dot{y}(t)+2\dot{\theta}\dot{x}(t)+\ddot{\theta}x(t)-\dot{\theta}^2y(t)+\omega_{n2}^2y(t)=f_{ey}(t) \end{cases} \tag{1}$$

where $x(t)$ and $y(t)$ are the displacement of drive mode (primary mode) and sense mode (secondary mode) respectively. Q_1 and Q_2 are the quality factors of two modes. ω_{n1} and ω_{n2} are the natural frequency of two modes. θ is the rotation angle. $f_{ex}(t)$ and $f_{ey}(t)$ are generalized electro-static force of drive mode and sense mode respectively.

In general case, a micromachined gyroscope is always working in primary close loop mode which the sensing mass is driven in its' resonant frequency of drive mode and the oscillation amplitude is equal to desired value A_0. In the primary close loop mode, the oscillation amplitude of primary mode is only influenced by the error of detection circuit. Therefore the oscillation of primary can be expressed by

$$x(t)=A_0(1+k_{e1})\sin\omega_{n1}t \tag{2}$$

where k_{e1} is the gain error of primary detection circuit.

In equation (1), assume that $2m\dot{\theta}\dot{y}(t)$, $m\ddot{\theta}y(t)$, $m\dot{\theta}^2x(t)$, $m\ddot{\theta}x(t)$ and $m\dot{\theta}^2y(t)$ are small enough and can be neglected. The error of secondary detection circuit is also taken into account. Then the scale factor of a micromachined gyroscope can be expressed by

$$SF\big|_O = \frac{2(1+k_{e1})(1+k_{e2})A_0\omega_{n1}}{\sqrt{\left(\omega_{n2}^2-\omega_{n1}^2\right)^2+\left(\omega_{n2}\omega_{n1}/Q_2\right)^2}} \tag{3}$$

where k_{e2} is the gain error of secondary detection circuit. $SF\big|_O$ is the scale factor in secondary open loop mode.

As the stiffness of structure is alternated by environment temperature due to the temperature dependency of Young's modulus, the nature frequency ω_{n1} and ω_{n2} are also changed with temperature change. The gain error k_{e1} and k_{e2} of detection circuit are influenced by the temperature dependency of the electronic components and the parasitic parameters. The temperature dependency of gas damping (also influenced by air pressure [5, 6]) and structure material damping can influence the quality factor Q_1 and Q_2. All the factors show that scale factor of a micromachined gyroscope is strongly related to environment temperature. In this case, a secondary close loop (also called force rebalance loop [4]) can be introduced to reduce the temperature influence of Q_2, $k_{e2}$and ω_{n2}. In force rebalance loop mode, the scale factor of a gyroscope can be simply expressed by

$$SF\big|_{C} = 2(1+k_{e1})A_0\omega_{n1} \tag{4}$$

3. Principal of online compensation for scale factor

According equation (4), when a gyroscope is working in primary close loop and force rebalance loop mode, the main errors of scale factor come from temperature drift of the drive detection circuit and nature frequency ω_{n1}. The key of online self-compensation is to construct a reference signal which can observe the drift of Yong's modulus and the detection circuit error. In simplified model, the movement of drive mode can be considered as a second order system. In this case, a reference sine signal is added to drive mode to force the sensing mass oscillation. Therefore the errors of scale factor can be observed. Assume expression of the reference oscillation is $x_r(t) = A_r \sin\omega_r t$, where A_r is the amplitude of reference sine oscillation and ω_r is the frequency of reference sine oscillation. Under the important assumption that the gain error of detection circuit to signals with frequency of ω_r and ω_{n1} is equal, then the amplitude of reference signal can be expressed by

$$A_r = \frac{(1+k_{e1})\mathrm{K}_{\mathrm{DRV}}V_r}{\sqrt{\left(\omega_{n1}^2-\omega_r^2\right)^2+\left(\omega_{n1}\omega_r/Q_1\right)^2}} \tag{5}$$

where V_r is the reference drive voltage. $\mathrm{K_{DRV}}$ is the conversion coefficient between drive voltage and drive force.

In a small area for similar and assuming $\omega_r \ll \omega_{n1}$ and Q>50, the expressions of A_r versus the temperature dependency of Yong's modulus and quality factor can be expressed by

$$A_r(T)\big|_{E} \approx \left[1+\kappa_{ET}\left(T-T_0\right)\right]\frac{(1+k_{e1})\mathrm{K}_{\mathrm{DRV}}V_r}{\omega_{n1}^2(T_0)} \tag{6}$$

$$A_r(T)\big|_{\Omega} \approx \left[1-\kappa_{BT}\left(T-T_0\right)\right]\frac{\omega_r^2(1+k_{e1})\mathrm{K}_{\mathrm{DRV}}V_r}{\omega_{n1}^4 Q_1^3(T_0)} \tag{7}$$

where κ_{ET} and κ_{BT} are the temperature coefficients of material Yong's modulus and system damping separately. $A_r(T)\big|_E$ and $A_r(T)\big|_{\Omega}$ are the amplitude expressions versus the temperature drift of Yong's modulus and damping separately.

Generally the gain error introduced by temperature dependency of damping is far small than the gain error introduced by temperature dependency of Yong's modulus, i.e. $A_r(T)\big|_{\Omega} \ll A_r(T)\big|_{\kappa}$. In this way, the amplitude's temperature drift of reference signal can be mainly expressed by equation (6) which shows the gain error of reference signal is directly proportion to the change of Yong's modulus and error of detection circuit. Therefore the amplitude change of the reference signal can be used to observe the temperature drift of scale factor.

The scheme of online self-compensation in drive mode is shown in figure 2. Beside the primary traditional control loop which is composed of phase-locked loop and AGC loop, the scheme has an additional online self-compensation loop which is include the following 3 steps

- Add a reference signal into the normal drive signal.
- Derive the amplitude of reference response signal through the phase sensitive detector by using reference signal and drive sense signal.
- According to the amplitude of reference response signal to adjust the desired value of AGC loop.

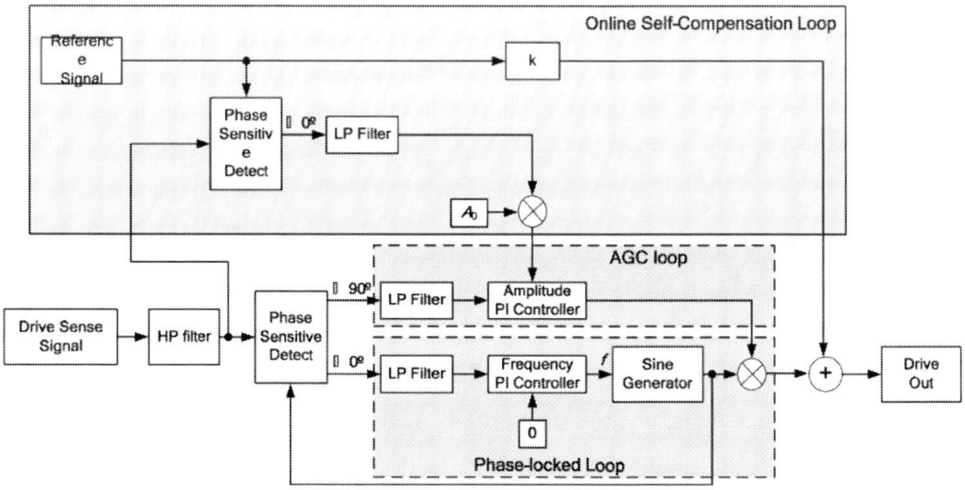

Figure 2. Schematic of Online Self-Compensation Loop.

4. Testing and results

As shown in figure 3, a micromachined gyroscope with digitalized control circuit has been implement to verify this online self-compensation method. The gyroscope is placed inside a temperature-controlled turntable..

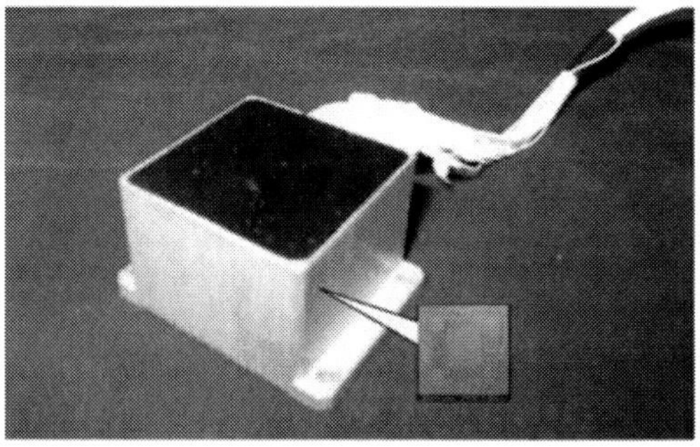

Figure 3. A testing sample of digital micromachined gyroscope.

The scale factor has been measured through -45℃ to +80℃. The test result without compensation is shown in figure 4. The result shows that the scale factor change is about -3.5% to 5.2% over full temperature range without compensation. The scale factor with online self-compensation is shown in figure 5. It shows that the scale factor change is only -0.009% to 0.15%.

The noise performance has also been measured. Without compensation the gyroscope output noise level is about $6.0\,°/\mathrm{h}/\sqrt{\mathrm{Hz}}$ while the noise level with self-compensation is about $6.1\,°/\mathrm{h}/\sqrt{\mathrm{Hz}}$.

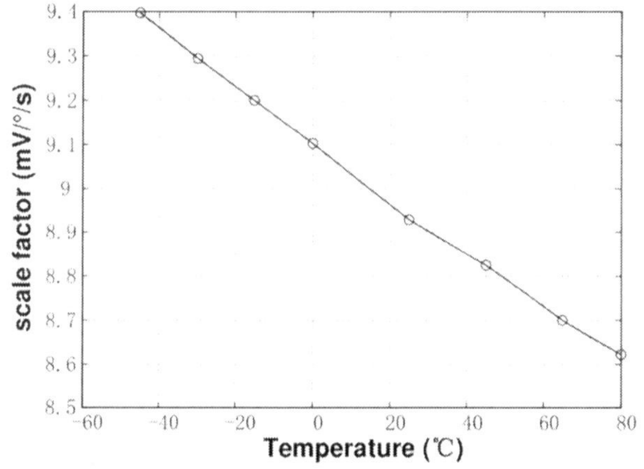

Figure 4. Scale factor versus temperature without compensation.

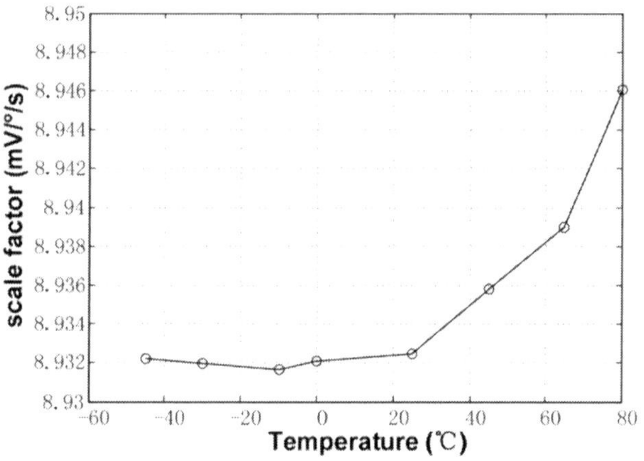

Figure 5. Scale factor versus temperature with online self-compensation.

5. Conclusions

An online self-compensation scheme for enhanced the scale factor stability has been developed. The effectiveness of this method has been verified by a digital micromachined gyroscope. Experiment results show that the self-compensation scheme tremendously reduce the temperature drift of gyroscope. The temperature hysteretic effect of output caused by temperature gradient is also significantly inhibited due to the compensation loop works in real-time. Additionally, due to the whole loop of self-compensation is implemented by using digital circuit, signal generation and algorithm calculation can be achieved without additional hardware cost. This self-compensation technique has been proved to be a no-cost and highly efficient way to improve the stability of scale factor. Future work will focus on the technology of bias compensation.

References

[1] Shkel Andrei M, Horowitz Roberto, Seshia Ashwin A, Park Sungsu and Howe Roger T 1999 Proceedings of the 1999 American Control Conference. **3** 2119
[2] Link T, ISimon, Trächtler M, Gaisser A, Braxmaier M, Manoli Y and Sandmaier H 2005 TRANSDUCERS '05. **1** 401
[3] Gaisser Alexander, Gao Zhongyu, Zhou Bin, Zhang Rong and Chen Zhiyong 2006 J. Tsinghua Science and Technology **11** 533

[4] Sung Woon-Tahk, Sung Sangkyung, Lee Jang Gyu and Kang Taesam 2007 Journal of Micromechanics and Microengineering. **17** 1939
[5] Acar Cenk, Shkel Andrei M, Costlow Lynn and Madni Asad M 2005 Proceedings of IEEE Sensors. **2005** 664
[6] Kulygin A, Schmida U and Seidel H 2007 J. Sensors and Actuators A: Physical. **145-146** 52

Journal of Physics: Conference Series **188** (2009) 012044 doi:10.1088/1742-6596/188/1/012044

Anticorrosive magnesium hydroxide coating on AZ31 magnesium alloy by hydrothermal method

Yanying Zhu[1,2], Guangming Wu[1 *], Qing Zhao[2], Yun-Hong Zhang[2], Guangjian Xing[1], Donglin Li[1]

[1]Beijing Institute of Petrochemical Technology, Beijing102617, China,

[2]Beijing Institute of Technology, Beijing 100081, China

Email address: wuguangming@bipt.edu.cn

Abstract. Magnesium alloys are potential biodegradable biomaterials in orthopedic surgery..However, the rapid degradation rate has limited their application in biomedical field. A great deal of studies have been done to improve the resistance of magnesium alloys. In this article, An anticorrosive magnesium hydroxide coating with a thickness of approximately 100μm was formed on an AZ31 magnesium alloy by hydrothermal method.The morphology of the coatings were observed by an optical microscope and SEM. And the samples were soaked in hank's solution (37℃) to investigate the corrosion resistance. Magnesium alloy AZ31 with magnesium hydroxide coatings present superior corrosion resistance than untreated samples.

1、Introduction

Metallic materials continue to play an essential role as biomaterials in orthopedic surgery. Previous studies have shown that magnesium alloys have potential as biodegradable materials. The fracture toughness of magnesium is greater than bioceramics, while the elastic modulus is closer to that of natural bone than other metallic implants [1]Moreover, magnesium ion plays an important role in many physiological activities of human body and it has been proved that the magnesium and its alloy are biocompatible[2,3,4].However, the poor corrosion resistance especially in the solution containing chloride ion, such as physiological environment of human body, inhibits their practical application as orthopedic implants. The mechanical integrity of the magnesium alloys are loosed before the tissue has healed. Fabricating protective coatings on the magnesium alloys are one of the most effective ways to overcome this drawback[5]. In this study, magnesium hydroxide coatings were deposited on magnesium alloy (AZ31) substrates by hydrothermal method. And so far, no report on increasing the

* Tel./Fax:+86-010-81293245

Email adrress: wuguanming@bipt.edu.cn

corrosion resistance of Mg alloy by magnesium hydroxide coating using hydrothermal method has been available. Furthermore, it is well documented that magnesium hydroxide coatings were non-toxic.

2、Experimental procedures

2.1 sample preparation

AZ31 magnesium alloy ingots were cut into discs of 2mm in thickness and 12 mm in diameter. Each sample was mechanically polished up to 2000 grit, cleaned in deionized water and ethanol, and then dried in open air.

2.2 magnesium hydroxide coatings formation

The $Mg(OH)_2$ coating was fabricated on the AZ31 alloy by hydrothermal method. The NaOH (6g) was added to 100ml deionized water, then the alkali solution was poured into a hydrothermal reaction vessel with a filling factor of 75%.The AZ31 samples were set in the vessel and the vessel was heated in an electric oven in order to prepare the anticorrosive coating. The temperature and the reaction time were 160°C and 3 h, respectively. The reactions in the hydrothermal reaction vessel may include the following reactions[6] :

$$Mg_{(s)} \rightleftharpoons Mg^{2+}_{(aq)} + 2e^- \quad (1)$$

$$2H_2O_{(aq)} + 2e^- \rightleftharpoons H_{2(g)} + 2OH^-_{(aq)} \quad (2)$$

$$Mg^{2+}_{(aq)} + 2OH^-_{(aq)} \rightleftharpoons Mg(OH)_{2(s)} \quad (3)$$

2.3 morphology observation

The surface morphology was observed using an optical microscope and SEM. Then an across-sectional view of the obtained coatings was observed by SEM.

2.4 The soaked experiment[7]

The degradation behavior of samples was investigated by immersion test in Hank's solution. The composition of the Hank's solution used was composed of 8 g/L NaCl + 0.4 g/L KCl+0.14 g/L$CaCl_2$ + 0.1 g/L $MgCl_2 \cdot 6H_20$ + 0.1 g/L $MgSO_4 \cdot 7H_20$ + 0.06 g/L KH_2PO_4 + 0.06 g/L $Na_2HPO_4 \cdot 2H_20$+ 0.35 g/L $NaHCO_3$+ 1.0g/L glucose. Before soaked in the solution, the dimensions and masses of samples were measured using a vernier caliper and an electric balance, respectively. The solution volume to specimen area ratio was 0.40 mL/mm^2.The containers were placed in the water bath and the temperature of the bath was kept at 37℃.The solution was renewed every two days. The samples were taken out after immersion 1, 3, 7, 15, and 31days, respectively. Then the samples were cleaned with deionized water and ethanol, and removed the corrosion products using chromic acid. Finally, the samples were put in the dryer for 24h, and then the masses were measured.

The corrosion rate was calculated as follows:

$$v = \frac{m_0 - m_1}{St} \quad (4)$$

where v is the corrosion rate , m_0 is the sample mass before soaked experiment, m_1 is the mass after soaked experiment, S is the surface area of an sample, t is the corrosion time.

3、Result and Discussion

3.1 Morphology of magnesium hydroxide costing

Figure 1(a) is the optical micrograph of the magnesium hydroxide coating. It can be seen that there are many cracks on the surface. Figure 1 (b) is the SEM micrograph of the coating. The grains of the magnesium hydroxide on the surface are about 20~80nm. Figure 1 (c) is SEM morphology of the cross-section of the coated AZ31 sample. It can be see that the coating is formed on the substrate of the AZ31 alloy and the thickness of the coating is about 100μm. The cross section graph of the sample also shows that coating is uniform and dense.

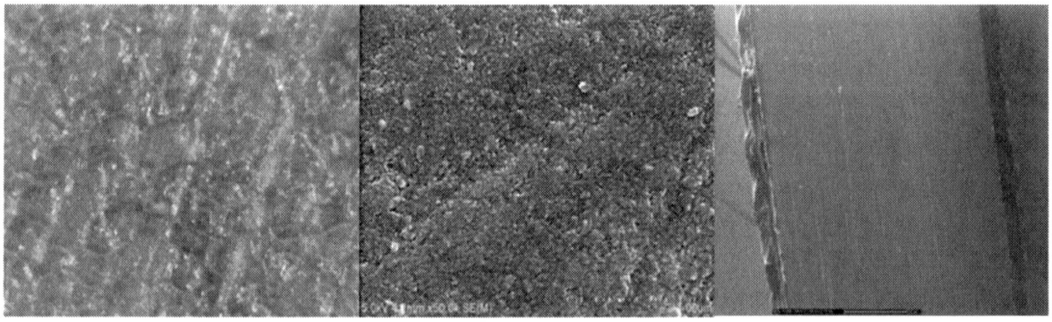

(a) (b) (c)

Figure 1 a、Optical micrograph of $Mg(OH)_2$ coating on AZ31 magnesium alloy
b、SEM micrograph of surface of micrograph
c、Cross-sectional SEM micrograph of $Mg(OH)_2$ coating on AZ31 magnesium alloy

Figure 2 shows the image of samples after immersion time 1,3,7,15 and 31 days in Hank's solution. It can be seen that after 1and 3 days, bare AZ31 magnesium alloy samples was slightly corroded, and the corrosion products covered the surface. The corrosion of bare samples became serious after 7, 15 and 31 days. The pitting holes appeared, developed and became enlarged. By comparison, the coated AZ31 has a better corrosion resistance than bare AZ31. The coatings were still basically intact after 31 days immersion in Hank's solution.

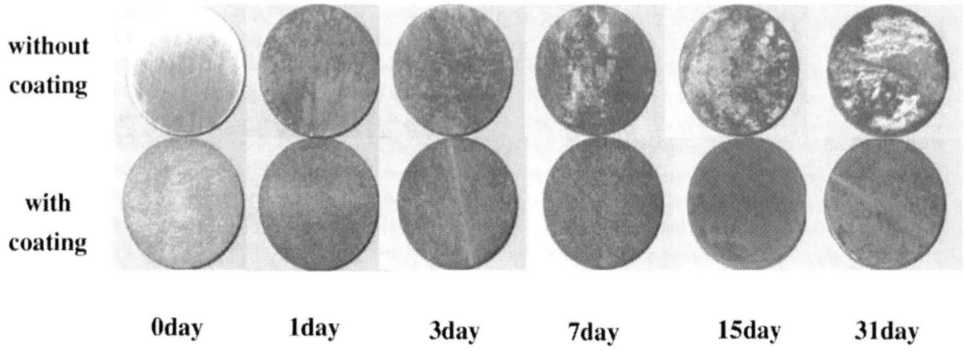

Figure 2 Appearance of the samples after 0,1,3,7,15,and 31days immersion in Hank's solution

The mass loss of the samples was measured after immersion 1,3,7,15,31 days in Hank's solution. The corrosion rates of the samples were calculated according to the equation (4), and the corrosion rate curves are showed by figure 3. It can be seen that the corrosion rates are decreased with the prolonging of immersion time. It is probably because the corrosion products covered the surfaces of samples, and decelerated the reaction. It was also showed that the coated samples showed better corrosion properties than that of bare AZ31 alloy samples. Therefore, the AZ31 magnesium alloy samples with magnesium hydroxide coating can decrease the corrosion rate effectively after implantation in the body.

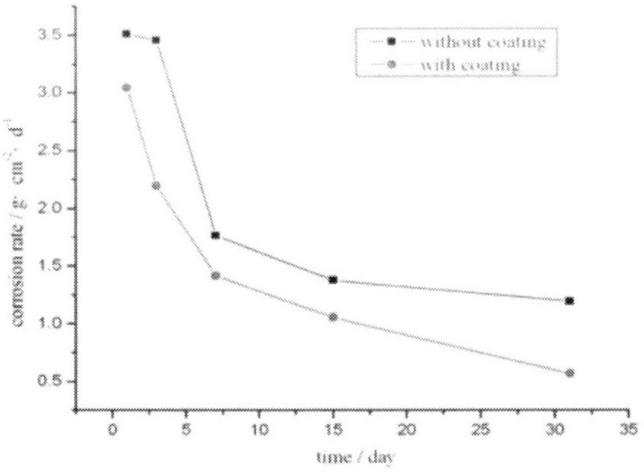

Figure 3 Variation of the corrosion rate of samples with coating and without coating with immersion time in Hank's solution.

4、Summary

The present study demonstrated that the magnesium hydroxide coating was successfully formed by hydrothermal method on the AZ31 magnesium alloy substrate and improved the corrosion resistance of the magnesium alloy effectively to compare with the bare alloy. Formation of magnesium hydroxide coating on the Mg alloy substrate can provide a potential material for the appropriate orthopedic surgery, and further studies will focus on the biocompatibility of the magnesium hydroxide coatings.

References

[1] M. P. Staiger et al. 2006 *Biomaterials* **27** 1728
[2] Saris NEL. 2000 *Clin Chim Acta* **294** 1
[3] Okuma T. 2001 *Nutrition* **17** 679
[4] Hartwig A. 2001 *Mutat Res/Fund Mol Mech Mutagen* **475** 113
[5] J. E Gray and B. Luan 2002 *Journal of Alloys and Compounds* **36** 88
[6] Frank Witte, Norbert Hort et al. 2009 *Current Opinion in Solid State and Materials Science* **12** 63
[7] American Society for Testing and Materials. 2004 ASTM G31-72 Standard Practice for Laboratory Immersion Corrosion Testing of Metals *American Society for Testing and Materials 2004 Annual Book of ASTM Standards* (Philadelphia Pennsylvania USA:ASTM vol 03.02) p5

Practical Integrated Design of a Condenser-objective Lens for Transmission Electron Microscope

Li Wen-ping [1], Wu Jian[1], Zhou Zhen[1], Gui Li-jiang[1], Han Li[2]

[1] Beijing University of Aeronautics and Astronautics, Beijing 100191, P.R.China

[2] Institute of Electrical Engineering, Chinese Academy of Sciences, Beijing 100190, P.R.China

liwp@buaa.edu.cn

Abstract. A condenser-objective lens is designed through combination of separating and integrating to consider the effect of the front condenser field on its objective performance. A practical lens model including magnetic pole piece, magnetic circuit and coil windings is built to optimize its rear field. The front field can be integrated into the rear one by simply adjusting the position of the specimen and the excitation on the condenser-objective lens. Optical performance of the integrated lens is researched as both a condenser lens and an imaging one. The total aberrations at the specimen plane are 0.01nm under STEM operation mode and its spherical aberration coefficient is 1.5mm when being an imaging objective lens, which can meet for high resolution microanalysis and TEM imaging.

1. Introduction

With the development of nanotechnology, it becomes important to improve the resolution of transmission electron microscope (TEM) in nano-metrology. The resolution of electron microscope is limited by the aberrations of its objective lens, so the popular so-called condenser-objective lens according to Riecke[1] has been the researching focus in this field[2-3]. Condenser-objective lens differs from conventional objectives in that the specimen is arranged in the middle of the pole piece gap. Its front field disposed above the specimen works as a condenser lens and the rear one disposed below the specimen as an objective lens. The condenser-objective lens is the most important lens in both the illumination system and the imaging system. However, most authors have concentrated their efforts on its objective performance either searching for optimized field distributions with minimum aberrations or considering the effect of the current density of coil windings on the objective lens design [3-5]. Chen Wen-xiong developed a program to calculate both condenser and objective performance of such a lens with a given model in 1988[6], but little attention has been paid to the effect of the front condenser field on its objective performance especially in modern field-emission TEM [3]. Furthermore, the specimen is disposed at very high field of the condenser-objective lens and its position is sure to affect its ultimate performance [6]. A more accurate simulation model is needed to obtain higher precise in its optical performance.

In this paper, a condenser-objective lens is designed through combination of separating and integrating the front and rear field. The practical lens model and its rear field are first optimized to achieve the resolution, then its front filed is designed to obtain better illumination. At last, the performance of the integrated lens is researched as both a condenser lens and an imaging lens.

2. Separate design of the condenser-objective lens

The condenser-objective lens works in TEM as shown in Fig.1. A beam path denoted by solid lines in Fig.1 represents a standard TEM illumination mode. A beam path denoted by dotted lines without activating the condenser lens in Fig.1 represents the STEM operation mode. The illumination system and the imaging system have different optical principles and different performance, so it is impractical to design the whole lens according to the united criterion.

2.1. Theoretical design of the rear field

The rear field is theoretical designed according to resolution in CTEM imaging. According to reference [7], for an optimum defocus value of $\Delta f_{opt} = -\sqrt{\frac{3}{2}}\Delta f_{Sch} = -1.22(C_s\lambda)^{1/2}$ (Δf_{Sch} is Scherzer defocus, C_s is the spherical aberration coefficient, λ is the wavelength of the electron) and optimum aperture size $\alpha_{opt} = 1.4(\lambda / C_s)^{1/4}$, the typical value of resolution is $\delta = 0.66C_s^{1/4}\lambda^{3/4}$. In our 200kV TEM, the spherical aberration coefficient should be smaller than 1.5mm to get the resolution better than 0.25nm.

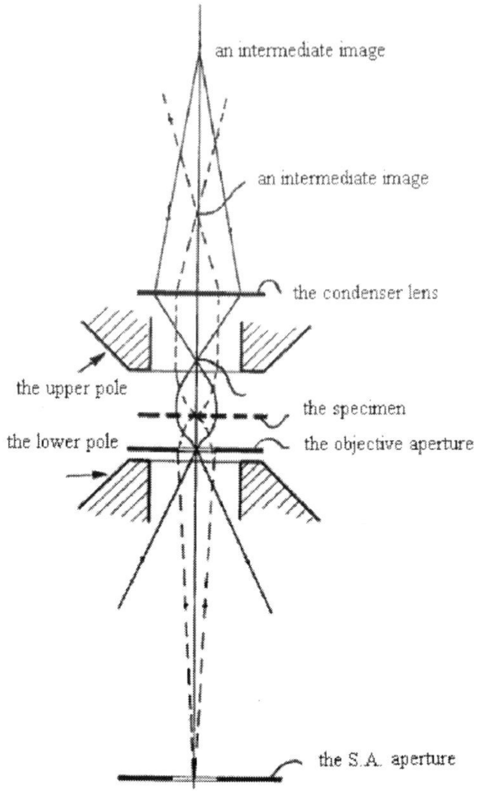

Fig.1 Principle of a condenser-objective lens

According to the conclusions in reference [3], we choose the gap S between the pole pieces as 6 mm and the bore diameter D as 3mm. When the excitation on the lens is 9976.212NI ($NI / \sqrt{V_r} = 20.4$, where V_r is the corrected voltage when taking into account relativistic effects), its spherical aberration coefficient will be 1.38mm with 1.8mm focal length.

2.2. Practical optimization of the lens model

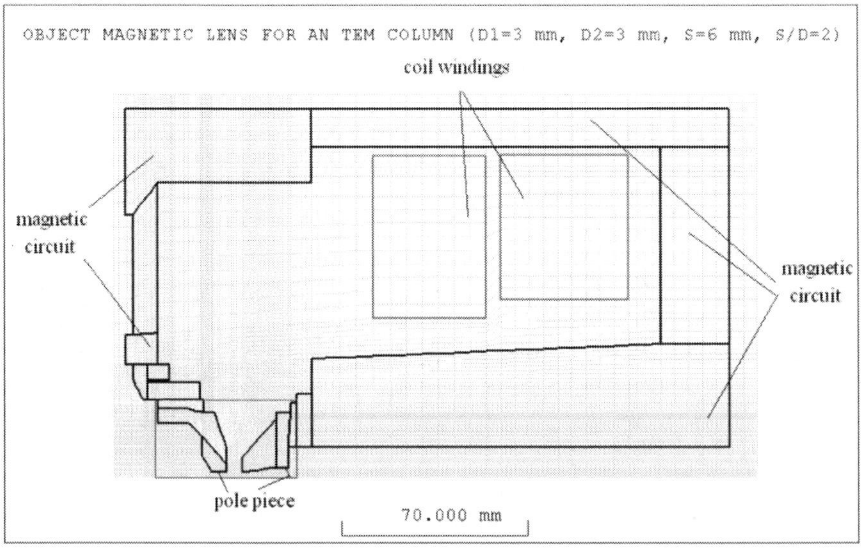

Fig.2 Optimized model of a condenser-objective lens

In order to obtain higher precise simulation, our lens model not only includes magnetic pole piece, magnetic circuit and coil windings but also consists of structures for mechanical assembling. Permendur alloys 1J22 and high purity electrical iron DT4C are chosen as the materials of the magnetic pole piece and magnetic circuit respectively. Better pole pieces structure means smaller fringe field beyond the pole gap. We refer to the conclusions in [8] when designing its magnetic circuit and coil windings and the optimized model is shown in Fig.2. We place the S.A. aperture below the centre of its pole pieces 200mm to leave enough space for other units. Only if the excitation on the lens is above 10377.7NI, the lens will have a spherical aberration coefficient of less than 1.5mm. Its spatial field distribution is shown in Fig.3 and the current densities of coil windings are 221.6 NI/cm^2. Judged from Fig.3, larger magnetic field (B>2.2T) is situated in the pole piece, so the magnetic circuit is unsaturated. The focal length of the lens is less than 1.92mm, the specimen is disposed below the centre of its pole pieces 0.15mm or farther and the magnification is more than 105.

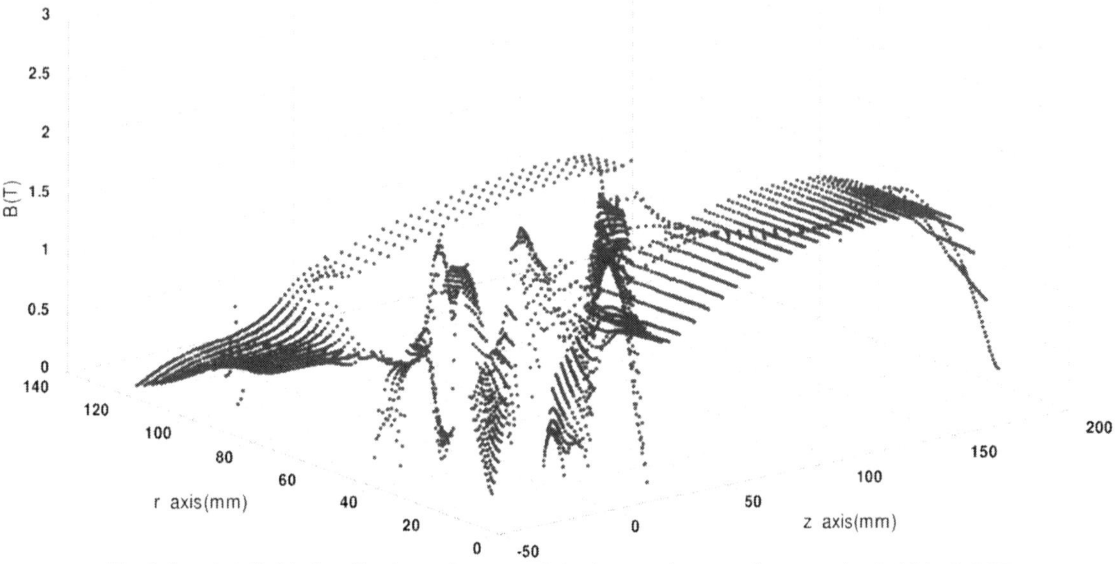

Fig.3 Spatial field distribution of our model: the maximum of magnetic field is 2.78T

3. Integrated design of the condenser-objective lens

The author gave an optimized ratio of the upper bore diameter to the lower one (D1/D2) in [6].However, the whole field will be changed greatly once we change the upper bore diameter. Then our above optimization of the rear field may become useless. Here we integrate the front condenser field into the rear one by optimizing the position of specimen and the excitation on the condenser-objective lens.

3.1. Integrating its front field into its rear one

In Fig.1, both the condenser lens and the front field work at the TEM illumination mode, but in STEM mode the front field works solely. The STEM mode is chosen to optimize the front field. The intermediate image plane at STEM mode should have a distance from the objective lens to supply enough space for auxiliary units in TEM. In addition, the aperture at the specimen may be larger than 2.5mrad being necessary for high resolution observation. Better illumination beam is obtained by placing the specimen below the gap centre 116.1μm. The excitation on the condenser-objective lens is 10382.0 NI (current densities of coil windings are 221.7NI/cm^2) and its spatial field distribution will be similar to that shown in Fig.3. The beam current density contours at the specimen is shown in Fig.4 if the energy spread of the field-emission gun is chosen 0.6eV. It can be seen from Fig. 4 that the total aberrations including spherical aberration and chromatic aberration are 0.01nm, which can be used in high resolution microanalysis.

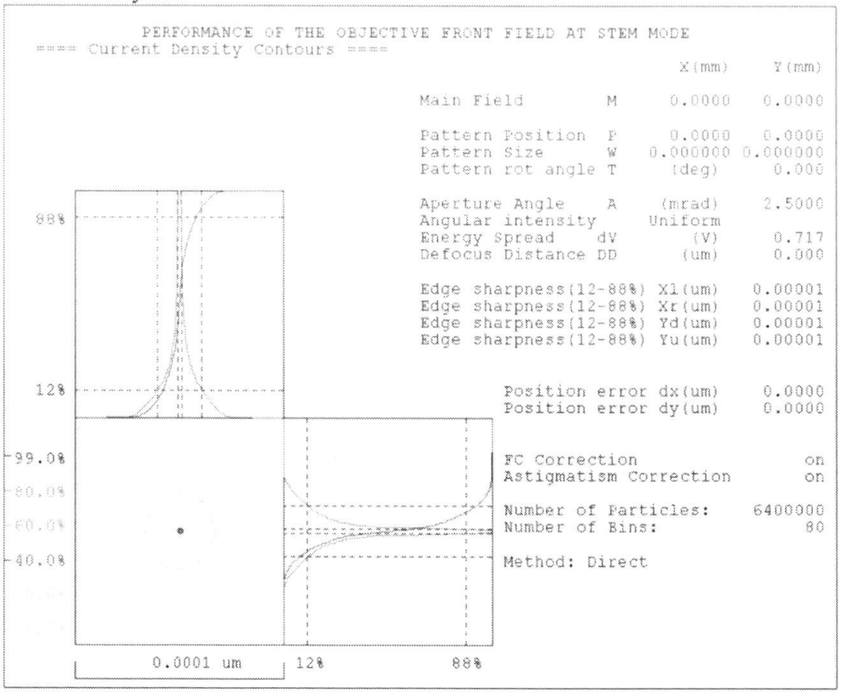

Fig.4 Beam current density contours at the specimen plane

3.2. Performance of the integrated lens

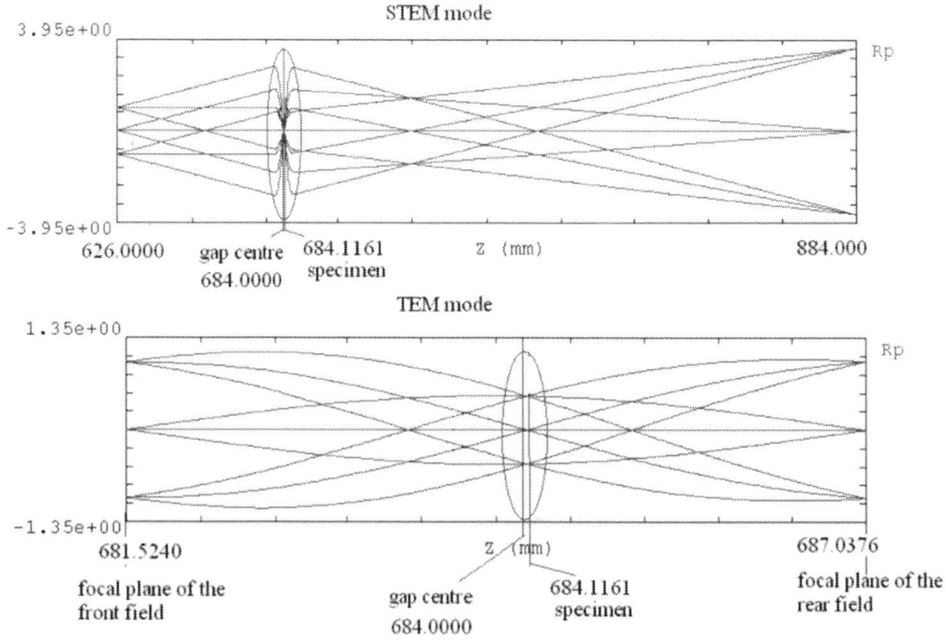

Fig.5 Electron trajectory under different mode

Under the STEM mode, electron trajectory is shown in the upper part of Fig. 5. Its illumination performance is detailed in part 3.1. The integrated lens has a spherical aberration coefficient of 1.5mm, a chromatic aberration coefficient of 1.4mm and a focal length of 1.91mm in the imaging system, which can meet for the ultimate resolution. The focal plane of the pre-field was studied before we get the electron trajectory under TEM mode as shown in the lower part of Fig. 5. In order to obtain the illuminated field, the lower part of Fig. 5 gives the electron trajectory when the focal plane of the pre-field is chosen as the object plane. The diameter of the illuminated field at the specimen plane is about 1.8um when the illumination aperture is 5 mrad. Compared to STEM mode, the same transmission electron trajectory is obtained under TEM mode as the excitation on the condenser-objective lens and the specimen plane keep fixed under both modes. Therefore, the same imaging performance has been obtained under TEM mode.

4. Conclusions

A condenser-objective lens is designed through combination of separating and integrating to consider the effect of the front condenser field on its objective performance. The practical model not only includes magnetic pole piece, magnetic circuit and coil windings but also consists of structures for future assembling, which improves the simulation precision greatly. Optical performance of the integrated lens shows that it can be used in high resolution TEM and STEM.

Acknowledgment

We would like to thank Munro's Electron Beam Software Ltd., London for supplying the finite element method software and optical properties computation software. We would like also to thank Dr. Liu Jun-biao at Institute of Electrical Engineering, Chinese Academy of Sciences for his help in designing the integrate lens. The research is supported by the Ministry of Sciences and Technology, P.R.China, under the Contract No. 2006BAK03A24.

References

[1] Wolfgang Dieter Riecke, Corpuscular Beam Microcope Apparatus, Patent:3560781,1967, Germany.
[2] Gerd Benner. Particle-optic illuminating and imaging system with a condenser-objective single field lens, Patent:6531698B1, 2003, Aalen.
[3] A.S.A. Alamir. Ultimate performance of objective magnetic lens,Ultramicroscopy, 2004,101: 241–246.
[4] K. Tsuno, D.A. Jefferson. Design of an objective lens pole piece for a transmission electron microscope with a resolution less than 0.1 nm at 200 kV, Ultramicroscopy ,1998,72: 31-39.
[5] O.L. Krivanek, G.J. Corbin, N. Dellby, B.F. Elston, R.J. Keyse, M.F. Murfitt, C.S. Own, Z.S. Szilagyi, J.W. Woodruff. An electron microscope for the aberration-corrected era. Ultramicroscopy, 2008, 108(3): 179-195.
[6] Chen Wen-xiong. How Does Polepiece Saturation Affect the Spherical and Chromatic Aberration Coefficient of the Magnetic Lenses, Journal of Chinese Electron Microscopy Society, 1(1988): 55-66.
[7] Mehmet Sarikaya. Resolution in Conventional transmission electron microscopy, Ultramicroscopy, 1992, 47: 145-161.
[8] Huang Lan You , Liu Xu Ping. Electron microscope and electron optics, 1991, Beijing: Science Press.

The anti-supercooling effect of surface-modified nano-scaled SiO_2 in hydrated salts phase transition system

Xiaolin Wu[1], Yihai Wang[1], Rong Sun[1]*, Maobai Lai[1], Ruxu Du[1,2] and Zhijun Zhang[3]

[1] Shenzhen Institute of Advanced Integration Technology, Chinese Academy of Sciences and The Chinese University of Hong Kong, Shenzhen 518067, China;

[2] The Chinese University of Hong Kong, Hong Kong SAR;

[3] Henan University, Henan 475004, China

Email: rong.sun@siat.ac.cn (Corresponding author)

Abstract. Phase change characteristics of hydrate salt were analyzed for a system consisting of three types of SiO_2 nano-particles doped phase change materials (PCMs). By using the three nano-particles of Aerosol SiO_2, RNS-A SiO_2 and Liquid phase SiO_2, the surface effect of the interaction between nano-particles and hydrate salts was investigated. The time-temperature curves and Differential Scanning Calorimeter (DSC) testing results showed that Aerosol SiO_2 had the most effective and stable performance as the nucleators for hydrate salts. The analysis of FT-IR showed that there were strong characteristic hydroxyl bonds on the surface of Aerosol SiO_2. And the designed hydroxyl controllable silica gel system could prove the effect of hydroxyl bonds on suppressing supercooling of hydrate salts. It was concluded that supercooling of hydrate salts could be easily suppressed by the nano-additives with high specific surface area and strong polar hydroxyls in the surface. The ion-exchange attraction between hydroxyls and hydrate salts might mainly result in the supersaturation of hydrate salts in the interface of nano-particles. And then the supercooling of hydrate salts could be suppressed. The theoretical analysing model based on interionic attraction is a novel approach for investigating the nucleation of hydrate salts. And this result might provide a potential low-cost approach for the applications of nano-additivs in building energy storage and coolant.

1. Introduction

Hydrate salts, a type of low cost phase change materials (PCMs), have attracted more attention owing to its high volumetric energy storage density. However, their applications are limited by supercooling.

There are two kinds of nucleation for hydrate salts: homogeneous nucleation which involves the formation of nucleation particles in the liquid phase from the PCMs itself and heterogeneous nucleation which involves the formation of nucleation particles on foreign particles or sites [1]. Homogeneous nucleation always has more serious supercooling than heterogeneous nucleation. So, heterogeneous nucleation is the main method to suppress supercooling. In order to solve the problem of supercooling, many studies about heterogeneous nucleation have been done. Telkes [2] introduced classic isomorphous and isotypic nucleators, which have crystal structures and lattice parameters nearly identical with, the substrate salts to control supercooling of hydrate salts. Lane [1] introduced epitaxial nucleation, which had different crystal morphology with substrate. It had preferred surface characteristics, which were suitable for deposition of the substrate crystal. However, there is little work about the effect of nano-additives on the phase change characteristics of hydrate salts.

In previous researches, nano-additives, such as nano-tube [3], nano-fibre [4], etc., which had very high surface to volume ratio, very low thermal resistance, great compatibility and light weight [5,6] were used in PCMs for improving the thermal conductivity [7-10] of organic PCMs, The measured values of latent heat for wax with SWCNT showed a good enhancement over the latent heat of pure wax. A maximum enhancement of approximately 13% was observed for the wax/SWCNT composite corresponding to 1% loading of SWCNT.

Some work had been done about the effect of nano-additives on organic PCMs. However, the effect of nano-additives with different surface characteristics on hydrate salts is not clear. In our present work, nano-SiO_2 powders with three kinds of surface characteristics were investigated to enhance thermal performance of hydrate salts. Different mass ratios of the three types of nano-scaled SiO_2 were added into saturated hydrate salts solution. The thermal characteristics of the new nanocomposites were tested to study the effect of the nano-scaled additives on hydrate salts. And a comparative study was performed to study the effect of nano-SiO_2 surface characteristics on the enhancement of the hydrate salt thermal performances. At the end, a novel theoretical analysing model based on interionic attraction was introduced to illuminate the nucleation of nano-particle doped hydrate salts. This result can provide a latent low cost solving method to suppress supercooling and improve thermal performances for the applications in building energy storage and coolant.

2. Experimental method

Three types of SiO_2 were used as the additives. They were Aerosol SiO_2 prepared by gas phase method with lots of active hydroxyl bonds in the surface, RNS-A SiO_2 which is a kind of surface modified SiO_2 with reactable bonds like double bonds, hydroxyl bonds, amidogen bonds, etc. in the surface and Liquid SiO_2 prepared by liquid phase method. The sizes of them were about 12nm, 40nm and 50nm, respectively. The phase change materials used contained $Na_2SO_4 \cdot 10H_2O$, $Na_2HPO_4 \cdot 12H_2O$, and $Na_2S_2O_3 \cdot 5H_2O$. The nano-additives doped hydrate salt samples were subjected to mixing by using mechanical stirring for about 5 min to obtain a well dispersed suspension.

We made a simple device to record the temperature in time. The samples were placed on test tubes with a thermal couple in the centre. The tubes were initially heated to 40 ℃ to reach the liquid phase. Then the tubes were immersed into a 10 ℃ water cooler. The temperature histories were recorded using a multipoint recorder (IDAQ-8018+) and the data were stored in a computer via an RS-232 port.

In order to analyze the phase transition in microscopic level, Differential Scanning Calorimeter (DSC) was used with nitrogen atmosphere. Samples were placed in a sealed aluminium pan with a mass of about 30.00 mg. The pan was placed in DSC and analyzed from -20 to 45 ℃ with a linear heating rate of 7 ℃ /min. The nitrogen flow rate was 50 ml/min. The accuracy of the DSC temperature measurements was ± 0.02 ℃. The latent heat was calculated as the total area under the peaks of solid-liquid transitions.

To study the microstructure of the samples, we used Transmission Electron Microscopy (TEM), x-ray Diffraction (XRD) and Fourier Transformation Infrared (FT-IR) Spectroscopy. For TEM analysis, the surface of sample was first coated by gold. TEM images were acquired on a JEM-2010 microscope at 22 ℃. For XRD curves, powder samples were put into the sample room directly. Then the microstructure characteristics could be gotten. For FT-IF spectroscopy, the samples were prepared by pressing potassium bromide troche KBr at 10 ℃.

3. Results and discussion

3.1. Structure of the three types of SiO_2 nano-particles

TEM and XRD were used to test the microstructure characteristics of the three nano-particles. Figures A-C in figure 1 are the TEM images of Aerosol SiO_2, Liquid SiO_2 and RNS-A SiO_2 nano-particles. No regular morphology of the three nano-particles could be seen from these figures. The size of Aerosol SiO_2 is smaller than the other two types of SiO_2. Figure D is the typical XRD result of these three types of nano-particles. It shows that these nano-particles have amorphous structures.

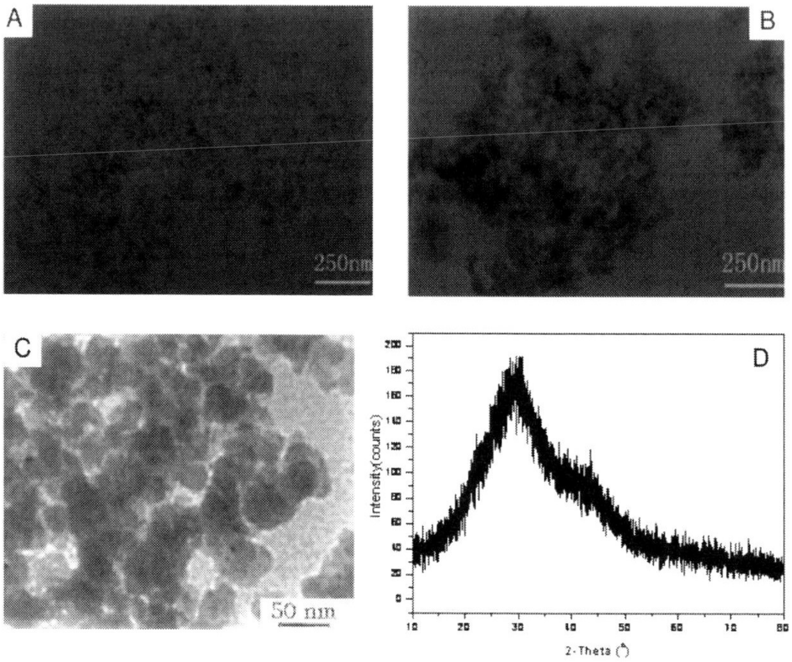

Figure 1. (A) TEM image of Aerosol SiO_2 nano-particles, (B) TEM image of Liquid SiO_2 nano-particles, (C) TEM image of RNS-A SiO_2 nano-particles, (D) XRD of SiO_2 nano-particles

3.2. Time-temperature curves of hydrate salts with three types of SiO_2 nano-particles

The three nano-SiO_2 particles mentioned above were used as additives of sodium sulfate decahydrate. With the adding of Liquid SiO_2 and RNS-A SiO_2, the solutions of hydrate salts were both chaotic suspensions. However, the hydrate salts with Aerosol SiO_2 was a colloid system, which was caused by lots of hydroxyl bonds in the surface of the SiO_2. Table 1 summarizes the experimental results of hydrate salts with different weight percentage of the three nano-particles.

As shown in table 1, with the increasing of weight percentage, the additives of Liquid SiO_2 and RNS-A SiO_2 can suppress supercooling of hydrate salts until their weight percentages are about 7% or more. But for Aerosol SiO_2, even 0.7% is enough to suppress supercooling of hydrate salts. This result indicates that the improvement of concentration caused by increasing the percentage of additive is not the key reason for suppressing supercooling of hydrate salts.

Table 1. Summaries of the experimental results: supercooling situation of hydrate salts with different weight percentage of the three types of nano-particles

Additives/weight percentage	0.7%	1.2%	3.3%	5%	7%
Liquid SiO_2	supercooling	supercooling	supercooling	supercooling	Phase change Temperature: 25℃
RNS-A SiO_2	supercooling	supercooling	supercooling	supercooling	Phase change Temperature: 29℃
Aerosol SiO_2	Phase change Temperature: 26~29℃	Phase change Temperature: 26~29℃	Phase change Temperature: 26~29℃	Phase change Temperature: 26~29℃	Phase change Temperature: 26~29℃

Figure 2 shows time-temperature (t-T) curves of hydrate salts without additives and with three nano-SiO_2 during cooling process. Figure 2 (a) corresponds to the t-T curve of hydrate salts without additives. There is serious supercooling with the phase change temperature about 28℃. Curves 1 to 3 in figure 2 (b) indicate the phase change of hydrate salts with RNS-A SiO_2, Aerosol SiO_2 and Liquid SiO_2 nano-additives. The weight percentages of them are all about 7%. From the three curves, there is only one successful phase change process without supercooling. The phase change temperatures are about 29℃, 28℃ and 25℃, respectively. Additives of RNS-A SiO_2 and Aerosol SiO_2 show better effect on hydrate salts.

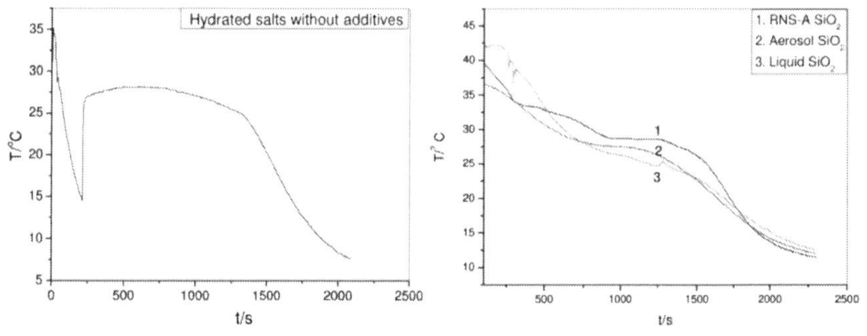

(a) (b)

Figure 2. t-T curves of sodium sulfate solution (a) without additives (b) with three nano-additives

From figure 2 (b), it still can be found that at the beginning process before phase change happening, the slope values of the curves are different. The slope values of the curves of hydrate salts with Liquid SiO_2 and RNS-A SiO_2 are nearly the same. They are larger than the slope value of hydrate salts with Aerosol SiO_2. That means the heat transfer rate of hydrate salts with Liquid SiO_2 and RNS-A SiO_2 nano-particles is higher than with Aerosol SiO_2 nano-particles.

3.3. DSC comparative results of hydrate salts with three types of the SiO_2 nano-particles

The precise DSC testing method was used to provide more detail phase behaviour in the thermal storage media. Figure 3 are the DSC curves of saturated sodium sulphate hydrate with and without nano-particles during heating process. Curve 1, 2 and 4 are the hydrate salts with the same weight percentage of additives of Liquid SiO_2, RNS-A SiO_2 and Aerosol SiO_2, respectively. Curve 3 is the hydrate salts without additives.

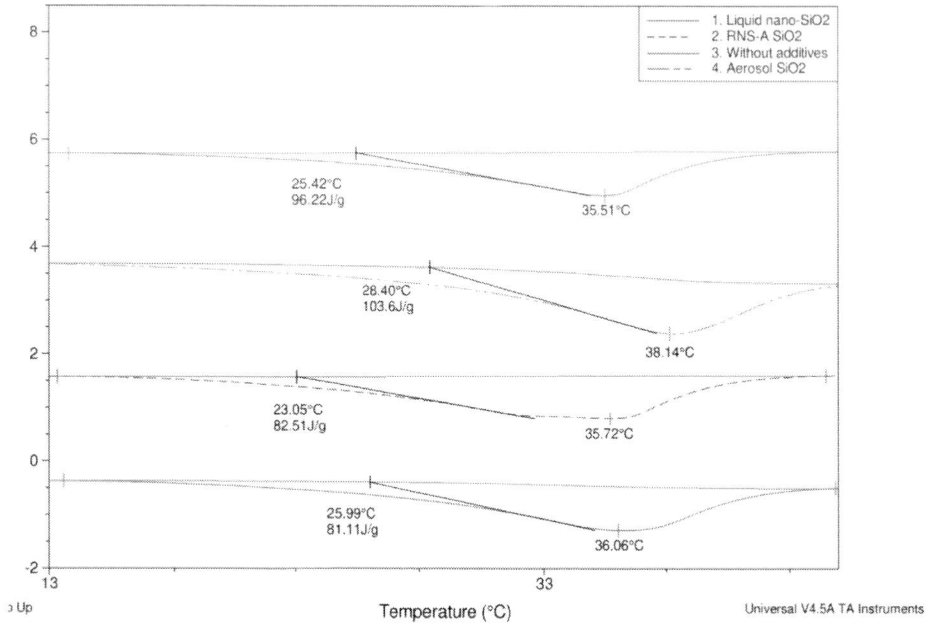

Figure 3. DSC curves of saturated sodium sulfate hydrate during heating process

In figure 3, the phase change temperature of hydrate salts with Aerosol SiO_2, Liquid SiO_2 and RNS-A SiO_2 are 28.40℃, 25.99℃ and 23.05℃ with the latent heat of 103.60J/g, 81.11J/g and 82.51J/g, respectively. The heat charging peak is caused by the melting of eutectic of sodium sulphate decahydrate-water [11]. Compared with curve 3 and connected with the analysis results in figure 2, it can be concluded that with the adding of Aerosol SiO_2, the latent heat of hydrate salts can be improved and the phase change process is stable. But for Liquid SiO_2 and RNS-A SiO_2, even they can increase the heat transfer of hydrate salts, their effects on latent heat of hydrate salts are small, even worse. And the phase change process is unstable.

3.4. Surface properties of the three types of nano-particles

Figure 4 gives the FT-IR spectrum of the three types of nano- SiO_2 particles. Curves 1 to 3 are the FT-IR spectrum of Aerosol SiO_2, Liquid SiO_2 and RNS-A SiO_2, respectively.

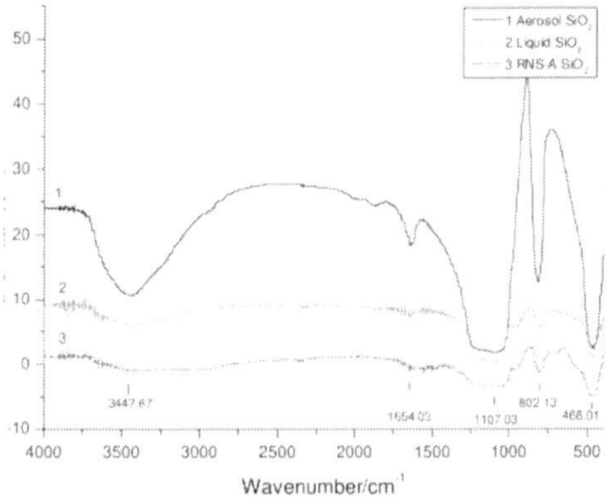

Figure 4. FT-IR spectrum of nano-particles:

1. Aerosol SiO_2, 2. Liquid SiO_2, 3. RNS-A

SiO_2

According to the results above, the characteristic peaks of the three types of nano-particles are in the same sites. The FT-IR spectrum of the nano-additives indicates the existence of three main absorption bonds at 468.01, 802.13 and 1107.03 cm^{-1} which can be assigned to bending vibration, symmetric and asymmetric stretching of SiO_2 [12,13]. The band at 1654.03 cm^{-1} is associated to the bending of the absorbed H_2O molecules [12]. The characteristic band for stretching (OH) groups is found at 3447.67 cm^{-1} [14]. It can conclude that the surface chemical bonds characteristics of the three nano-particles are nearly the same. For Aerosol SiO_2, there are strong characteristic hydroxyl bonds on its surface.

3.5. Effects of nano-additive Aerosol SiO_2 on other saturated hydrate salts

Figure 5 are the t-T curves of (a) $Na_2HPO_4 \cdot 12H_2O$ and (b) $Na_2S_2O_3 \cdot 5H_2O$ with about 2% weight percentage of Aerosol SiO_2. There is no supercooling during their solidifying process. That means the nano-additives of Aerosol SiO_2 is still suitable for some other hydrates to suppress supercooling.

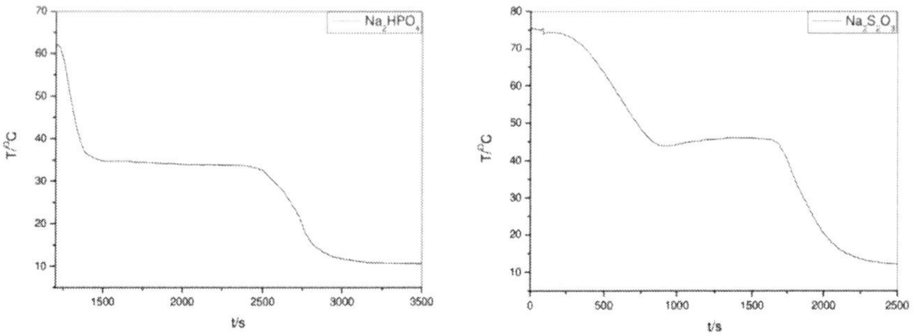

(a) (b)

Figure 5. t-T curves of (a) $Na_2HPO_4 \cdot 12H_2O$ (b) $Na_2S_2O_3 \cdot 5H_2O$ with

Aerosol SiO_2

3.6. A possible mechanism of anti-supercooling in our systems

To conclude, the nucleation processes of hydrate salts with nano-additives belong to heterogeneous nucleation. The improvement of heat transfer and concentration of hydrate salts should not be the most effective parameter for suppressing supercooling. According to the studies of nano-additive effects on organic PCMs [3,4], the interaction between the surfaces of nano-additives and PCMs should be surface dependent. The three nano-additives are all amorphous structures with nearly the same surface chemical bonds. However, Aerosol SiO_2 has higher specific surface area with more and stronger active hydroxyl bonds on the surface. We first presumed that the effect of hydroxyl bonds was the key effective parameter.

To prove the effect of hydroxyl bonds, the amorphous silica gel system was designed. Saturated hydrate salts were confined in it by in-situ polymerization. The changing of its microstructure controlled the situation of hydroxyls in the surface of the gel.

Figure 6 (a) shows the molecular structural characteristics of silica gel with hydrate salts by FT-IR spectrum. The absorption bond at 1099.43 cm^{-1} is stretching groups of Si-O. The band at 1618.22 cm^{-1} is associated to the bending of the absorbed H_2O molecules. At 2103.77cm^{-1}, the absorption bond may be the vibrating groups of Si=O. The characteristic band for stretching (OH) groups is found at 3477.66 cm^{-1}. The characteristic bond is wide. It indicates that free hydroxyls and associated hydroxyls exist at the same time. Employing NMR techniques, Sindorf and Maciel [15,16] have shown that the single hydroxyl group is likely to be the most prolific.

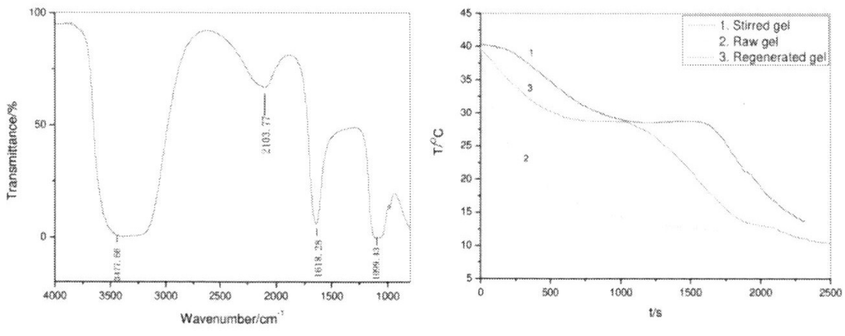

(a) (b)

Figure 6. (a) FT-IR spectrums and (b) t-T curves of silica gel with hydrate salts

In this solid-gel system, the supercooling of hydrate salts could not be suppressed. That could be proved by curve 2 of the t-T curves in figure 6 (b). But after the solid gel stirred homogeneously, it changed into white liquid homogenous suspending solution of gel particles. Then the supercooling of hydrate salts could be suppressed like curve 1 in figure 6 (b). After staying the suspending solution for some time, it changed into solid gel again and it still could suppress supercooling of hydrate salts like curve 1 in figure 6 (b). During the stirring process, a lot of Si-O-Si bonds of silica gel were broken up and changed into lots of single or more silanol groups on the surface of silica gel [15,16]. The re-gelling of silica gel was caused by the physical hydrogen bond absorption instead of lots of Si-O-Si bonds prepared by condensation reaction. So, there were still a lot of free hydroxyl bonds on the surface. All the analysis above indicates that the effect of hydroxyl bonds is the key effective parameter.

From the information of solid surface absorption and surface characteristics of silica gel [17], the absorption relationship between hydroxyl bonds on kinds of SiO_2 surface and salt solution belongs to polar absorption. The surface silanol groups are ionized in aqueous solutions of electrolytes:

$$\equiv Si - OH \Leftrightarrow \equiv Si - O^- + H^+ \tag{1}$$

The metal cations can be absorbed on the surface:

$$\equiv Si-OH+M^{n+}\Leftrightarrow\equiv Si-OM^{(n-1)+}+H^{+} \quad (2)$$

Another possible mechanism of hydroxyl bonds effect can be complex surface-induced hydrolysis:

$$\equiv Si-OH+M^{n+}+H_2O\Leftrightarrow Si-OM^{(n-1)+}OH^{-}+2H^{+} \quad (3)$$

So, the main reason that nano- SiO_2 can suppress supercooling of hydrate salts should be caused by the ion-exchange absorption relationship between strong polar functional -OH bonds and ions of hydrate salts. The magnitude of the free hydroxyl bonds should be enough to provide sufficient congregation of hydrate salts on the interface. For the three nano-additives, Aerosol SiO_2 has higher specific surface area and lots of free hydroxyls on the interface, which are more suitable for the congregation of ions of sodium sulphate. So, only small weight percentage of it can suppress supercooling of hydrate salts successfully.

4. Conclusion

By using three different types of SiO_2 nano-partiles as the anti-supercooling agents, we investigated the surface effect of the interaction between nano-particle and hydrate salts during the phase changing process. Aerosol SiO_2 showed the most effective and stable performance as the nucleators for hydrate salts. Supercooling of hydrate salts can be easily suppressed by the nano-additives, which have high specific surface area and strong polar hydroxyl bonds in the surfaces. Through the ion-exchange absorption between the ions of nano-additivs and hydrate salts, the supersaturation of hydrate salts in the interface of nano-additives can be obviously changed. Then the phase changing properties were improved by increasing phase transition efficiency of hydrate salts. These results might provide a potential low-cost approach for the applications of nano-additives in building energy storage and coolant.

ACKNOWLEDGEMENT: The presented research is partially supported by a research grant from Shenzhen/ Hong Kong Innovation Circle Fund.

References

[1] Lane G A 1992 *Solar energy materials and solar cells* **135-160** 27
[2] Telkes M 1952 *Ind. Eng. Chem.* **1308-1310** 44
[3] Shaikh S, Lafdi K and Hallinan K 2008 *Journal of Applied Physics* **094302-1-6** 103
[4] Elgafy A and Lafdi K 2005 *Carbon* **3067 - 3074** 43
[5] Xavier P Y, Regis O and Sylvain M 2001 *Int. J. Heat Mass Transfer* **2727 - 2737** 44
[6] Mesalhy O, Lafdi K and Elgafy A 2006 *Carbon* **2080-2088** 44
[7] Bugaje I M 1997 *Int. J. Energy Res.* **759-766** 21
[8] Erk H F and Dudukovic M P 1996 *AIChE J.* **791-808** 42
[9] Hasnain M S 1998 *Energy Conversion and Management* **1127 - 1138** 39
[10] Jianfeg W, Ouyang Y and Chen G 2001 *Int. J. Energy Res.* **439 - 447** 25
[11] Ashbee E and Ashbee G H K 1985 *Journal of Materials Science Letters* **249-250** 4
[12] Bruni S, Cariati F, Casu M, Lai A, Musinu A, Piccaluga G and Solinas S 1999 *Nanostruct. Mater.* **573–586** 11
[13] Innocenzi P 2003 *J. Non-Cryst. Solids* **309–319** 316
[14] Moreno M E, Zayat M, Morales P M, Serna J C, Roig A and Levy D 2002 *Langmuir* **4972–4978** 18
[15] Sindorf W D and Maciel E G 1983 *J. AM. Chem. Soc.* **1487-1493** 105
[16] Sindorf W D and Maciel E G 1983 *J. Phys. Chem.* **5516-5521** 87
[17] Gordeeva G L, Gubar V A, Plyasova M L, Malakhov V V and Aristov Y A 2005 *Kinetics and Catalysis* **736-742** 46

Amperometric Immunosensor for α-Fetoprotein Antigen in Human serum Based on Co-immobilizing Dinuclear Copper Complex and Gold Nanoparticle Doped Chitosan Film

Ning Gan, Ling Hua Meng and Feng Wang

The State Key Laboratory Base of Novel Functional Materials and Preparation science, Faculty of Material Science and Chemical Engineering of Ninbo University, Ningbo, 315211, China

ganning@nbu.edu.cn

Abstract. A sensitive amperometric immunosensor for α-fetoprotein (AFP), a tumor marker for the diagnosis of hepatocellular carcinoma (HCC), was constructed , The immunosensor is prepared by co-immobilizing $[Cu_2(phen)_2Cl_2]$ $(\mu\text{-}Cl)_2$ (CuL), nano-Au/Chitosan(Chit) composite, horseradish peroxidase（HRP）and AFP antibody(anti-AFP) on a glassy carbon electrode (GCE). Firstly ,CuL was irreversibly absorb on GCE electrode through π-π stacking interaction; then nano-Au/Chit composite was immobilized onto the electrode because of its excellent membrane-forming ability, finally HRP and anti-AFP was adsorbed onto the surface of the gold nanoparticles to construct GCE | CuL/nanoAu-chit/HRP/anti-AFP immunosensor. The preparation procedure of the electrode was characterized by electrochemical and spectroscopy method.The results showed that this immunosensor exhibited an excellent electrocatalytic response to the reduction of hydrogen peroxide (H_2O_2) without the aid of an electron mediator, offers a high-sensitivity (1710 nA • ng^{-1} • ml^{-1}) for the detection of AFP and has good correlation for detection of AFP in the range of 0.2 to 120.0 ng/ml with a detection limit of 0.05 ng/ml. The biosensor showed high selectivity as well as good stability and reproductivity.

1. Introduction

In recent years, a variety of nanomaterials have been defined as promising candidates for biomedical applications because of the unique characteristics inherent to the nanoscale[1]. High surface area, small size, novel magnetic and optical effects allow nanomaterials to enhance both diagnostic and therapeutic techniques in the coming years [1]. Among these nanomaterials, nanometer-sized gold (NG) particles have been used widely to immobilize protein because of unique physical and chemical features: large specific surface area, good biocompatibility and high surface free energy and so on [2,3].NG particles with small size may allow more freedom in orientation for the protein molecules anchored to maximize the utilization of their bioactive sites .Chitosan(chit) as one kind of biopolymer is a polysaccharide derived by deacetylation of chitin. It possesses many advantages such as excellent membrane-forming ability, high permeability towards water, good adhesion, biocompatibility, nontoxicity, high mechanical strength and susceptibility to chemical modification due to the presence of abundant primary amino groups [4]. It is well known that some functional groups such as mercapto (–SH), and amino ($-NH_2$) groups have a high affinity for Au. Therefore, the nano-Au/Chit composite was prepared via covalent bonds between gold nanoparticles (nano-Au) and the $-NH_2$ groups of the chitosan. This material combined the advantages of inorganic nanoparticles and organic polymer[5].

α-fetoprotein (AFP) is the most widely used tumor marker for the diagnosis of hepatocellular cancer (HCC) [6, 7]. Application of AFP detection to the monitoring of HCC treatment is well accepted in patients with increased AFP levels prior to therapy, and is recommended by the European Association for the Study of the Liver (EASL), the European Group on Tumor Markers (EGTM)[8],

Hepatocellular carcinoma (HCC) has been recognised as the fifth most common cancer and the third most common cause of cancerrelated death worldwide [9]. In the recent years ,there have been as many as 500,000 new cases diagnosed annually. An incidence of 15/100,000 in population per year has been observed [10, 11] in Asia and Africa. More importantly, some patients are often asymptomatic with early HCC [12], resulting in late diagnosis and high death rate . Therefore, the early detection for HCC is highly desirable. AFP is specific for HCC and is widely used in diagnosis of HCC [13]. Compared with conventional immunoassay methods for AFP quantitative determination, such as radioimmunoassay (RIA) [14], immunoradiometric assay (IRMA) [15], and fluoroimmunoassay [16,17]. Electrochemical detection of the label has several advantages such as simple,direct detection techniques,high sensitivity and the low cost of the resulting sensors and instrumentation[5].

H_2O_2 has emerged as an important by-product of enzymatic reactions in the field of biosensing. Thus, through detection and quantification of H_2O_2 could obtain the quantification of immune reaction between antigen and antibody. Many copper complexes with aromatic ligand have been examined as potential catalyst for the reduction of H_2O_2 and can irreversibly absorb on glassy carbon electrode (GCE) through π-π stacking interaction [18]. We synthesized one novel binuclear copper centre complex $[Cu_2(phen)_2Cl_2]$ $(\mu\text{-}Cl)_2$ (CuL) with 1,10-phenanthrolin as ligand , We found that CuL could adsorb on GCE surface firmly to form a stable CuL modified electrode(GCE|CuL), and exhibited electro-catalytic activity toward reduction of H_2O_2. It has been reported that the electron transfer would be blocked when the antigen reacts with its antibody to form an immune complexes. The detection of AFP level was performed by detecting the decrease of amperometric response of the immunosensor to H_2O_2 after the immunoreaction was performed[19].

In this work, we combine the advantages of a co-immobilized CuL and HRP immunosensor to design a simple, fast and direct detection immunoassay method. In order to improve the sensitivity of the immunosensor, the glassy carbon electrode (GCE) was modified with nano-Au/Chit composite to form gold nanoparticles bilayer to increase the amount of the antibody loaded.

2. Experimental section

2.1. Reagents and materials

The ELISA kits were consist of a series of AFP standard solutions with different concentrations of 20,50,100,200 ng /ml, and a solution of horseradish peroxidase conjugated monoclonal anti-AEP antibody. Bovine serum albumin (BSA, 96–99%), chloroauric acid ($HAuCl_4$), sodium citrate(Fluka co ltd,USA). $CuCl_2{\cdot}2H_2O$, 2,2'-Bipyridine, All chemicals and solvents used were of analytical grade. Double distilled water was used throughout all experiments.

2.2. Apparatus

Cyclic voltammetric measurements were carried out with a CHI 660B electrochemistry workstation (Shanghai CH Instruments Co. China). A three-electrode system contained a platinum wire auxiliary electrode, a saturated calomel reference electrode (SCE) and the GCE |CuL /nanoAu-chit/HRP/anti-AFP electrode as working electrode. Easysizer20 laser particle analyzer(OMEC Co. USA), Model X650 Scan electron Microscope (Hitachi corp, JAPAN).

2.3. The synthesis of CuL,nano-Au colloid and nano-Au/Chit composite material

CuL was synthesized according to literature[20] and the nano-Au colloid were prepared according to the literature [21] by adding 2 ml of 1% (w/w) sodium citrate solution into 50 ml of 0.01% (w/w) $HAuCl_4$ boiling solution. The Au colloid was characterized by Easysizer20 laser particle analyzer, the average size of NG particles was about 20 nm. nano-Au/Chit composite were prepared according to the literature[5], Appropriate amount of nano-Au colloid was mixed with 0.1% of chitosan (0.05 mmol/L acetic acid). The mixture was stirred for 2 h, and then it was placed at 4 °C overnight.

2.4. The preparation of GCE |CuL /nano-Au/HRP/anti-AFP immunosensor

GCE was polished on chamois leather with alumina slurry, washed in an ultrasonic bath for 5 min and dried in air. Then it was dipped in 20 mmol/L CuL for 24 h to form GCE | CuL. According to literature [22], dipping 25 μl nano-Au/Chit composite solution onto the surface of the GCE | CuL to form the modified electrode (GCE |CuL /nanoAu-chit) . Thereafter 10 μL 2mg/mL HRP and 10 μl anti-AFP were successfully dripped on the surface of GCE | CuL/nano-Au/chit and dried in air. At last, the electrode was immersed in the 5% BSA solution for 5 h in order to block possible remaining active sites of the electrode, The procedures used for construction of the immunosensor were shown in Fig 1. The experiments was maintained at an atmosphere of N_2.

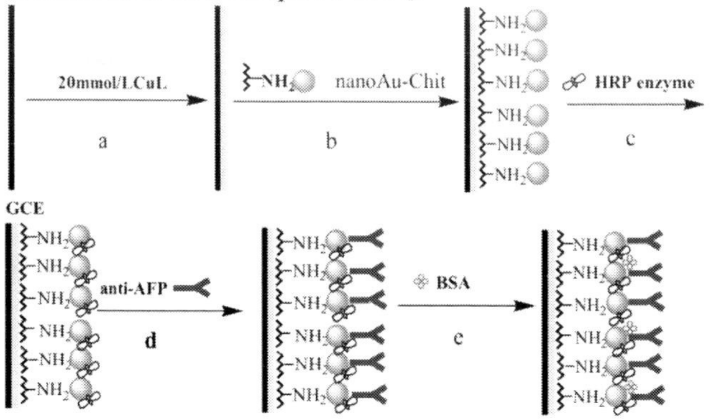

Figure 1. The fabrication process of the immunosensor.

3. Results and Discussion

3.1. Characterization of the immunosensor

Electrochemical impedance spectra (EIS) was employed to monitor the assembly process of the electrode. In EIS, the semicircle part at higher frequencies corresponds to the electron-transfer limited process. Its diameter corresponds to the electron transfer resistance, in the presence of 5.0 mmol/L $Fe(CN)_6^{3-/4-}$ as electrochemical probe. The results were shown in Fig 2. At bare GCE, only a small semicircle could be observed, When CuL,nanoAu/chit,HRP and anti-AFP were successively modified on the electrode surface, the diameter of the semicircle increased until to the maximum, which implyed that the deposited membrane obstructs the electron transfer of the electrochemical probe. The above results could imply the success of the assembly of the electrode after each modified step.

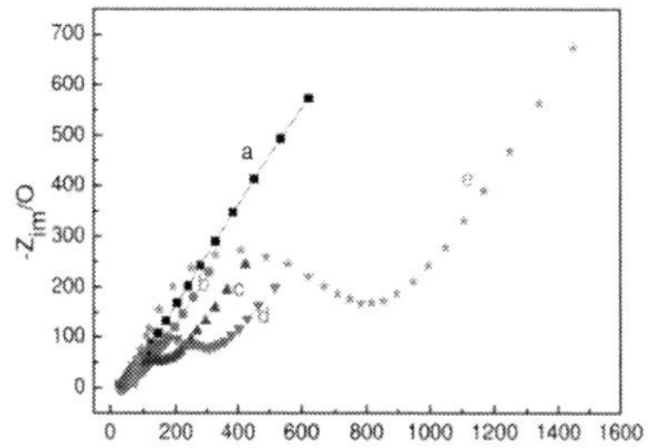

Figure 2. AC impendence spectrum of (a) bare GCE, (b) GCE | CuL, (c) GCE | CuL/nano-Au/chit, (d) GCE | CuL/nano-Au/chit/HRP, (e) GCE | CuL/nano-Au/chit/HRP/anti-AFP, in the presence of 5 mmol/L $Fe(CN)_6^{3-/4-}$ with 0.5 mol/L KCl solution. The frequency range was from 1 to 1×10^5 Hz.

3.2. Electrochemical behavior of different modified electrodes

The electrochemical behavior of different modified electrodes was studied by cyclic voltammetry (CV) and the results were shown in Fig 3. It could be seen that the electrodes at different stages all exhibited good electrochemical activity . In pH 7.0 PBS buffer, the CuL film modified electrode(GCE | CuL) showed a couple of characteristic Cu(II)/Cu(I) redox peaks(E_{pa}^{1}=98mV, E_{pc}^{1}=-220mV) of CuLat a sweep rate of 100mVs^{-1} (Fig.3a). After the GCE | CuL was modified with nanoAu-chit to form GCE | CuL/nano-Au/Chit, the cathodic peak current and the anodic peak current decreased (Fig 3b).It was attributed to the Chit membrane obstruct electron transfer between CuL and electrode. When GCE | CuL/nano-Au/Chit was immersed in 1mmol/L H_2O_2, The cathodic peak increased and anodic peak decreased which showed that the electrode can catalyze the reduction of H_2O_2 (Fig 3c). After HRP and anti-AFP were immobilized on the surface of GCE | CuL/nano-Au/Chit to form the immunosensor, the CV Fig in H_2O_2 showed a slightly decrease of cathodic current responses (Fig 3d). When the immunosensor was incubated in the 50 ng ml^{-1} AFP solution for 30 min, the cathodic peak current decreased obviously (Fig 3e). This is attributed to the formation of anti-AFP/AFP immunocomplex, which acts as the inert electron blocking layer. The reduction magnitude of cathodic peak current was linear to the concentration of AFP. In the scan rate range of 50–250mVs^{-1}, the peak currents were proportional to the scan rate which means the electron transfer of CuL is controlled by its absorption on electrode(Fig 4). The surface coverage (Γ) of CuL in the GCE electrode was calculated as 1.2×10^{-10}mol/cm^2 by the function of $I_p=n^2F^2\upsilon A\Gamma/4RT$ [12] at scan rate of 10mV/s. The value is correspondent with that of mono-molecular layer film ($\Gamma_{mono}=2.34\times10^{-10}$ molcm^{-2})[13]

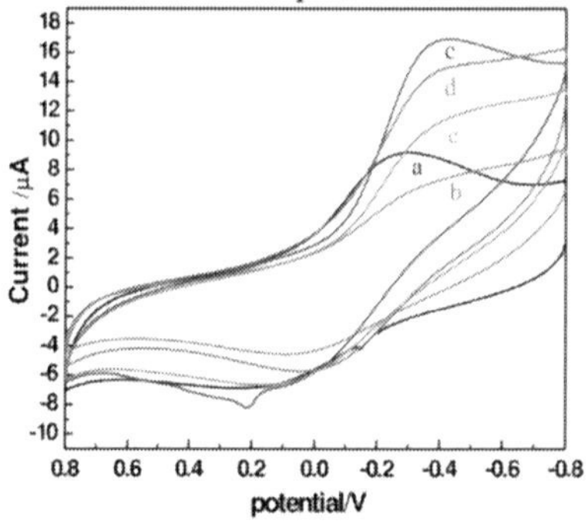

Figure 3. Cyclic voltammograms of the electrode at different stages: (a) GCE | CuL, (b) GCE | CuL/nano-Au/Chit, (c)b+H_2O_2(1mmol/L) (d) GCE | CuL/nanoAu/Chit/HRP/anti-AFP+ H_2O_2(1mmol/L), (e) the electrode of d incubated in the solution containing 50 ng/ml AFP for 30 min+H_2O_2(1mmol/L); scan rate, 100 mV/s.

Cyclic voltammograms of different electrode in PH 7.0 PBS with 1mmol H_2O_2 showed that the peak current responses difference(ΔI) between GCE/CuL and GCE/CuL/anti-AFP was calculated as 6.3μA(Fig 5a,b), when nano-Au was modified on the electrode, the ΔI was 8.6μA(Fig 5c,d).which indictated the nano Au modified on the elctrode adsorbed more AFP antibody because of its large specific surface area, and high surface free energy.

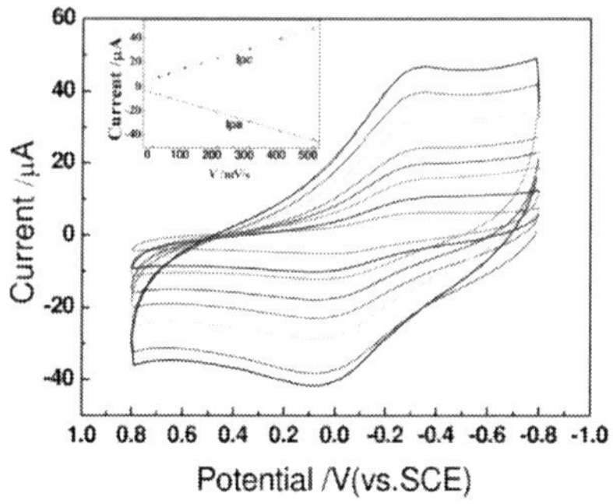

Figure 4. Cyclic voltammograms of immunosensor in pH 7.0 PBS at 10, 50, 100, 150, 200, 300, 400, and 500 mV/s (from lowest to highest peak currents). Inset: Plots of peak current vs. scan rate.

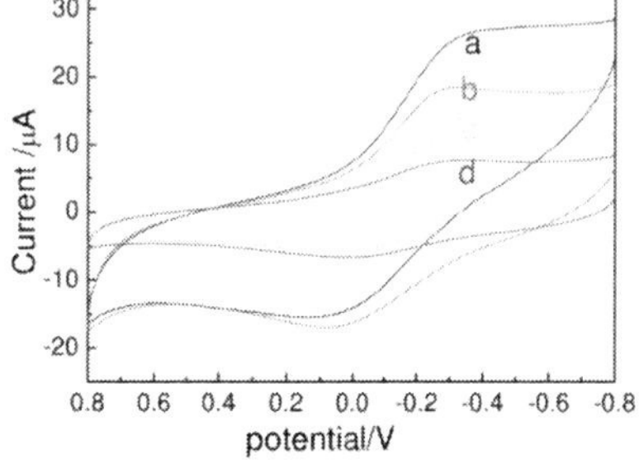

Figure 5. Cyclic voltammograms of different electrode in pH 7.0 PBS with 1mmol H_2O_2: (a)GCE | CuL, (b)GCE | CuL/anti-AFP, (c)GCE | CuL/nano-Au/Chit, (d)GCE | CuL/nano-Au/Chit/HRP-anti-AFP.

3.4. Optimization of measurement variables.

The experimental variables, which can affect the amperometric response of the immunosensor, include the pH, the incubation temperature,the incubation time and the concentration of H_2O_2.

The effect of incubation temperature was studied in the temperature range from 20 to 50 °C (not shown). The current responses increased with increasing temperature up to 35 °C, while temperatures over 40 °C resulted in a deterioration of current response. The results showed that the high temperature caused an irreversible behavior (denaturation of proteins) involved in the process. Thus, the temperature of 35 °C was selected as a compromise. The dependence of the electrochemical behavior of the immunosensor on the pH of the working solution containing 1.0 mmol/L H_2O_2 was studied between 4.5 and 8.5. As shown in Fig 6, the current response had a maximum value at pH 7.0. Hence, all the experiments were carried out in pH 7.0 PBS solution throughout this study. The effect of incubation time on the amperometric responses of the immunosensor was investigated. The decrease value ΔI_0 increased with increasing incubation time and tended to a maximum value at 30 min, and at this time immune response on the electrode surface can completely finish. the amperometric response increased linearly with the H_2O_2 concentration were in ranged from 0.1 to 1.0mmol/L.When the H_2O_2 concentration was higher than 1.0 mmol/L, the ΔI tended to a constant value. Therefore, 1.0 mmol/L H_2O_2 was chosen for the whole experiment.

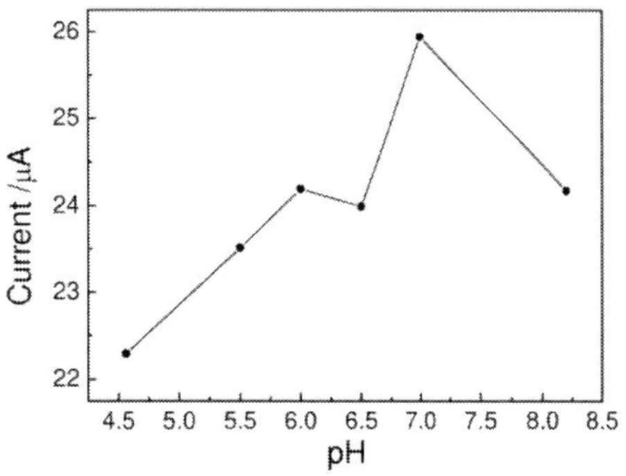

Figure 6. The effect of the immunosensor on the pH of the working solution.

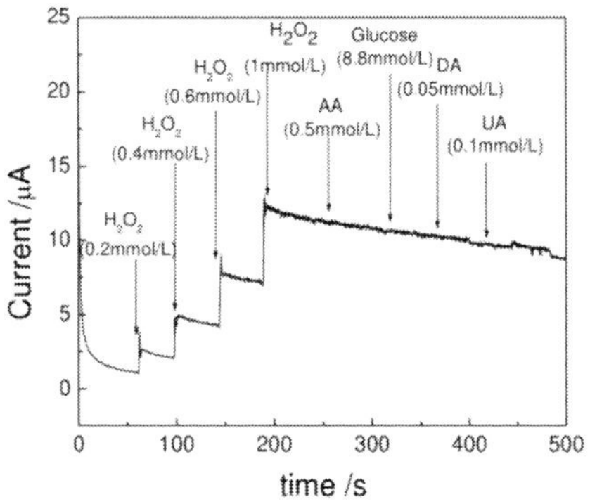

Figure 7. interfering experiment of the immunosensor.

3.5. Selectivity and stability of the immunosensor

In order to test the selectivity of this biosensor, four common interference substances with usual concentration in human body solution (8.8 mmol/L Glucose, 0.1mmol/L uric acid(UA), 0.05 mmol/L dopamine (DA)and 0.5 mmol/L ascorbic acid(AA))were added respectively in the solution containing 1 mmol/L H_2O_2. We found that all signals' change of the sensor were no more than 10% which meaned it could eliminate their interference. The immunosensor also showed good stability. After 2 weeks storage in PBS at pH 7.0 in 4℃, it retained 96% of its initial activity.

4. Conclusions

In this work, a novel amperometric immunosensor for the rapid determination of AFP is prepared by co-immobilizing $[Cu_2(phen)_2Cl_2]$ $(\mu\text{-}Cl)_2$ (CuL) ,nano-Au/Chit composite ,horseradish peroxidase （HRP） and anti-AFP on a glassy carbon electrode (GCE). Based on the catalysis of CuL and HRP to H_2O_2, besides, the nano-Au/Chit composite was used to increase the amount of the antibody loaded, which improve the sensitivity of the immunosensor. after incubation with AFP solution, the immunosensor could directly detects AFP concentrations in the incubation solution without the addition of electron transfer mediator. The immunosensor shows good accuracy, selectivity and storage stability, wider linear range and less assay time. All of these indicated that it was a convenient, economical, and superior method in the clinical analysis.

References

[1] Yuanzhe Piao, Andrew Burns, Jaeyun Kim, Ulrich Wiesner and Taeghwan Hyeon 2008 Adv. *Funct. Mater* **18** 3745–58
[2] Xu S Y and Han X Z 2004 *Biosens Bioelectron* **19** 1117–20
[3] Liu SQ, Leech D and Ju HX 2003 *Anal Lett* **36** 1–17
[4] Muzzarelli RAA. Chitin. Oxford and Pergamon 1977 p. 140–1
[5] Xiulan He, Ruo Yuan , Yaqin Chai and Yintao Shi J 2008 *Biochem. Biophys. Methods* **70** 823–9
[6] L Zhou, J Liu and F Luo 2006 *World J Gastroenterol* **12** 1175
[7] Koteish A, Thuluvath P J and Vasc J 2002 *Interv. Radiol* **13** S185
[8] Qianyun Zhang, XuWang, Zhenjia Li and Jin-Ming Lin 2009 *Analytica Chimica Acta* **631** 212–7
[9] Llovet JM, Burroughs A and Bruix J 2003 *The Lancet* **362** 1907–17
[10] Bosch X, Ribes J, Borras J 1999 *Semin Liver Dis* **19** 271–85
[11] Di Bisceglie AM, Rustgi VK and Hoofnagle JH et al 1988 *AnnIntern Med* **108** 390–401.
[12] Kew MC, Dos Santos HA and Sherlock S. 1971*Br Med* J **4** 408–11
[13] Xu Wang, Qian-Yun Zhang, Zhen-Jia Li, Xi-Tang Ying and Jin-Ming Lin 2008 *Clinica Chimica Acta* **393** 90–4
[14] Christensen RL, Rea MRKG and Crane JP et al 1986 *Clin Chem* **32** 1812–7
[15] Nomura M, Imai M and Takahashi K, et al 1983 *J Immunol Methods* **58** 293–300
[16] Wang WK, Ho LT and Chiang Y 1993 *Clin Chem* **39** 1659–61
[17] Eduardo VN, Eulália CDC and Armando F 2005*Clin Chim Acta* **360** 151–9
[18] Cai C X, Xue K H and Xu X Y et al 1997 *J. Appl. Electrochem* **27** 793-8
[19] Zong Dai, Feng Yanb, Hua Yua, Xiao Hu and Huang xian Ju, 2004*Journal of Immunological Methods* **287** 13– 20
[20] Katz E and Willner I 2003 *Electroanalysis* **15** 913-47
[21] Frens G 1973 *Phys Sci* **241** 20–2
[22] He X L, Yuan R, Chai Y Q and Shi Y T 2008 *J. Biochem. Biophys. Methods* **70** 823–9

A study of the crystal phase separation of mLLDPE/EVA blend

Z H Deng

Institute of Chemical Materials, China Academy of Engineering Physics,

Mianyang 621900, China

E-mail: prodzh@sohu.com

Abstract. The crystal phase separation of mLLDPE/EVA (40/60) (wt/wt) blend was investigated by FTIR micro-spectroscopy in a sequential cooling process. To reveal microcosmic causation and details during the process, the changes of absorbance ratio corresponding to high and low vibrational states, and the apparent enthalpy differences of vibration energy states transformation (ΔH_v) of characteristic groups in pure mLLDPE, pure EVA and mLLDPE/EVA blend were calculated based on FTIR spectra data. By analyzing ΔH_v values of individual characteristic group relating to the exothermal peaks examined in the cooling process, it has been found that methylene groups play an important role in the crystal phase separation. Besides, the energy analysis of the corresponding methyl band at A1369.4/A1367.7cm^{-1} in blend indicated that the chain segments of EVA were involved in the phase separation and crystallization process of mLLDPE taking the form of cocrystallization, which further illuminate the results of DSC analysis. Simultaneously, the worse miscibility of ester and alkyl groups was proved by the absence of enthalpy changes at 110°C.

Keywords: Metallocene linear low density polyethylene; Ethylene/vinyl acetate copolymer; Crystal phase separation

1. Introduction

The research interest in polyolefin blends has lasted for several decades both in academic and industrial scopes [1-3]. Concerning the transition from melting to crystalline state, the compatibility between the components, the liquid-liquid and liquid-solid phase separation, as well as the crystal structure formed by the balanced competition among the kinetic, relaxation factors and the thermodynamic equilibrium not only have considerable influence on the melting process and the rheological properties, but also directly affect the final properties of the products [4-5]. Crystal phase separation has decisive influence on the morphology and properties of polyolefin blends, and the effect of the constitutive changes on the crystal phase separation process could reflect the interactions and the dispersion state among the components. However, owing to the similarity in the structure and properties of the components, the phase separation or phase behavior of many commercial binary polyolefin blends could not be directly discerned in the melt state, e.g., HDPE/LDPE blend. As the crystal kinetic process could provide relevant information on phase separation, differential scanning calorimetry (DSC) analysis has been exercised to investigate the crystal phase separation of polyolefin blends via the measurement of the energy changes in the phase transition process, along with the profile change of exothermic crystallizing peaks [6-8]. However, the microcosmic causation or details concerning crystal phase separation are still unclear by DSC analysis.

In view of the congregation structure formed in non-isothermal crystallization, the macromolecular chain segments in the crystal region of polyolefin blends possess special conformational characteristics and form three-dimensional ordered structure. Owing to the strong intermolecular interactions, the crystalline phase will behave distinctly different physical properties from the melting state. As clarified by large numbers of research reports or monographs [9-13], IR spectroscopy displays prominent advantages concerning the investigation on the changes of characteristic groups during phase transition process, and the effectivity and sensitivity of this method has been widely testified. During the crystallizing process, the infrared spectra of polymers will manifest notable changes in the intensity, frequency, and the profile of absorption peaks, which might be ascribed to two factors: (1) the force constant of interior chemical bonds and the intermolecular interactions of polymer chain segments are changed by heat shrinkage, accordingly resulting in changed characters of the molecular vibration dipole moment; (2) inside the macromolecular chains, the changes of infrared absorbing concentration of characteristic groups would arise as the temperature decreases. Our previous work has proven that the changes of absorbance ratio corresponding to different vibrational states, and that the apparent enthalpy differences of vibration energy states transformation (ΔH_v) of characteristic groups could be analyzed via the collection of FTIR spectra data [14-15]. In this case, particular information about the crystallization process of the blends, accompanying the energy change of corresponding groups could be obtained via FTIR spectral measurement, thus obtaining more detailed information compared to DSC measurement.

In the present study, the morphology of mLLDPE/EVA (40/60) (wt/wt) blend was firstly observed by transmission electron microscopy (TEM), which evidently indicated the formation of individual rich domains. Afterward, by comparison of the infrared spectra of pure metallocene linear low density polyethylene (m-LLDPE), pure ethylene/vinyl acetate copolymer (EVA), and mLLDPE/EVA blend during a cooling process, the phase separation along with the formation of crystal structure in mLLDPE/EVA blend were discussed, with the intention of supplying a new insight into the thermodynamics investigation on crystal phase separation of polyolefin blends. At the same time, the combination of FTIR with DSC analysis was carried out to give more information.

2. Experimental

2.1. Materials

The mLLDPE used was supplied by Exxon Chemical Company as Exceed350D65 grade. EVA was provided by Beijing Organic Chemical Plant (China), with vinyl acetate content of 14%. The characteristics of the two materials are listed in Table 1, in which the molecular parameters of materials, *Mn, Mw, Mw/Mn,* were measured on a PL-210 GPC instrument.

2.2. Preparation of mLLDPE/EVA Blend

The blend was prepared by dissolving the two species (40wt % mLLDPE and 60wt % EVA,

Table 1 Characteristics of materials used

Samples	Density (g/cm^3)	MI (g/10 min)	$\overline{M}_n$	$\overline{M}_w$	$\overline{M}_w/\overline{M}_n$
mLLDPE	0.917	1.0	43529	113030	2.60
EVA	0.935	0.7	23068	95704	4.15

weighed 2g in all) in 200ml of *p*-xylene, refluxing for 2 hours at constant temperature, and by rapidly precipitating into 400ml of methanol at 0 °C. The sample was placed for 12 hours at room temperature, then filtrated and washed with methanol, and dried in vacuum at 50 °C.

2.3. TEM observation

0.04 g of mLLDPE and 0.06 g of EVA were dissolved in 50 ml of *p*-xylene, refluxing for 2 hours at constant temperature. Afterward, the solution was droplet taken using a sucker and dripped onto phosphonic acid surface at 150 °C to form thin film, then cooled to the ambient temperature at a rate of - 2°C /min. The film was put over copper mesh and dried after washed by de-ionized water for many times, then dyed using RuO_4. The morphology of corresponding samples was observed with a Hitachi-800 TEM.

2.4. FTIR micro-spectroscopy

The FTIR measurements were performed in 10kHz rapid-scan at a resolution of $4cm^{-1}$ on a UMA500 microscope with a Bio-Rad FTS6000 FTIR spectrometer equipped with a MCT detector, and 16 scans were co-added in each scanset. The sample as a clear KBr pellet was laid between KBr windows in ICL Basics™ microscope compression stage equipped with the temperature controller in precision better than ±1°C over entire temperature range. The samples was heated from ambient to 150 °C at a rate of 80 °C /min, held at 150 °C for 5 min, then a cooling rate of - 2°C/min was adopted from 150°C to 67°C.

2.5. DSC analysis

The DSC thermograms were measured using a Perkin-Elmer DSC-2C calorimeter. The temperature and heat flow were calibrated before the measurement. All samples were heated from room temperature to 150 °C at a rate of 80 °C /min in nitrogen atmosphere, held at 150 °C for 5 min, then cooled down to 50 °C at a rate of -2 °C /min.

3. Results and discussion

Fig.1 represents the morphology of mLLDPE/EVA (40/60) (wt/wt) blend via TEM observation. As indicated in the micrograph at low magnification, distinct phase separation has occurred, resulting in the formation of two individual rich domains. Obviously, the dark area represents EVA rich domain, whilst the bright area corresponds to mLLDPE rich domain. More interestingly, it can be seen from the micrograph at high magnification that bundle form configuration has formed in the crystallization process. It might be reasonably supposed that the liquid-liquid and liquid-solid phase separation, as well as the formation of crystal structure in the blend are very complicated. As to the formation of the radial sheet crystals, the circumstances concerning the diffusion, migration and regular stacking of the two species of macromolecular chains need more experimentation results to elucidate, which will be discussed in a further study. Besides, it is noteworthy that the sheet crystals penetrate into each other, at the interface, indicating that there is partial compatibility between the two components.

Fig. 2 depicts the FTIR spectra of pure mLLDPE, pure EVA and mLLDPE/EVA blend recorded during the cooling process, from 150 °C to 67 °C at 5 °C intervals. It can be seen that the intensity of most absorption peaks showed enhanced uptrend as the temperature decreased to a certain extent, indicating that the corresponding characteristic groups are correlative to the crystal and/or the conformation in the macromolecular congregation structure. Therefore, whether for the two individual components or mLLDPE/EVA blend, the investigation on the crystallization process could essentially be transferred to a study on the conversion from liquid amorphous state to solid crystalline structure, simultaneously accompanied by the conformation changes of macromolecular chain segments. Apparently this process will lead to incontinuous changes of the vibration mode of correlative groups in FTIR spectra, which might be employed to determine the crystallization temperature. In addition, as mentioned earlier, FTIR micro-spectroscopy possesses the advantages of providing intrinsic and detailed information on a microcosmic molecular level, by way of inspecting the structural changes during the crystallization process.

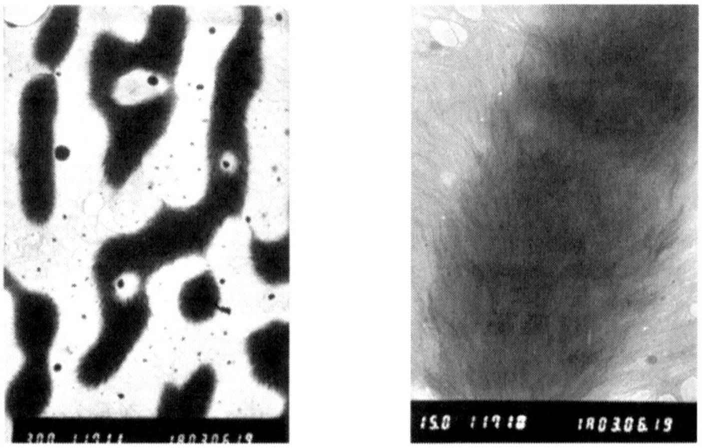

×3000 ×15000

Fig.1. Transmission electron micrograph of mLLDPE/EVA (40/60) (wt/wt) blend.

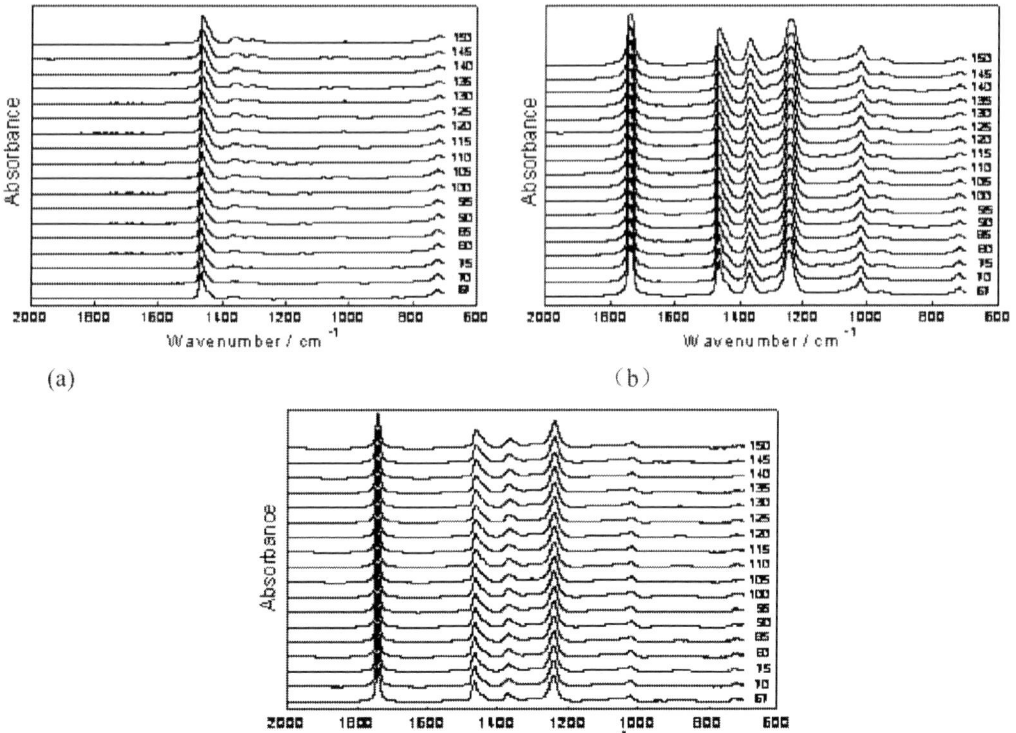

(a) (b)

(C)

Fig.2. Infrared spectra of (a) pure mLLDPE, (b) pure EVA and (c) mLLDPE/EVA blend from 150℃ to 67℃ at 5℃ intervals.

The FTIR spectra of pure mLLDPE, pure EVA and mLLDPE/EVA (40/60) (wt/wt) blend at 67°C is shown in Fig. 3. As to pure mLLDPE, it was found that after the slowly cooling course, the bending distortion vibration band corresponding to methyl and methylene groups

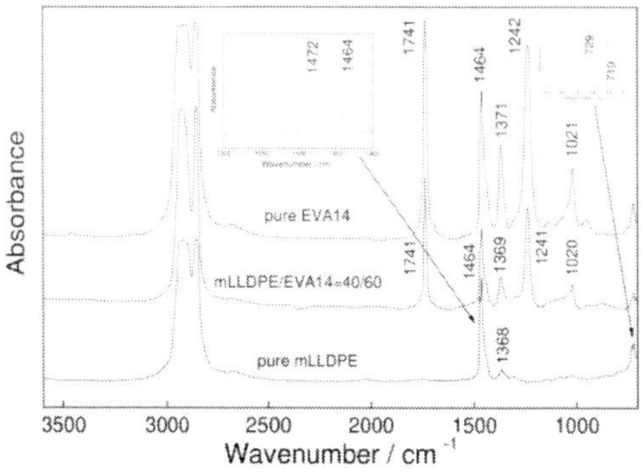

Fig.3. Infrared spectra of pure mLLDPE, pure EVA and mLLDPE/EVA blend at 67℃.

at 1461 cm^{-1} and the rocking vibration band corresponding to methylene around 720 cm^{-1} all became sharper and divided into double peaks, manifesting the comparatively higher crystallinity formed. The appearance of double peaks is due to the function of crystal force field, which induces the coincidence of molecular vibration and crystal lattice vibration. Once there are two or more macromolecular chains get across one crystal lattice, the split of crystal band might happen. However, similar phenomenon was not found in pure EVA or mLLDPE/EVA blend investigated.

If we regard 150°C as the start-point of the energy changes in the crystallization process, and that 67°C as the end-point, it can be seen that for both the blend and the pure polymers the infrared spectra of the characteristic groups investigated all take on gradual changes in peak intensity, peak profile and absorption frequency with the decrease of temperature. This transformation is considered to be an equilibrium achieved between the two energy states of specific characteristic groups in same vibration mode. According to Ref [14-15], we can obtain the apparent enthalpy differences of vibration energy

$$\Delta H_v = -R\partial[\ln(A_H/A_L)]/\partial(1/T) \qquad (1)$$

states transformation ΔH_v using the following equation, namely where A_H and A_L are the peak absorbance corresponding to high and low energy states for one single characteristic group, T represents the temperature. In this way, the enthalpy differences relating to one specific vibration mode of characteristic groups could be acquired via the necessary calculation by determining the peak absorbance ratio at different temperatures, and it provides the possibility to obtain some indications of the contributions made by individual characteristic groups in the crystallization process. Specifically, as to the present study, the crystal phase separation of mLLDPE/EVA blend could be analyzed on a

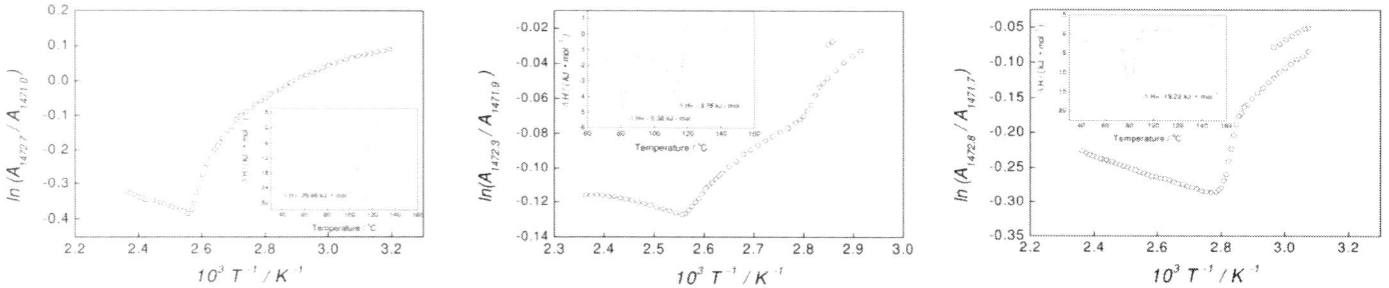

(a) (b) (c)

Fig.4. The typical plots of $\ln(A_L/A_H)$ *vs.* reciprocal temperature represented methylene in (a) mLLDPE, (b) mLLDPE/EVA blend and (c) EVA.

Table 2 Grouped apparent enthalpy of vibration energy states transformation in mLLDPE, mLLDPE/EVA blend, EVA and the corresponding band assignments.

Group	mLLDPE/EVA=40/60 blend			Pure mLLDPE		Pure EVA		IR assignments
	A_L/A_H	ΔH_{80}/(kJ· mol^{-1})	ΔH_{110}/(kJ· mol^{-1})	A_L/A_H	ΔH_{113}/(kJ·mol^{-1})	A_L/A_H	ΔH_{80}/(kJ·mol^{-1})	
-CH_2-	2930.0/2941.4	-2.37	-1.52	2935.9/2939.2	-0.80	2936.2/2941.5	-0.64	ν_{as} (CH_2)
-CH_2-	2857.7/2859.8	-2.06	-0.86	2856.0/2859.4	-0.72	2853.7/2861.6	-1.01	ν_s (CH_2)
-CH_2-	2847.6/2849.0	-1.74	-1.47	2848.4/2851.3	-1.35	2849.1/2849.6	-0.91	ν_s (CH_2)
-CH_2-	1472.3/1471.9	-5.38	-3.78	1472.7/1471.0	-29.70	1472.8/1471.7	-19.22	δ (CH_2)
-CH_2-, -CH_3	1464.0/1462.9	-2.30	-2.77	1463.6/1463.2	-1.04	1464.1/1462.4	-4.58	δ (CH_2)
-CH_3	1369.4/1367.7	-2.71	-1.67	1368.7/1364.6	-4.10	1371.2/1369.4	-1.63	δ_s (CH_3)
-$OOCCH_3$	1741.4/1742.0	-0.62	N	—	—	1741.4/1742.4	-0.39	ν (C=O)
-$OOCCH_3$	1241.1/1239.8	-1.42	N	—	—	1242.4/1239.7	-0.75	ν (-CO-O-)
-$OOCCH_3$	1020.1/1019.6	-0.76	N	—	—	1020.8/1019.3	-0.65	ν (-O-CH-)

molecular level with this method.

The typical plots of $\ln(A_L/A_H)$ *vs.* reciprocal temperature representing methylene bending vibrations in pure mLLDPE, pure EVA and mLLDPE/EVA blend are shown in Fig. 4, subsequently relating ΔH_v values with the temperature. In addition, the grouped apparent enthalpy differences of vibration energy states transformation and band assignments in the samples are listed in Table 2. It can be seen from Fig. 4 that with the given cooling rate, for whether pure mLLDPE and EVA, or mLLDPE/EVA blend, the vibrational absorption peak of the corresponding groups all present sudden changes during the cooling course, showing typical one-stage phase transition in the thermodynamics essence. As indicated from Table 2, the ΔH_v values of characteristic groups all manifest minus values, which is indicative of an exothermal process towards ordered structure. Note that as to EVA, the vibration modes of ester groups only showed sudden changes in the vicinity of 80°C, and the ΔH_v values relating to 1741/1742-cm^{-1}, 1242/1240-cm^{-1} and 1021/1019-cm^{-1} bands are comparatively small, indicating that the vibration modes of methyl and methylene bearing on the main chain play a dominant role during the crystallization process. Especially, the ΔH_v values concerning the bending vibration of methylene near 1472 ± 1/1471 ± 1-cm^{-1} manifest prominent changes in all the samples, showing that methylene is sensitive to the formation of the congregation structure in this process.

Fig. 5 showed the DSC thermograms of pure mLLDPE, pure EVA and mLLDPE/EVA (40/60) (wt/wt) blend in the cooling course. It can be seen that there are two evident exothermic peaks on the curve of mLLDPE/EVA blend, corresponding to the apex values of 80.3°C and 110.3°C, respectively. However, under the same cooling procedure, there is no enthalpy changes around 80°C for pure mLLDPE, as well as the absence of enthalpy changes around 110°C for pure EVA. The corresponding apex temperature of enthalpy changes for pure mLLDPE and EVA occurs at 111.1°C and 78.6°C, respectively. The above results indicate that with a slow cooling rate, liquid-liquid and liquid-solid phase separation arise in the blend, which act as a driving power for the macromolecular chain segments of mLLDPE and EVA to move towards their individual rich domains in the crystallization process, subsequently form the crystal structure via the regular arrangement of chain segments. Accordingly, there are two crystallization peaks come into being in the blend as the temperature decreasing, in which the one at low temperature relates to EVA-rich domain, whereas the other corresponds to mLLDPE-rich domain.

According to Kohji Tashiro *et al* [16], the crystallization temperature and peak profile of polymer blends might be influenced by the lamellar size and the interaction at the boundary between the two lamellae of the different species. In addition, the lamellae of one species might be surrounded not only

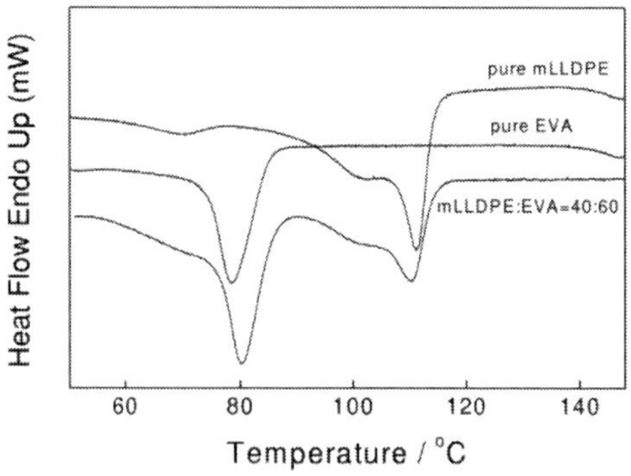

Fig.5. DSC curves of mLLDPE, EVA and mLLDPE/EVA blend in the cooling course.

by the amorphous phase of itself but also by the amorphous chains of the other species, for these reasons the crystallization behaviors of each component in the blend system were influenced. As seen in Fig. 5, owing to the insertion of EVA chain segments, the apex temperature relating to mLLDPE-rich domain in the blend was lower than that of pure mLLDPE; likewise, the peak temperature relating to EVA-rich domain in the blend is a little higher than that of pure EVA, owing to the insertion of mLLDPE chain segments. That is to say, although the two characteristic peaks are entirely separated, the corresponding apex values come nearer to each other as compared to their individual species, implying that there is a small quantity of cocrystallization in the blend. Accordingly, the changes of intermolecular interactions could be induced by the interlacement or entanglement among the chain segments of two different species, which might be reflected in FTIR spectra.

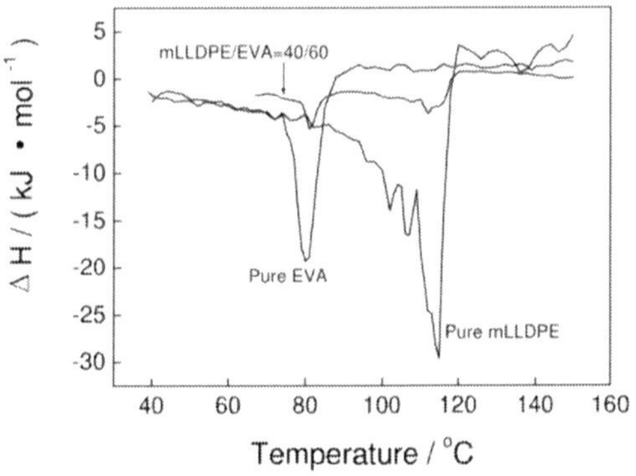

Fig.6. ΔH_v curves of mLLDPE, EVA and mLLDPE/EVA blend represented methylene bending vibration.

Fig.6 depicts the apparent enthalpy differences of vibration energy states transformation corresponding to the bending vibration of methylene in pure mLLDPE, pure EVA and mLLDPE/EVA blend. It can be seen that Fig. 5 and Fig. 6 present much resemble form, manifesting that thermal analysis using FTIR micro-spectroscopy could receive consistent results with DSC measurement. Moreover, it is noteworthy that as indicated in Fig. 6, both pure mLLDPE and EVA manifest much larger ΔH_v values compared to the blend investigated, showing that in the cooling course, larger changes in absorbance ratio arose in the two pristine polymers than those in mLLDPE/EVA blend. This result further proves the existence of certain compatibility between mLLDPE and EVA in the blend system, which is consistent with the TEM observation mentioned above.

4. Conclusions

TEM observation indicated the occurrence of phase separation during the crystallization process of mLLDPE/EVA (40/60) (wt/wt) blend, and there was compatibility between the two components. In comparison of the ΔH_v values of various characteristic groups calculated from FTIR spectra data, it was found that for mLLDPE/EVA blend, the bending vibration mode of methylene made prominent contribution to the liquid-liquid and liquid-solid phase separation, as well as the following migration and regular arrangement of mLLDPE and EVA chain segments to form crystal structure in the cooling course. The cocrystallization of mLLDPE and EVA arose during the phase separation process.

References

[1] Krishnamoorti R, Graessley WW, Dee GT, Walsh DJ, Fetters LJ, Lohse DJ. Macromolecules 1996; 29(1): 367-376.

[2] Economou IC. Macromolecules 2000; 33(13): 4954-4960.

[3] Wignall GD, Alamo RG, Ritchson EJ, Mandelkern L, Schwahn D. Macromolecules 2001; 34(23): 8160-8165.

[4] Li YY, Liu CY, Sun X, He JS. Acta Polymerica Sinica 2000; 6: 707-710.

[5] Na B, Zhang Q, Wang Y, Du RN, Fu Q. Polymer 2003; 44:5737-5747.

[6] Morgan RL, Hill MJ, Barham PJ. Polymer 1999; 40 (2): 337-348.

[7] Marcilla A, Sempere FJ, Reyes-Labarta JA. Polymer 2004; 45: 4977- 4985.

[8] Marcilla A, Reyes-Labarta JA, Sempere FJ. Polymer 2001; 42: 5343- 5350.

[9] O'Reilly JM and Mosher RA. Macromolecules 1981; 14: 602-608.

[10] Coleman MM, Lee KH, Skrovanek DJ, Painter PC. Macromolecules 1986; 19: 2149-2157.

[11] Wu G, Yoshida T, Cuculo JA. Polymer 1998; 39 (25): 6473-6482.

[12] Koenig JL, Spectroscopy of polymer, Washington DC: American Chemical Society, 1992.

[13]Schrader B. Infrared and raman spectroscopy: methods and applications, New York: VCH, 1995. pp: 684-687.

[14] Wu Q, Li CX, Li XC, Li YH. Chemical Journal of Chinese Universities (Chinese) 2001; 22 (12): 2123-2125.

[15] Wu T, Li Y, Wu Q, Wu G, Tan HM. J Appl Polym Sci 2005; 96(2): 261-267.

[16] Tashiro K, Stein RS, Hsu SL. Macromolecules 1992; 25:1801-1808.

[17] Xue Q. The spectroscopy method in the study of macromolecular structure, Beijing: Higher education press, 1995.

[18] Su ZQ, Zhao Y, Xu YZ, Zhang XQ, Zhu SN, Wang DJ, Han CC, Xu DF. Polymer 2004; 45: 3577-3581

Effect of biphasic calcium phosphate nanocomposite on healing of surgically created alveolar bone defects in beagle dogs

Lanlei Wang[1], Aizhong Guan[2], Han Shi[1], Yangxi Chen[1] and Yunmao Liao[1]
[1] State Key Laboratory of Oral Diseases, Sichuan University, Chengdu, Sichuan Province,610041,China
[2]Linyi People's Hospital, Linyi, Shandong, 276000, China
Correspondent author: Yangxi Chen, E-mail: 406975817@163.com

Abstract. The aim of the present study was to investigate the effect of porous biphasic calcium phosphate nanocomposite (nanoBCP) scaffolds bioceramic. Alveolar bone defects were surgically created bilaterally at the buccal aspects of the upper second premolar in fourteen beagle dogs. After root conditioning with ethylenediaminetetraacetate (EDTA), nanoBCP was randomly filled in the defects and nothing was put into the contralaterals as controls. Dogs were killed at the 12th weeks. Histological observations were processed through a light microscopy. The results revealed that a great amount of functional periodontal fissures formed in the defects in the nanoBCP groups while minimal bone took shape in the controls. In this study, nanoBCP has proved to work well as a biocompatible and osteoconductive scaffold material to promote periodontal regeneration effectively.

1. Introduction

Regenerative periodontal therapy aims at the restitution of periodontal tissues that have been lost following inflammatory periodontal disease. This type of healing should result in regeneration of cementum, a functionally oriented periodontal ligament, and the alveolar bone [1, 2]. At present, the hotspot in the area of periodontal regeneration focuses on the biomaterial scaffold graft on alveolar bone defects to provide adequate space for periodontal regeneration. Both limited availability of autografts and potentiality of infection of allografts produce an increasing demand on synthetic bone grafts. Among various artificial bone materials, calcium phosphate bioceramic is similar to the animal bone in chemicophysics compositions, which is thus noticeable [3]. Due to a good many advantages including its excellent bioactivities, biocompatibility, osteophony and osteoinductivity, NanoBCP is expected to be an ideal tissue engineering scaffold material [4-6].

2. Material and method

2.1 Biomaterial preparation

In our study, the nanoBCP bioceramic was supplied by the Material Lab of West China College of Stomatology Sichuan University (Figure.1a). The proportion of hydroxyapatite(HA) / β -tricalcium phosphate is 4/6 with 40% NanoHA.The 3-dimentional images from μCT (μCT80, Scanco Medical AG, Basserdorf, Swiss) scan and reconstruction revealed the porosity of the material was 85% and the macropores were interconnected. The diameter of macropore ranged from 116 to 515 μm (Figure. 1b, c). The observation of nanoBCP by SEM revealed that the distribution of macropores was uniform and micropores was distributed widely in the wall of macropores. The size of micropores ranged from 100 nm to 10μm.

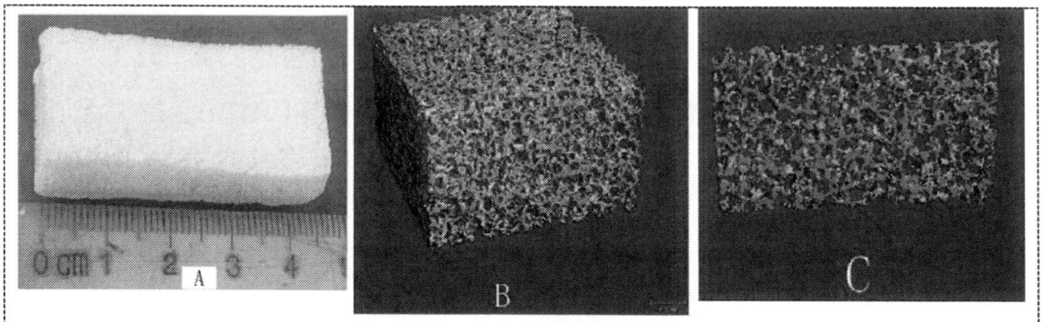

Figure 1. The structural profile of the improved nanoBCP. (a) The aspect of nanoBCP sample; (b) the 3D reconstruction image of nanoBCP by μCT, Bar scales: 1 mm; (c) the reconstruction transect image of nanoBCP by μCT, Bar scales: 1 mm.

2.2 Animal preparation

Fourteen healthy male beagle dogs of 1.5 years old with the weight of 9.4± 1.5kg were selected in this experiment. They were offered by the Experimental Animal Center of Sichuan University (Chengdu, China). The experimental protocol was approved by the Animal Ethical Committee for Animal Research of Sichuan University (Chengdu, China).

2.3 Surgical procedure

The dogs were anesthetized with an intravenous injection of 3% pentobarbital sodium (Shanghai, China, 30 mg/kg). Following mesial vertical incisions, a full-thickness mucoperiosteal flap was raised to expose the buccal mesial aspect of maxillary second premolars, where a rectangular piece of the bone, about 4 mm long, 3 mm wide and deep into the root surface was removed with an osteotome. Reference notches indicating the bottom of the defects were made on the respective root surfaces with a high-speed handpiece under saline irrigation (Figure.2a). Periodontal ligaments were scaled thoroughly. Root surfaces in defects had been conditioned with EDTA for two minutes. The surgical sites were rinsed thoroughly with sterile saline. NanoBCP was trimed into appropriate patches or particles and immersed into sterile saline. Then it was randomly filled in the defects and was impacted lightly to eliminate any gap or void that might appear at the interfaces (Figure.2b). The contralaterals were filled with void as controls. Finally, the flap was repositioned and sutured. Sutures were removed 2 weeks later. During the entire experiment chemical plaque control (0.2% chlorhexidine rinsing twice a week) was performed. The beagle dogs were fed with soft-food diet and water thoughout the experiment.

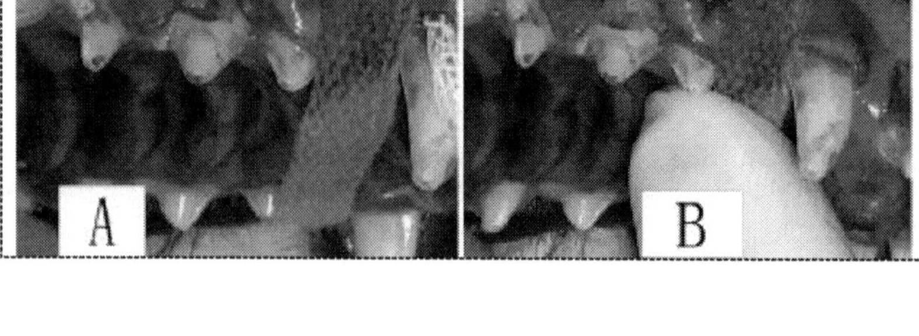

Figure 2. (a) The alveolar bone block about 3mm wide, 4mm high and deep into the root surface was removed from the buccal mesial aspect of the upper second premolars. The horizontal notch was made on the bottom of the root surface in bone defects; (b) nanoBCP was filled in the right bone defect.

2.4 Specimen obtainment

Fourteen dogs were euthanized with an overdose of 3% pentobarbital sodium at the 12th week postoperatively. The experimental teeth, together with their surrounding hard and soft tissues,were harvested and trimmed to the appropriate size. The specimens had been fixed in 70% ethanol for 3 days, dehydrated with increasing concentrations of ethanol,and then embedded in methylmethacrylate without decalcification. Buccal-lingual sections were cut parallel to long axis of the teeth with a microtome (Leica SM 2500E, Germany) set at 5 μm. The sections were stained with hematoxylin-eosin (H&E) staining and observed under light microscope. The images were captured by a digital camera connected with the light microscope.

3.Experimental results

3.1 Clinical observations

All surgeries went smoothly and the clinical incision healing was uneventful in all the animals. No complications, such as allergic reactions to material, abscesses and exposure of nanoBCP, occurred throughout the experimental period.

3.2 Histological obverations

NanoBCP group: alveolar bone defects had about 60-86% bone-filling. The new bone possessed a normal structure. There was no significant difference between the new alveolar bone and the host one near the notch, which showed nanoBCP has excellent biocompatibility. The cementum mixed with inserted collagen fibers formed on the root surface of the defect. The width of new PDL was normal and uniform. One end of the newly formed periodontal ligament (PDL) fiber was inserted into the newborn cementum, the other end was embedded in the new adjacent alveolar bone. Thus, a new functional periodontal supporting tissue formed (Figure. 3 a,b).

In the control group: alveolar bone defects had about 10-26% bone-filling, which just covered the notch partly or fully at most. The limited new cementum deposited someplace on the local root surface of the notch, which had no collagen fibers embedded in (Figure. 3 c,d). The newly formed fiber bundles were thin and little, which were parallel to the root surface. It means the healing was still a periodontal repair via a long junctional epithelium instead of a new attachment.

4.Discussion

There are mainly two kinds of materials used as the tissue engineering scaffold at present: organic polymer material and inorganic material. The polymer materials, such as collagen, fibrin, polylactic acid, chitosan and alginate, have been widely applied as scaffolds [7, 8]. Although they meet the requirements of bone tissue engineering in many ways, they still have many defects, including poor mechanical properties, acid products of degradation in vivo and aseptic inflammation, to affect growth and reproduction of cells and tissues. Among inorganic materials bioceramic has a positive effect as the tissue engineering scaffold material. However, its high brittleness and slow degradation in vivo affect both bone ingrowth and followed reconstruction. As a result, its application has been restricted in some fields.

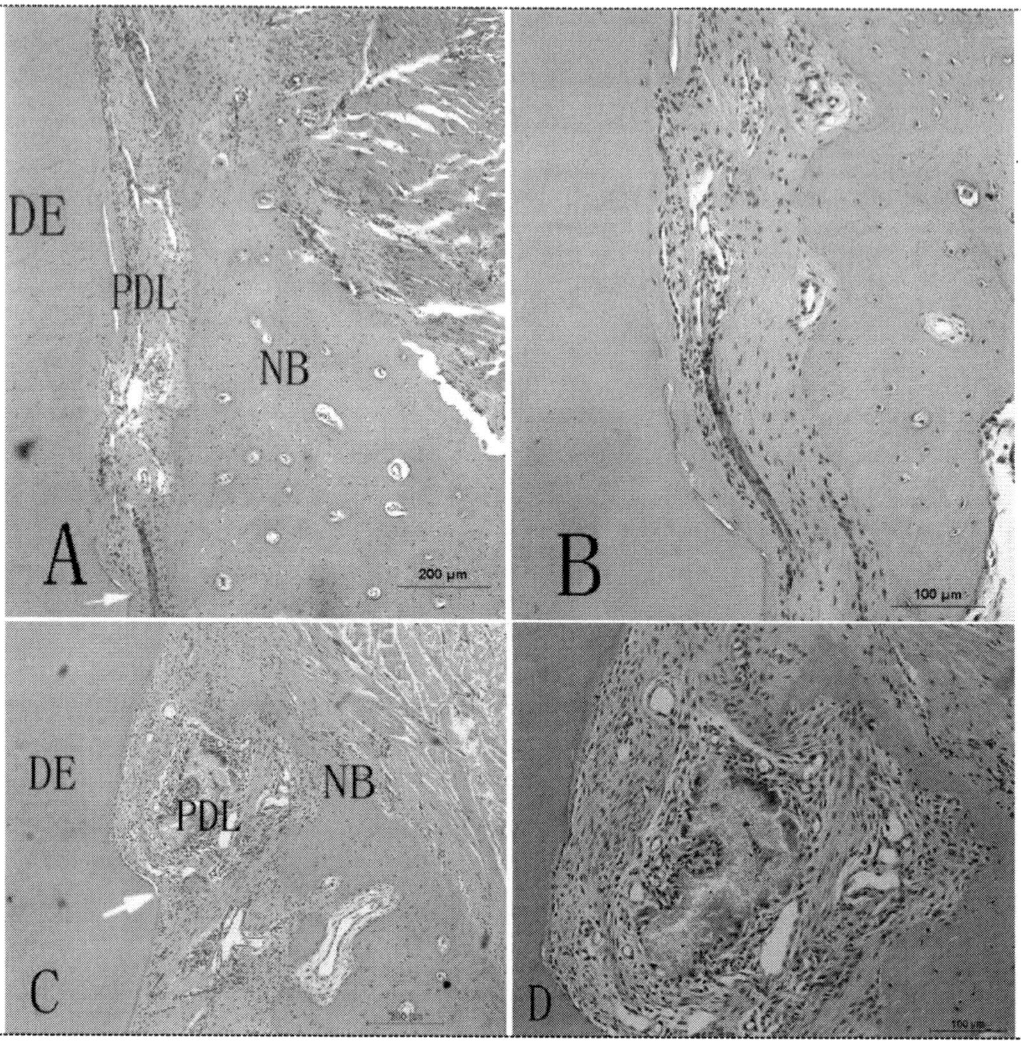

Figure 3. Photomicrograph of healing. (Yellow arrow pointed to the notch, NB new bone, DE dentin, PDL periodontal ligament) (**a**) nanoBCP group: healing resulted in abundant periodontal regeneration and there was no apparent difference between the new bone and the normal one（×100）; (**b**) magnified view of(a), (×200); (**c**) Control group: the healing was characterized by formation of a long junctional epithelium along the root surface all the way down to the notch. There were limited new alveolar bones formed in the defect (×100); (**d**) Magnified view of (c) (×200).

In this study, nanoBCP is a kind of new scaffold which has a porous and 3D structure. The porosity of the scaffold was 85% and the pore size was 116-515 μm; the proportion of hydroxyapatite / β -tricalcium phosphate was 4/6 with 40% NanoHA. Because of interconnected macropores and numerous micropores on macropore walls, the large surface area of the material is beneficial to acceleration of interface reaction. The good interconnected pore structure provides larger space for proliferation and attachment of osteoblastic cells and favors both permeation of nutrition composition and formation of vascularization[9]. Due to a good many advantages including its excellent bioactivities, biocompatibility, osteophony and osteoinductivity, nanoBCP is expected to be an ideal tissue engineering scaffold material[10,11]. With the small-size effect and the surface effect, nano-ceramic material shows different characteristics from traditional ceramic. Increase of its specific surface area enhances its guidance ability of histiocytes and its biodegradation. Research has shown that microstructure and sub-microstructure on the surface of tissue engineering scaffold have a great influence on cell adhesion and growth.

In this study,histological analysis reveals abundant and homogeneous new alveolar bones formed in the defects of nanoBCP group while little in the controls. The findings of the experiment provide evidence that treatment of acute intrabony defects with nanoBCP can enhance periodontal regeneration. It shows that nanoBCP as a promising bone grafting might provide a favorable choice for periodontal therapy.

References

[1] Caton JG, Greenstein GG 1993 *Periodontol 2000.* 9–15
[2] Brunsvold MA, Mellonig JT 1993 *Periodontol 2000.* 80-91
[3] Shirtliff VJ, Hench LL 2003 *J Mater Sci.* 4697–4707
[4] Racquel Zapanta LeGeros, PhD. 2002 *Calcium Phosphates, Clinical orthopaedics and related research.* 81–98
[5] Damien Le Nihouannen, Afchine Saffarzadeh,Olivier Gauthier 2008 *J Mater Sci:Mater Med.* 667–675
[6] Balasundaram G, Sato M, Webster TJ 2006 *Biomaterials.* 2798-2805
[7] Meyer U, Büchter A, Hohoff A, et al. 2005 *Int J Oral Maxillofac Implants.* 882-890
[8] Ren T, Ren J, Jia X, et al. 2005 *J Biomed Mater Res A.* 562-569
[9] Julia Will, Reinhold Melcher, Cornelia Treul, Nahum Travitzky 2008 *J Mater Sci: Mater Med.* 2781–2790
[10] Hassna R. R. Ramay and M. Zhang 2004 *Biomaterials.* 5171-5180

A green access to highly pure single-walled carbon nanotubes by taurocholate-assistant dispersion and centrifugation

Yongfu Lian
Key Laboratory of Functional Inorganic material Chemistry (Heilongjiang University), Ministry of education, Harbin, 150080, China

E-mail: chyflian@hlju.edu.cn

Abstract. Raw single-walled carbon nanotubes produced by arc discharge were oxidized in the air to eliminate amorphous carbon, and then dispersed in the aqueous solution of sodium taurocholate supersonically. Thus obtained stable dispersion was subjected to centrifugation, and the metal catalysts and varying carbon impurities were separated with carbon nanotubes. The efficiency of the above procedure was confirmed by scanning electron microscope observation, thermogravimetry, and optical absorption and Raman spectroscopic analyses. The advantage of this procedure lies in easiness, high purity, and no pollution to environment.

1. Introduction

Since discovered by Iijima in 1991 [1], carbon nanotubes have been investigated actively and widely. Because of their unique one-dimension structure, high surface area, electrical conductivity, and extremely high strength, single-walled carbon nanotubes (SWNTs) are proposed to have many promising applications in nanoelectronics, nanoprobes, nanostructural composites, field-emission displays, chemical sensors, etc. [2,3] To realize these potentials and to study their intrinsic properties, high quality SWNTs are required. Unfortunately, the available methods for large-scale production of SWNTs generate significant amounts of carbonaceous impurities, such as amorphous carbon, fullerene, turbostratic graphite, and polyhedral carbon nanoparticles, along with transition metals that were introduced as catalysts in the growth of nanotubes [4,5]. Normally, amorphous carbon is removed by acidic treatment of starting raw material or burning them in air under a relatively high temperature [6,7]. The oxidation of SWNTs has been studied previously [8-13], and strong oxidizing agents such as $KMnO_4$, HNO_3, and O_3 are applied. These chemicals could damage SWNTs, offer little selectivity to remove metal catalysts, and be environment unfriendly.

In this paper, raw SWNTs prepared by arc discharge are purified by a biological reagent assisted dispersion and then subjecting to centrifugation. The high purity of thus treated SWNTs is checked by scanning electron microscope observation, thermogravimetry, and optical absorption and Raman spectroscopic analyses. In sharp contrast to those purification processes reported previously, a simple and efficient method is developed, in which application of strong acid or strong base and toxic solvents such as THF, DMF are avoided.

2. Experimental

Raw SWNTs are prepared by arc-discharge using Tb/Ni powers as catalysts, and FeS as promoter. In a typical experiment, 100 mg raw SWNTs are first oxidized in the air at 400℃ for 4 hours to remove amorphous, and the remaining material is then dispersed in an aqueous solution of 1% (wt%) sodium taurocholate [14,15] by ultrasonic agitation. Thus obtained stable dispersion is subjected to centrifugation for 3 hours at 15000 rpm. The upper suspension is decanted and filtrated with 1μm poly-(tetrafluoroethylene) (PTFE) membrane. After drying for1 h at 100 °C in a vacuum, the solid sample is heated at 200 °C for 20 min in dry air to remove sodium taurocholate.

Scanning electron microscope (SEM) observation was carried out with a JEOL JSM-6700FT field emission electron microscope (accelerating voltage: 5.0 kV; beam current: 10 ÌA); Thermogravimetric analysis (TGA) data were recorded with a SII Seiko Instruments EXSTAR 6000 thermogravimetric analyzer; Raman spectra were measured with a Jasco NRS-2100 spectrophotometer using laser excitation at 457.9 nm; The solution-phase optical absorption data were recorded with a Shimadzu UV-3150 spectrophotometer using a quartz cell with a path length of 10 mm.

3. Results and discussion

From figure 1(left), the SEM image of raw SWNTs, it can be observed that some amorphous carbon is on the surface of tube bundles, and SWNTs entangle with carbon and metal catalyst particles. Figure 1(right) shows the SEM image of purified SWNTs, in which high-density SWNTs are observed with few impurities. It is obvious that the purified SWNTs show cleaner surfaces and larger bundles than those observed in figure 1(left).

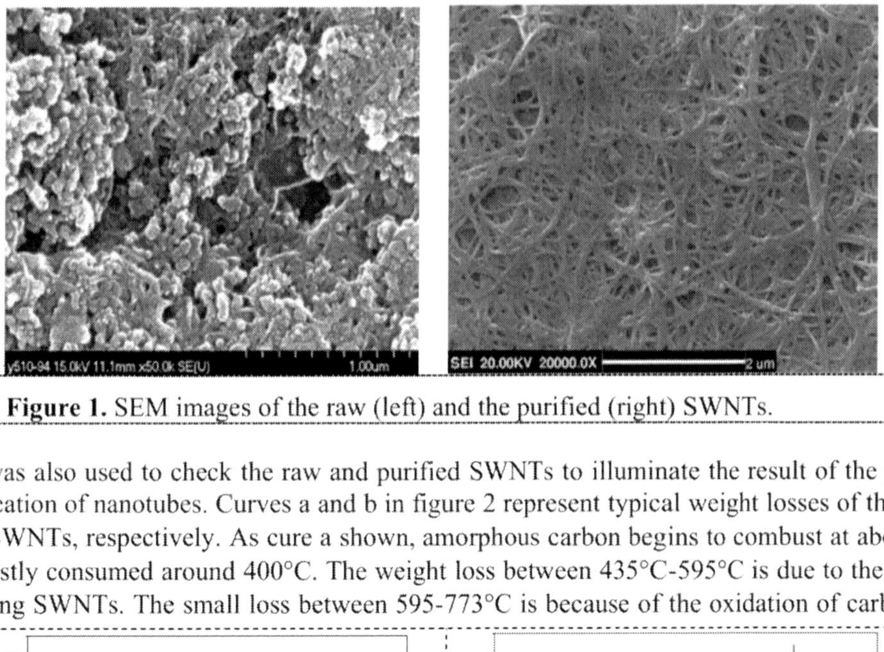

Figure 1. SEM images of the raw (left) and the purified (right) SWNTs.

TGA was also used to check the raw and purified SWNTs to illuminate the result of the procedure for purification of nanotubes. Curves a and b in figure 2 represent typical weight losses of the raw and purified SWNTs, respectively. As cure a shown, amorphous carbon begins to combust at about 350°C and is mostly consumed around 400°C. The weight loss between 435°C-595°C is due to the oxidation of surviving SWNTs. The small loss between 595-773°C is because of the oxidation of carbonaceous

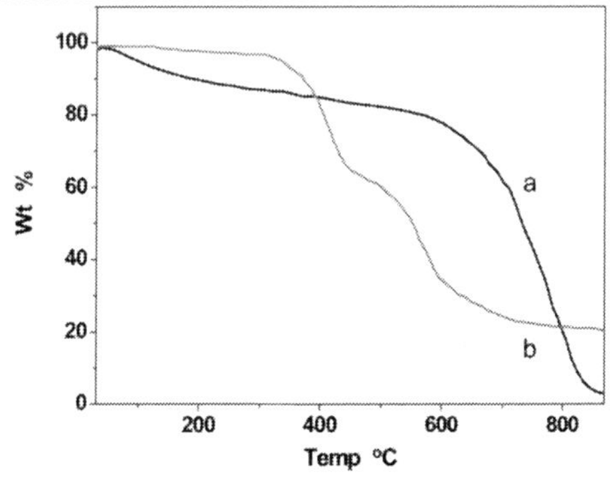

Figure 2. TGA curves of the raw (a) and the purified (b) SWNTs.

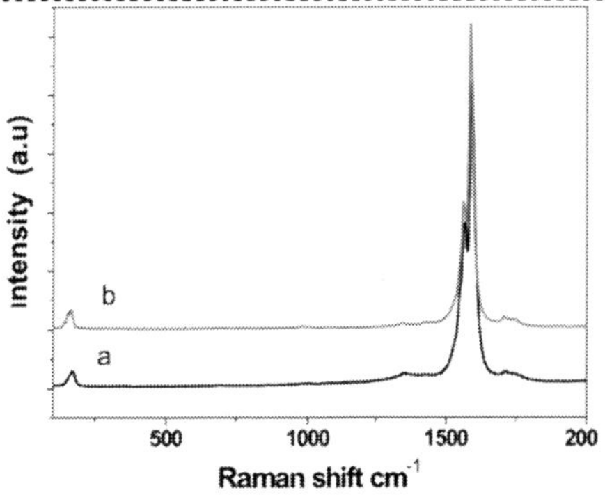

Figure 3. Raman spectra of the raw (a) and the purified (b) SWNTs.

nanoparticles that have higher thermal stability than SWNTs. The prominent weight loss observed for curve b starts at 600°C and ends at 810°C, which could be ascribed to the burning of SWNTs. It is obvious that most of the carbonaceous impurities have been removed. On the other hand, the residual metal oxides after burning of carbonaceous materials decrease from 23% in the raw SWNTs to 3% in the purified ones. Furthermore, the combustion temperature of SWNTs increases from 435-595℃ to 600-810 ℃, which is mainly owning to the removal of metal catalysts. Thus, it is concluded that the purification procedure is efficient to remove both carbonaceous impurities and metal catalysts.

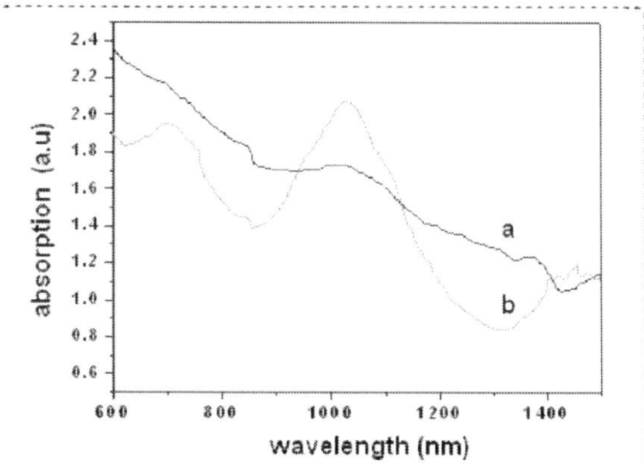

Figure 4. UV-vis-NIR spectra the raw (a) and the purified (b) SWNTs dispersed in an aqueous solution of 1% (wt%) sodium taurocholate.

It is suggested that the area ratios of the Raman peaks (G/D) are proportional to the in-plane crystallite size and inversely proportional to the quantity of "unorganized" carbon in graphitic materials [16,17]. Figure 3 shows the Raman spectra of the raw (a) and the purified (b) SWNTs, respectively. The integrated ratio of the Raman peaks (G/D) increase from 21 for the raw SWNTs to 50 for the purified ones. Moreover, it is also noted that the RBM mode of the purified SWNTs has a higher intensity than that of the raw materials. Thus, the purification procedure leads to a dramatic improvement in the purity of SWNTs. It is worth mentioning that the RBM mode has an obvious blue shift for the purified SWNTs in comparison with that of the raw materials. Such shift is most probably due to the increase in bundle's size, implying that the purification process generates SWNTs with good crystallization. This result is consistent with that observed in figure 1(right).

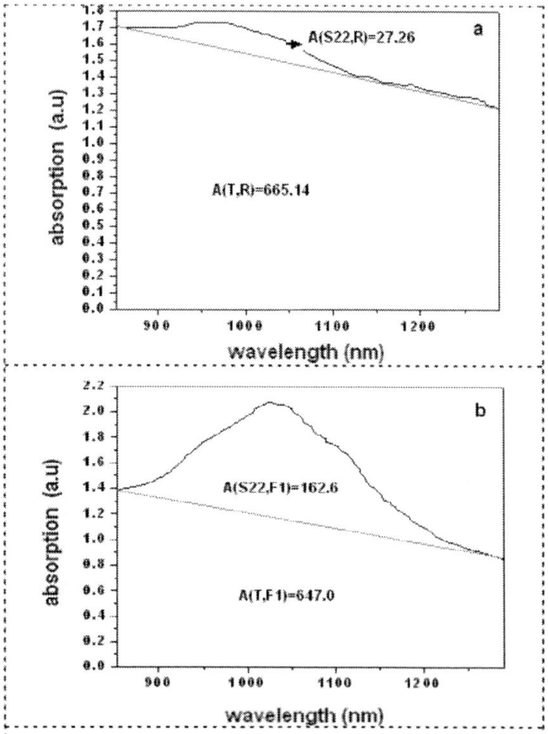

Figure 5. Absorption spectrum of the raw (a) and the purified (b) SWNTs in the range of 850-1290 nm. A(T, R), A(T, F_1) and A(S_{22}, R), A(S_{22}, F_1) stand for the integrated areas under the curve before and after baseline subtraction, respectively.

In the optical spectra of SWNTs by arc-discharge, three characteristic absorption bands are usually observed at approximately 1800, 1000, and 700 nm, respectively, superimposed on a broad background [18]. Figure 4 shows the absorption spectra of the raw and the purified SWNTs, respectively. It is obvious that the characteristic absorptions at 1000 and 700 nm of the purified SWNTs are much stronger than those of raw ones, indicating a large improvement in the purity of SWNTs.

Haddon et al [19,20] proposed a spectroscopic approach to quantitatively evaluate the purity of SWNTs. Even though a 100% pure reference is unavailable and it could not generate absolute purity of SWNTs. However, it could be applied to estimate the efficiency of the present purification procedure. Figure 5 shows the solution-phase NIR absorption spectra of the raw and purified SWNTs, respectively, in the range of 850-1290 nm. From figure 5a, it is estimated that the integrated areas ratio under the curve after and before baseline subtraction is about 0.241 ($A(S_{22},F1)=A(S_{22}, R_2)/A(T, R_2)$ =162.66/674.01) for the purified SWNTs. In contrast, it is about 0.0411 ($A(S_{22}, R)/A(T, R)$= 27.2635/665.14) for the raw SWNTs. This means that the proposed purification procedure could generate SWNTs with purity about 5.8 times of the raw SWNTs.

4. Conclusion

In summary, we developed a simple and efficient procedure for purification of raw SWNTs, which takes advantage of oxidization at 400℃ in dry air, dispersion in an aqueous solution of sodium taurocholate and centrifugation. From the results of SEM, Raman spectra, optical absorption, and TGA analyses, high quality of SWNTs are obtained. In addition, the present purification procedure is environment friendly since it does not involve strong acid, strong base and toxic solvents.

Acknowledgements. This project is sponsored by the National Natural Science Foundation of China under grant No. 50472022 and by the Scientific Research Foundation for the Returned Overseas Chinese Scholars, Department of Personnel Heilongjiang.

References:

[1] Iijima S 1991 *Nature* **354** 56.
[2] Valentini F, Amine A, Orlanducci S, Terranova M L and Palleschi G 2003 *Anal Chem.* **75** 5413.
[3] Baughman R H, Zakhidov A A and Heer W A 2002 *Science* **297** 787.
[4] Thess A, Lee R, Nikolaev P, Dai H, Petit P, Robert J, Xu C, Lee Y H, Kim S G, Rinzler A G, Colbert D T, Scuseria G E, Tomanek D, Fisher J E and Smalley R E 1996 *Science* **273** 483.
[5] Bronikowski M J, Willis P A, Colbert D T, Smith K A and Smalley R E 2001 *J. Vac. Sci. Technol.* A **19** 1800.
[6] Dillon A C, Gennett T, Jones K M, Alleman J L, Parilla P A. and Heben,M A 1999 *Adv. Mater.* **11** 1354.
[7] Moon J M, An K H, Lee Y H, Park Y S, Bae D J and Park G S 2001 *J. Phys. Chem.* B **105** 5677.
[8] Mawhinney D B, Naumenko V, Kuzenetsova A, Yates J T J, Liu J and Smalley R E 2000 *Chem. Phys. Lett.* **324** 213.
[9] Nagasawa S, Yudasaka M, Hirahara K, Ichihashi T and Iijima S 2000 *Chem. Phys. Lett.* **328** 374.
[10] Yang Y, Zou H, Wu B, Li Q, Zhang J and Liu Z 2002 *J. Phys. Chem.* B **106** 7160.
[11] Zhang J, Zou H, Qing Q, Yang Y, Li Q, Liu Z, Guo X and Du Z J 2003 *J. Phys. Chem.* B **107** 3712.
[12] Hu H, Zhao B, Itkis M E and Haddon R C 2003 *J. Phys. Chem.* B **107** 13838.
[13] Simmons J M, Nichols B M, Baker S E, Marcus M S, Castellini O M, Lee C S, Hamers R J. and Eriksson M A 2006 *J. Phys. Chem.* B **110**, 7113.
[14] Barone P W, Baik S, Heller D A and Strano M S 2005 *Nat. Mater.* **4** 86-92.
[15] O'Connell M J, Bachilo S M, Huffman C B, Moore V C, Strano M S, Haroz E H, Rialon K L, Boul P J, Noon W H, Kittrell C, Ma J P, Hauge R H, Weisman R B and Smalley R E 2002 *Science* **297** 593.
[16] Tuinstra P and Koenig J 1970 *J. Chem. Phys.* **53** 1126.
[17] Elkund P C, Holden J M and Jishi R A 1995 *Carbon* **33** 959.
[18] Kataura, H, Kumazawa Y, Maniwa Y, Umezu I, Suzuki S, Ohtsuka Y and Achiba Y 1999 *Synth.* **103** 2555.
[19] Itkis M E, Perea D E, Niyogi S, Rickard S M, Hamon M A, Hu H, Zhao B. and Haddon R C 2003 *Nano Lett.* **3** 309.
[20] Haddon R C, Chen Y, Fang S, Rao A M, Eklund P C, Lee W H, Dickey E C, Grulke E A, Pendergrass J C, Chavan A, Haley B E and Smalley R E 1998 *J. Mater. Res.* **13** 2423.

Journal of Physics: Conference Series **188** (2009) 012051 doi:10.1088/1742-6596/188/1/012051

High Throughput Exfoliation of Graphene Oxide from Expanded Graphite with Assistance of Strong Oxidant in Modified Hummers Method

T Chen[1], Baoqing Zeng[1, 3], J L Liu[1], J H Dong[1], X Q Liu[1], Z Wu[1], X Z Yang[1] and Z M Li[2]

[1] Vacuum Electronics National Laboratory, School of Physical Electronics, University of Electronic Science and Technology of China, Chengdu 610054, P. R. China

[2] State Key Laboratory of Polymer Materials Engineering, College of Polymer Science and Engineering, Sichuan University, Chengdu, Sichuan, 610065,P. R. China

E-mail: bqzeng@uestc.edu.cn

Abstract. The large-quantity (Grams) of graphene oxide (GO) was prepared by modified Hummers method, which the purifying process of GO suspension was much faster by filtering dried GO agglomeration dispersed into deionized water than usually used directed filter. The GO could be easily reduced to graphene.

1. Introduction

As a one-atom-thick two-dimensional crystal, graphene has been considered a basic building block for all sp^2 carbons including fullerene, carbon nanotubes and graphite [1]. Since Novoselov peeled a few graphene sheets from highly crystalline graphite by a scotch tape method in 2004 [2], some unique electronic properties of this conceptual matter [3,4] have been found, which indicate potential applications in quantum devices [5–7], nanocomposites with various matrixes [8, 9], and ultrathin membrane materials [10, 11]. Since stable suspension of graphene oxide (GO) can be obtained by ultrasonic treatment of GO platelets in water [12, 13], much effort has been made to assemble these well dispersed oxidized or chemically reduced (under controlled conditions, still water-dispersible) graphene nanosheets into membrane-shaped ordered macrostructures, through flow-directed assembly by filtration and self-assembled free-Standing [10, 11, 14-16]. Recently, molecular templates [17], Langmuir–Blodgett assembly [18, 19], and direct chemical vapor deposition [20] have also been employed to obtain graphene-based or graphene-oxide-based membranes on selected substrates.

Though the GO has been used in various investigations and applications, the application range, such as

[3]Authors to whom correspondence should be addressed.

conductive fillers in polymer, barrier material and coating material, is limited because of small-quantity of GO. If the chemically reduced GO is regarded as conductive filler, large-quantity (Grams) of GO may be needed by an experiment of conductive polymer composite [21]. Dongxing Yang has sacrificed the GO purity for gaining large-quantity of GO platelets [22, 23]. The GO platelets contain a few acids, so it is not an ideal nano-material. A new method that could dramatically increase quantity of GO becomes more and more necessary. In the present work, the large-quantity of graphene oxide (GO) was prepared by modified Hummers method, which the purifying process of GO suspension was much faster by filtering dried GO agglomeration dispersed into deionized water than usually used directed filter. The GO could be easily reduced to graphene.

2. Experimental

The expanded graphite powder (20 g) is put into concentrated H_2SO_4 (460 mL). $KMnO_4$ (60 g) is added gradually with stirring and cooling, so that the temperature of the mixture is not allowed to reach 20 °C. The mixture is then stirred at 35 °C for 2 h, and deionized water (920 mL) is added. In 1 h, the reaction is terminated by the addition of a large amount of deionized water (2.8 L) and 30% H_2O_2 solution (50 mL), causing violent effervescence and an increase in temperature to 100 °C, after which the color of the suspension changes to bright yellow. The suspension washed with 1:10 HCl solution (5 L) in order to remove metal ions by filter paper and funnel. The paste collected from the filter paper is dried at 60 °C, until it becomes agglomeration. The agglomeration is dispersed into deionized water in static state for 2~3 hours and slightly stirred by glass bar. The suspension is washed with much deionized water at 5~7 times for two days by filter paper and funnel, until the PH is nearly 7. The paste collected on the filter paper is dispersed into water by ultrasonication. The obtained brown dispersion is then subjected to 30 mins of centrifugation at 4000 r.p.m. to remove any unexfoliated GO using centrifuge with a rotor radius of 14 cm. The GO platelets are obtained by dehydration at 60 °C in air.

The GO hydrosol is obtained by the GO platelets dispersed into water by ultrasonication and centrifugation at 4000 r.p.m. for 2 hours to remove any unexfoliated GO. Then Atomic force microscopy (AFM) shows that the resulting GO sheets that are cast on a silicon wafer are flat.

3 Results and Discussion

Figure 1. (a) Powder of expended graphite. (b) Expended graphite oxidized. (c) GO suspension with acids and metal ions. (d) Paste of GO washed with HCl solution and deionized water.

Figure 1 (a) shows that the expended graphite is dark squamous powder with a little metal shine. It can be known in figure 1 (b) that the expended graphite oxidized by concentrated H_2SO_4 and $KMnO_4$ is dark paste. The paste can cling on to the glass rod. The paste mixture with a mass of water and a little

H_2O_2 becomes a brown-yellow suspension in figure 1 (c). If the GO suspension with a lot of acids and metal ions is laid for long time, GO will sink to the bottom and become floccules. The figure 1 (d) indicates that the GO paste without metal ions and acids is dark brown.

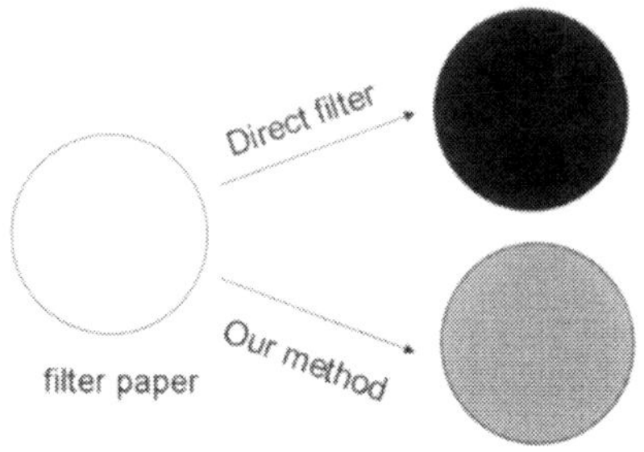

Figure 2. The model of filter

In Hummers method [23], little GO could be prepared by laboratory. The most important influencing factor of GO preparation is very slow filtering rate. In other words, the residual acids are hardly wiped away. The suspension with metal ions and acids is washed with 1:10 HCl solution in order to remove metal ions. However the GO paste with acids is dried at vacuum drying oven [22], a few acids remain in the GO. If the GO is continually washed with deionized water, the PH will increase. The suspension gradually becomes colloid. The filtering rate becomes slower and slower and then finishes in the end. The residual acids are hardly completely wiped away. The higher is PH of GO suspension, the larger is the zeta potential of GO as a function and the electrostatic force among GO sheets [14]. If the electrostatic force among GO sheets is greater than or equal to the total of gravity and van der waals force, the GO hydrosol can be stabilize forever. The model in Figure 2 shows that the GO suspension with acids and metal ions is directly filtered, and then the surface of filter will be quickly covered with a layer compact film which could block the water passing. The large-quantity of GO is hardly obtained. For obtaining large-quantity of GO, the purifying process of GO suspension should be modified.

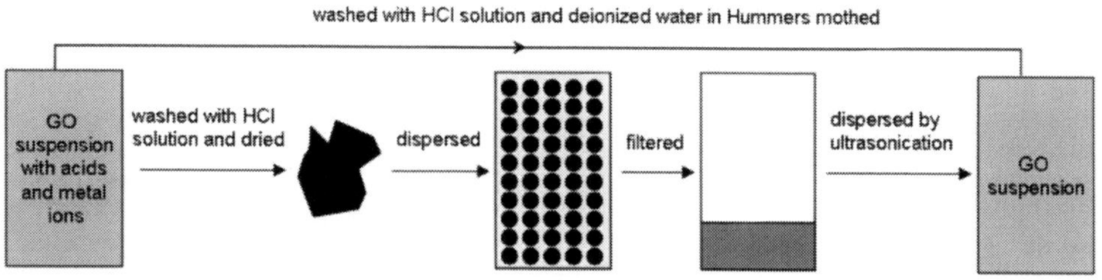

Figure 3. The different process between modified Hummers method and Hummers method.

The large-quantity of GO has been obtained by modified Hummers method. Figure 3 shows that the purifying process of modified Hummers method is different from Hummers method. In Hummers method, the GO suspension is obtained from GO suspension with acids and metal ions that is washed with HCl solution and deionized water. In modified Hummers method, the GO suspension with acids and metal ions is washed with HCl solution and dried at 60 °C. The GO agglomeration is dispersed into deionized water, and then the suspension is washed with much deionized water. The GO suspension is obtained from the paste which is collected on the filter paper and dispersed into water by ultrasonication. The dispersion of the GO agglomeration in static state is very slow. Figure 2 shows that few colloids are dispersed into deionized water in the beginning, so the surface of filter paper is not covered a layer compact film which blocks the water passing. The water with residual acids could pass through the interstice of GO film. As the GO is washed with water, the film of GO becomes more and more compact. The filtering rate becomes slower and slower. The residual acids are already wiped away before the compact film formed.

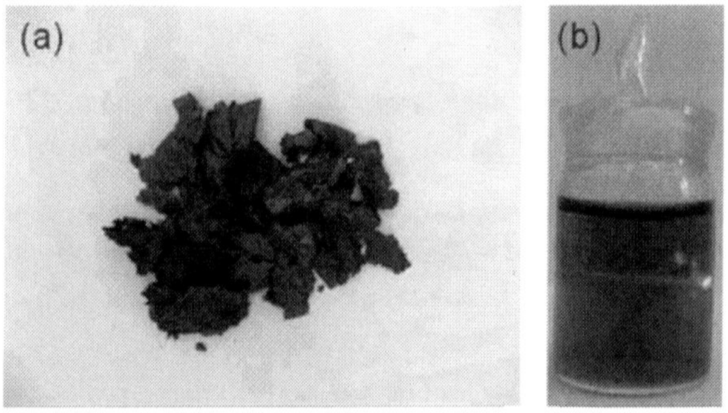

Figure 4. (a) GO platelets. (b) The Tyndall effect confirming the colloidal nature of the GO dispersions.

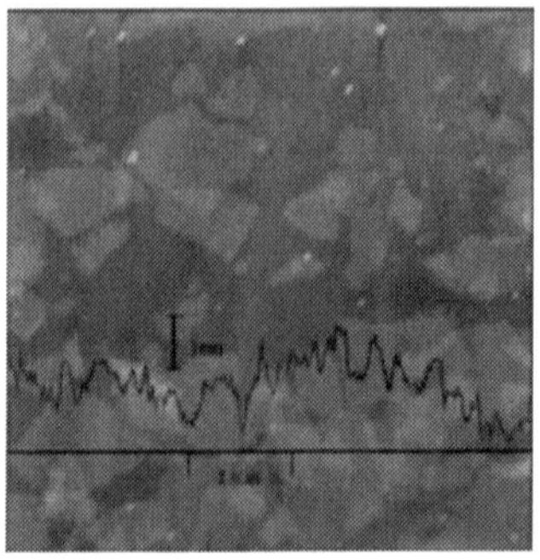

Figure 5. The AFM image of GO sheets on the Si slice

Figure 4 (a) shows that the GO platelets are dark and flexibility. The GO hydrosol obtained by the GO platelets dispersed into water by ultrasonication and centrifugation at 4000 r.p.m. to remove any unexfoliated GO is brown yellow. The GO hydrosol laid several months in static state has no obvious deposition. The Tyndall effect confirms the colloidal nature of the GO dispersions in figure 4 (b).
Figure 5 is an atomic force microscopy (AFM) image of the GO hydrosol. It offers immediate evidence for peeled-off single GO sheets. The thicknesses of the GO layers are about 0.8 nm, as indicated in Figure. As reported previously [24, 25], these GO layers should be mostly monolayer. The sheets are a little bumpier than predicted, which is possibly due to the existence of abundant functional groups, such as epoxy and hydroxyl groups, bonded to both sides of the graphene sheets, which disrupts the original conjugation and introduces lattice defects to result in folds and distortions on the sheets [26]. The AFM image observations indicate that most graphene sheets are of a relatively large size (several hundred nanometers to several micrometers).

4. Conclusion

In modified Hummers method, the purifying process of GO suspension has been studied, which can obtain the large-quantity of GO. More increasing quantity of GO is the next center work. In addition, the device, filter paper and funnel, is inexpensive and simple. The large-quantity of GO prepared in industry may be enlightened by this work.

Acknowledgements

The work is supported by NSFC (Grants No.60071043 and 60532010), the Doctor Station Foundation of the Ministry of Education of China (Grant No. 200806140007), pre-study foundation of electronic science academic, and National Laboratory for Vacuum Electronics.

References

[1] Geim A K, Novoselov K S 2007 *Nat. Mater.* **6** 183
[2] Novoselov K S, Geim A K, Morozov S V, Jiang D, Zhang Y, Dubonos S V, Grigorieva I V, Firsov A A 2004 *Science* **306** 666
[3] Zhang Y B, Tan Y W, Stormer H L, Kim P 2005 *Nature* **438** 201
[4] Novoselov K S, Geim A K, Morozov S V, Jiang D, Katsnelson M I, Grigorieva I V, Dubonos S V, Firsov A A 2005 *Nature* **438** 197
[5] Wu J, Pisula W, Mullen K 2007 *Chem. Rev.* **107** 718
[6] Lin Y M, Avouris P 2008 *Nano Lett.* **8** 2119
[7] Eda G, Lin Y Y, Miller S, Chen C W, Su W F, Chhowalla M 2008 *Appl. Phys. Lett.* **92** 233305
[8] Stankovich S, Dikin D A, Dommett G H B, Kohlhaas K M, Zimney E J, Stach E A, Piner R D, Nguyen S T, Ruoff R S 2006 *Nature* **442** 282
[9] Ramanathan T, Abdala A A, Stankovich S, Dikin D A, Alonso M H, Piner R D, Adamson D H, Schniepp H C, Chen X, Ruoff R S, Nguyen S T, Aksay I A, Prud'homme R K, Brinson L C 2008 *Nat. Nanotechnol.* **3** 327
[10] Dikin D A, Stankovich S, Zimney E J, Piner R D, Dommett G H B, Evmenenko G, Nguyen S T, Ruoff R S 2007 *Nature* **448** 457
[11] Xu Y, Bai H, Lu G, Li C, Shi G 2008 *J. Am. Chem. Soc.* **130** 5856
[12] Stankovich S, Dikin D A, Piner R D, Kohlhaas K A, Kleinhammes A, Jia Y, Wu Y, Nguyen S T, Ruoff R S 2007 *Carbon* **45** 1558
[13] Gilje S, Han S, Wang M, Wang K L, Kaner R B 2007 *Nano Lett.* **7** 3394
[14] Li D, Muller M B, Gilje S, Kaner R B, Wallace G G 2008 *Nat. Nanotechnol.* **3** 101
[15] Chen H Q, Müller M B, Gilmore K J, Wallace G G, Li D 2008 *Adv. Mater.* **20** 3557
[16] Chen C M, Yang Q H, Yang Y G, Lv W, Wen Y F, Hou P X, Wang M Z, Cheng H M 2009 *Adv. Mater.* **21** 1–5
[17] Wei Z Q, Barlow D E, Sheehan P E 2008 *Nano Lett.* **8** 3141
[18] Li X, Zhang G, Bai X, Sun X, Wang X, Wang E, Dai H 2008 *Nat. Nanotechnol.* **3** 538
[19] Cote L J, Kim F, Huang J X 2009 *J. Am. Chem. Soc.* **131** 1043
[20] Kim K S, Zhao Y, Jang H, Lee S Y, Kim J M, Kim K S, Ahn J H, Kim P, Choi J Y, Hong B H 2009 *Nature* **457** 706
[21] Xu X B, Li Z M, Shi L, Bian X C, Xiang Z D 2007 *Small* **3** 408
[22] Yang D X, Velamakanni A, Bozoklu G, Park S, Stoller M, Piner R D, Stankovich S, Jung I, Field D A. Ventrice J C A, Ruoff R S 2009 *Carbon* **47** 1 145
[23] Hummers W S, Offeman R E 1958 *J. Am. Chem. Soc.* **80** 1339
[24] Jung I, Vaupel M, Pelton M, Piner R, Dikin D A, Stankovich S, An J, Ruoff R S 2008 *J. Phys. Chem. C* **112** 8499
[25] Ni Z H, Wang H M, Kasim J, Fan H M, Yu T, Wu Y H, Feng Y P, Shen Z X, 2007 *Nano* Lett. **7** 2758
[26] Schniepp H C, Li J L, McAllister M J, Sai H, Alonso M H, Adamson D H, Prud'homme R K, Car R, Saville D A, Aksay I A, 2006 *J. Phys. Chem. B* **110** 8535

Molecular dynamics simulation for mechanical properties of CNT/Polyethylene composites

Liu Jia Yang Qingsheng [1]
Dept. of Engineering Mechanics, Beijing University of Technology, Beijing 100124, China,

E-mail: jiajia2007@emails.bjut.edu.cn

Abstract. The pull-out process of the carbon nanotube from polyethylene was simulated by molecular dynamics method. A model of a carbon nanotube in polyethylene was established. In the simulation, Adaptive Intermolecular Reactive Empirical Bond Order(ARIEBO) potential was adopted to describe the interaction of C-C and C-H in the carbon nanotube and polymer, and Lennard-Jones pair potential was used to describe the interaction between the carbon nanotube and polymer; NVT ensemble was adopted in the whole simulation and Nose-Hoover method was used to control the temperature at absolute zero, which avoided the influence induced by thermal activation; Verlet algorithm was used to solve molecular dynamics equations in the procedure of simulation. The deformation and forces on interfaces between the carbon nanotube and polymer was analyzed by simulating the process of pulling-out of the carbon nanotube from polyethylene.
Keywords. carbon nanotube, molecular dynamics, potential function, polyethylene.

1. Introduction

Carbon nanotube has typical layered hollow structure and light weight as one-dimensional nano-material. Its hexagonal structure is perfectly connected and it has quite high slenderness ratio, usually greater than 1000. The unique structure makes it have excellent mechanical, electrical, thermal and chemical performances. Once discovered, carbon nanotube caused comprehensive interest and concern of researchers in academia and industry field because of its unique performance. Comprehensive concern is provoked on the increment of carbon nanotube synthetic ratio, the purification, opening, cutting, filling and the polymerization between carbon nanotube and polymer. Since 1994 Ajayan[1] mixed carbon nanotubes as inorganic filler into polymer matrix and to prepare polymer composite materials, a lot of work is done to make carbon nanotube evenly scattered in the polymer matrix by nano-composite technology which is in order to produce newly high-property and multi-functioned polymer matrix nano-composite materials. Composites in which carbon nanotube is the reinforcement reflect extensive application prospects on information materials, biomedical materials, high-performance structural materials, multi-functional materials and so on. Great progress has been made in the mechanical properties, thermal properties and function of CNT/ polymer composites.
Schadler[2] et al researched tension and compression property of MWNT/epoxy resin composites, a fact that compression modulus of composite materials was larger than tensile modulus was found, which was because only the outer of carbon nanotube transferred load when it was tensile, while both inner and outer woke when compression. Cooper[3] et al did pull-out experiment on a single carbon

[1] E-mail: qsyang@bjut.edu.cn

nanotube which crossed the hole of matrix by atomic force microscope and directly found interface strength between multi-wall carbon nanotube and epoxy resin is 35-376MPa. Biercuk[4] et al found that thermal conductivity of CNT/epoxy resin composite material increased about 120% when 1% percent of carbon nanotubes were added. Wei[5] et al found that heat exchange and diffusion coefficient of composites was improved after adding carbon nanotube by using molecular dynamics method. Mechanical properties of CNT/Si composites increased two times when volume fraction ratio of carbon nanotubes in composites was 0.01 found by Dalton[6] et al. Lei[7] et al tested the mechanical properties of CNT/epoxy resin composites at normal and elevated temperature. Sharon[8] et al studied the mechanical properties of CNT/polystyrene and how irradiation affected the surface structures under argon irradiation. Zheng[9] et al studied interface adhesion of different kinds of CNT/polymer using molecular mechanics and molecular dynamics.

Because of special mechanical properties of carbon nanotube reinforced polymer composites, a model of a carbon nanotube in polyethylene was built and the process of pulling out of carbon nanotube from polyethylene was simulated by using molecular dynamics method in this article. Then polymer interfaces were analyzed from microcosmic angle and the force on carbon nanotube was calculated.

2. Model and method of simulation

2.1. Build the model

Carbon nanotube has periodicity in axial direction. The configuration of carbon nanotube can be shown by (m, n) which decides the diameter, helix angle, lattice constant and physical properties exclusively. Polyethylene is made of a great many of ethylene chains. The whole model is built in Materials Studio in this article. Firstly, carbon nanotube of (6, 6) is built. Its radius is 0.407nm and the length is 1.476nm . Then, polyethylene chains are built. At last, carbon nanotube is placed in polyethylene, and the polyethylene chains in carbon nanotube are deleted (Figure 1.).

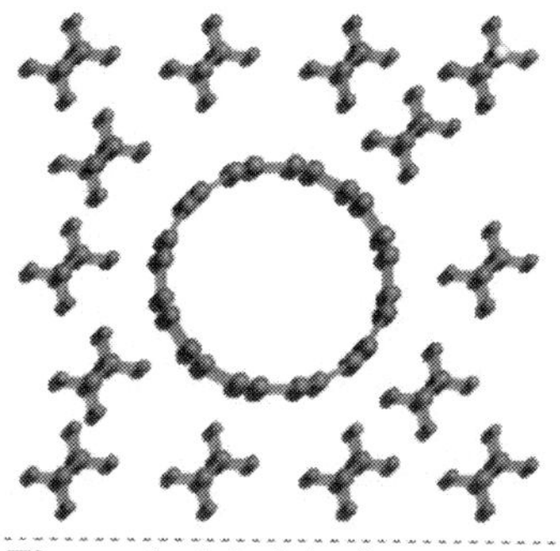

Figure 1. Model of composite

Coordinate of every atom will be written in a file in order to simulate by molecular dynamics programme.

2.2. Principle of the molecular dynamics simulation

The steps of molecular dynamics method: 1) setting of a model, which is the selection of potential function; 2) giving initial conditions; 3) tending to Balance calculation.

The movement of atoms on nano-scale can still be described by classical Newton motion equation in the simulation of molecular dynamics

$$m_i \frac{\partial^2 \boldsymbol{r}_i}{\partial t^2} = \boldsymbol{F}_i \tag{1}$$

m_i is the quality of an atom, $\boldsymbol{r}_i$ is the coordinate of an atom, $\boldsymbol{F}_i$ is the composition of forces acting on atom i from the atoms surrounding atom i. In molecular dynamics method, there is

$$F_i = -\frac{\partial V}{\partial r_i} \tag{2}$$

V is potential function among atoms. In order to obtain the movement of atoms, kinds of finite difference algorithm can be used.

Common algorithms are: Verlet algorithm, Beeman algorithm and Leapfrog algorithm. The research object of molecular dynamics is multi-particle system. The number of particles in system which can be simulated is finite because of the impact of external conditions. But the rule of statistical physics is still tenable, so suitable ensemble should be selected when simulating.

The master ensembles used in molecular simulation are Microcanonical Ensemble (NVE), Canonical Ensemble (NVT), Isobaric-isothermal Ensemble (NPT) and Pressure-enthalpy Ensemble (NPH). NVT is adopted in simulation in this article.

Particle number, volume, temperature and total momentum is conservation in NVT. The total energy of a system is not a conserved quantum at a constant temperature. Energy exchange must be occurred between the system and the external to maintain the temperature unchanged. The adopted method usually makes the heat balance between the system and the heat bath of external. A variable which bewrite the interaction between the system and the external is imported. Then Eq.(1) is changed into

$$\ddot{r} = \frac{F_i}{m} - \eta \dot{r}_i \tag{3}$$

Where

$$\dot{\eta} = \frac{1}{\tau^2}(\frac{T}{T_C} - 1) \qquad \tau = \sqrt{\frac{Q}{3Nk_B T_C}}$$

τ is the relaxation time of hot bath, determining the rate of tending to constant of system temperature, which can be evaluated equal or several times as time step. $k_B=1.3806\times10^{-23}$J/K is Boltzmann constant.

The main numerical solutions of molecular kinetic equation are Verlet algorithm, Beeman algorithm and Leapfrog algorithm. The most extensively used Verlet algorithm is adopted in this article.

Suppose the positions of atoms in molecular system $r(t)$ when the time is t, then the positions of atoms can be obtained by Taylor expansion when the time is $t+\delta t$

$$r(t+\delta t) = r_i(t) + \delta t v_i(t) + \frac{1}{2}\delta t^2 a_i(t) + \cdots \tag{4}$$

and when time is $t-\delta t$, there has

$$r(t-\delta t) = r_i(t) - \delta t v_i(t) + \frac{1}{2}\delta t^2 a_i(t) - \cdots \tag{5}$$

The Eq.(6)can be gotten by adding the Eq.(4) and Eq.(5)

$$r(t+\delta t) = r2_i(t) - r_i(t-\delta t) + \delta t^2 \frac{F_i(t)}{m_i} \tag{6}$$

and then

$$v(t) = [r(t+\delta t) - r(t-\delta t)]/2\delta t \tag{7}$$

Because models are made of carbon nanotube and polyethylene, there are only C and H atom. Adaptive Intermolecular Reactive Empirical Bond Order (AIREBO) potential aiming at C and H atom is adopted in this article. The function between carbon nanotube and polyethylene is simulated by Lennard-Jones pair potential.

AIREBO potential[11] can be written into

$$E = \frac{1}{2}\sum_i \sum_{j\neq i}[E_{ij}^{REBO} + E_{ij}^{LJ} + \sum_{k\neq i,j}\sum_{l\neq i,j,k} E_{kijl}^{TORSION}] \tag{8}$$

E_{ij}^{REBO} gives the model its reactive capabilities and only describes short-ranged C–C, C–H and H–H interactions ($r < 2\overset{\circ}{A}$); E_{ij}^{LJ} adds longer-ranged interactions ($2\overset{\circ}{A} < r < r_{cut}$) using a form similar to the standard Lennard-Jones potential. $E_{kijl}^{TORSION}$ is an explicit 4–body potential that describes various dihedral angle preferences in hydrocarbon configurations.
Lennard-Jones potential can be written into

$$E = 4\varepsilon[(\frac{\sigma}{r})^{12} - (\frac{\sigma}{r})^{6}] \quad r < r_{cut} \tag{9}$$

Different parameters are given to ε and σ according to different atoms in the formula above.

3. procedure and result of simulation

In the simulation, carbon nanotube is supposed to be rigid material, which is distortion would not happen in the process of simulation. AIREBO potential is adopted in carbon nanotube and polyethylene, and the function between carbon nanotube and polyethylene is simulated by Lennard-Jones pair potential. The whole system adopts NVT ensemble and Nose-Hoover method is used to control the temperature at absolute zero, which avoided the influence brought by thermal activation. Verlet algorithm is used to solve molecular dynamics equations in the procedure of simulation. First, model will relaxed in 10000 time steps, so the energy is the smallest. Give a displacement of 0.01nm to carbon nanotube along to the direction of length. Relax 1000 time steps, and the time is 10^{-16}s. This will circulate until the carbon nanotube is divorced from polyethylene.
Figure 2 is the deformation of composite. From it we can see that some polyethylene chains will move along with carbon nanotube and the polyethylene chains that are far from carbon nanotube will keep the position when the displacement of carbon nanotube is small enough. Polyethylene chains will move slowly because of the increase of the displacement of carbon nanotube and the function of the immovable polyethylene chains. Last carbon nanotube divorces from polyethylene.

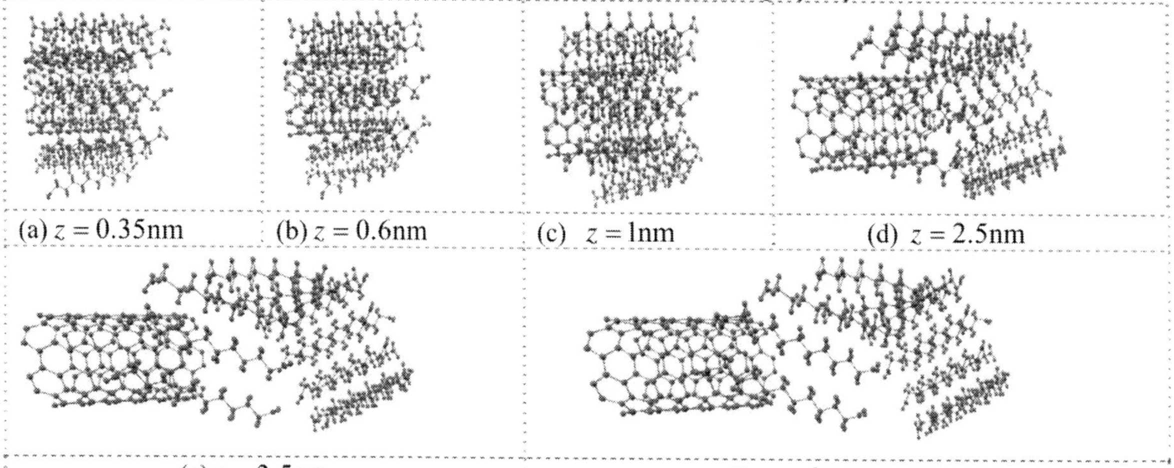

(a) $z = 0.35$nm (b) $z = 0.6$nm (c) $z = 1$nm (d) $z = 2.5$nm

(e) $z = 3.5$nm (f) $z = 4$nm

Figure 2. Deformation of composite

Figure 3 expresses the relationship of average potential energy and displacement of carbon nanotube. The potential energy arrives at minimum when the displacement is 0.35nm, as shown in Figure 2(b), because that the deformation of polyethylene is not obvious. From Figure 2., we can see that when the displacement in 0.35nm-0.62nm, carbon nanotube connects closely with polyethylene and mean stress of carbon nanotube achieves maximum along the direction of displacement. As the

displacement increases, the distance between carbon nanotube and polyethylene augments and mean stress decreases. It is shown as in the figure 4.
In the program used in this article, mean stress can be expressed as

$$S_{ab} = -[mv_a v_b + \frac{1}{2}\sum_{n=1}^{N_p}(r_{1a}F_{1b} + r_{2a}F_{2b}) + \frac{1}{2}\sum_{n=1}^{N_b}(r_{1a}F_{1b} + r_{2a}F_{2b}) + \frac{1}{3}\sum_{n=1}^{N_a}(r_{1a}F_{1b} + r_{2a}F_{2b} + r_{3a}F_{3b}) + \frac{1}{4}\sum_{n=1}^{N_d}(r_{1a}F_{1b} + r_{2a}F_{2b} + r_{3a}F_{3b} + r_{4a}F_{4b}) + \frac{1}{4}\sum_{n=1}^{N_i}(r_{1a}F_{1b} + r_{2a}F_{2b} + r_{3a}F_{3b} + r_{4a}F_{4b}) + \sum_{n=1}^{N_f} r_{ia}F_{ib})] \tag{10}$$

The first term is a kinetic energy contribution for atom i. The second term is a pairwise energy contribution where n loops over the N_p neighbors of atom i, r_1 and r_2 are the positions of the two atoms in the pairwise interaction, and F_1 and F_2 are the forces on the two atoms resulting from the pairwise interaction. The third term is a bond contribution of similar form for the N_b bonds which atom i is part of. There are similar terms for the N_a angle, N_d dihedral, and N_i improper interactions atom i is part of. Finally, there is a term for the N_f fixes that apply internal constraint forces to atom i.
Stress in Eq.(10) is the stress of every atom. Mean stress is the average of sum of stresses of all the atoms, it can be obtained by

$$P = \frac{Nk_B T}{V} + \frac{\sum_i^N r_i \cdot f_i}{dV} \tag{11}$$

The maximal mean stress can be computed out from the program when the displacement is 0.62nm. The numerical value is 17159.4Mpa.
In continuum mechanics, there has $F = \sigma A$, and $A = 2\pi rl$. The maximum force of the carbon nanotube is 4.994×10^{-7} N when the displacement is 0.62nm.
The interaction between carbon nanotube and polymer is the most obvious when displacement force is 0.35nm-0.62nm and then carbon nanotube gradually divorces from polyethylene.

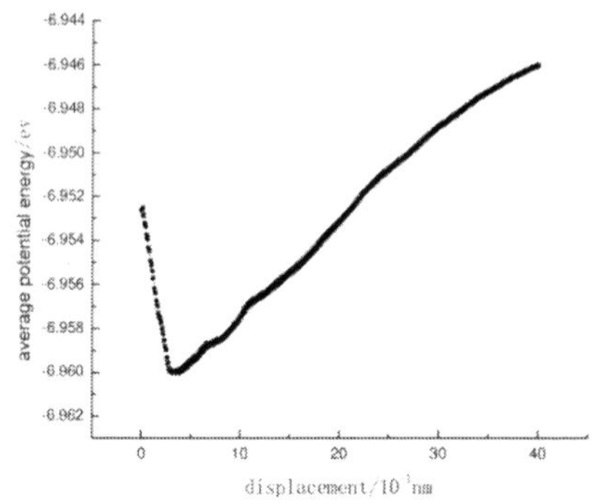

Figure 3. displacement-average potential energy curve

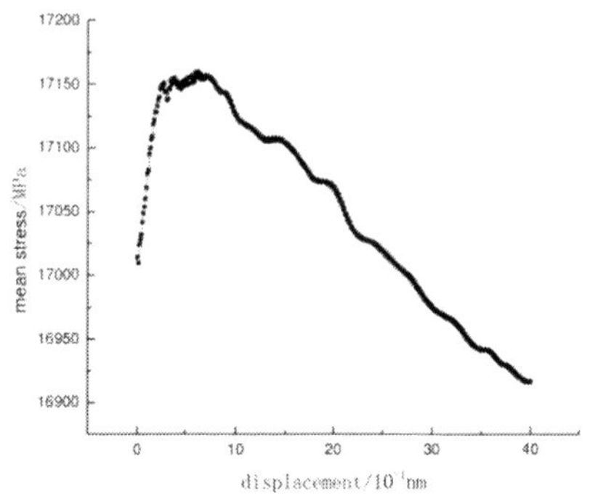

Figure 4. displacement-mean stress curve

4. Conclusion

The process of pulling out of the carbon nanotube from polyethylene was simulated by inlaying the carbon nanotube into polyethylene in Materials Studio and using a molecular dynamics program. The results showing the relationship of load displacement versus average potential energy and mean stress are obtained. Average potential energy reduces with the increase of displacement and then gradually increases at a specific displacement. Axial force of the carbon nanotube was obtained at different displacement. The interaction between the carbon nanotube and polyethylene when pulling out and the potential changes of the carbon nanotube are obtained. The mechanical properties of polymer matrix composite materials are obtained on atomic scale.

References

[1] Ajayan P M, Stephan O, Collies C, Trauth D. Aligned carbon nanotube arrays formed by cutting a polymer resin-nanotube composite. 1994, *Science*, 265:1212-1214.
[2] Schadler L S, Gianngaris S C, Ajayan P M. Load transfer in carbon nanotube epoxy composites. 1998, *App l Phys Lett*, 73 (26):3842-3844.
[3] Cooper C A, Young R J, Halsall M. Investigation into the deformation of carbon nanotubes and their composites through the use of Raman spectroscopy. 2001, *Composites: PartA*, 32 (3-4):401-411.
[4] Biercuk M J, Llaguno M C, Radosavljevic M. Carbon nanotube composites for thermal management. 2002, *Applied Physics Letters*, 80(15):2767-2769.
[5] Wei C, Srivastava D, Cho K. Thermal expansion and diffusion coefficients of carbon nanotube-polymer composites. 2002, *Nano Letters,* 2(6):647-650.
[6] Dalton A B, Ferraris J P, Baughman R H. Continuous carbon nanotube composite fibers: properties, potential applications, and problems. 2004,*J. Mater. Chem,* 14:1-4.
[7] Lei Z K, Qiu W, Li Q, Kang Y H, Pan X M. Mechanical Properties of Carbon Nanotube Polymer Composites. 2008, *Polymer Materials Science & Engineering*, 12(24):134-140.
[8] Sharon K Pregler a, Byeong-Woo Jeong, Susan B Sinnott. Ar beam modification of nanotube based composites using molecular dynamics simulations. 2008, *Composites Science and Technology*, 68: 2049–2055.
[9] Zheng Qingbin, Xia Dan, Xue Qingzhong, Yan Keyou, Gao Xili, Li Qun. Computational analysis of effect of modification on the interfacial characteristics of a carbon nanotube/ polyethylene composite system. 2009, *Applied Surface Science,* 255, 3534–3543.
[10] Zhang Y, Gu J H, Shang J X, Ma Y. The base of computational materials science. 2007, *Beijing University of Aeronautics and Astronautics Press.*
[11] Fan J H. Multiscale analysis for deformation and failure of materials. 2008, *Science Press.*

Quasicontinuum Study of Nanoindentation into Nanocrystalline Aluminum Thin Film

JI Zhong*, HOU Shuai

(School of Materials Science and Engineering, Shandong University, Jinan 250061, China)

ABSTRACT: Quasicontinuum simulations are performed to examine the mechanical response of nanocrystalline aluminum thin film under nanoindentation. The simulations are conducted by driving the flat indenter into the (111) face of one grain in the aluminum film, two load drops have been found during the indentation process, evidence indicates that these load drops are related to the stacking fault emitted from the grain boundary. The effect of indenter width on nanoindentation response is also studied through three simulations with different indenter sizes. A rule is discovered that larger load is needed to initiate the first load drop for the wider indenter; however, this rule is not followed by the second load drop.

KEYWORDS: nanoindentation; quasicontinuum method; nanocrystalline materials; aluminum

0 Introduction

Nanocrystalline materials show unique mechanical properties under applied load. Many experiments and simulations have been carried out to study the mechanical properties of nanocrystalline materials. Nanoindentation, a small-scale contact experiment, was established to test the hardness, elastic modulus, strain hardening exponent and other important mechanical parameters of small volumes of materials[1]. And it is very suitable to investigate the plastic deformation of nanocrystalline metals. The indenter size can be selected to be either significantly smaller or significantly larger than the average grain diameter so different aspects of the deformation can be revealed[2]. Yang and Vehoff observed that when the indenter size was smaller than the grain size, the dislocations nucleated below the indenter tip interact directly with the surrounding grain boundaries (GBs) for grains below 900nm[3]. Dupont and Sansoz's quasi-2D atomistic simulation indicated that when the indenter was several times larger than the grain size, the GB networks cause softening effects at the tip/surface interface and the deformation mechanisms are GB movement as well as deformation twins[4].

In this study, the simulation of nano- indentation into nanocrystalline aluminum under zero temperature was performed using a quasicontinuum (QC) model. The width of the indenter was kept much less than the average grain diameter. We obtained the load-displacement curve for the nanocrystalline aluminum and we compared it with the nanoindentation response of single crystalline aluminum. Simulations using another two indenters of different size were also performed in order to find out how the mechanical response depends on the indenter width.

1 Computational model

The quasicontinuum method is a multi-scale method combining atomistic and continuum simulations [5]. In a quasicontinuum simulation, the critical region subjected to large deformation gradients are treated by molecular dynamics (MD) and the other regions are treated using the finite element method (FEM). Thus the computational effort is significantly reduced. This technique is performed by energy minimization under zero temperature, and the result is provided as the equilibrium atomic configurations.

A 2D quasicontinuum model with the dimensions of 200nm×100nm was created. The crystalline orientation of the crystal is illustrated in Fig.1(a). The flat knife-like indenter was assumed rigid and it had no interfacial interaction with the sample. The bottom of the sample was fixed and a periodic boundary condition was applied to the out-of-plane direction. The other sides were kept free. Regions under the indenter and the GBs were refined down to the atomic scale, while the other part of the crystal was coarsely meshed. A closer look of the region under the indenter was shown in Fig.1(b). The sample was considered to contain only five grains: the indented grain, its three neighbors and the left part of the crystal. The left part of the crystal was treated as a single crystal grain because the indenter was relatively small and was thought to merely affect the indented grain and its neighbors. The grains were numbered as shown in Fig.1(b). Grains 2, 3 and 4 have geometry of regular hexagon with the edge length of 4.33nm. Grain 1, named indented grain, is a half hexagon. The grain boundaries between grain 1 and grains 2, 3, 4 were all pure <110> tilt GBs. The orientations of grains 1 to 4 were set as follows: Grain 1 has the same orientation as the global crystal orientation shown in Fig.1(a); grains 2 to 4 were rotated along the $[1\bar{1}0]$ direction from the global crystal orientation to 5 degrees, 5 degrees and -3 degrees respectively.

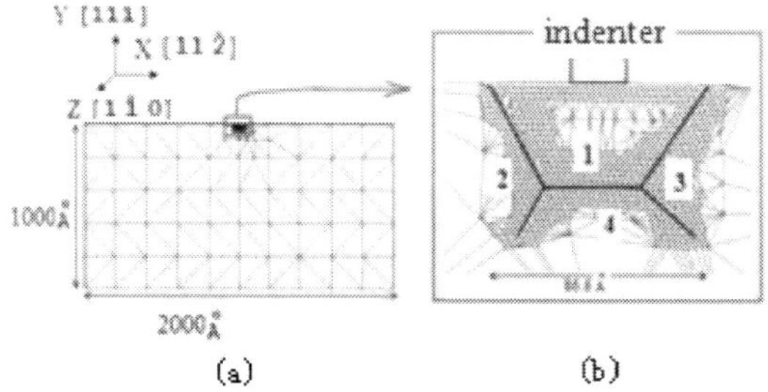

Fig.1 Quasicontinuum model of a nanocrystalline aluminum thin film indented by a flat indenter.
(a) Full view of both finite element and atomistic regions
(b) Close view of the region under the indenter. Bold lines were added to highlight the GBs.

The sample was first relaxed to accomplish minimized energy. Then the indenter was driven into the (111) face of the sample with the step of 0.2 Å until the indentation depth reached 8 Å. Before every step, energy minimization was carried out to adjust to the new boundary conditions. The embedded atom method (EAM) potential for FCC Al used in our simulation was developed by Ercolessi and Adams [6]. Within the EAM model, the total energy E for a system can be written as,

$$E=\sum_{i}^{N}\left(F_i(\rho_i)+\frac{1}{2}\sum_{j\neq i}^{N}Vi_j(R_{ij})\right) \tag{1}$$

where the summations extend over the total number of atoms N in the system, F_i is the embedding function, ρ_i is the electron density at atom i, V_{ij} is a pair interaction function and R_{ij} is the distance between atoms i and j.

2 Nanoindentation Response of Polycrystalline Aluminum

The flat indenter used here was 20 Å. The computed load-displacement curve is presented in Fig.2 in a solid line. The load was calculated by dividing the force exerted on the nods by the interplanar spacing of the $[1\bar{1}0]$ planes. The curve starts from a negative value because of the initial relaxation [7]. Then the curve rises in a linear pattern until point A. At point A the load drops from 12.94N/m to 11.52N/m, and then the curve rises nonlinearly to point B where the load drops again from 16.97N/m to 11.65N/m. After point B, the curve continues going up until the indentation depth reaches 8 Å.

The result for the single crystalline aluminum under the same condition is also shown in Fig.2 in a dashed line. This single crystal simulation was initially carried out by Tadmor and Miller [7], and we repeated the simulation by changing the indenter width to 20Å. The load applied to the single crystal increases linearly with the increasing indentation depth, and is larger as compared to the polycrystal simulation at any indentation depth.

It is indicated in previous literature that in this particular crystal orientation, the dominant deformation mechanism of single crystalline aluminum thin film is deformation twinning [8]. In our study of nanocrystalline aluminum thin film, deformation twinning as well as dislocation movement were both observed.

Fig.3 shows the atomic configuration at point A. Two phenomena were observed. First, a twin (Fig.3(a)) was formed near the surface. Its twinning plane is $(11\bar{1})$, and the twinning direction is [112]. The existence of the twin is further evidenced by plotting the atomic displacement along the Y and Z axes (Fig.4). Second, a stacking fault (Fig.3(b)) with the dislocation core at approximately (15,-30) was found near the grain boundary of grain 1 and grain 3. After careful observation, this stacking fault is believed to be separated from the grain boundary and moved leftward to the place where it was found.

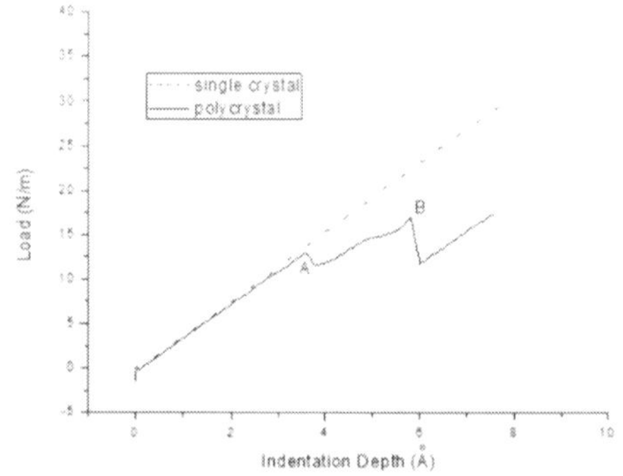

Fig.2 Load-displacement curve for nanoindentation into a (111) face of single crystalline aluminum (dashed line) and polycrystalline aluminum (solid line).

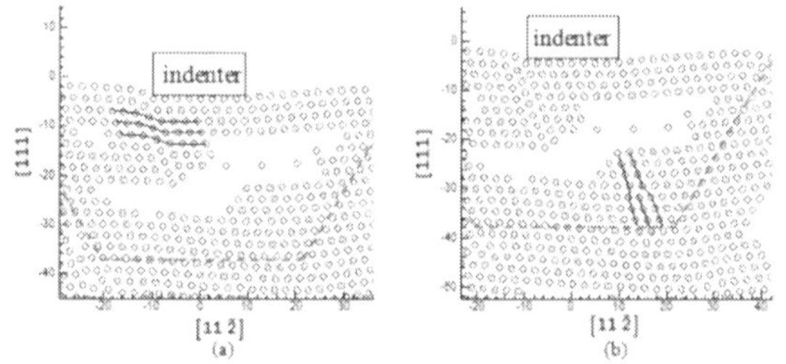

Fig.3 Atomic configuration at point A. Red dashed lines indicate boundaries of grain 1.
(a) Twin formed near the surface. (b) Stacking fault near the grain boundary of grains 1 and 3.

Fig.5 shows the mechanism of the second load drop at point B. Fig.5(a) is the atomic configuration just before point B. The twin formed at point A propagated along the [112] direction and its front arrived at approximately (0,-30). The stacking fault mentioned above traveled leftward with its core reached nearly the same position. Fig.5(b) demonstrated the result of the interaction of the twin and the stacking fault. The twin disappeared and another stacking fault was form with its core near (0, -25).

On the basis of the above, the load drop is related to the grain boundary. The stacking fault separated from the grain boundary can interact with twins and thus cause change in mechanical response.

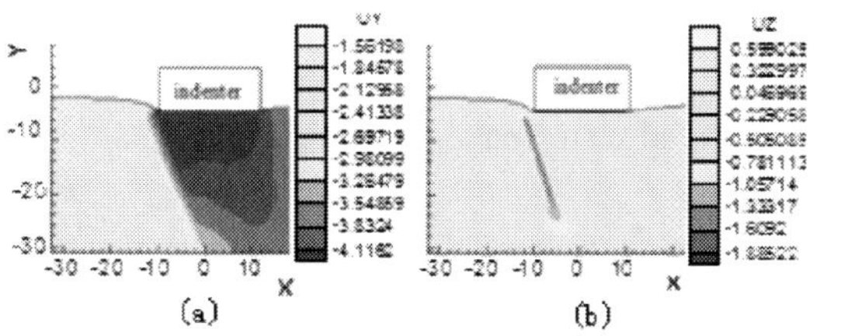

Fig.4 Displacement experienced by the atoms along the Y and Z axes
(a) UY contour plot. (b) UZ contour plot

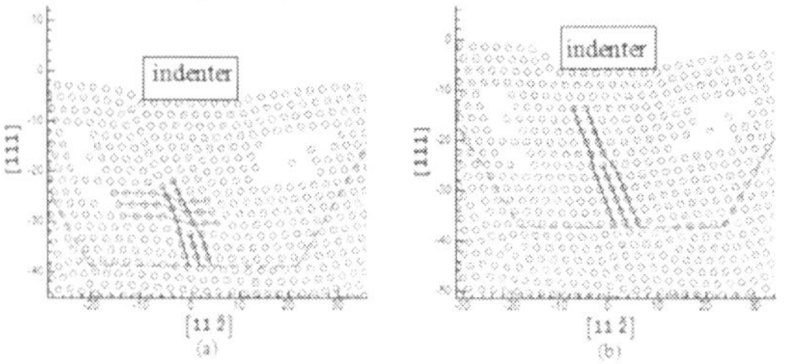

Fig.5 Twin-stacking fault interaction around point B
(a) Before point B, twin and stacking fault co-existed
(b) At point B, twin disappeared and new stacking fault was formed.

3 Effects of Indenter Width on Nano- indentation Response

Three indenters with different widths of 10Å, 15Å and 20Å were used in this study. The results were plotted as shown in Fig.6. It is indicated in the figure that the mechanical response under nanoindentation is very sensitive to the indenter width. As the indenter becomes wider, the first load drop occurs at a larger indentation depth and the corresponding load is also larger. However, when the hardness (mean contact pressure) instead of force is considered, the widest indenter is corresponding to the smallest hardness. Hardness of 10.13GPa, 7.68GPa and 6.47GPa were obtained for the 10Å, 15Å and 20Å indenters respectively. This is consistent with the literature that significant increase in hardness was observed with reduced indenter size.

However, the second load drop does not obey the same rule as found for the first load drop. The loads to initiate the second load drop are 13.66N/m, 9.95N/m and 16.97N/m for the 10Å, 15Å and 20Å wide indenters respectively. No monotonicity is observed here.

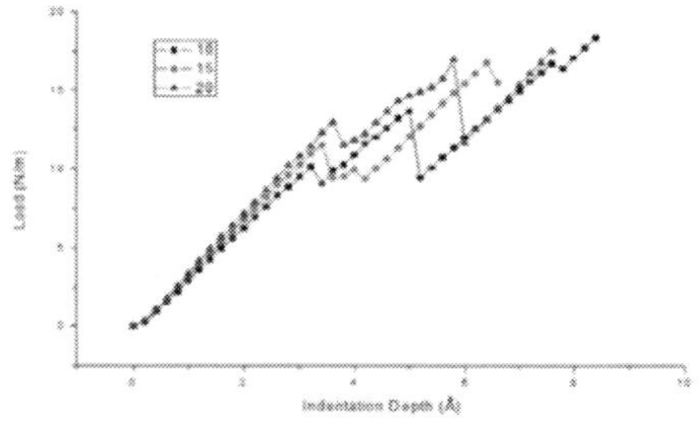

Fig.6 Indenter width effect on nanoindentation into (111) face of polycrystalline aluminum.

4 Conclusions

Quasicontinuum simulations of nano- indentation into the (111) face of one grain in nanocrystalline aluminum thin film were performed. The different nanoindentation response of polycrystalline and single crystalline aluminum was studied. And the effect of indenter width on nanoindentation response was observed through three simulations with different indenter sizes. The major conclusions of this work can be summarized as follows:

(1) At the same indentation depth, the force needed to hold the indenter is smaller for the polycrystal as compared to the single crystal.

(2) Two load drops were observed on the nanoindentation response curve for nanocrystalline aluminum thin film.

(3) As the first load occurs, both mechanical twinning and stacking fault were found.

(4) The second load drop is attributed to the twin-stacking fault interaction.

(5) The first load drop is sensitive to the indenter size. A rule is found out that, as the indenter becomes wider, larger indentation depth and larger load is required to initiate the first load drop. A smaller hardness was observed with the wider indenter. However, the second load drop does not follow the same rule.

Acknowledgements

This work is sponsored by the National Natural Science Foundation of China (No. 50775131), the Natural Science Foundation of Shandong Province (No.Y2007F55) and the Scientific Research Foundation for the Returned Overseas Chinese Scholars, State Education Ministry.

References

[1] Fischer-Cripps A C. Nanoindentation[M]. New York: Springer, 2004

[2] Dupont V, Sansoz F. Quasicontinuum study of incipient plasticity under nanoscale contact in nanocrystalline aluminum[J]. Acta Mater. 2008, 56: 6013-6026

[3] Yang B, Vehoff H. Dependence of nanohardness upon indentation size and grain size - A local examination of the interaction between dislocations and grain boundaries[J]. Acta Mater. 2007, 55 (3): 849-856

[4] Dupont V, Sansoz F. Grain boundary structure evolution in nanocrystalline Al by nanoindentation simulations[C]. Mater. Res. Soc. Symp. Proc. 2006, 903: 61-66

[5] Miller R E, Tadmor E B. The Quasicontinuum Method: Overview, applications and current directions[J]. J. Comput.-Aided Mat. Des. 2002, 9 (3): 203-239

[6] Ercolessi F, Adams J B. Interatomic potentials from first-principles calculations - the force-matching method[J]. Europhys. Lett. 1994, 26: 583.

[7] Tadmor E B, Miller R, Phillips R. Nanoindentation and incipient plasticity[J]. J. Mater. Res. 1999, 14 (2): 2233-2250.

Authors:

JI Zhong (1966-), male, professor of Shandong University. Research area: micro plastic forming and multiscale numerical simulation. Email: zhongji@umich.edu

HOU Shuai (1988-), male, graduate of Shandong University. Research area: nanomaterials

Journal of Physics: Conference Series **188** (2009) 012054 doi:10.1088/1742-6596/188/1/012054

Correlation between photoluminescence and oxygen vacancies in In_2O_3, SnO_2 and ZnO metal oxide nanostructures

Ping Wu,[1] Qiang Li, Xingquan Zou, Wende Cheng, Danli Zhang, Chuanxi Zhao, Lingfei Chi, Tan Xiao
Department of Physics, Shantou University, Shantou 515063, People's Republic of China
E-mail: pwu@stu.edu.cn (P. Wu)

Abstract. In_2O_3, SnO_2 and ZnO nanostructure have been synthesized by thermal evaporation method and their photoluminescence have been investigated to explore the light emission mechanism. All of the three as-grown metallic oxide nanostructures have a strong blue or green photoluminescence at room temperature. However the blue or green light emission disappears after the as-grown materials are oxidized in air at high temperature, and then reappears again after the oxidized products are deoxidized in forming gas of argon and H_2 at high temperature. The common feature is referred to that the blue or green emission originates from the oxygen vacancies.

1. Introduction

In recent years semiconducting oxides nanostructures, such as ZnO[1,2], In_2O_3[1,3], SnO_2[1,3,4], SnO [4], Ca_2O_3 [5], MgO [6] and so on, have been synthesized and the growth mechanism have been discussed. In particularly, metal oxide nanostructures (e.g. ZnO, In_2O_3, SnO_2) have been studied for applications ranging from optoelectronic devices [7-10] field-effect transistors [11], ultra-sensitive nanoscale gas sensors [12] and field emission cathode emitters [13]. Among them, the wide-band-gap semiconducting oxide nanostructure with room-temperature photoluminescence (PL) in blue-light and ultraviolet laser play an important role in laser printing, information storage and nanoscale optoelectronic devices. The photoluminescence of these three oxide nanostructures attract more research interests, with many research findings reported. It is known that the bulk In_2O_3 can't emit light at room temperature [14]. However, a strong and broad PL emission spectrum recorded from the In_2O_3 nanofibers is reported with its maximum intensity centered at 470 nm [15]. While it is reported that the In_2O_3 nanowires have PL peaks centered at 425, 429, 442, 460 nm [16], 398 nm [17], 416 and 435 nm [18]. The nanocubes have a strong PL peak centered at 450 nm [19]. The authors [15-18] thought the photoluminescence peaks in the visible emission would be attributed to the oxygen vacancies. The emission thus resulted from the radioactive recombination of a photo-excited hole with an electron occupying the oxygen vacancies.

The energy gap of bulk SnO_2 is 3.62 eV. However for different morphologies of SnO_2 nanostructure,

[1] Author to whom any correspondence should be addressed.

the room temperature photoluminescence are different too. For SnO_2 nanowires the PL peaks centered at 602 nm, which is contributed to that oxygen vacancies interact with interfacial tin vacancies and lead to the formation of a considerable amount of trapped states within the band gap, resulting in a strong PL signal [20]. In the as-grown SnO_2 nanoblade, an intensive blue emission centered at 445 nm was observed, which can be attributed to oxygen-related defects that have been introduced during growth [21]. And the 500 nm PL emission was obtained in SnO_2 nanoribbles [22]. Luo et al [23] demonstrated that the 480 and 600 nm emission peaks originate from defect electronic states formed by surface oxygen vacancies and 435 nm luminescence emission peaks from the in-plan oxygen vacancies.

For different morphologies of nanostructure ZnO, the different room temperature PL emissions were reported too. For ZnO nanowires, the photoluminescence spectra showed a ultra-violet (UV) emission at 3.26 eV and a green emission at 2.44 eV, which are attributed to near band-edge transition and radial combination of a singly ionized oxygen vacancies with a photo-induced hole, respectively [24]. The UV emission peaks at 380 nm and green emission at 419 nm [25] have also been reported for nanowires. The ZnO nanowires have a weak UV peak at 380 nm and a strong green band peak at 520 nm depending upon the processing conditions [26]. According to that samples annealed in vacuum showed a stronger green band peak the author indicated that the deep-band green light emission is due to the radiative recombination of the photo-generated holes with electrons belonging to the oxygen vacancy of the surface [26]. The nanorods have PL emission centered at 384 nm [27] and nano-tetrapod ZnO have a 491 nm [28], 495 nm and 520 nm [29]. Usually the photoluminescence of ZnO includes near-band-edge ultraviolet emission and deep-level emission (usually broad green, blue peaks), where the latter is relate to structure defects and impurity [30]. Vanheusden [31] demonstrated that the green emission bands are attributed to radial combination of a singly ionized oxygen vacancy with a photo-induced hole. Vanheusden et al. measured oxygen-vacancy density, free-carrier concentration and the PL spectrum in the research of green photoluminescence in ZnO phosphor powder, confirming that oxygen vacancy is responsible for the visible emission in metal oxide structures.

For ZnO, In_2O_3, SnO_2 and many other nanostructures, such as ZrO_2 nanowires [32], there are so many authors speculate that the origin of strong emission in visible emission is due to oxygen vacancies produced in the growth of nanomaterial. [15-21,23,24,26,31] For the wide-gap oxide nanomaterials, oxygen vacancy in the crystalline will induce the formation of new energy levels in the energy gap. Although there are so many authors contribute the origin of photoluminescence in visible range to the radiative recombination of a photo-generated hole with an electron occupying the oxygen vacancy, the theoretic or experimental supporting is few. The theoretic calculation identifies electron-hole radiative recombination at the V_o center as the source of the green luminescence in nano-ZnO [33]. Vanheusden et al [31] confirm that oxygen vacancy is responsible for the visible emission in metal oxide structures. For SnO_2 nanowires, the indirect experimental result is from Fourier Transform Infrared spectrum, obtained from as-grown and annealing of SnO_2 nanowires [23]. The peak at 581 cm^{-1} in Raman spectrum shows the existence of oxygen deficiency [34]. However, some authors have different ideas. Roy et al [35] fabricated ZnO nanorods in different conditions and found that significant differences in the ratio of UV and visible emission and the visible emission cannot be fully explained by a single type of defect. Djurišić et al [36] found that the transition between a single charged oxygen vacancy and photo-excited hole does not explain the green emission in all ZnO samples.

So the direct or even the indirect proof is needed, to support or not support the oxygen vacancy. In this paper, three nanostructures, including In_2O_3 nanowires, SnO_2 nanoblade and ZnO nanotetrapod, were synthesized by thermal evaporation method. The photoluminescence of as-grown, oxidized, and deoxidized ZnO, In_2O_3, SnO_2 nanostructure were investigated and the results provided an indirect proof supporting the oxygen vacancies.

2. Experimental details

Our experimental set-up for thermal evaporation consists of a furnace system equipped with a proportional-integral-differential (PID) controller and a horizontal quartz tube of 7 cm in diameter and 150 cm in length. The raw materials were a mixture of active carbon powder and analytical reagent In_2O_3 powder, SnO powder or zinc powder to synthesize In_2O_3, SnO_2 and ZnO nanostructure, respectively. A clean p-type Si substrate covered on the raw materials with a 1 cm separation to collect the product. The raw material and the substrate were placed in the constant temperature zone at the middle of the tube. High purity argon (99.99% purity) gas was used as a carrier gas. Argon gas was used first to purge the reactor for 30 min before the furnace was heated to set-up temperature with a constant elevating rate. Then the temperature was kept at that set-up temperature for a certain time to grow nanostructures. Finally those products cooled down to the room temperature under the protection of argon flow. For the growth of In_2O_3, SnO_2 and ZnO nanostructures, the carrier gas flow rates are 200, 200 and 240 sccm, and corresponding temperature elevating rates 20, 25 and 50 ^{o}C/min, respectively. The growth temperatures of In_2O_3, SnO_2 and ZnO are 1050 ^{o}C, 1100°C and 750 ^{o}C, with 120 min, 240 min and 30 min, respectively.

Scanning electron microscope (SEM, JOESL JSM-6360LA) equipped with the Energy Dispersive X-ray Detector and high resolution transmission electron microscope (TECNAI F30) were employed to characterize the morphologies and composition of the products. X-ray diffractometer (XRD, Panalytical X-pert, CuKα) were employed to analyze the crystal structures. The photoluminescence spectrum was measured at room temperature by using the fluorescence spectrometers (Edinburgh Instruments FLS 920).

3. Results and discussion

Figure 1 shows the morphology, crystalline and composition of as-grown In_2O_3 nanowires. Figure 1 (a) is SEM image of as-grown In_2O_3 nanowires, with several hundreds of nanometer in diameter and several tens of micrometer in length. Figure 1 (b) is the X-ray diffraction pattern of the product. All diffraction peaks can be indexed to a body-centered cubic (bcc) structure with a lattice constant of *a* =1.011 nm (PDF card No. 06-0416). These peaks are almost the same as that of bulk In_2O_3, except that the strongest diffraction peak comes from (400) rather than (222). This difference suggests that the as synthesized products have a preferential growth direction along <100>. [37] In figure 1 (c), the energy dispersion spectrum shows that the compositions of the nanowires are In and O with slight deficiency of oxygen compared with the stoichiometric In_2O_3.

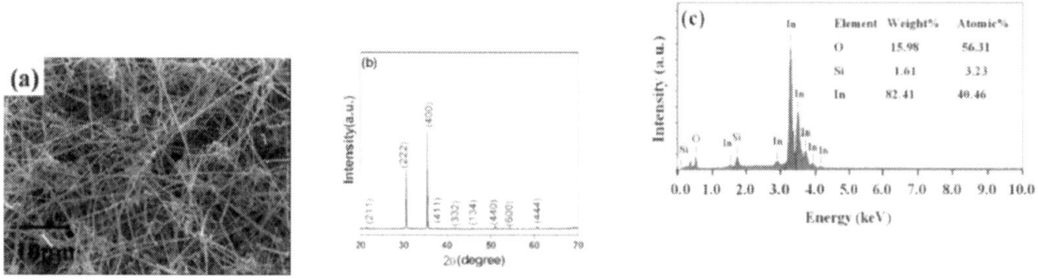

Figure.1 The morphology, crystalline and composition of as-grown In_2O_3 nanowires: (a) SEM image; (b) XRD pattern; (c) EDX spectrum.

The SEM image of synthesized SnO_2 nanoblades is shown in figure 2(a) and the XRD pattern is shown in figure 2 (b). The SnO_2 nanobelts are rutile structure with all the diffraction peaks indexed (PDF card No.88-0287). The EDX analysis shown in figure 2 (c) indicates slight deficiency of oxygen compared with the stoichiometric SnO_2.

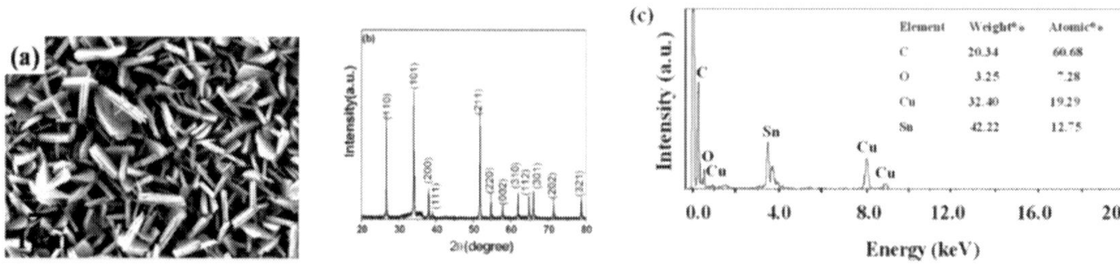

Figure.2 The morphology, crystalline and composition of as-grown SnO_2 nanoblades: (a) SEM image; (b) XRD pattern; (c) EDX spectrum.

For the ZnO nanostructure, SEM image, XRD pattern and EDX spectrum are shown in figure 3 (a), (b) and (c), respectively. The as-grown nano-ZnO consists of four legs, linked together at one end to form tetrapodlike morphology. The diameter of legs varies from 100 nm to 300 nm with a smaller end tip (Figure 3 (a)). XRD pattern is shown in figure 3 (b). The diffraction peaks can be indexed as the hexagonal wurtzite structure of ZnO (PDF card No. 89-1397) with the lattice parameters of a = 0.3249 nm and c = 0.5206 nm. The as-prepared products have a pure single phase of hexagonal wurtzite structure. EDX spectrum shown in figure 3 (c) gives the deficient of oxygen compared with the stoichiometric ZnO.

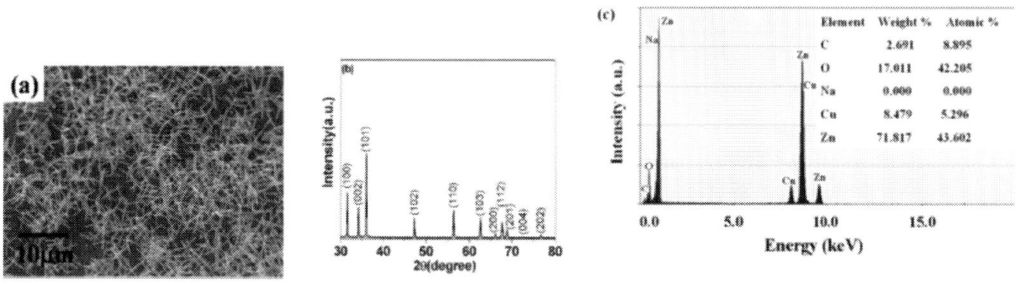

Figure.3 The morphology, crystalline and composition of as-grown ZnO tetrapods: (a) SEM image; (b) XRD pattern; (c) EDX spectrum.

The photoluminescence spectra of the as-grown, oxidized and deoxidized of the In_2O_3, SnO_2 and ZnO nanostructures were measured under the excitation at 325 nm. Figure 4 (a) shows the PL spectrum obtained from the as-grown In_2O_3 nanowires. The peak centered at 394 nm must be originated from the free exciton emission around 3.14 eV from the wide-band-gap In_2O_3 nanowires. The stable blue light was also observed at 423 nm, 436 nm and 487 nm. After the as-grown In_2O_3 nanowires was oxidized in air at 850℃ for 3h, the blue emission peaks disappeared, shown in figure 4 (b). If we deoxidized the oxidized In_2O_3 nanowires in the forming gas of 300 sccm argon and 30 sccm H_2 at 750℃ for 1 hour, the blue emission appeared again, with a slight stronger blue emission than that of the as-grown In_2O_3 nanowires. The result is shown in figure 4 (c).

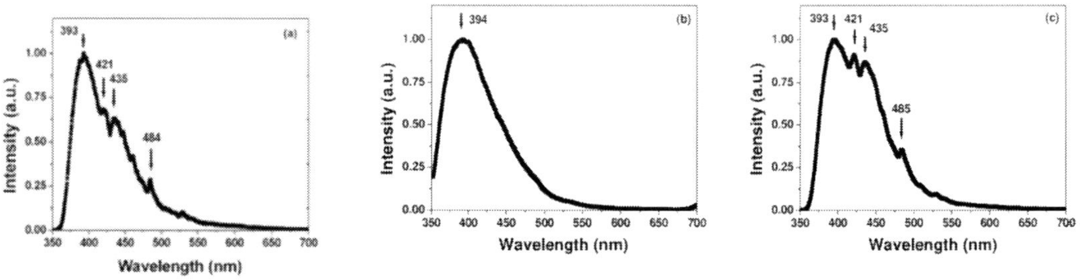

Figure.4 The room-temperature PL spectrum of In_2O_3 nanowires measured from: (a) as-grown In_2O_3 nanowires ; (b) oxidized in air at 850℃ for 3h ; (c) deoxidized the oxidized-In_2O_3 nanowires in the forming gas of 30 sccm H_2 and 300 sccm argon at 750℃ for 1 h.

The PL spectra of the as-grown, the oxidized and deoxidized of the oxidized- SnO_2 nanoblades are shown in figure 5(a), 5(b) and 5(c), respectively. Here again, the blue and green emission obtained from the as-grown SnO_2 nanoblades (Fig. 5(a)) disappeared after the as-grown SnO_2 nanoblades annealed in air at 900℃ for 3h (Fig. 5(b)). When the oxidized-SnO_2 nanobelts were deoxidized in the forming gas of 300 sccm argon and 30 sccm H_2 at 750℃ for 30 min, the blue and green emission reappeared again (Fig. 5(c)).

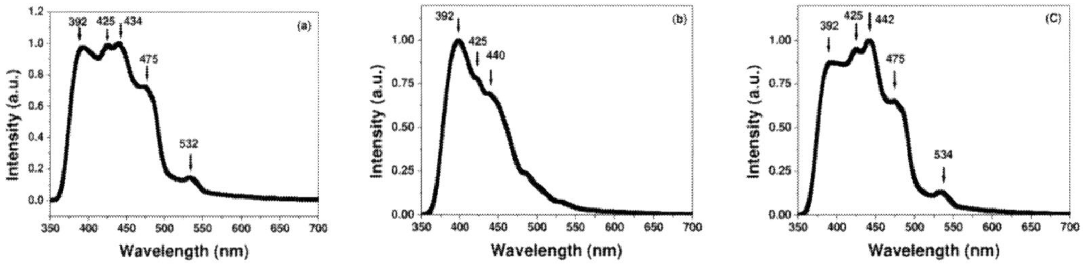

Figure 5. The room-temperature PL spectrum of SnO_2 nanobelts measured from: (a) as-grown SnO_2 nanobelts ;(b) oxidized in air at 900℃ for 3h ; (c) deoxidized oxidized-SnO_2 nanobelts in the forming gas of 30 sccm H_2 and300 sccm argon at 750℃ for 30 min.

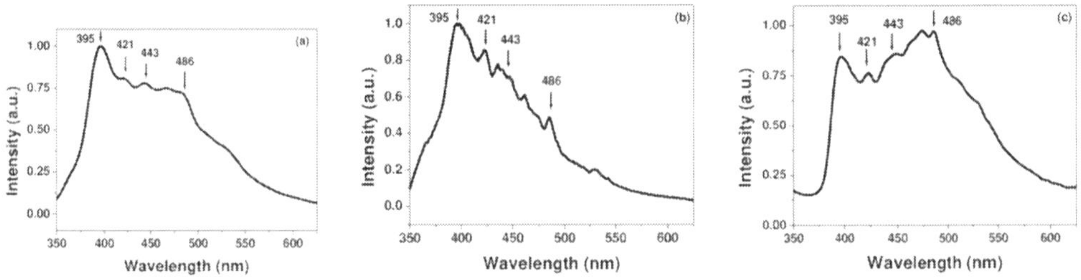

Figure 6. The room-temperature PL spectrum of ZnO nanostructure measured from: (a) as-grown nano-ZnO; (b) oxidized in air at 700℃ for 1 hour; (c) deoxidized the oxidized-ZnO nanostructure in the forming gas of 30 sccm H_2 and 150 sccm argon at 700℃ for 1 hour.

The PL spectra of the as-grown, the oxidized and deoxidized of the oxidized-ZnO nanostructure are shown in figure 6(a), 6(b) and 6(c), respectively. Once more, the blue and green emission in the as-grown ZnO nanostructure (Fig. 6(a)) disappeared after the as-grown ZnO nanostructure annealed in air at 700℃ for 1 hour (Fig. 6(b)). When the oxidized-ZnO nanostructure deoxidized in the forming gas of 30 sccm H_2 and 150 sccm argon at 700℃ for 1 hour, the blue and green emission reappeared again (Fig. 6(c)).

The photoluminescences of these three metallic oxide nanomaterial consist of the near band edge of the ultraviolet (UV) and visible emission. Obviously, the changes of the luminescence in the visible must be originated from the surface state change. Annealing in oxygen-rich atmosphere or reducible atmosphere is an easy way to change the concentration of surface oxygen vacancy. Oxidation in air has the effect of physically removing oxygen vacancy sites. Since the oxidization reduces partial surface oxygen deficiencies and leads to a higher stoichiometric surface. In other words, the decrease of the oxygen vacancy sites results in the decrease of the defect states localized in the energy gap. The decrease of the defect states directly results in the disappearance or diminishment of the emission peaks in visible range. From the experimental disappearance or diminishment of the luminescence in the oxidized samples, it is reasonably inferred that these luminescence are caused by surface oxygen vacancies. On the other hand, deoxidized in the reduction forming gas removes surface oxygen and then destroys the surface stoichiometric structure and increases the structure disorder. The more disorder leads to an increase of defect states in the energy gap and then the enhancement of the emission intensity. That's why the photoluminescence peaks in all of the deoxidized samples exhibit reappearance or even enhancement.

4. Conclusion

The In_2O_3 nanowires, SnO_2 nanoblade and ZnO nanotetrapods have been synthesized by thermal evaporation high purity of In_2O_3 powder, SnO powder and metal zinc powder, respectively. The photoluminescence of the as-grown nanomaterial was compared with that sample annealed in air, and then in forming gas of argon and H_2 at high temperature. All of the three as-grown metallic oxide nanostructures have a strong visible photoluminescence at room temperature. As expected, the visible emission disappears after the as-grown materials are oxidized in air at high temperature, and then reappears again after the oxidized products are deoxidized in forming gas of argon and H_2 at high temperature. The common PL feature of the nanomaterials with different crystal and morphologies suggests that the visible emission originates from the oxygen vacancies.

Acknowledgments

Financial supports from the Natural Science Foundation of China (Grant No. 10572079 and 10772102), the Natural Science Foundation of Guangdong Province, China (Grant No.07008103) and Research & Development Foundation of Shantou University (130-934062) are gratefully acknowledged.

References

[1] Pan Z W, Dai Z R and Wang Z L 2001 Science **291** 1947
[2] Kong X Y, Ding Y, Yang R and Wang Z L 2004 Science **303** 1348
[3] Wang Z L, 2003 Adv. Mater. **15** 432
[4] Dai Z R, Gole J L, Stout J D and Z. L. Wang 2002 J. Phys. Chem. B **106** 1274
[5] Zhang H Z, Kong Y C, Wang Y Z, Du X, Bai Z G, Wang J J, Yu D P, Ding Y, Hang Q L and Feng S Q 1999 Solid State Commun, **109** 677
[6] Yang P and Lieber C M 1997J. Mater. Res. **12** 2981
[7] Kind H, Yan H, Messer B, Law M and Yang P 2002 Adv. Mater. **14** 158
[8] Park W I and Yi G.-C.2004 Adv. Mater. **16** 87
[9] Könenkamp R, Word R C and Schlegel C 2004 Appl. Phys. Lett. **85** 6004
[10] Keem K, Kim H, Kim G.-T., Lee J. S.2004 Appl. Phys. Lett. **84** 4376
[11] Arnold M S., Avouris P, Pan Z W and Wang Z L.2003 J. Phys. Chem. B **107** 659
[12] Wang H T, Kang B S, Ren F, Tien L C, Sadik P W, Norton D P, Pearton S J and Lin J 2005 Appl. Phys. Lett. **86** 243503
[13] Zhou J, Xu N.-S., Deng S.-Z., Chen J, She J.-C. and Wang Z.- L.2003 Adv. Mater. **15** 1835
[14] Ohhata Y, Shinoki F and Yoshida S 1979 Thin Solid Films **59** 255
[15] Liang C, Meng G, Lei Y, Phillipp F, and Zhang L.2001Adv. Mater. **13** 1330
[16] Zheng M J, Zhang L D, Li G H, Zhang X Y and Wang X F 2001 Appl. Phys. Lett. **79** 839
[17] Cao H, Qiu X, Liang Y, Zhu Q and ZhaoM 2003 Appl. Phys. Lett. **83** 761
[18] Wu X C, Hong J M, Han Z J, Tao Y R 2003 Chem. Phys. Lett., **373** 28
[19] Tang Q, Zhou W, Zhang W, Ou S, Jiang K, Yu W and Qian Y 2005 Crystal Growth & Design, **5**, 147
[20] He H Jr., Wu T H, Hsin C L, Li K M, Chen L J, Chueh Y L, Chou L J and Wang Z L 2006 Small **2** 116
[21] Hera Y.-C., Wu J.-Y., Lin Y.-R. and Tsai S.-Y 2006 Appl. Phys. Lett. **89** 043115
[22] Hu J Q , Bando Y, Golberg D 2003 Chem. Phys. Letters **372** 758
[23] Luo S, Chu P K, Liu W, Zhang M and Lin C, 2006 Appl. Phys. Lett. **88** 183112
[24] Lyu S C, Zhang Y, Ruh H, Lee H.-J., Shim H.-W., Suh E.-K. Lee C J 2003 Chem. Phys. Letters **363** 134
[25] Huang M H, Wu Y. Feick H, Tran N, Webber E and Yang P 2001 Adv. Matt. **13** 113

[26] Banerjee D, Lao J Y, Wang D Z, Huang J Y, Ren Z F, Steeves D, Kimball B and Sennett M 2003 Appl. Phys. Lett. **83** 2061
[27] Wang X, Summers C J and Wang Z L 2004 Nano Lett. **4** 423
[28] Yu W D, Li X M and Gao X D 2004 Appl. Phys. Lett. **84**, 2658
[29] Roy V A L, Djurišić A B, Chan W K, Gao J, Lui H F and Surya C 2003 Appl. Phys. Lett. **83** 141
[30] Dingle R 1969 Phys. Rev. Lett. **23** 579
[31] Vanheusden K, Seager C H, Warren W L, Tallant D R and Voigt J A 1996 Appl. Phys. Lett. **68** 403
[32] Cao H Q, Qiu X, Luo B, Liang Y, Zhang Y, Tan R, Zhao M, Zhu Q 2004 Adv. Funct. Mater. **14** 245
[33] Zhang S B, Wei S H and Zunger A 2001 Phys. Rev. B. **63** 075205
[34] Wang X, Li Q, Liu Z, Zhang J, Liu Z, and Wang R 2004 Appl. Phys. Lett., **84** 4941
[35] Roy V A L, Djurišić A B, Chan W K, Gao J, 2003 Appl. Phys. Lett. **83** 141
[36] Djurišić A B, Leung Y H, Choy W C H, Cheah K W, and Chan W K 2004 Appl. Phys. Lett. **84** 2635
[37] Gao T and Wang T 2006 J. Crystal Growth **290** 660

Enhanced Efficiency of a Fluorescing Nanoparticle with a Silver Shell

Wallace C. H. Choy[1]*, Xue-Wen Chen[1,2], and Sailing He[2]
1. Department of Electrical and Electronic Engineering, University of Hong Kong, Pokfulam Road, Hong Kong
2. Centre for Optical and Electromagnetic Research, Zhejiang University, Zhijingang campus, Hangzhou 310058, China.

*E-mail: chchoy@eee.hku.hk

Abstract. Spontaneous emission (SE) rate and the fluorescence efficiency of a bare fluorescing nanoparticle (NP) and the NP with a silver nanoshell are analyzed rigorously by using a classical electromagnetic approach with the consideration of the nonlocal effect of the silver nano-shell. The dependences of the SE rate and the fluorescence efficiency on the core-shell structure are carefully studied and the physical interpretations of the results are addressed. The results show that the SE rate of a bare NP is much slower than that in the infinite medium by almost an order of magnitude and consequently the fluorescence efficiency is usually low. However, by encapsulating the NP with a silver shell, highly efficient fluorescence can be achieved as a result of a large Purcell enhancement and high out-coupling efficiency (*OQE*) for a well-designed core-shell structure. We also show that a higher SE rate may not offer a larger fluorescence efficiency since the fluorescence efficiency not only depends on the internal quantum yield but also the *OQE*.

1. Introduction

Fluorescent nanomaterials have been the subject of intensive research in recent years for their vast applications ranging from biomedical therapeutics and diagnostics to information storage and optoelectronics [1-3]. For fluorescence based applications, fluorescence efficiency, i.e. the external quantum efficiency (*EQE*) of the emitter, is an important issue. Due to the existence of pronounced nonradiative decay of excitons in the nanoscale structure, low *EQE* is an often-observed feature. Most of the strategies so far employed aim to reduce the nonradiative decay rate [4-6] for improving the fluorescence efficiency. The direct and effective approach to improve *EQE* is to increase radiative decay rate[7] and out-coupling efficiency (*OQE*) since *EQE* is the product of internal quantum yield (i.e. internal quantum efficiency) (*IQE*) and *OQE*. The enhancement of radiative decay rate results in the Purcell enhancement of *IQE* [8,9]. Here, we will address this issue.

The radiative decay rate, i.e. spontaneous emission (SE) rate, can be enhanced by utilizing the surface plasmon effect [10-12] or the microcavity effect[13]. According to the near field nature of surface plasmon, the SE rate in a dielectric nanoparticle (NP) encapsulated with a metallic shell can be greatly enhanced. However, the metallic shell introduces extra absorption loss and may result in a small *OQE*. Consequently, fluorescence efficiency may still be low although the SE rate is greatly enhanced. Thus a rigorous theoretical study on the SE rate and *OQE* of emitters in nanoscale structure is desirable for getting a better physical understanding and optimal design of the nano-structure to

Journal of Physics: Conference Series **188** (2009) 012055 doi:10.1088/1742-6596/188/1/012055

achieve high-efficiency fluorescence. In this paper, the SE rates and *EQE* of emitters in a spherical NP with and without a silver nanoshell are rigorously investigated through a classical electromagnetic approach with the consideration of the nonlocal effect of the silver nanoshell [14-15].

Different from the SE rate in a dielectric particle with a diameter around several wavelengths [16], the SE rate in a NP with a diameter below 100nm is considerably slower than that in free space by almost an order of magnitude. Our results show that the fluorescence of a bare NP is inefficient since nonradiative decay outweighs radiative decay. By encapsulating the NP with a silver nano-shell, the SE rate can be enhanced by orders of magnitude due to the resonant excitation of surface plasmons. However, our results also show that *OQE* is usually low for very high SE rate.

2. Result and Discussion

In this section, the SE properties of emitters in a bare NP and the NP with a silver nanoshell are theoretically studied. The dependences of the fluorescence efficiencies on the structural parameters of NP are investigated for two initial *IQE*, i.e. *IQE* =0.25 and *IQE* =0.75. NP under study here is a dielectric sphere and its material is not restricted here and in principle can be any type of fluorescent nanomaterials. In our model, the refractive index of NP is set to be 2.0 over the wavelength of interest. The dielectric function of bulk silver [17] is used in the calculation.

2.1 SE in a bare NP

SE properties of the emitters in a bare spherical NP with a radius r is studied and discussed in this subsection. Fig.1 (a) plots the wavelength dependence of the normalized SE rate of emitters located at the center of the NP with a radius of 20nm. One sees that the SE rate is nearly one tenth of the SE rate in free space and decreases gradually as the wavelength increases. The size dependence of the normalized SE rate at the wavelength of 500nm is shown in Fig.1 (b) as the solid line with filled squares. The SE rate decreases as the radius of the NP decreases. As the radius further decreases, the SE rate has a limiting value around 0.11. This value depends on the refractive index of the NP and becomes smaller as the refractive index increases. Fig.1 (b) also shows *EQE* for the cases of *IQE* =0.25 and *IQE* =0.75 at the wavelength of 500nm. Due to the low SE rate, *EQE* are much smaller than their *IQEs* that the power radiated to farfield U equals averaged total radiation power (F) in the case of a bare NP. The degradation of *EQE* for *IQE* =0.25 is more pronounced than that for *IQE* =0.75. This is because the ratio of radiative decay rate to nonradiative decay rate for *IQE* =0.25 drops faster as F decreases. But even for *IQE* =0.75, η_{ext} drops to about 20% at r = 20nm. Consequently, a bare NP usually shows inefficient fluorescence.

2.2 SE in the NP with a silver nanoshell

2.2.1 Thickness effect of the silver shell

The thickness effect of the silver shell on the SE properties is studied for the silver encapsulated NP with various shell thickness s and a fixed core radius r of 30nm as shown in Fig. 2(a). As compared with a bare NP, the present case shows some remarkable features. First of all, the spectrum of the SE rate shows a resonant structure and the SE rate at the resonant wavelength is several orders of magnitude larger than the SE rate in a bare NP due to the resonant excitation of surface plasmons. The large increment of the SE rate will result in a great enhancement of *IQE* due to the Purcell enhancement. Secondly, another distinct feature of Fig.2 (a) is that the resonant wavelength of the spectrum of the SE rate shows a blue shift as s increases. For the shells thinner than 5nm, the resonant wavelength shifts out of the visible wavelength range. Meanwhile, for the shells thicker than 30nm, the resonant wavelength shows no obvious shift and the spectrum of the SE rate flattens out as s further increases. In addition, the peak value of the normalized SE rate (hereafter named "the peak SE rate") is reduced as s increases. To understand the reduction of the peak SE rate, we define an effective radius as $R=2\pi r/\lambda_r$, where λ_r is the resonant wavelength, to characterize the strength of surface plasmon effects. The surface plasmon effect decreases exponentially as R increases. The inset of Fig.2(a) shows the dependence of the effective radius on the shell thickness. It can be observed that the effective radius firstly increases as s increases. Thus in Fig.2 (a) the peak SE rate is reduced and the resonant spectrum broadens as s increases from 5nm to 30nm. In another word, the reason why the peak SE rate is larger for thinner silver shell is that the strength of surface plasmon effect is larger since the effective radius R is smaller. For thicker silver shell (s > 30nm), the microcavity effect of the silver shell gradually dominates and the resonant spectrum sharpens again as s further increases.

Fig.2 (b) shows the dependence of the fluorescence efficiency at the resonant wavelength on the shell thickness. As compared with Fig. 1(b) for a bare NP, the fluorescence efficiency of the NP encapsulated with a suitable thickness of silver shell is much larger. Moreover, the fluorescence efficiency for the core-shell structure with r = 30nm and s = 25nm is much larger than the initial *IQE*, particularly for *IQE* =0.25 where a nearly 100% increment is achieved. This is due to a combination of the Purcell enhancement of *IQE* and a high *OQE*. From Fig.2 (a) and (b), one sees that for the NP with ultra-thin silver shell although the SE rate is very large the fluorescence efficiency is still low since *OQE* is small, i.e. most of the emitted power is absorbed in the silver shell. As shown in Fig. 2(b), *EQE* for the cases of *IQE* =0.25 and *IQE* =0.75 increases as s increases from 5nm to 25nm. As s further increases fluorescence decreases gradually since the absorption loss in the Ag shell increases again. Consequently, from Fig.2 (a) and (b), one sees that the fluorescence efficiency can be maximized through simultaneously achieving a large Purcell enhancement and high *OQE* by carefully designing the Ag thickness.

2.2.2 Size effect of the NP

The size effect of the NP on the SE properties is studied for the silver encapsulated NPs with various core radii r and an optimized shell thickness s of 25nm as shown in Fig. 2(c). The resonant wavelength of the SE rate spectrum shows a red-shift as r increases. For r < 70nm, the peak SE rate in Fig.2(c) decreases as the radius of the NP increases. To explain this, the effective radius R is calculated and its dependence on the core size is plotted in the inset of Fig. 2(a). We see that even through the resonant wavelength increases with r, the effective radius R still increases with r. In other words, the strength of surface plasmon effect decreases as r increases and thus the peak SE rate is reduced and the resonant spectrum is broadened as r increases. For r>70nm, the microcavity effect of the silver shell gradually goes up and the spectrum of the SE rate becomes narrower and the peak value is increased as r increases.

Fig.2 (d) plots the size dependence of the fluorescence efficiency for *IQE* =0.25 and *IQE* =0.75 at the resonant wavelength. For r>120nm and r<15nm, the result is not shown here since the resonant wavelength is out of the visible range. As compared with Fig. 1(b), large enhancement of the fluorescence efficiency is obtained due to the Purcell enhancement of the *IQE* and a high *OQE*. For the NPs with a smaller radius although the SE rate is much higher the fluorescence efficiency is low since the *OQE* is low. The fluorescence efficiency at the resonant wavelength increases as the radius of NP increases.

3. Conclusion

In this paper, we have formulated a general approach for calculating the SE rate and *EQE* of the emitters in a spherically multilayered structure with arbitrary permittivity. In the presence of silver nano-shell, nonlocal effect has been taken into account. By using the method, the SE rates and fluorescence efficiencies of the emitters in a bare NP and the NP with a silver nanoshell are investigated. The SE rate in a bare NP is only one tenth of the SE rate in free space and consequently the fluorescence is usually inefficient. The SE rate can be enhanced by orders of magnitude through encapsulating the NP with a silver nanoshell due to the resonant excitation of the surface plasmon. The spontaneous emission properties of the emitters in the NP with a silver nanoshell are studied separately for two cases, i.e. (i) the size effect of the NP and (ii) the thickness effect of the silver shell. Through a proper design of the core radius and shell thickness, highly efficient fluorescence can be achieved as a result of the combination of the large Purcell enhancement of *IQE* and a high *OQE*.

References:

[1] P. N. Prasad, Introduction to Biophotonics, John Wiley & Sons, New York, 2003

[2] Y.P. Leung, W.C.H. Choy, I. Markov, G.K.H. Pang, H. C. Ong and T.I. Yuk, 2006 Appl. Phys. Lett., 88, 183110.
[3] J.P. Zimmer, S.W. Kim, S. Ohnishi, E. Tanaka, J.V. Frangioni and M. G. Bawendi, 2006 J. Am. Chem. Soc., 128, 2526.
[4] M.A. Hines and P. Guyot-Sionest, 1996 J. Phys. Chem. 100, 468.
[5] S. Haubold, M. Haase, A. Kornowski and H. Weller, 2001 Chemphyschem, 2, 331.
[6] N. Gaponik, D.V. Talapin, A.L. Rogach, K. Hoppe, E.V. Shevchenko, A. Kornowski, A. Eychmuller and H. Weller, 2002 J. Phys. Chem. B 106, 7177.
[7] J.R. Lakowicz, 2001 Anal. Biochem. 298, 1.
[8] E. M. Purcell, 1946 Phys. Rev., 69, 681.
[9] W. Lukosz, 1980 Phys. Rev. B 22, 3030.
[10] K. Okamoto, I. Niki, A. Shvartser, Y. Narukawa, T. Mukai, and A. Scherer, 2004 Nat. Mater. 3, 601.
[11] A. Moroz, 2005 Chem. Phys.317, 1.
[12] J. Enderlein, 2002 Appl. Phys. Lett. 80, 315.
[13] X.W. Chen, W.C.H. Choy, S. He and P. C. Chui, 2008 J. Appl. Phys. 103, 024301,.
[14] A.R. Melnyk, M.J. Harrison, 1970 Phys. Rev. B 2, 835 .
[15] P.T. Leung, 1990 Phys. Rev. B 42 7622.
[16] H. Chew, 1988 Phys. Rev. A 38, 3410.
[17] E. D, Palik, Handbook of Optical Constants of Solids Academic press, Boston, 1985

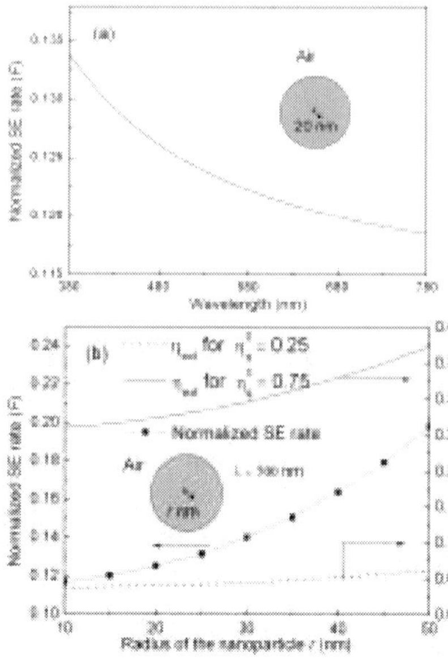

Fig.1 (a) Wavelength dependence of the normalized SE rate for a fixed nanoparticle r = 20nm (b) Variance of the normalized SE rate and the fluorescence efficiency for IQE=0.25 and IQE=0.75 with the increase of the size of the nanoparticle at the wavelength of 500nm.

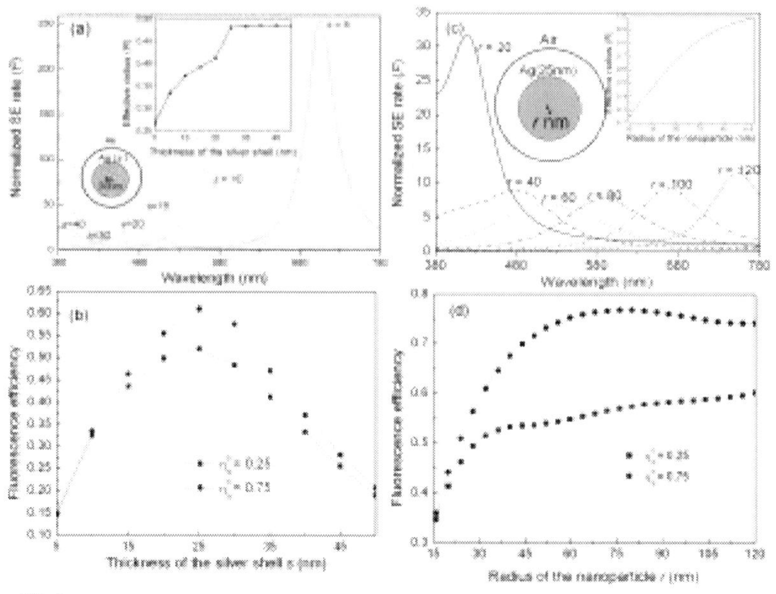

Fig2. (a) Wavelength dependence of the normalized SE rate in silver encapsulated nanoparticles of various Ag shell thicknesses s (nm) and fixed core radius; inset shows the variance of the effective radius with the increase of the shell thickness. (b) Variance of the fluorescence efficiency with the increase of Ag shell thickness and fixed core radius. (c) Wavelength dependence of the normalized SE rate in silver encapsulated nanoparticles of various core radii r (nm) and fixed Ag shell thickness; inset shows the variance of the effective radius with the increase of the core radius (d) Variance of the fluorescence efficiency with the increase of core radius and fixed Ag shell thickness.

Room-temperature Synthesis of $MnMoO_4 \cdot H_2O$ Nanorods by the Microemulsion-based Method and Its Photocatalytic Performance

Yan Mi, Zaiyin Huang,* Zeguang Zhou, Feilong Hu, Qiufeng Meng

College of Chemistry and Ecological Engineering, Guangxi University for Nationalities, Nanning 530006, PR China

E-mail: hzy210@yahoo.cn (Z Y Huang)

Abstract. Manganese molybdate hydrates ($MnMoO_4 \cdot H_2O$) nanorods have been synthesized at room temperature by a facile water-in-oil reverse microemulsion method. This technique was carried out in the reverse microemulsion of OP-10 (Po1yoxyethylene octylphenol ether)-n-octanol-water-cyclohexane with a water/surfactant molar ratio $\omega = 10$. Field-emission scanning electron microscopy (SEM) and transmission electron microscopy (TEM) images showed that the diameters of these formed nanorods about 70 nm and lengthe up to 4 μm, respectively. High-resolution transmission electron microscopy (HRTEM) results showed that each nanorod was formed by serveral nanobelts which are stacked by a layer-by-layer process. These unique nanorods demonstrate good photocatalytic properties.

1. Introduction

One-dimention (1-D) inorganic nanomaterials, such as nanotubes, nanowires, nanorods, and nanobelts, have become a research focus owing to their distinctive geometries, peculiar chemical and physical properties, and widely potential applications in nanodevices [1-4]. These 1D materials offer fundamental scientific opportunities for investigating the size- and shape-dependent optical, magnetic, and eletronic properties. [5-9]

Great efforts have been made to synthesize the 1D nanomaterials with uniform size and aspect ratios employing various strategies, such as microwave irradiation method [10-12], One-step γ-irradiation route [13], chemical vapor deposition method [14-15], reverse micelles [16-18] and hydrothermal approach [19-22]. Previous publications have shown that the facile reverse microemulsion solution process is a versatile strategy for the synthesis of inorganic nanomaterials with different sizes and shapes. A variety of 1D marterials including BaF_2 [23], ZnO [24], CdS [25], $BaCO_3$ [26], $PbSO_4$ [27], $CaSO_4$ [28] and $BaMoO_4$ [29] nanorods or nanowires have been successfully synthesized in the reverse microemulsion system.

Metal molybdates have received much attention in the past years due to their potential applications in photoluminescence, humidity sensors, scintillator materials, catalysis, and magnetic properties [30]. Zhang et al have synthesized $SrMoO_4$ nanowires form polyoxometalates via a facile hydrothermal process [31]. Li et al. reported the synthesis of single crystalline $BaMoO_4$ nanofibers in CTAB reverse microemulsions [29]. The solution synthesis of $CaMoO_4$ [32] and lanthanum molybdate nanoflakes [33] was also reported. Hydrothermal synthesis of transition metal molybadte hydrates $MMoO_4{\cdot}nH_2O$ (M = Co, Ni, Mn, n = 0, 3/4, 1) nano/microcrystals [20] was also investigated. Recently, our group fabricated high aspect-ratio $MnMoO_4{\cdot}H_2O$ nanorods with diameters of about 70 nm and lengths up to 4 μm using a simple reverse microemulsion method. The molar ratio of water : surfactant : oil is crucial for controlling the morphology of the products. Furthermore, such prepared $MnMoO_4{\cdot}H_2O$ nanorods showed good photocatalytic performance in the photodegradation of the methyl orange upon irradiation of UV light.

2. Experimental

In a typical preparation process, the reverse microemulsion with the surfactant-to-oil concentration of 0.4 M and $[H_2O]/[surfactant]$ molar ratio $\omega = 10$ were prepared by dissolving an aqueous solution of 0.1 M $Mn(CH_3COO)_2$ or 0.1 M Na_2MoO_4 in a solution of OP-10 in cyclohexane and n-octanol and stirring for 30 min, respectively, and then we got the two transparent microemulsions. Equal volumes of the above two microemulsions were then mixed slowly and stirred for another 5 minutes. Subsequently, such prepared mixture was incubated without stirring for 2 days at room temperature, resulting in the formation of white precipitates. The powder samples were collected by a centrifuge at 10000 rotation /min for 10 min, and washed several times with a mixed solution of acetone and ethanol (volume ratio 1:1) to remove the remaining organic. Pure deionized water and ethanol were subsequently used to rinse these products for 3 times, respectively. Finally, the products were dried at room temperature in vacuum for 12 hours.

3. Results and discussion

The X-ray powder diffraction (XRD) pattern of the as-prepared product is shown in Fig. 1. All the reflection peaks can be indexed as a pure, anorthic wolframite structure $MnMoO_4{\cdot}H_2O$ with cell parameters a = 5.776, b = 5.964, and c = 6.992Å, which agrees well with the standard data (JCPDS Card No. 78-0220). No obvious impurity phases were detected in the XRD pattern.

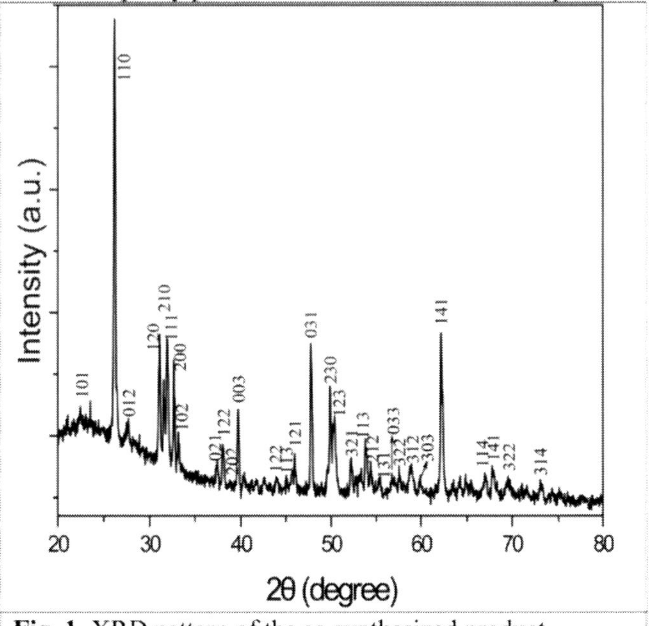

Fig. 1. XRD pattern of the as-synthesized product.

FESEM shows that there is a large quantity of $MnMoO_4{\cdot}H_2O$ nanorods with clean and smooth surface (see Fig. 2a). These nanorods have the length up to 4 μm and diameters of about 70 nm (Fig.

2b). More careful observation shows that these nanorods tended to spontaneously align side-by-side to form raftlike (Fig. 2c). Each nanorod was assembled by several nanobelts which are stacked by a layer-by-layer form (ellipse marked area in Fig. 2c), which was further investigated by HRTEM image (Fig. 2d). HRTEM image also indicated that these nanobelts are monocrystalline. The lattice fringe has a spacing of 3.3962 Å, [which can be indexed to the (110) plane of $MnMoO_4{\cdot}H_2O$ nanorods], implying that these nanorods grow preferentially along <110> direction. The single-crystalline structure and preferential growth orientation of the $MnMoO_4{\cdot}H_2O$ nanorods is also mirrored in the corresponding SEAD image shown in the inset of Figure. 2d.

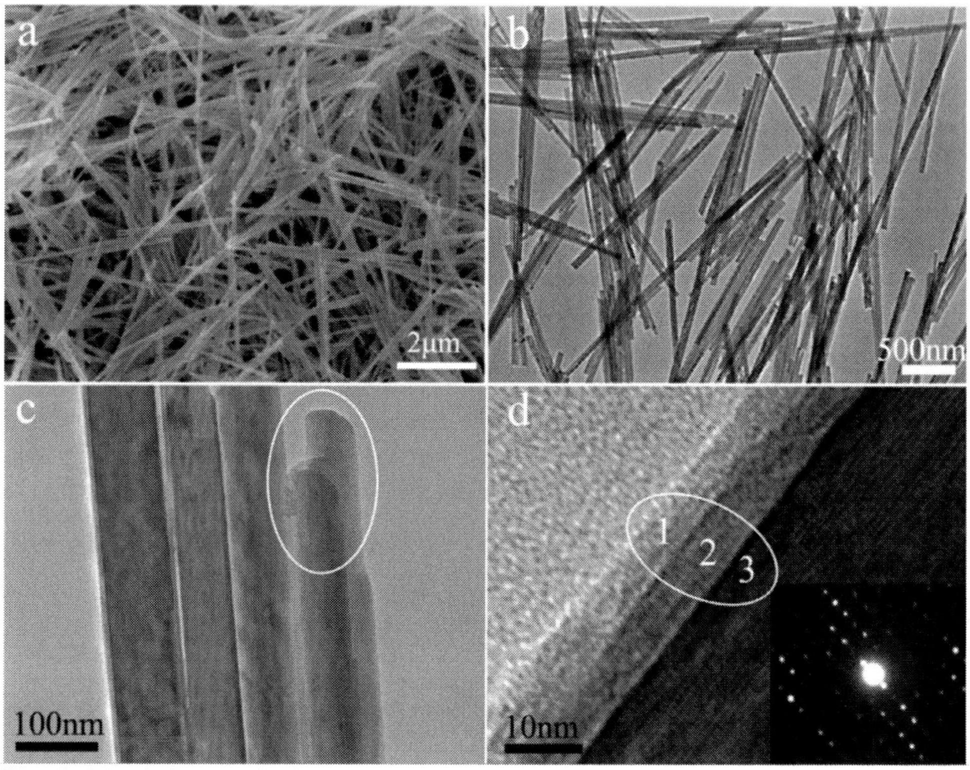

Fig. 2. Morphology and microstructure of the product (a) FESEM, (b) TEM, (c) mediate magnification TEM, and (d) high resolution TEM images, Inset in (d) shows the corresponding electron diffraction patterns.

It was found that the value of ω played an important role in the formation of the $MnMoO_4{\cdot}H_2O$ nanorods. As shown in Fig. 3a, the product obtained when ω = 5 was composed of irregular particles and some much shorter nanorods. Increasing ω to 20, three-dimensional honeycomb nanostructures (Fig. 3b) were obtained. Other factors such as the molar ratio between Mn^{2+} and MoO_4^{2-}, and reactant concentration exhibited a less pronounced effect on the fabrication of such nanorods. For example, when the initial reactant concentration was reduced to 0.05 M, the lengths and diameters of nanorods were about 1.5 μm and 100 nm, respectively. When it was increased to 0.2 M, the products were still nanorods with lengths of 1 μm and diameters of 135 nm (not shown here). Therefore the present reverse microemulsion method is a very effective route in fabricating the uniform $MnMoO_4{\cdot}H_2O$ nanorods. A possible mechanism is that the reverse microemulsion adopted a highly ellipsoidal microdroplet structure due to high concentration of surfactant, and the surfactants preferentially adsorb on the faces parallel to the c axis of the $MnMoO_4{\cdot}H_2O$ nanorods. However, the detailed growth mechanism of the $MnMoO_4{\cdot}H_2O$ nanorods is still unclear and need further confirmation.

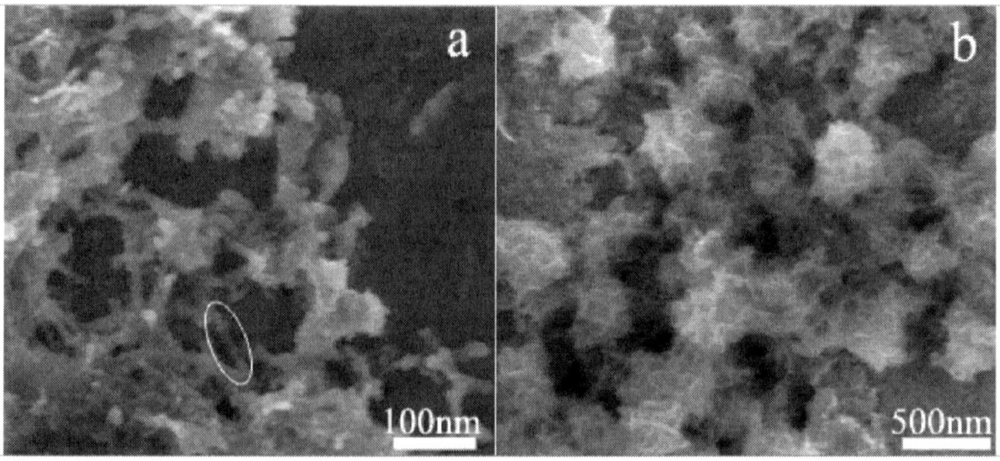

Fig. 3 FESEM images of $MnMoO_4 \cdot H_2O$ nanoparticles obained when (a) ω= 5 and (b) ω= 20.

To explore the potential application of the as-prepared $MnMoO_4 \cdot H_2O$ nanorods, the photocatalytic activities of the nanorods were investigated by choosing the photocatalytic degradation of methyl orange under 40 W UV light with continuous maganetic stirring. Fig. 4 shows the time-dependent UV-vis absorption spectra of the solution with initial concentration of 2.0×10^{-5} M (20 mL) methyl orange in the presence of 10 mg products under exposure to UV light at different times. The typical peak at 465 nm weakened and the color of the solution gradually lightened as the exposure time extended (Fig 4a), 95% of methyl orange was degraded after 3 hours of irradiation (Fig 4b). The concentration change versus time clearly shows the result that the as-prepared $MnMoO_4 \cdot H_2O$ nanorods showed high photocatalytic capability.

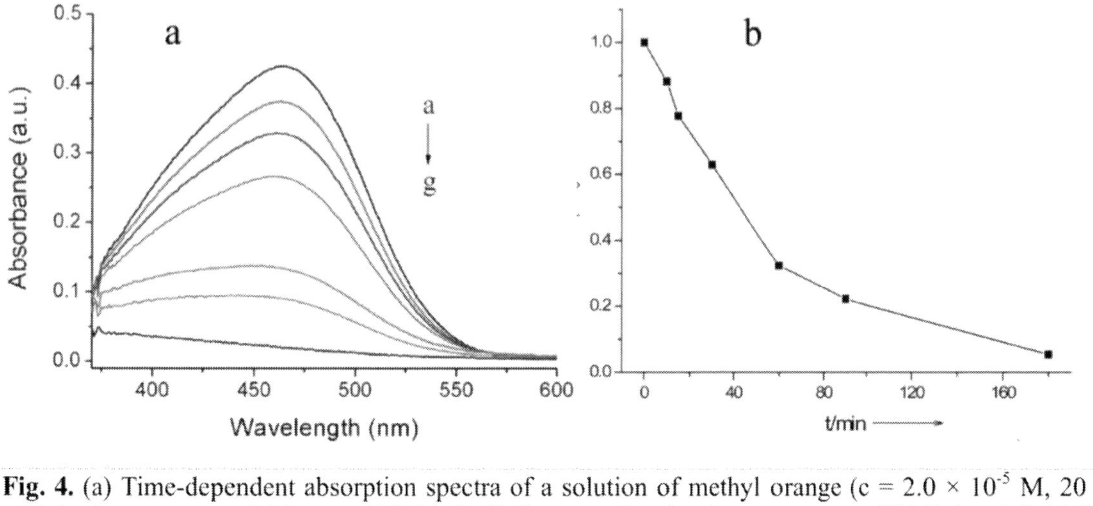

Fig. 4. (a) Time-dependent absorption spectra of a solution of methyl orange ($c = 2.0 \times 10^{-5}$ M, 20 mL) in the presence of 10 mg of $MnMoO_4 \cdot H_2O$ nanorods after its exposure to UV light, a) Initial solution; b-g) after exposure to UV light for 10, 15, 30, 60, 90 and 180 min; (b) Concentration evolution as a function of time (c_0, ct are concentrations at times 0 and t).

4. Conclusion

In summary, high aspect-ratio, single crystalline $MnMoO_4 \cdot H_2O$ nanorods have been synthesized in a selected reverse microemulsion system, and each nanorod was fabricated by serveral nanobelts which are stacked by a layer-by-layer form. The size of these nanorods can be tuned by surfactant concentration and the morphology is very sensitive to the ratio between water and oil. It is found that these nanorods showed a high photocatalytic activity. Further studies of the growth mechanism and properties of the nanorods are underway.

Acknowledgments. The authors thank the financial support of the Guangxi Natural Science Foundation of China No. 0575030 and No. 0832078.

References

[1] Wang Z L and Song J H 2006 *Science* 312 242
[2] Xia Y N, Yang P D, Sun Y G, Wu Y Y, Mayers B and Gates B 2003 *Adv Mater* 15 353
[3] Li L, Fang X S, Chew H G, Zhang F, Liew T H and Xu XJ 2008 *Adv Funct Mater* 18 1080
[4] Bavykin D V, Friedrich J M and Walsh F C 2006 *Adv Mater* 18 2807
[5] Habas S E, Yang P D and Mokari T 2008 *J Am Chem Soc* 130 3294
[6] Fang X S, Ye C H, Zhang L D and Xie T 2005 *Adv. Mater* 17 1661
[7] Li L, Yang Y W, Li G H and Zhang L D 2006 *Small* 2 548
[8] Fang X S, Bando Y, Ye C H, Shen G Z and Golberg D 2007 *J Phys Chem C* 111 8469
[9] Gantam U K, Fang X S, Bando Y, Zhan J H and Golberg D 2008 *Acs Nano* 2 1015
[10] Limaye M V, Gokhale S, Acharya S A and Kulkarni S K 2008 *Nanotechnology* 19 415602.
[11] Godinho M, Ribeiro C, Longo E and Leite E R 2008 *Cryst Growth Des* 8 384
[12] Panda A B, Glaspell G and El-shall M S 2006 *J Am Chem Soc* 128 2790
[13] Hu Y, Chen J F, Xue X, Li T W and Xie Y2005 *Inorg Chem* 44 7280
[14] Zhang H, Zhang Q, Tang J and Qin L C 2005 *J Am Chem Soc* 127 8002 .
[15] Lensch-Falk J L, Hemesath E R, Lopez F J and Lauhon L J 2007 *J Am Chem Soc* 129 10670 .
[16] Kwan S, Kim F, Akana J and Yang P D2001 *Chem Commun* 5 447
[17] Shi H T, Qi L M, Ma J M and Cheng H M 2002 Chem Commun 16 1704
[18] Yu T, Park J, Moon J, An K, Piao Y and Hyeon T 2007 *J Am Chem Soc* 129 14558
[19] Yu S H, Antonietti M, Cölfen H and Giersig M 2002 *Angew Chem Int Ed* 41 2356
[20] Ding Y, Wan Y, Min Y L, Zhang W and Yu S H 2008 *Inorg Chem* 47 7813
[21] Zhu H L, Yang D R, Yu G X, Zhang H, Jin D L and Yao K H 2006 *J Phys Chem B* 110 7631
[22] Wang J M, Khoo E, Lee P S and Ma J 2008 *J Phys Chem C* 112 14306
[23] Cao M H, Hu C W and Wang E B 2003 *J Am Chem Soc* 125 11196
[24] Zhang J, Sun L D, Jiang X C, Liao C S and Yan C H 2004 *Cryst Growth Des* 4 309
[25] Sinnons B A, Li S C, John V T, McPherson G L, Bose A and Zhou W L2002 *Nano Lett* 2 263
[26] Li L L, Chu Y, Liu Y, Dong L H, Huo L and Yang F Y 2006 *Mater Lett* 60 2138
[27] Xiang J H, Yu S H, Geng X, Liu B H and Xu Y 2005 *Cryst Growth Des* 5 1157
[28] Rees G D, Evans-Gowing R, Hammond S J and Robinson B H 1999 *Langmuir* 15 1993
[29] Li Z H, Du J M, Zhang J L, Mu T C, Gao Y N and Han B X 2005 *Mater Lett* 59 64
[30] Yu S H, Liu B, Mo M S, Huang J H, Liu X M and Qian Y T 2003 *Adv Funct Mater* 13 639
[31] Zhang Y M, Yang F D, Yang J, Tang Y and Yuan P 2005 *Solid State Commun* 133 759
[32] Chen D, Tang K B, Li F Q and Zheng H G 2006 *Cryst Growth Des* 6 247
[33] Xu Y P, Jiang D Y, Bu W B and Shi J L 2005 *Chem Lett* 34 978

Journal of Physics: Conference Series **188** (2009) 012057 doi:10.1088/1742-6596/188/1/012057

Hexavalent Chromium Removal from Water Using Chitosan-Fe^0 Nanoparticles

LI Tielong[1], GENG Bing[2], ZHANG Na[1], JIN Zhaohui[1*], QI Xinhua[1]

1College of Environmental Science and Engineering/Tianjin Key Laboratory of Environmental Remediation and Pollution Control/Ministry of Education Key Laboratory of Environmental Processes and Criteria, Nankai University, Tianjin 300071, China;2 Institute of Environment and Sustainable Development in Agriculture, The Chinese Academy of Agricultural Sciences, Beijing 100081 P R China.

jinzh@nankai.edu.cn

Abstract. Chitosan-Fe^0 nanoparticles (chitosan-Fe^0) were prepared using nontoxic and biodegradable chitosan as a stabilizer. Laboratory batch and column experiments were conducted to investigate the feasibility of using chitosan-Fe^0 for in situ reductive removal of Cr(VI) from water. Batch kinetic tests indicated that the overall disappearance of Cr(VI) may include both physical adsorption of Cr(VI) onto the chitosan-Fe0 surface and subsequent reduction of Cr(VI) to Cr(III). The rate of reduction of Cr(VI) to Cr(III) can be expressed by a pseudo-first-order reaction kinetics and the rate constants increase with the increase in iron loading. Column experiments indicated that the chitosan-Fe^0 were highly deliverable in the column. When the column was treated with Cr(VI)-contaminated deionized water, the removal capacity of chitosan-Fe^0 was found to be about 32 mg Cr/g Fe(0). A 25% drop in the Cr(VI) removal capacity was found when treated with Cr(VI)-contaminated surface water.

1. Introduction

Chromium is widely detected in surface water and groundwater at sites associated with industrial and military activities (Palmer and Wittbrodt, 1991). In natural waters, chromium exists in both Cr(VI) and Cr(III) valence states. Cr(VI) anions, including chromate (CrO_4^{2-}) and dichromate ($Cr_2O_7^{2-}$), are highly soluble in aquatic systems and are severe contaminants to environment due to their carcinogenic, mutagenic, and teratogenic features in biological systems (Fendorf et al., 2000). Recently, uses of Fe^0 nanoparticles as reactive media in situ subsurface environment remediation for trichloroethene (TCE), polychlorinated biphenyl (PCB) (Choe et al., 2001), benzoic acid (Joo et al., 2005), Ni(II) (Li and Zhang, 2006), As(III) (Kanel et al., 2005), As(V) (Kanel et al., 2006), Cr(VI) and Pb(II) (Ponder et al., 2000) have also been extensively investigated. Similarly, using a Fe^0 nanoparticle system for the removal of nitrate in aquatic environment has also been investigated in this laboratory and elsewhere (Choe et al., 2000; Wang et al., 2006). However, these Fe^0 nanoparticles were found to be lack in stability in water. This arises from the forces holding the Fe^0 together is either much too weak to hold particles together or too strong to disperse Fe^0 in water (He and Zhao, 2005; He et al., 2007).

---

[*] To whom any correspondence should be addressed.

Normally, Fe^0 nanoparticles are prepared by borohydride reduction of Fe(II) or Fe(III) in aqueous solution (Zhang, 2003). Fe^0 nanoparticles prepared in this way tend to agglomerate in water through direct interparticle interactions such as van der Waals forces and magnetic interactions leading to poor particle size suitable for any reactions (He et al., 2007). A stabilizer can reduce agglomeration of Fe^0 nanoparticles through (a) electrostatic repulsion (Coulombic repulsion caused by the accumulation of charged) and (b) steric hindrance (existence of sterically bulky groups) (Chen et al., 2004; He and Zhao, 2005, 2007; He et al., 2007). Thus, a proper stabilizer must be cheap, widely available and can well disperse the metals to form right particle size and, at the same time, can stimulate the reaction on its surface. Chitosan (β-[1-4]-2-amino-2-deoxy-D-glucopyranose) is one of the important natural polymers composed by layers of crustaceans and cell walls of many fungi. Chitosan has also been widely used in the green synthesis for the nanoscale materials due to its excellent biocompatibility, biodegradability, and lack of toxicity. Previous investigations have demonstrated that chitosan nanoparticles of silver, gold, platinum and palladium nanoparticles are stable in aqueous solutions (Wei and Qian, 2008; Wu et al., 2008). To date, only limited studies have been reported on Fe^0 nanoparticles stabilized by chitosan (Zhu et al., 2006).

The specific objectives of this research were to: (1) prepare chitosan-Fe^0, and (2) test the feasibility of using chitosan-Fe^0 for reductive removal of Cr(VI) from contaminated water through batch and column experiments.

2. Materials and Methods

2.1. Materials and Chemicals

Potassium dichromate ($K_2Cr_2O_7$) was obtained from Fuchen Chemical Reagent Manufactory, Tianjin, China. 1,5-diphenylcarbohydrazide and acetone were obtained from Third Chemical Reagent Manufactory, Shanghai, China. Ferrous sulfate heptahydrate ($FeSO_4{\cdot}7H_2O$) and potassium borohydride (KBH_4) were purchased from First Chemical Reagent Manufactory, Tianjin, China. Chitosan with a molecular weight of 10^5 g mol^{-1} with a degree of deacetylation of 85% was purchased from Zhejiang Jinke Marine Biochemistry Co., Ltd., Zhejiang, China. All the chemicals used in this research were analytical grade or higher and used as received.

2.2. Preparation and Characterization of Chitosan-Fe^0

Chitosan was dissolved in 0.05 M HNO_3 to make the final concentration of 0.5% by weight. Finally, chitosan solution was stirred overnight and was filtered through 0.22μm syringe filters to remove any suspensions. Chitosan-Fe^0 was prepared in situ by reducing Fe(II) with KBH_4 in the presence of chitosan as a stabilizer. To ensure all the Fe(II) were reduced, excess of KBH_4 (6 times) over the Fe(II) was used. The detailed procedure was as follows: 10 mL of solution containing 0.2978 g of $FeSO_4{\cdot}7H_2O$ was first mixed with 3 mL of 0.5% chitosan solution. The mixture was stirred for 30 min under nitrogen gas. Then, to the mixture, 5 ml of freshly prepared aqueous solution containing 0.3467 g of KBH_4 was added dropwise. At this stage, gas was evolved vigorously and black precipitation was formed. Again, the mixture was stirred for another 90 min. The resulted black precipitate was collected and washed by deoxygenated water for three times to get rid of the excess chemicals. The whole process was carried out in a nitrogen atmosphere. The prepared chitosan-Fe^0 nanoparticle power was left at room temperature waiting for characterizations. Controls were prepared simply by dissolving chitosan in 0.05 M HNO_3 only.

Transmission electron micrographs were obtained using a Phillips Tecnai F20 super-twin transmission electron microscope (TEM). A 40 μL of aqueous sample containing the nanoparticles was placed onto the amorphous carbon-coated 200 mesh copper grid. The droplet was then allowed for full contact/spreading on the grid, followed by drying the sample at ambient temperature for 5 min. Then the residual nanoparticles attached to the grid were photographed using the TEM.

2.3. Batch Experiments

Kinetic batch experiments for the reduction of Cr(VI) in water were carried out in polytetrafluoro ethylene (PTFE)- sealed serum bottles covered with aluminum foil at atmospheric pressure. To each bottle, 100 mL Cr(VI) solution prepared with deionized water and a certain amount of Fe^0 nanoparticles were added. The initial pH was adjusted prior to the experiment. All bottles were shaked on a shaker at a rate of 220 rpm during the experiment. Sampling was made at certain time-interval and the samples were filtered through a 0.22 μm filter.

2.4. Column Experiments

Glass columns (30 cm long and 3.2 cm in internal diameter) were packed with quartz sand (20 mesh) and 2 g of chitosan-Fe^0 in four layers. Synthetic Cr(VI)-contaminated deionized water was prepared by dissolving reagent-grade potassium dichromate ($K_2Cr_2O_7$) into deionized water. The pH was then adjusted to 6.0 by adding 0.5 mol/L sodium hydroxide (NaOH) and 0.5 mol/L hydrochloric acid (HCl). Synthetic Cr(VI)-contaminated surface water was prepared by dissolving reagent-grade potassium dichromate ($K_2Cr_2O_7$) into river water (Weijin River, Tianjin China). This two kinds of synthetic contaminated water were fed in an up-flow manner at a water velocity of 0.12 mL/min by using a multi-channel precision peristaltic pump. Usually, the seepage velocity is ranged from 0.5 mL/min to 0.03 mL/min (Lai and Lo, 2002). The required time of every pore volume (PV) for the columns was 2.5 h. The experimental set up of the columns is shown in Fig. 1.

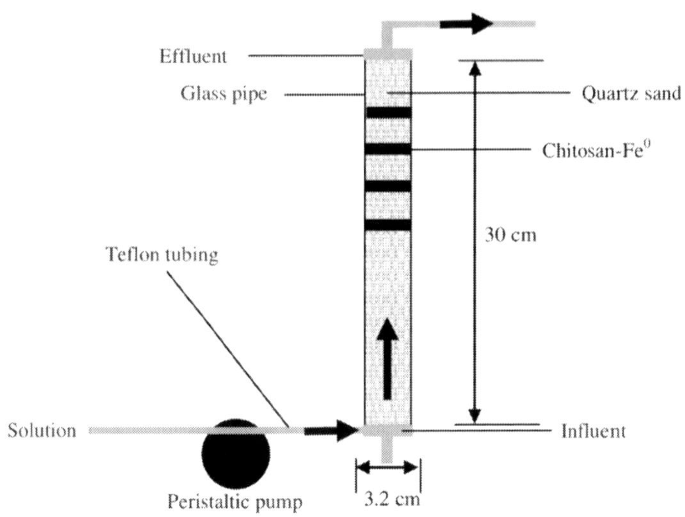

Fig. 1 Experimental set up of columns.

3. Results and Discussion

3.1. TEM Characterization

Fig. 2 shows that Fe^0 nanoparticles and chitosan polymers are arranged in a homocentric layered structure with both Fe^0 nanoparticles and chitosan polymers on top of each other. Excess chitosan in the solution may form polymer nanoparticles in different morphology. Similar phenomena are also found in other metallic nanoparticles such as silver, gold, platinum, and palladium (Ding et al., 2006; Wei and Qian, 2008). The distribution of particle size of chitosan-Fe^0 is widespread ranging from 20 to 150 nm with a mean diameter of 82.4 nm. This is consistent with figures obtained by Sreeram et al. (2004) (chitosan-Fe(III), 78.1 nm) and Boddu et al. (2003) (chitosan-Cr(III), 71.1 nm). Particles with mean diameter of less than 100 nm for a binary chitosan-Pd-Fe system were reported by Zhu et al.

(2006). Interestingly, due to its high chemical reactivity and large surface area, Fe^0 nanoparticles are easily oxidized by oxygen or ignite spontaneously when exposed to air. Introduction of oxygen to Fe^0 nanoparticles causes a strongly exothermic redox reaction. Ponder et al. (2000) mentioned in their preparation for the Fe^0 nanoparticles that precautions must be taken in limiting oxygen to enter into the systems and, at the same time, heat produced during the preparation must be carefully dissipated. However, in our work, similar problem is never occurred even the chitosan-Fe^0 directly exposed to the open atmosphere. It is believed that in the chitosan-Fe^0 system, Fe^0 nanoparticles is better protect by chitosan and oxidation is hindered. The chitosan-Fe^0 is precipitated out as a powder and can be dispersed into aqueous solutions.

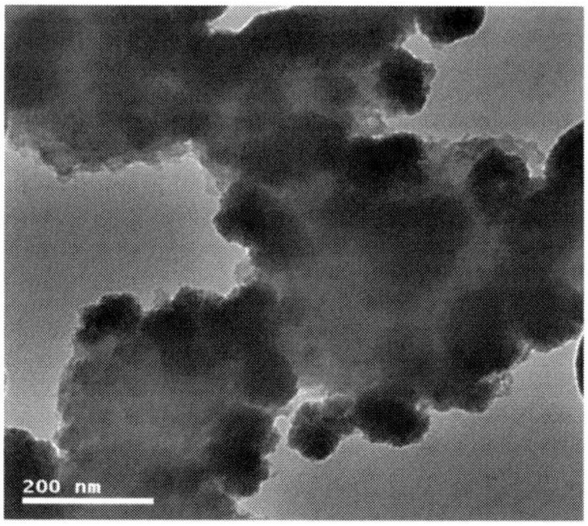

Fig. 2 TEM image of chitosan-Fe^0 (The scale bar represents 200 nm).

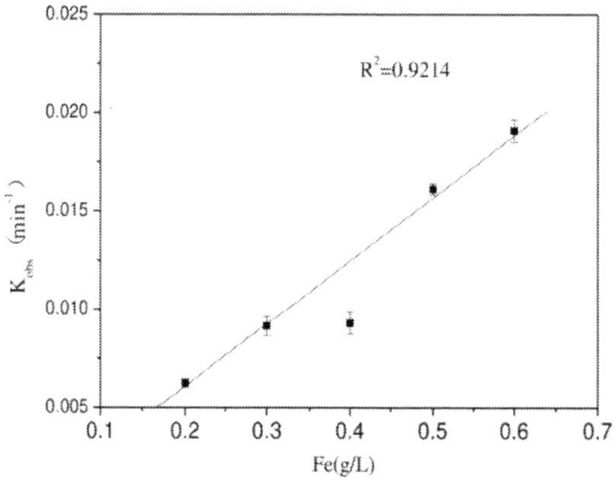

Fig. 3 Pseudo-first-order rate constant (k_{obs}) for Cr(VI) reduction by chitosan-Fe^0 using different iron loadings.

3.2. Batch Experiments

In this study, it is found that reduction of Cr(VI) is proceeded in two steps. During the first 2 min of the experiment, decrease of Cr(VI) is much dramatic with an average rate of about 10 mg L^{-1} min^{-1}. After the first 2 min till the end of the experiment, the reaction can be expressed with a pseudo-first-order kinetics. The reductions of Cr(VI) in other Fe^0 nanoparticles system have been reported as pseudo-first-order (Ponder et al., 2000; Xu and Zhao, 2007).

$$\frac{d[Cr(VI)]}{dt} = -K_{obs}[Cr(VI)] \qquad (1)$$

where [Cr(VI)] is the concentration of Cr(VI) in the aqueous phase (mg/L) at time t (min), K_{obs} is the pseudo-first-order rate constant (min^{-1}). Integration of eq 1 results in:

$$[Cr(VI)] = [Cr(VI)]_0 e^{-K_{obs}t} \qquad (2)$$

Pseudo-first-order rate constant (K_{obs}) were determined by linear regression of ln normalized concentration versus time. However, when a support is introduced into the reduction of Cr(VI), complicated kinetic behavior is noticed in our study and also in the investigations by other researchers (Deng et al., 1999). The two-step kinetics have been explained as reduction of Cr(VI) under the existence of Fe^0 followed by a faster physical or a chemical adsorptions (Deng et al., 1999; Ponder et al., 2000; Kanel et al., 2005). The overall reaction kinetics may possibly be overlays of different first order kinetics or reactions with different orders (WÜst et al., 1999).

Fig. 3 shows that rates of reduction of Cr(VI) were increased as the loadings of iron were increased. This phenomenon is attributed to the increase in the active sites on the chitosan-Fe^0 surface. In all cases, the regression line did not go through the origin, indicating that physical adsorption maybe played a major role in the Cr(VI) dissipation. Similar phenomena have been observed by the other researchers (Burris et al., 1995; Su and Puls, 1999). In this experiment, we found that the mass-normalized rate constant for chitosan-Fe^0 is about twice as fast as that reported by Xu and Zhao. (2007). It has been proposed that Cr(III) can be co-precipitated with Fe(III) in the form of Fe(III)-Cr(III) hydroxides on the Fe^0 nanoparticles surface. This co-precipitate will reduce the electron transfer from Fe^0 to Cr(VI) and accordingly eliminated Cr(VI) reduction (Li et al., 2008). However, when a support like chitosan is introduced into the system, chitosan can hinder the formation of Fe(III)-Cr(III) precipitate by virtue of its coordinating capability. This is because of the presence of the $-NH_2$ and $-OH$ groups on the bulky chitosan surface. These groups can be behaved either as a mono- or bidentate ligands to coordinate with most of the transition metals, such as Fe(III), Cr(III) and Cr(VI) (Boddu et al., 2003; Varma et al., 2004). Recent Mössbauer studies of Fe(III) ions in a chitosan-Fe system demonstrated that Fe(III) is either penta- or hexa-coordinated and more than one polymer chains were found in these complexes (Bhatia and Ravi, 2003).

3.3. Column Experiments

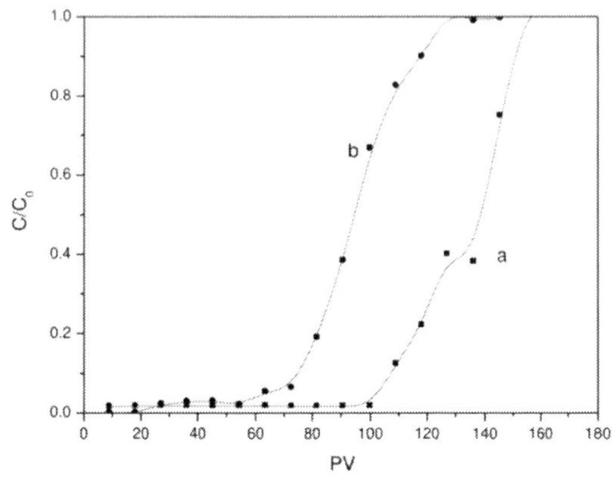

Fig. 4 Breakthrough curve of columns treated with Cr(VI)-contaminated deionized water (a) and surface water (b).

Many researchers have reported that the removal of Cr(VI) by Fe(0) is achieved by a coupled reduction-oxidation reaction in the system in which Fe(0) donates the electrons to reduce the Cr(VI) to

Cr(III). The reaction equation can be written as (Naftz et al., 2002; Powell et al., 1995; Eary and Rai, 1988; Gandhi et al., 2002):

$$CrO_4^{2-} + Fe^0 + 4H_2O \rightarrow Cr^{3+} + Fe^{3+} + 8OH^- \quad (3)$$

When both Cr(III) and Fe(III) were in a proximity, a thin film of Fe(III)-Cr(III) hydroxides on the Fe^0 nanoparticles surface was formed (Powell et al., 1995). The thin film was believed to increase the resistance for the electron transfer from Fe^0 to Cr(VI) and thus to retard the reduction of Cr(VI) to Cr(III). Recent investigations further confirmed that the actual Fe-Cr surface structure was a mixture of $Cr_{0.667}Fe_{0.333}OOH$ and $(Cr_{0.667}Fe_{0.333})(OH)_3$ (Li et al., 2008). This important phenomenon suggests that Fe(0) can be exhausted after reacting with a certain amount of Cr(VI). Moreover, the exhaustion of Fe(0) results in the migration of Cr(VI) fronts through the columns, and the columns are eventually exhausted if a continuous loading of Cr(VI) is applied. The breakthrough curve, illustrated in Fig. 4, shows the Cr(VI) migration pattern in columns treated with Cr(VI)-contaminated deionized water and surface water. The Cr(VI) breakthrough was observed at 100 PV then attained 100% concentration breakthrough after 160 PV as illustrated in Fig. 4.

The method developed by Lo et al. (2006) was used to estimate the Cr(VI) removal capacity of each column. This value is used as a reference value for comparison of the reactivity of chitosan-Fe^0 on the removal of Cr(VI) under various geochemical conditions. The estimated removal capacity of column treated with Cr(VI)-contaminated deionized water is 32 mg Cr/g Fe(0) based on the data in Fig. 4a. This value is consistent with previously work with a groundwater or DI water (Ponder et al. 2000; Cao and Zhang 2006; Li et al. 2008). There was about 25% drop in the Cr(VI) removal capacity of chitosan-Fe^0 in the presence of contaminated surface water compared with in the presence of contaminated deionized water throughout the experiment.

4. Conclusions

Highly reactive Fe^0 nanoparticles was synthesized using nontoxic and biodegradable chitosan as a stabilizer. Laboratory batch and column experiments were conducted to investigate the feasibility of using chitosan-Fe^0 for in situ reductive removal of Cr(VI) from water. Batch kinetic tests indicated that the overall disappearance of Cr(VI) may include both physical adsorption of Cr(VI) onto the chitosan-Fe^0 surface and subsequent reduction of Cr(VI) to Cr(III). The rate of reduction of Cr(VI) to Cr(III) can be expressed by a pseudo-first-order reaction kinetics and the rate constants increase with the increase in iron loading. Column experiments indicated that the chitosan-Fe^0 were highly deliverable in the column. When the column was treated with Cr(VI)-contaminated deionized water, the removal capacity of chitosan-Fe^0 was found to be about 32 mg Cr/g Fe(0). A 25% drop in the Cr(VI) removal capacity was found when treated with Cr(VI)-contaminated surface water.

Acknowledgments

This work was supported in part by Tianjin Natural Science Foundation of China under grant No. 07JCZDJC01800 and Doctoral Foundation for Young Scholar from Ministry of Education of China under grant No. 20070055053.

References

[1] Bhatia, S.C., Ravi, N., 2003. A Mössbauer study of the interaction of chitosan and D-glucosamine with iron and its relevance to other metalloenzymes. Biomacromolecules 4, 723-727.

[2] Boddu, V.M., Abburi, K., Talbott, J.L., Smith, E.D., 2003. Removal of hexavalent chromium from wastewater using a new composite chitosan biosorbent. Environ. Sci. Technol. 37, 4449-4456.

[3] Burris, D.R., Campbell, T.J., Manoranjan, V.S., 1995. Sorption of trichloroethylene and tetrachloroethylene in a batch reactive metallic iron-water system. Environ. Sci. Technol. 29, 2850-2855.

[4] Chen, S.S., Hsu, H.D., Li, C.W., 2004. A new method to produce nanoscale iron for nitrate removal. J. Nanopart. Res. 6, 639-647.
[5] Choe, S., Chang, Y., Hwang, K., Khim, J., 2000. Kinetics of reductive denitrification by nanoscale zero-valent iron. Chemosphere 41, 1307-1311.
[6] Choe, S., Lee, S., Chang, Y., Hwang, K., Khim, J., 2001. Rapid reductive destruction of hazardous organic compounds by nanoscale Fe^0. Chemosphere 42, 367-372.
[7] Chuang, F.W., Larson, R.A., Wessman, M.S., 1995. Zero-valent iron-promoted dechlorination of polychlorinated biphenyls. Environ. Sci. Technol. 29, 2460-2463.
[8] Deng, B., Burris, D.R., Campbell, T.J., 1999. Reduction of vinyl chloride in metallic iron-water systems. Environ. Sci. Technol. 33, 2651-2656.
[9] Ding, Y., Xia, X., Zhang, C., 2006. Synthesis of metallic nanoparticles protected with N,N,N-trimethyl chitosan chloride via a relatively weak affinity. Nanotechnology 17, 4156-4162.
[10] Dupont, L., Guillon, E., 2003. Removal of hexavalent chromium with a lignocellulosic substrate extracted from wheat bran. Environ. Sci. Technol. 37, 4235-4241.
[11] Fendorf, S., Wielinga, B.W., Hansel, C.M., 2000. Chromium transformations in natural environments: The role of biological and abiological processes in chromium(VI) reduction. Int. Geol. Rev. 42, 691-701.
[12] Fiedor, J.N., Bostick, W.D., Jarabek, R.J., Farrell, J., 1998. Understanding the mechanism of uranium removal from groundwater by zero-valent iron using X-ray photoelectron spectroscopy. Environ. Sci. Technol. 32, 1466-1473.
[13] He, F., Zhao, D., 2005. Preparation and characterization of a new class of starch-stabilized bimetallic nanoparticles for degradation of chlorinated hydrocarbons in water. Environ. Sci. Technol. 39, 3314-3320.
[14] He, F., Zhao, D., 2007. Manipulating the size and dispersibility of zerovalent iron nanoparticles by use of carboxymethyl cellulose stabilizers. Environ. Sci. Technol. 41, 6216-6221.
[15] He, F., Zhao, D., Liu, J., Roberts, C.B., 2007. Stabilization of Fe-Pd nanoparticles with sodium carboxymethyl cellulose for enhanced transport and dechlorination of trichloroethylene in soil and groundwater. Ind. Eng. Chem. Res. 46, 29-34.
[16] Joo, S.H., Feitz, A.J., Sedlak, D.L., Waite, T.D., 2005. Quantification of the oxidizing capacity of nanoparticulate zerovalent iron. Environ. Sci. Technol. 39, 1263-1268.
[17] Kanel, S.R., Greneche, J., Choi, H., 2006. Arsenic(V) removal from groundwater using nano scale zero-valent iron as a colloidal reactive barrier material. Environ. Sci. Technol. 40, 2045-2050.
[18] Kanel, S.R., Manning, B., Charlet, L., Choi, H., 2005. Removal of arsenic(III) from groundwater by nano scale zero-valent iron. Environ. Sci. Technol. 39, 1291-1298.
[19] Kendelewicz, T., Liu, P., Doyle, C.S., Brown, G..E., Nelson, E.J., Chambers, S.A., 2000. Spectroscopic study of the reaction of aqueous Cr(VI) with Fe_3O_4(111) surfaces. Surf. Sci. 469, 144-163.
[20] Kieber, R.J., Helz, G.R., 1992. Indirect photoreduction of aqueous chromium(VI). Environ. Sci. Technol. 26, 307-312.
[21] Laidler, K.J., 1965. Chemical Kinetics. McGraw-Hill, New York.
[22] Li, X.Q., Cao, J., Zhang, W.X., 2008. Stoichiometry of Cr(VI) immobilization using nanoscale zerovalent iron (nZVI): A study with high-resolution X-ray photoelectron spectroscopy (HR-XPS). Ind. Eng. Chem. Res. 47, 2131-2139.
[23] Li, X.Q., Zhang, W.X., 2006. Iron nanoparticles: The core-shell structure and unique properties for Ni(II) sequestration. Langmuir 22, 4638-4642.
[24] Li, X.Q., Zhang, W.X., 2007. Sequestration of metal cations with zerovalent iron nanoparticless-a study with high resolution X-ray photoelectron spectroscopy (HR-XPS). J. Phys. Chem. C 111, 6939-6946.
[25] Manning, B.A., Kiser, J.R., Kwon, H., Kanel, S.R., 2007. Spectroscopic investigation of Cr(III)- and Cr(VI)-treated nanoscale zerovalent iron. Environ. Sci. Technol. 41, 586-592.

[26] Martin, J.E., Herzing, A.A., Yan, W., Li, X.Q., Koel, B.E., Kiely, C.J., Zhang, W.X., 2008. Determination of the oxide layer thickness in core-shell zerovalent iron nanoparticles. Langmuir 24, 4329-4334.

Sol-gel synthesis and characterization of nanocrystalline $(Bi_{0.5}Na_{0.5})TiO_3$ powders from the poly vinyl alcohol evaporation route

Liu Haitao[1,2], Wang xiaohui[2] and Li Longtu[2]
[1]Col lege of Chemistry and Material Engineering, Wenzhou University, Wenzhou , 325035 P.R. China
[2]State Key Laboratory of New Ceramics and Fine Processing , Department of Material Science and Engineering, Tsinghua University, Beijing, 100084 P.R. China

E-mail: liuhaitao05@tsinghua.org.cn

Abstract: Nanocrystalline pure perovskite phase bismuth sodium titanate $(Bi_{0.5}Na_{0.5})TiO_3$(BNT) powders have been prepared by a sol-gel method from the poly vinyl alcohol evaporation route. bismuth nitrate [$Bi(NO_3)_3$], tetra-butyl titanate [$Ti(OC_4H_9)_4$] and sodium acetate(CH_3COONa) were used as raw materials and poly vinyl alcohol(PVA) as the sol-gel forming solvent. Stoichiometric amounts of the individual raw materials were thoroughly mixed in accordance with the BNT composition to form the stock solution. The pH of the stock was adjusted to 1-3 by adding acetate. Aqueous solution of PVA[10%(w/v)] was then added to the cationic mixture with thorough stirring. The weight ratio of the cationic mixture to the PVA aqueous solution was maintained at 2:1. The resulting viscous liquid was then heated over a hot crucible up to form a fluffy dry gel. The fluffy dry gels were calcined at different temperatures and times and then cooled to room temperature naturally. The X-ray powder diffraction(XRD) patterns of the heat-treated powders were obtained using a Bruker D8 Advance X-ray diffractometer with Cu Kα radiation and nickel filter. Scanning electron microscope (SEM) studies of the NBT powders were performed using a JSM-6700F electron microscope. Phase-pure BNT powders were obtained at 550℃ for 2-2.5h which is relatively lower calcination temperature than other reports. The BNT powders consists of phase-pure perovskite nanocrystals with an average size of 100-200nm.

1. Introduction

Lead has recently been expelled from many commercial applications and materials owing to concerns regarding its toxicity. Ferroelectric lead zirconate titanate ceramics (PZT) are extensively used for many industrial purposes due to their high dielectric and electromechanical properties. However, the toxicity of lead oxide and its high vapor pressure during the sintering process not only causes environmental pollution but also generate unstability of composition and electrical properties of products. Therefore, it is necessary to develop environment-friendly lead-free piezoelectric ceramics to replace the PZT-based ceramic, which has become one of the main trends in present development of piezoelectric materials[1].

Lead zirconate titanate ceramics(PZT) are high performance piezoelectric materials which contain more than 60 weight percent lead. As a consequence much of the current research is oriented towards more environmentally friendly lead-free materials. Although there has been a concerted effort to develop lead-free piezoceramics, no effective alternative to PZT has yet been found. Bismuth sodium titanate $(Bi_{0.5}Na_{0.5})TiO_3$ (abbreviated to BNT) has been considered to be an excellent candidate as a key material of lead-free piezoelectric ceramics and has attracted considerable attention in the past few decades. It has been studied continually by several researchers[2-4] in recent years. Conventional solid-state method for synthesizing BNT ceramic powder often form large grains which are difficult to disperse and affect the sintering properties of BNT . Many chemical methods have recently been developed by many researchers for the synthesis of NBT powders due to the merits of easy composition control, better homogeneity and low processing temperature in comparison to the conventional solid-state reaction process.

In this work, fine crystalline pure phase bismuth sodium titanate $(Bi_{0.5}Na_{0.5})TiO_3$(BNT) powders have been prepared by a sol-gel method from the poly vinyl alcohol evaporation route. Phase-pure BNT powders were obtained at 550℃ for 2-2.5h which is relatively lower calcination temperature than other reports. The BNT powders consists of phase-pure perovskite nanocrystals with an average size of 100-200nm. This synthetic process is easily controlled and convenient in comparison with other methods.

2. Experimental

All starting materials are analytical grade and used without further purification. Bismuth nitrate $[Bi(NO_3)_3]$, tetra-butyl titanate $[Ti(OC_4H_9)_4]$ and sodium acetate(CH_3COONa) were used as raw materials and poly vinyl alcohol(PVA) as the sol-gel forming solvent. Stoichiometric amounts of the individual raw materials were thoroughly mixed in accordance with the BNT formula of $Na_{0.5}Bi_{0.5}TiO_3$ to form the stock solution. The purity of a reagent was taken into account in weighing. During the mixing process, electromagnetic agitating was needed. The pH value of the stock was adjusted to 1-3 by adding acetate to obtain a transparent sol. Aqueous solution of PVA[10%(w/v)] was then added to the cationic mixture with thorough stirring . The weight ratio of the cationic mixture to the PVA aqueous solution was maintained at 2:1. The resulting viscous liquid was then heated over a hot crucible up to form a fluffy dry gel. The fluffy dry gels were calcined at different temperatures and times, from 450 to 550℃ for 0.5-2.5h and then cooled to room temperature naturally.

The X-ray powder diffraction(XRD) patterns of the heat-treated powders were obtained using a Bruker D8 Advance X-ray diffractometer with Cu Kα radiation and nickel filter. Scanning electron microscope (SEM) studies of the BNT powders were performed using a JSM-6700F scanning electron microscope.

The flow diagram of the sol-gel method employed in this study is shown in Fig.1.

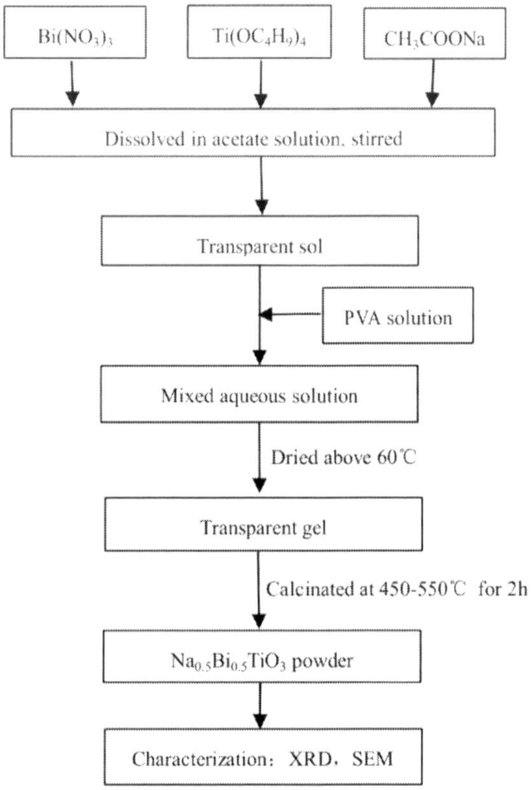

Fig.1 Flow diagram of the sol-gel method

3. Results and discussion

The polymer (PVA) has been employed to adjust the sol-gel property. The poly vinyl alcohol is widely used as a binder for green shaping of ceramic materials and decomposing exothermally at low ignition temperatures (500°C), leaving behind hardly any residue. This characteristic, together with water-soluble nature and gel ability has motivated the present study.

Fig.2 shows the XRD patterns of Na0.5Bi0.5TiO3 powders calcined at different temperatures for 2h. On comparison of the XRD patterns in Fig.2, It can be seen that the BNT phase started to form at a temperature of as low as 450 ℃ for 2h, while phase-pure pervoskite NBT powders were obtained at 550℃ for 2h.

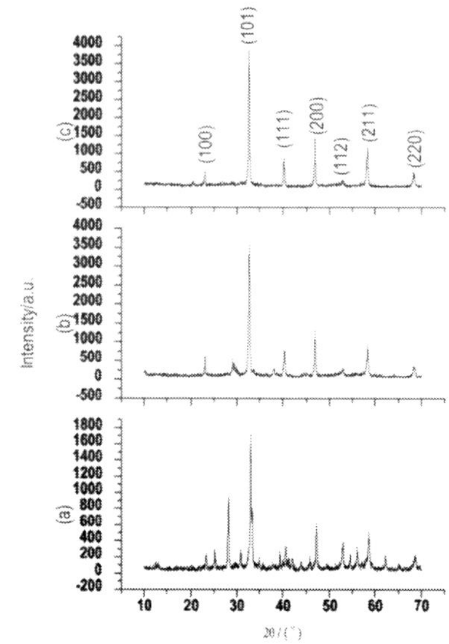

Fig.2 XRD patterns of Sol-gel powders synthesized at (a)450℃2h,(b) 500℃2h,(c) 550℃2h

Fig.3 shows the XRD patterns of Na0.5Bi0.5TiO3 powders calcined at 550℃ for 0.5-2.5h. On comparison of the XRD patterns in Fig.3, we can observed that the BNT phase started to form at 550℃ for 0.5h and transformed to phase-pure pervoskite BNT powders at 550℃ for 2h. Further increasing calcination time up to 2.5h, phase-pure pervoskite NBT powders still maintained unchangeable and no impurity phase was found.

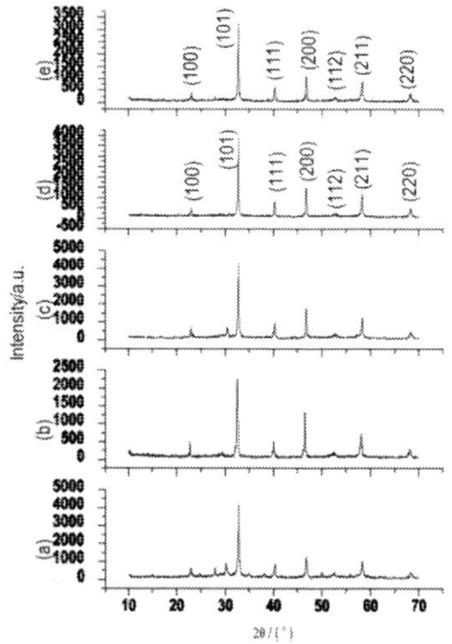

Fig.3 XRD patterns of sol-gel powders synthesized at (a)550℃0.5h,(b) 550℃1h,(c) 550℃1.5h,(d) 550℃2h, (e) 550℃2.5h

The particle sizes and morphologies of BNT powders corresponding Fig.3(d) and (e) were examined by SEM,.which were shown in Fig.4(a) and (b). At Fig.4(a), all BNT particles were homogeneous cubic grains with an average size of 100–200nm. The formation of homogeneous cubic grains might be due to that the combustion of the poly vinyl alcohol presented in the gel, along with the evolution of gases. At Fig.4(b) NBT grains are much larger and unhomogeneously than at Fig.4(a), which suggests that the grain grows unnormally with calcining time increasing.

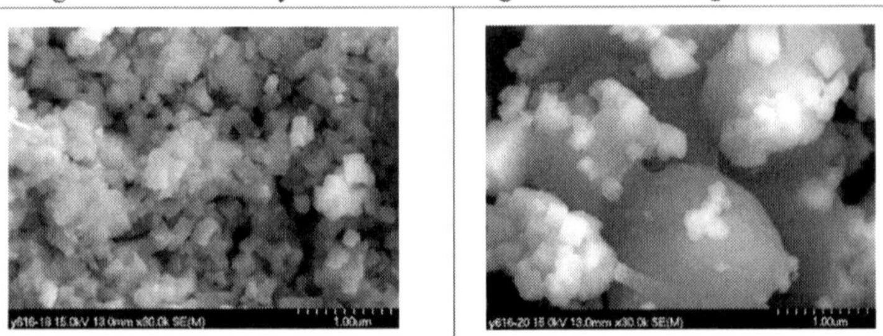

(a) (b)

Fig.4 SEM micrograph of synthesized powders calcined at (a)550℃2h, (b) 550℃2.5h

4 Conclusion

Single-phase, good dispersing, homogeneity and phase-pure perovskite Nanocrystalline NBT powders with an average grain size of 100-200nm have been synthesized via a sol-gel method by employing the poly vinyl alcohol evaporation process. BNT phase started to form at 450℃ for 2h, while phase-pure NBT powders were obtained at 550℃ for 2-2.5h, which is relatively lower calcination temperature than other reports.

Acknowledgements

This study was supported by a research grant from 973-project of the ministry of science and technology of China(2002CB613301).

References

[1] Yasuyoshi Saito, Hisaaki Takao and Toshihiko Tani 2004 *Nature.* **432** 84-87
[2] Junjie Hao, Xiaohui Wang and Renzheng Chen 2005 *J. Materials chemistry and physics.* **90** 282-285
[3] P. Marchet, E. Boucher and V. Dorcet 2006 *J. of the European Ceramic Society.* **26** 3037–41
[4] Guisheng Xu, Ziqing Duan and Xiaofeng Wang. 2005 *J. Journal of Crystal Growth.* **275** 113–19

Journal of Physics: Conference Series **188** (2009) 012059 doi:10.1088/1742-6596/188/1/012059

Synthesis of nano-$BaTiO_3$ by solid-state reaction at 80℃

S W Ding[1], Y Y Zhang, C Y Feng and H J Zhang
College of Chemistry and Environmental Science, Hebei University, Baoding Hebei 071002, China
Correspondence should be addressed to S W Ding

E-mail:dingsw@ hbu.edu.cn

Abstract. Nano-$BaTiO_3$ powder was synthesized by solid-state reaction at 80℃.The method is high productive and can be easily operated.XRD pattern demonstrates that the powder is pure cubic phase.And TEM photograph of the powder shows that the particles are uniform and spherical with average size of 50nm in diameter.Furthermore we confirmed that chemical reaction is the rate-determining step.

1. Introduction

The outstanding dielectric, ferroelectric and piezoelectric properties of $BaTiO_3$ make it the desirable material for a variety of applications including multilayer capacitors,thermistors and electrooptic devices[1].Because of its properties,many methods were explored to synthesize $BaTiO_3$.Traditional solid-state reaction needs high temperature up to 1200℃,which not only makes the grains bulky,but also easily introduce some impurities into the product[2].And sol-gel method has attracted many researcher's attention, but this method can't also be for application because its original material price is too expensive.We have used hydrothermal method[3,4],but it makes environment polluted and needs higher quality for equipment.In our present work,a new method called solid-state reaction at low temperature[5,6,7]has been explored to synthesize $BaTiO_3$.This method has many advantages,such as no solvents,little pollution and high productive.

2. Experimental

2.1 Chemicals and Instruments

$TiCl_4$,$Ba(OH)_2{\cdot}8H_2O$ and $NH_3{\cdot}H_2O$ are analytical reagents made in China.All experiments were carried out in the deionized water.

Y-2000 X-ray Diffractometer (Dandong),JEM-100SX Transmission Electrical Microscope (Japan),etc.

2.2 Synthesis of $BaTiO_3$ Powder

A suitable amount of $TiCl_4$ was slowly dropped to 100 mL water.The pH value of the solution was adjusted to 6~8 by dropping 1:1 $NH_3 \cdot H_2O$ to form $HiTiO_3$(or $TiO_2 \cdot H_2O$).Then it was filtered and washed with deionized water until the Cl^- could not be detected.Subsequently it was mixed with needed $Ba(OH)_2 \cdot 8H_2O$(molar ratio is Ti:Ba=1:1).The mixture was milled for 60min at room temperature,then dried at 80℃ for 5.5h.

3. Results and Discussion

3.1 XRD Analysis of $BaTiO_3$ Powder

Comparing XRD pattern of synthetical $BaTiO_3$ powder with JCPDS Card (31-0174),it can be seen that synthetical $BaTiO_3$ powder is pure cubic phase.(shown in figure 1)

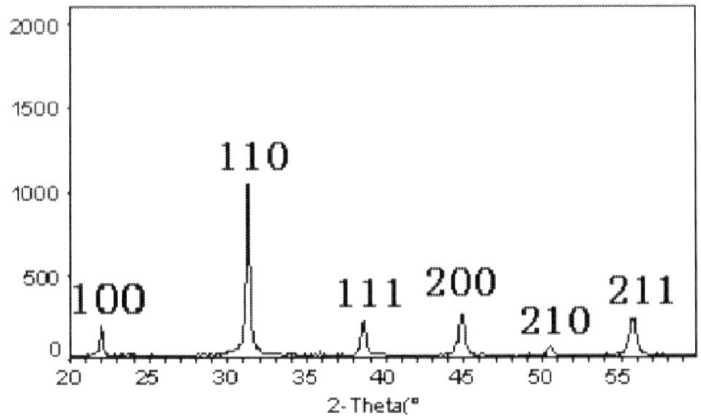

Figure 1 XRD pattern of synthetical $BaTiO_3$

3.2 TEM Analysis of $BaTiO_3$ Powder

TEM photograph of the powder shows that the particles are uniform and spherical with average size of 50nm in diameter.(shown in figure 2)

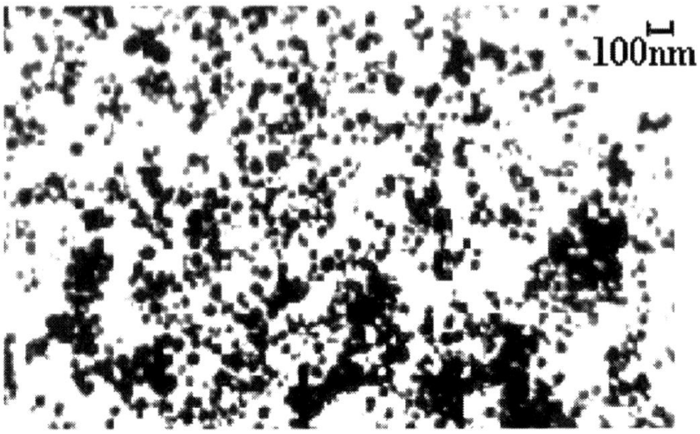

Figure 2 TEM photograph of sample

3.3 Influence of Reaction Temperature

The mixture was devided into severn parts,then dried respectively at room temperature,40℃,60℃,70℃,80℃,100℃ and 120℃ for 6d,48h,34h,28h,24h,5.5h and 4h in order to be dried completely. XRD patterns of the dried powder are shown in figure 3.

In figure 3,1,2,3,4,5,6,7 are respectively XRD pattern of dried powder at room temperature,40℃,60℃,70℃,80℃,100℃ and 120℃.

It is can be seen that dried powder at room temperature has no $BaTiO_3$ phase,only has some impurities,such as $BaCO_3$,$Ba(OH)_2{\cdot}8H_2O$ and $Ba(OH)_2$.As temperature rising, $Ba(OH)_2{\cdot}8H_2O$ disappears at 40℃,$Ba(OH)_2$ and $BaCO_3$ decrease gradually.At 80℃,the main phase is $BaTiO_3$ with little impurities.So we confirmed the reaction temperature is 80℃.

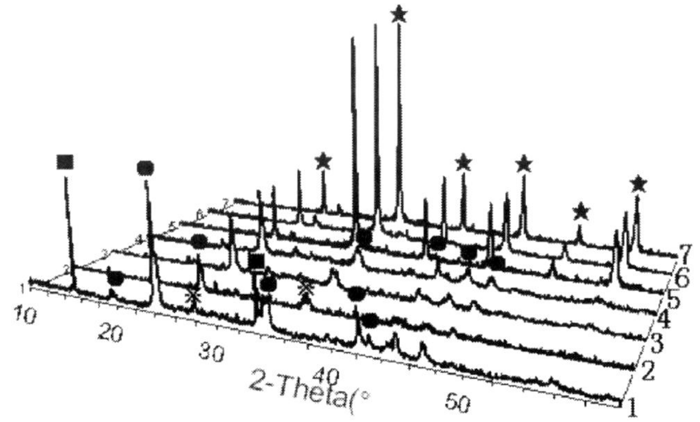

Figure 3 XRD patterns of powders dried at different temperatures
★$BaTiO_3$, ●$BaCO_3$, ■$Ba(OH)_2{\cdot}8H_2O$, ※$Ba(OH)_2$.

3.4 Influence of Reaction Time

In order to definite how long the mixture needs to form $BaTiO_3$ at 80℃,the mixture react respectively for 1h,2h,3h,4h,7h,10h,13h under 80℃,and then dried at room temperature.Their XRD pattrerns are shown in figure 4.

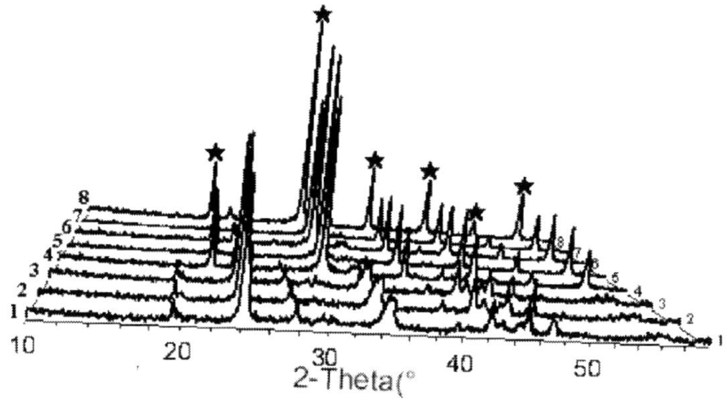

Figure 4 XRD patterns of powders reacting different time under 80℃
★$BaTiO_3$， other peaks are ascribed to $BaCO_3$.

In figure 4,1, 2, 3, 4, 5, 6, 7 respectively indicates XRD pattern of powders reacting for 1h, 2h, 3h, 4h, 7h, 10h, 13h under 80℃ then dried at room temperature.Curve 8 indicates the XRD pattern of the powder dried under 80℃.

It can be seen that when the reaction time is 1h,2h,and 3h, the sample is $BaCO_3$, no $BaTiO_3$ occures. When the reaction time is 4h,$BaTiO_3$ phase starts to form,and the intensity of its diffraction peaks increases with the reaction time prolongation,while that of $BaCO_3$ gradually decrease.

3.5 Reaction Mechanism

Diffusion,reaction,nucleation and growth are the four steps of the solid-state reaction.In the past, diffusion or nucleation is considered to be the rate-determining step of the high temperature solid-state reaction. From figure 3, it can be seen that the strength of the $Ba(OH)_2$ diffraction peaks is very small from the start to the end,so we can know the diffusion from $Ba(OH)_2$ to H_2TiO_3 is very quick.In figure 3,reaction under 80℃ just exhibits the $BaTiO_3$ diffraction peaks and in figure 4, reaction for 4h under 80℃ just begins to exhibit the $BaTiO_3$ diffraction peaks with high strength.From the above,we draw a conclusion that diffusion, nucleation and growth is quick.So,we confirmed that reaction is the rate-determining step of the solid-state reaction at low temperature.

4. Conclusions

(1) It is the first time to use solid-state reaction at low temperature to synthesize cubic $BaTiO_3$.The method is high productive and partly process achieving economic atom reaction according to the ideas of green chemistry.

(2) In the solid-state reaction at low temperature, chemical reaction is the rate-determining step.

Acknowlegements: This work was supported by the Natural Science Foundation of Hebei Province,China (Grant No. B2008000574)

References

[1] Liu J B, Wang Z M, Zhen C P, Ji L F, Xu L H and Li W C 2000 The humidity sensitivity of La-doped $BaTiO_3$-based nanocrystalline ceramic sensor *J.Jfmd.* 6 106-110

[2] Fu W Y and Haribala 2005 Preparation of nanocrystalline bariumtitanate by microwave-hydrothermal method *J.Jounal of inner mongolia university for nationalities.* 20 43-46

[3] Ding S W, Wang J, Qin J L,and Li X M 2001 Hydrothermal synthesis ,structure and property of nano-$BaTiO_3$-based dielectric materials *J.Scichina B.* 44 657-662

[4] Ding S W, Zhai Y Q, Qin J L, Wang J and Li X M 2001 Hydrothermal synthersis,structure and property of $Ba_{1-x}Sr_xTi_{1-y}Sn_yO_3$ nanometer materials *J.Acta Chimica Sinica.* 59 1121-24

[5] Ding S W , Chai J and Feng C Y 2005 Synthesis,structure and dielectric property of $BaTiO_3$ by solid-state reaction at room temperature *J.Chem J On Internet.* 7 15-18

[6] Liu M L，Zhou H L and Chen Y R 2005 Room temperature solid–solid reaction preparation of iron boron alloy nanoparticles and mossbauer spectra *J.Materials Chemistry and Physics.* 89 289-294

[7] Chen Y 1997 Low-temperature oxidation of ilmenite($FeTiO_3$) induced by high energy ball milling at room temperature *J.Journal of Alloys and Compounds.* 257 156-160

Study on methane conversion to synthesis gas over nano Pt/MgO catalysts III. Influence of the precursors on the physico-chemical properties of the catalysts

Min Yang[1,2], **Helmut Papp**[2]

[1] School of Environmental and Chemical Engineering, Dalian Jiaotong University, Dalian 116028, China; [2]Institute of Technical Chemistry, University of Leipzig, 04103 Leipzig, Germany

E-mail: minyang@djtu.edu.cn

Abstract. Using different precursors, two series of nano Pt/MgO catalysts (Pt-II and Pt-IV, respectively) loaded with Pt in a wide range from 0.1 wt% to 2.2 wt% were prepared by using wet impregnation methods. The physico-chemical properties of catalysts were characterized by using BET, XRD, TPD-CO_2, and other techniques. The results revealed that the precursor and content of Pt, especially the acid/base properties of the precursor solution, had a big influence on the structure, and surface properties of the catalysts. The loaded Pt is in metallic state in all the catalysts calcined at 1073 K. The Pt-II catalysts had larger surface areas and relatively narrow pore size distribution range. The MgO crystals in the calcined Pt-II-3 catalyst had more regular structure than that in the calcined Pt-IV-3 catalyst.

1. Introduction

Methane is the principal constituent of natural gas. Conversion of methane to more useful and easily transportable chemicals (liquids) has therefore been given high priority by chemical scientists. Since Sabatier and Senderens [1] discovered the formation of synthesis gas from methane by steam reforming, different routes have been developed to convert methane to synthesis gas [2].

The conversion of methane into value-added products is a challenging task and the main reason is the high CH_3–H(g) bond dissociation energy (439.3 kJ/mol). Nickel and other transition metals are active for methane conversion but quickly deactivation due to deposits of carbon limits their utilization [3–5]. Certain noble metals (Rh, Ru, Pt, Pd) exhibit much higher intrinsic activities for dry reforming than Ni and are much less susceptible to carbon formation, and become more attractive [6-8]. Our earlier work showed that Pt supported on MgO has highly catalytic activity and stability for methane conversion to syngas due to the high stability of dispersed Pt and the low carbon deposition [9-11]. In this paper, we report about the comparison of the characterization results of two series of catalysts with different Pt content.

2. Experimental

2.1 Catalysts Preparation

Two series of nano Pt/MgO catalysts were prepared by the wet impregnation method using different precursors of Pt, i.e. $Pt(NH_3)_4(NO_3)_2$ and H_2PtCl_6, for Pt-II and Pt-IV, respectively. The catalyst was dried at 393 K for 24 h, then, calcined in air at the desired temperature (473 ~ 1073 K) for 6 h. The Pt content of the catalysts calcined at 1073 K were detected by ICP-AES and showed in Table 1.

Table 1. Pt content of the catalysts calcined at 1073 k detected by ICP-AES.

Catalyst	Pt-II-1	Pt-II-2	Pt-II-3	Pt-II-4	Pt-IV-1	Pt-IV-2	Pt-IV-3	Pt-IV-4
Pt content, wt%	*0.11*	*0.56*	*1.05*	*2.19*	*0.09*	*0.47*	*0.84*	*1.51*

2.2 Characterization techniques

The textural analysis of catalysts was tested by N_2 adsorption at 77 K on the automatic adsorption analyzer ASAP 2000 (Micrometric, USA). The specific surface area, and the pore size distribution were calculated from the adsorption isotherm according to the BET model.

X-ray powder diffraction (XRD) was carried out as described in earlier paper [11].

The AMI 100 system (Zeton Altamira, USA) equipped with a TCD detector was used to carry out the temperature programmed desorption of CO_2 (TPD-CO_2). For each experiment, the first step was the activation of the catalyst in a flow of He (25 ml/min) from room temperature to the desired temperature at a rate of 10 K/min, and kept for 60 min. After activation, the temperature was cooled down to 308 K, and then, a test of TPD-CO_2 was performed, the TCD signal was recorded and saved by the computer.

3. Results and Discussion

3.1. Textural analysis of the catalysts

The BET surface areas for all Pt-II and Pt-IV catalysts calcined at 1073 K and the pore size distributions of the Pt-II-3 and Pt-IV-3 catalysts are presented in figure 1. It is shown that the two series of catalysts have different surface behavior compared to the support MgO. There was a big difference between the Pt-II samples and the support MgO. After loading Pt, the surface area showed large increases, between about two to four times. The surface area decreased with an increase of the Pt content. For the Pt-IV samples, the amount of Pt content had little influence on the surface area.

For the Pt-II-3 catalyst, most of the pores had diameters of 10 ~ 100 nm, different from the support MgO. But Pt-IV-3 catalyst had similar pore size distribution to the support MgO. The pore size distributions of other catalysts give the similar results. This is the first difference in pore size distributions between the two catalyst series. Another difference lies in the influence of Pt content. For Pt-II catalysts, the amount of Pt content did not influence the pore size distribution. For Pt-IV catalysts, the amount of Pt content had no influence on the pore size distribution (not showed here) when it was below 1.0 wt%. When the Pt content reached 1.5 wt%, the pore size distribution showed a relatively narrow range (10 ~ 100 nm).

The above results indicate that the two precursors of Pt had different influence on the texture of the support MgO. $Pt(NH_3)_4(NO_3)_2$ had a big influence, whereas H_2PtCl_6 had only a small influence. The reason for this may be the acid-base properties of the impregnation solutions. The impregnation solution of H_2PtCl_6 is acidic (pH = 6~7), whereas the impregnation solution of $Pt(NH_3)_4(NO_3)_2$ is basic (pH = 7~8).

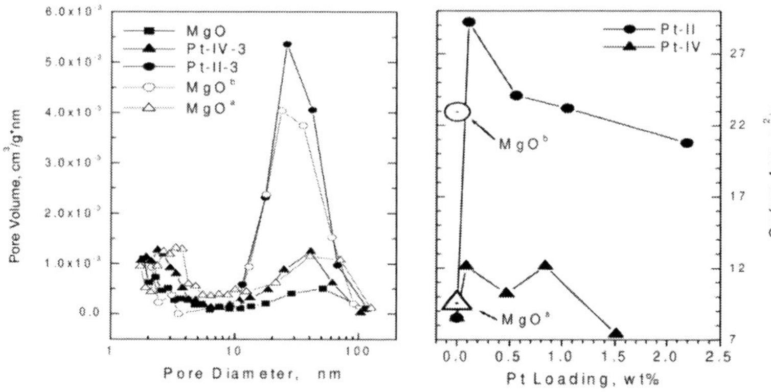

Figure 1. BET surface areas of all the catalysts (right) and pore size distribution of Pt-II-3 and Pt-IV-3 catalysts (left).

Journal of Physics: Conference Series **188** (2009) 012060 doi:10.1088/1742-6596/188/1/012060

In order to verify the above idea, the support MgO was treated with acidic and basic water solutions at the equivalent pH values as the impregnation solutions and under similar condition as impregnation, which were named MgO^a and MgO^b, respectively, and their pore size distributions and surface areas are showed in figure 1 also. The BET results of MgO^a and MgO^b show that the pH value of the impregnation solution influenced the texture of MgO. After treating with basic solution, MgO had similar BET surface and pore size distribution as the Pt-II-3 catalyst. MgO^a showed similar BET surface and pore size distribution as the Pt-IV-3 catalyst. This supports above idea that the textural differences between the Pt-II and Pt-IV catalysts mainly result from the acid-base properties of the impregnation solution.

3.2. Phase composition of the catalysts

As showed in figure 2, all the patterns have similar XRD patterns as the support MgO, but, new peaks appear in the patterns of catalysts having high Pt content. These new peaks at 2θ=40, 46, 67, 81 and 86 belong to Pt crystallites. For the Pt-II series, these peaks appeared when the Pt content is more than 0.5 wt%, and their intensities increased with an increasing Pt content. For the Pt-IV catalysts, only Pt-IV-4 (1.5 wt% Pt) showed Pt diffraction peaks. This means that the dispersed Pt on the surface of the support MgO existed in the metallic state after calcination at 1073 K in both catalyst series.

The phase composition of the dried catalysts was complicated. The XRD patterns (not showed here) of the dried Pt-II-3, Pt-IV-3, and, Pt-IV-4 catalysts show that the supports consisted of $Mg(OH)_2$ and MgO. The relative amount of the two phases depended on the acid/base properties of the impregnation solution, as supposed. The impregnation solution for Pt-II-3 was basic, and was weakly acidic for Pt-IV-3. The XRD patterns of both catalysts show that a large amount of $Mg(OH)_2$ existed in these catalysts. Whereas, the impregnation solution for the Pt-IV-4 catalyst was relatively strong acidic, about two times as that of the Pt-IV-3 catalyst. The XRD pattern of Pt-IV-4 catalyst shows that there was only a small amount of $Mg(OH)_2$ in it. In these dried catalysts, there was no indication of a phase of Pt. XPS results show that Pt existed in the Pt^{2+} state in the dried Pt-II-3 catalyst, and, in the dried Pt-IV-3 catalyst, Pt may exist in the halide state.

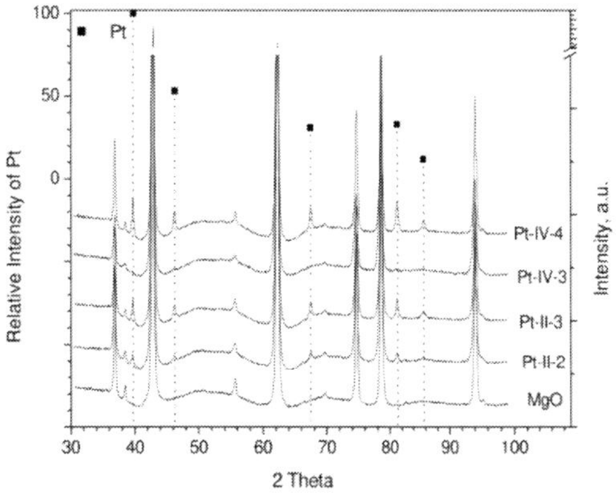

Figure 2. XRD patterns of MgO and Pt/MgO catalysts

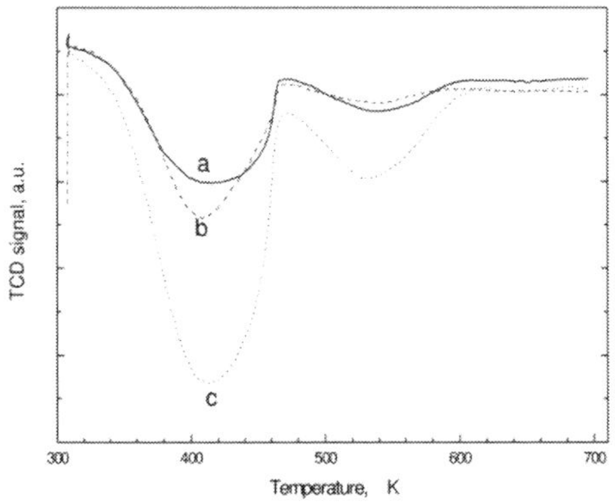

Figure 3. CO_2-TPD profiles of MgO (a), Pt-IV-3 (b) and Pt-II-3 (c) catalysts

When the dried catalysts were calcined at high temperatures (>393 K), the phase transformation from $Mg(OH)_2$ to MgO took place. After calcination at 573 K, there was no $Mg(OH)_2$ detected. Although the phase transformation from $Mg(OH)_2$ to MgO was completed after calcination at 573 K, higher calcination temperatures (> 573 K) still had influence on the MgO phase, i.e. the MgO diffraction intensities increased with the temperature increase in the XRD patterns.

Pt appeared as phase at different temperatures for the Pt-II-3 and Pt-IV-4 catalysts. The Pt phase appeared when the Pt-II-3 catalyst was calcined at 473 K, and higher calcination temperatures (>473 K) had only a small influence on it. For the Pt-IV-4 catalyst, the Pt phase began to appear after calcination at 573 K, and the XRD peak intensities of the Pt phase increased with an increase in the calcination temperature. The above results indicate that there was difference in the formation of the Pt phase between the two catalysts. In the Pt-II-3 catalyst, the Pt phase was formed completely after calcination at 473 K, whereas in the Pt-IV-4 catalyst, the Pt phase was formed step by step above a calcination temperature of 573 K.

3.3. Basicity of the catalysts
The basicity of the calcined catalysts was investigated by TPD-CO_2. Figure 3 shows that the there were two CO_2 desorption peaks in the TPD-CO_2 profile of support MgO. One was at 413 K, and the other at 540 K. This indicates that there were two kinds of basic sites on the surface of the support MgO, but they were only weak basic sites as judged from the low desorption temperatures. For the Pt-II-3 catalyst, the TPD-CO_2 profile had also two CO_2 desorption peaks, and the desorption temperatures were almost identical to that in the TPD-CO_2 profile of the support MgO, but the amount of desorbed CO_2 at both temperatures increased greatly after loading of Pt. This means that the loading of Pt did not change the basicity of the support MgO, but increased the amount of the basic sites, resulting in the increase of the amount of desorbed CO_2.

Different to the TPD-CO_2 profile of Pt-II-3, Pt-IV-3 catalyst had similar TPD-CO_2 profile as that of the support MgO, with respect to both the basicity and the amount of the desorbed CO_2. Therefore, for Pt-IV catalyst, the loading of Pt did not change the basic property of the support MgO.

The TPD-CO_2 results show that both series of catalysts had only two kinds of weak basic sites. But the two series of catalysts exhibit a big difference in the number of the basic sites. The number of the basic sites in all the Pt-II catalysts were about 2 ~ 3 times higher than in the support MgO, whereas all Pt-IV catalysts had a similar amount of basic sites as the support MgO.

As shown in figure 1, the surface area of the Pt-II catalysts were two times of that of the support MgO, but there was only small difference in these parameters between the Pt-IV catalysts and the support MgO. These results are in accordance with the difference in the TPD-CO_2 results. So, the big difference in the number of the basic sites may be caused by the difference in the textural properties of the two series of catalysts.

4. Conclusion
All catalysts calcined at 1073 K showed the typical XRD pattern of the support MgO. The precursor, especially the acid/base properties of the precursor solution, had a big influence on the structure of the catalysts. The Pt-II catalysts had larger surface areas than the Pt-IV catalysts. The MgO crystals in the calcined Pt-II-3 catalyst had more regular structure than that in Pt-IV-3 catalyst. Both series of catalysts had a similar basicity as the support MgO, but the Pt-II catalysts had more weak basic centers than the Pt-IV catalysts.

5. Acknowledgments
Financial supports from National Science Foundation of China (Contract No. 20873013), and Scientific Research Foundation for the Returned Overseas Chinese Scholars, State Education Ministry are greatly appreciated.

References
[1] Sabatier P and Senderens J B 1902 *Compt. Rend.* **134** 689
[2] Rostrup-Nielsen J R and Sehested J 2002 *Adv. Catal.* **47** 65
[3] Tomishige K, Yamazaki O, Chen Y, Yokoyama K, Li X and Fujimoto K 1998 *Catal. Today* **45** 35
[4] Wei J M, Xu B Q, Cheng Z X, Li J L and Zhu Q M 2000 *Stud. Surf. Sci. Catal.* **130** 3687
[5] Ascroft S T, Cheetham A K , Green M L H and Vernon P D F 1991 *Nature* **352** 225
[6] Souza M M V M, Aranda D A G and Schmal M 2001 *J. Catal.* **204** 498
[7] Bradford M C J and Vannice M A 1999 *Catal. Rev. Sci. Eng.* **41** 1
[8] Bitter J H, Seshan K and Lercher J A 1997 *J. Catal.* **171** 279
[9] Yang M and Papp H 2006 *Catal. Today* **115** 199
[10] Yang M and Papp H 2007 *Chin. J. Chem. Physi.* **20** 690
[11] Yang M and Papp H 2008 *Chin. J. Catal.* **29** 228

AUTHOR INDEX

AUTHOR INDEX

Institute of Physics
Dirac House, Temple Back
Bristol BS1 6BE UK

proceedings.com